Soft Ground Tunnel Design

Soft Ground Tunnel Design

Benoît Jones

CRC Press
Taylor & Francis Group
Boca Raton London New York

CRC Press is an imprint of the
Taylor & Francis Group, an **informa** business

First edition published 2022
by CRC Press
6000 Broken Sound Parkway NW, Suite 300, Boca Raton, FL 33487-2742

and by CRC Press
2 Park Square, Milton Park, Abingdon, Oxon, OX14 4RN

Library of Congress Cataloging-in-Publication Data

Names: Jones, Benoît, author.
Title: Soft ground tunnel design / Benoît Jones.
Description: First edition. | Boca Raton, FL : CRC Press, [2022] | Includes index.
Identifiers: LCCN 2021013258 (print) | LCCN 2021013259 (ebook) | ISBN 9780367419592 (hbk) | ISBN 9781482254679 (pbk) | ISBN 9780429470387 (ebk)
Subjects: LCSH: Tunnels--Design and construction. | Tunneling. | Soil stabilization.
Classification: LCC TA815 .J66 2022 (print) | LCC TA815 (ebook) | DDC 624.1/93--dc23
LC record available at https://lccn.loc.gov/2021013258
LC ebook record available at https://lccn.loc.gov/2021013259

ISBN: 978-0-367-41959-2 (hbk)
ISBN: 978-1-4822-5467-9 (pbk)
ISBN: 978-0-429-47038-7 (ebk)

DOI: 10.1201/9780429470387

Typeset in Sabon
by KnowledgeWorks Global Ltd.

Contents

2 Undrained stability   45

3 Drained stability   83

Preface

Dear Reader,

This textbook covers the design of tunnels and other underground spaces in soft ground. 'Soft ground' means soil, as opposed to rock. The philosophies of rock and soft ground tunnel design are different enough to merit separate treatment.

In 2011, when I was setting up the MSc in Tunnelling and Underground Space at the University of Warwick, I found that there was no comprehensive textbook available on soft ground tunnel design. So I had to synthesise the literature and my own experience to find the best ways to present this subject to my students. Over several years the lectures and example problems were developed and improved. This book aims to extend and enhance these teaching materials and make them more widely available.

Readers who will gain most from this book already have a civil engineering degree or similar and understand the basic principles of soil mechanics and structural engineering, but have little or no experience of tunnel design. My target audience is MSc students, graduate tunnel engineers and experienced engineers who are new to soft ground tunnelling but I hope that even for experienced tunnel design experts, there is something new or interesting for you here and that you will find it a useful reference work, compilation of methods and summary of the state of the art.

There are already several reasonably good (and very detailed) books available on tunnel construction methods so I have not unnecessarily duplicated that subject matter here. If you do not already know the basics of how TBMs, shotcrete or segmental linings work, there is a book list below that I would urge you to make use of as preparation for and in combination with this textbook. There are also plenty of videos and other resources available online.

There are many aspects of design that are not covered in this book, including sustainability (encompassing health, safety and the environment), design for watertightness, fire, durability, spaceproofing and operational considerations. Site investigation and geological characterisation are also not covered explicitly. These are all very important subjects and their absence is purely due to space constraints.

This is first and foremost a textbook: you won't just *know* a lot at the end but will be able to *do* a lot. Each chapter begins by listing the knowledge, understanding and skills you should have gained by the end of it. The use of worked examples throughout this book and exercises at the end of most chapters will give you an in-depth understanding and the confidence to apply that understanding.

The three principal themes are stability, prediction of ground movements, and structural design of the tunnel lining. These themes are the key to understanding how tunnels work and are the basis for choices of construction method and lining type. The reader is guided through the basic principles of soil-structure interaction, the three-dimensional effects of construction sequence and the effects of construction on other surface or subsurface structures, in steps of gradually increasing complexity from basic principles to sophisticated design. As the calculations become more complex, it is more and more important that they remain grounded in an in-depth understanding of how real tunnels behave. For this reason, the very first chapter starts with case studies of real tunnel behaviour, to provide a conceptual framework of what happens to the tunnel lining and what goes on in the ground around a tunnel in terms of displacements, stresses and strains.

My final aim in this book is to collate and clearly present all the most commonly used design methods from first principles. Even in the original papers where these methods were first introduced, and more so in subsequent papers that cite and expand on them, there are sometimes missing steps in the derivations, missing information, implicit assumptions that are not explained, and even errors. It is alarming to consider how often these errors have been inherited, used and disseminated in the intervening years. I have made every effort to ensure that in this book these design methods and their derivations are now set out in full; transparently and correctly. I hope I've made it as easy as possible for you to spot any errors that remain and draw them to my attention. Contact me, we will add them to the corrigendum online at www.inbye.co.uk/soft-ground-tunnel-design and get them fixed in the second edition.

Warmest Regards,
Benoît Jones

Books on tunnel construction methods

The following books may be useful to gain an appreciation of tunnel construction methods.

INTRODUCTION TO TUNNEL CONSTRUCTION

This is the only book intended as an introductory text for tunnel construction and as such is probably the one best suited to beginners.

Chapman, D., Metje, N. & Stärk, A. (2010). *Introduction to tunnel construction*. Abingdon: Spon Press.

Another excellent reference that all tunnelling engineers should read is the British Standard on health and safety in tunnelling.

BS 6164: 2019. *Health and safety in tunnelling in the construction industry – Code of practice*. London: British Standards Institution.

TBM TUNNELLING

'Mechanized tunnelling' means tunnelling using a tunnel boring machine (TBM). This book covers all the different types of TBMs and how to manage and control their use, particularly in urban areas:

Guglielmetti, V., Grasso, P., Mahtab, A. & Xu, S. (2008). *Mechanized tunnelling in urban areas – design methodology and construction control*. London: Taylor & Francis Group.

The following are essentially handbooks for practitioners. They are big and heavy and cover almost every aspect of TBM use, but have very little to say about design.

Maidl, B., Herrenknecht, M., Maidl, U. & Wehrmeyer, G. (2012). *Mechanised shield tunnelling*, 2nd Edition. Berlin: Ernst & Sohn.

Maidl, B., Thewes, M. & Maidl, U. (2013). *Handbook of tunnel engineering I – structures and methods*, English edition translated by David Sturge. Berlin: Ernst & Sohn.

Maidl, B., Thewes, M. & Maidl, U. (2014). *Handbook of tunnel engineering II – basics and additional services for design and construction*, English edition translated by David Sturge. Berlin: Ernst & Sohn.

SPRAYED CONCRETE

Sprayed Concrete for Ground Support is published by a product supplier but is nevertheless a very good explanation of the use of sprayed concrete in tunnelling, covering equipment, chemical admixtures, mix design and application.

BASF (2014). *Sprayed concrete for ground support*. BASF Construction Chemicals Europe Ltd.

Alun Thomas's book is both an introduction to sprayed concrete tunnelling and a reference guide for experienced practitioners. The second edition has recently been released.

Thomas, A. H. (2019). *Sprayed concrete lined tunnels*, 2nd Edition. Abingdon: Taylor & Francis.

Acknowledgements

First of all I would like to thank my civil partner Anna, who read the entire first draft with a ruthless eye for anything unnecessary, confusing or inelegant. She proposed myriad improvements, the vast majority of which I adopted. Her patience and craftsmanship improved the book on almost every page.

Secondly, I would like to thank everyone at Taylor & Francis, particularly Frazer Merritt and Tony Moore, who have been incredibly patient as deadline after deadline flew by.

Many friends and colleagues provided constructive comments on draft chapters, in particular Ben Swatton, James Boswell, Dr Giorgia Giardina and Sotiris Psomas.

My derivation of the Curtis-Muir Wood equations benefited from access to handwritten notes from John Curtis himself, which explained some (but not all!) the intermediate steps between equations in the published papers. These were given to me by Nick Tucker when I was working at Mott MacDonald.

My derivation of the wedge-prism method was aided by discussion with and assistance from Dr Tiago Dias of the University of Southampton, and correspondence with Prof. Dr Georgios Anagnostou of ETH Zurich. Prof. Vicky Henderson of the University of Warwick helped me solve the differential equations.

Dr Giorgia Giardina of Delft University of Technology gave advice and made many useful comments on the building damage assessments chapter. Dr Stana Zivanovic of the University of Exeter helped me to understand shear deflection and warping. Jamie Bradley showed me the importance of the unit width assumption.

To all those others that corresponded with me to provide papers, data or clarifications, a sincere thank you. These include Dr Alec Marshall of the University of Nottingham, Prof. Daniel Dias of Polytech' Grenoble, Prof. Adam Bezuijen of Ghent University, Dr Jiang Su of Ramboll and John Greenhalgh and Benoît de Rivaz of Bekaert.

I tried for several years to write this book while working full-time in very demanding jobs, but progress was slow. It has only been finished because I was able to work part-time on a contract basis for an extended period. I would like to thank London Bridge Associates and OTB Engineering for providing me with the work to stay solvent during this time, in particular Ken Spiby and Phil Astle. It was a bonus that much of the work they gave me was very interesting and I learnt a lot while doing it. This book literally wouldn't have been possible without them.

Much of the bringing together of disparate sources to achieve the coherent synthesis of the state-of-the-art presented in this book was done between 2011 and 2015 to prepare teaching materials for the MSc in Tunnelling and Underground Space at the University of Warwick, where I was Course Director. I would like to thank all my students for their enthusiasm, patience and feedback.

This synthesis was developed further in a series of 32 articles for Tunnelling Journal between 2012 and 2017. Thank you to Tris Thomas for giving me that platform, and all his readers who gave me comments and encouragement.

The British Tunnelling Society and the International Tunnelling Association do so much to further the education and skills of their members, and also to provide a forum for exchanging ideas and building friendships and support networks. It is hard to imagine what a career in tunnelling would be like if they didn't exist. The writing of a book like this would certainly be much harder.

I have benefited, over a career of more than 20 years now, from the friendship, knowledge, experience and guidance of a large number of managers, supervisors and colleagues – too many to name.

Finally I would like to thank the rest of my family – Christophe, Mum, Dad, Damien and all the rest of you – for your love and support.

With heartfelt gratitude to all,

Benoît
Coventry, August 2021

Author

Dr Benoît Jones has two decades of experience in tunnelling as a designer, researcher, contractor, academic, teacher, inventor and consultant. He is a Chartered Civil Engineer and is the founder and Managing Director of Inbye Engineering, a technology and consulting company established in 2014.

Benoît graduated from the University of Bristol in 2000 with a First Class Honours degree in Civil Engineering then joined Mott MacDonald as a Graduate Tunnel Engineer. Between 2002-2006 he undertook research for an EngD at the University of Southampton supervised by Professor Chris Clayton. After working for Mott MacDonald on Crossrail MDC3, he moved in 2007 to Morgan Est (now Morgan Sindall) as Section Engineer on the King's Cross Underground Station Redevelopment and later as Engineering Manager on the Stoke Newington to New River Head Thames Water Ring Main Extension Tunnel. In Autumn 2010 he joined OTB Engineering, a small consultancy specialising in tunnelling and geotechnics, working on temporary works designs for Tottenham Court Road Station Upgrade and for the Crossrail Eastern Running Tunnels (C305). From 2011 to 2014 he set up and then ran the MSc in Tunnelling and Underground Space at the University of Warwick. Since 2015 Benoît has devoted his time to Inbye Engineering, a company he set up in 2014 to deliver Strength Monitoring Using Thermal Imaging (SMUTI) and provide engineering/tunnelling design and consultancy services.

Benoît has led multiple research projects from within both industry and academia and is the author of more than 60 articles and papers. He won the ICE Telford Premium Prize for outstanding contribution to the civil engineering literature in 2009 and was invited to give the ICE Geotechnical Engineering Lecture in 2010 and 2012. He won two NCE Tunnelling Awards and an ITA Austria Innovation Prize for SMUTI in 2016, followed by an International Tunnelling Association award in 2017.

Chapter 1

Real tunnel behaviour

Even the most sophisticated design model is an idealised representation of reality. In a model, it is neither possible nor desirable to include all the complexity and uncertainty of ground and tunnel lining behaviour and their interactions. Therefore, it is important to understand reality before attempting to model it.

At different depths and in different geological and hydrogeological conditions, and with different sizes of tunnels constructed using different methods, different aspects of real behaviour will be more or less important. It is a matter of experienced judgement, and of trial and error, to decide which aspects of real behaviour are important enough to include in the design model. The aim is to remove superfluous complexity so that the design model outputs can be more easily interpreted and validated, while retaining a design that is realistic. A designer who understands what reality looks like will be better equipped to interrogate their model's outputs and produce a reliable design.

This chapter describes what happens when real tunnels are constructed: the ground movements, the tunnel lining movements, the effective stresses and pore pressures in the ground, and the stresses in the tunnel lining, using data from real projects. The objective of this chapter is to 'upload a conceptual model to your necktop' (to paraphrase Daniel Dennett), so you know what results to expect and you are able to achieve efficient and realistic designs.

After working through this chapter, you will understand:

- how the ground moves towards the face of an advancing tunnel
- what happens to pore pressures and effective stresses within the ground as the tunnel excavation passes
- the factors affecting stability and what causes collapse or blow-out failure
- how the ground and groundwater apply loads to the tunnel lining, how these loads develop as the tunnel advances and what happens in the long-term
- how all these aspects depend on the ground conditions, the depth of the tunnel and the tunnel construction method

DOI: 10.1201/9780429470387-1

1.1 IN SITU STRESS STATES

Before a tunnel is excavated, there are existing 'in situ' stresses in the ground. The principal stresses are usually in the vertical and horizontal directions because they depend on gravity and the history of geological deposition and erosion. Where there are steep slopes or tectonic stresses, this may no longer be true, but this would rarely be the case for a tunnel in soft ground.

Total stress in the ground is made up of two components: the pore pressure and the effective stress, such that:

$$\sigma = \sigma' + u \tag{1.1}$$

σ is the total stress in kPa
σ' is the effective stress in kPa
u is the pore pressure in kPa

The vertical stress in the ground is determined by the bulk unit weight of the soil using the following equation:

$$\sigma_v = \gamma z \tag{1.2}$$

σ_v is the vertical total stress in kPa
γ is the bulk unit weight of the soil, which is the weight of the soil grains and pore water in kN/m^3
z is the depth in m

Therefore, we can easily calculate the vertical total stress from basic knowledge of the stratigraphy of the site and the bulk unit weight of the soil. If the ground is made up of several layers with different values of bulk unit weight, then they can be added together as follows:

$$\sigma_v = \gamma_1 z_1 + \gamma_2 z_2 + \gamma_3 z_3 + \ldots + \gamma_n z_n \tag{1.3}$$

$\gamma_1, \gamma_2, \gamma_3 \ldots \gamma_n$ are the bulk unit weights of soil layers 1, 2, 3 ... in kN/m^3
$z_1, z_2, z_3 \ldots z_n$ are the thicknesses of soil layers 1, 2, 3 ... in m

The vertical total stress at the tunnel axis level is sometimes referred to as the 'full overburden pressure'.

In situ stresses are rarely equal in the vertical and horizontal directions. The vertical stress is determined by the weight of the soil, but horizontal stress depends on other factors as well. The relationship between horizontal and vertical effective stress is described by the coefficient of earth pressure at rest, K_0, which is given by:

$$K_0 = \frac{\sigma'_h}{\sigma'_v}$$
(1.4)

K_0 is the coefficient of earth pressure at rest
σ'_h is the horizontal effective stress in kPa
σ'_v is the vertical effective stress in kPa

In 'normally consolidated' soils, where the maximum vertical effective stress in the soil's history is the current value, the horizontal stress will be lower than the vertical stress. In this case, the value of K_0 may be given by 'Jaky's formula':

$$K_0 = 1 - \sin\phi'$$
(1.5)

ϕ' is the soil's angle of friction in a drained condition

Overconsolidated soils, where the soil has experienced a higher effective stress at some point in its past than at present, perhaps because overlying soils have subsequently been eroded or the water table has risen, can have a higher horizontal effective stress than Jaky's formula would predict and they can even be significantly larger than the vertical effective stress, i.e. K_0 may be greater than unity.

1.2 OVERVIEW OF TUNNEL BEHAVIOUR

During excavation of a tunnel, a void is formed in the ground. In an open-face tunnel, the stress normal to the exposed surface of the excavation has to be zero for there to be equilibrium. When a closed face tunnel boring machine (TBM), such as an earth pressure balance (EPB) or slurry TBM, is used, or when the tunnel face is pressurised using compressed air, the stress normal to the excavated surface can be greater than zero and will be equal to the applied pressure.

If the pressure applied to the surface of the excavation is less than the in situ stress, then the ground will be unloaded, and will relax towards the excavation. Therefore, tunnel construction nearly always induces ground movements towards the tunnel, both radial displacements around the tunnel and longitudinal displacements towards the face. There will also be a tendency for there to be more ground movement downwards than upwards, due to gravity. An example of displacements at two different transverse sections of a sprayed concrete tunnel in London Clay is shown in Figure 1.1, from a paper by Deane & Bassett (1995). London Clay is a stiff to very stiff

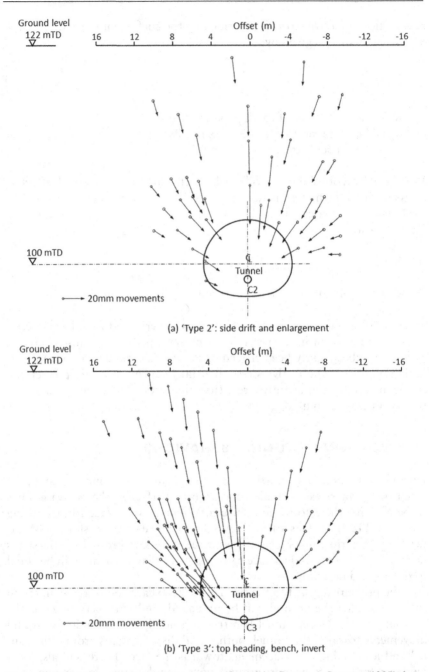

(a) 'Type 2': side drift and enlargement

(b) 'Type 3': top heading, bench, invert

Figure 1.1 Subsurface displacement vectors inferred by Deane & Bassett (1995) from inclinometers and extensometers installed at two transverse sections about 30 m apart to monitor the Heathrow Express Trial Tunnel construction. Displacement vectors are exaggerated. (a) 'Type 2' side drift and enlargement, (b) 'Type 3' top heading, bench, invert.

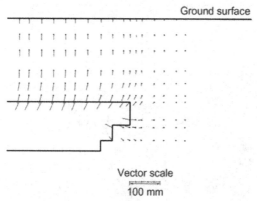

Vector scale

100 mm

(a) Displacements on a vertical plane through the tunnel centreline

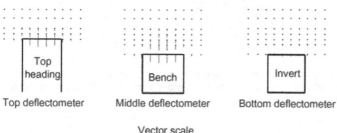

Top deflectometer | Middle deflectometer | Bottom deflectometer

Vector scale

100 mm

(b) Plan view of longitudinal displacements towards the top heading, bench and invert

Figure 1.2 Patterns of ground displacements above and ahead of the advancing Heathrow Express Terminal 4 Station Concourse Tunnel (from van der Berg et al., 2003). (a) Displacements on a vertical plane through the tunnel centreline, and (b) Plan view of longitudinal displacements towards the top heading, bench and invert.

overconsolidated clay. Figure 1.1 also shows the foci of the displacements, marked by circles 'C2' and 'C3'. In these cases, the foci lie somewhere between the axis and the invert of the tunnel.

Figure 1.2 shows displacement vectors around an advancing tunnel face from van der Berg et al. (2003). This was a shotcrete-lined tunnel excavated in a top heading, bench, invert sequence, also in London Clay. Advances were 1 m long for the top heading and bench, and 2 m for the invert.

Diagram (a) in Figure 1.2 is a long section through the tunnel and shows how displacements developed in front of and then behind the face. Behind the face, displacements continued to occur, but once the full ring had been sprayed by closing the invert, displacements did not visibly increase further. In diagram (b), plan sections show horizontal displacements at different levels – the top deflectometer was in the top heading 1 m above axis level,

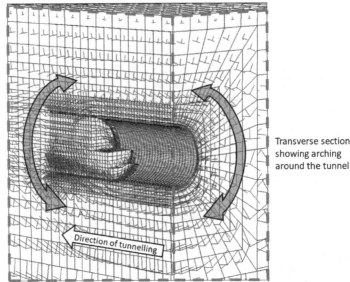

Long section showing arching around the face

Transverse section showing arching around the tunnel

Direction of tunnelling

Figure 1.3 Principal stress tensors showing rotation of principal stress around the tunnel and around the face (background image from numerical modelling by Thomas, 2003).

the middle deflectometer was at axis level and the bottom deflectometer was 2.3 m below axis level. The strongly three-dimensional (3D) nature of ground movements is evident – the longitudinal component of displacements is as important as the radial. This is important to remember when we approximate tunnel construction using two-dimensional (2D) models.

The displacements in Figures 1.1 and 1.2 occur because the ground around the tunnel heading is trying to find a new equilibrium. If it cannot, then there will be a collapse. When a new equilibrium is found, it will look something like Figure 1.3, which is a visualisation of principal stress magnitude and orientation in the ground around an advancing tunnel, in a block cut from a larger 3D numerical model. In the transverse section on the right hand side, we can see that as the ground has moved radially towards the tunnel, the principal stresses have rotated and the major principal stresses are now circumferential. This is referred to as 'arching'. We can also see, in the longitudinal section on the left hand side of the figure, that arching also occurs around the face, which I like to call 'front-to-back arching'.

The redistribution of ground stresses around a tunnel during its construction means that ground movements are inevitable. These ground movements will have an impact on existing surface and subsurface structures. By minimising the redistribution of stresses, either by applying an internal pressure to the face or installing a stiff lining as early as possible, we can minimise ground movements.

1.2.1 Undrained soil behaviour

The displacements shown in Figures 1.1 and 1.2 are 'constant volume'. This means that although the ground has deformed, the volume of the soil has not changed. This has happened because these displacements are in the short-term and the London Clay is behaving in an undrained manner. 'Undrained' means that pore water in the clay cannot flow during this time-frame (that would take years or decades, whereas these readings have been taken during construction over days or weeks). Therefore, virtually all the changes in mean total stress caused by excavation are experienced as pore water pressure changes, because water has a much higher bulk modulus (i.e. it is much less compressible) than the soil particles. Since the pore water is experiencing the change in stress, and water can be considered incompressible at these stress levels, volume change will be negligible.

This undrained behaviour may be illustrated by the conceptual model in Figure 1.4, representing Terzaghi's principle of effective stress (Terzaghi, 1936). The spring represents the effective stress (the average of contact stresses between soil particles), the water represents the pore water, and the thin tube represents the permeability of the soil. As the loading changes, the change in stress is immediately applied to the pore water. Over time, depending on the diameter of the tube, water will flow in or out of the container and gradually the change in stress will transfer from the water to the spring.

As we have said, water has a high bulk modulus and is very difficult to compress, but on the other hand, it cannot resist shear. Therefore, it is useful to think of soil behaviour in terms of separate 'shear deformation' and 'volumetric deformation', governed by a shear modulus and a bulk modulus respectively. For an undrained soil, the shear modulus will depend on the soil particles and the bulk modulus will depend on the pore water.

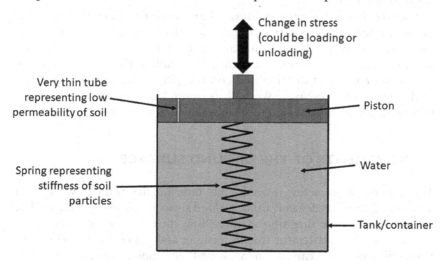

Figure 1.4 Conceptual model illustrating undrained behaviour.

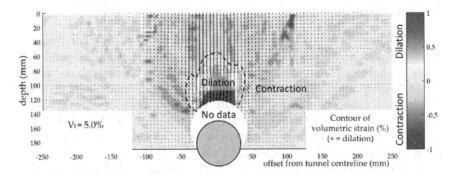

Figure 1.5 Volumetric strain contours from a centrifuge test by Marshall (2009), showing zones of contraction and dilation. Note no data available in white area around tunnel.

1.2.2 Drained soil behaviour

When a tunnel is constructed in a drained material, such as a sand or gravel, which allows pore water to flow and pore water pressures to reach either equilibrium or a steady-state within the timescale of construction, then the soil can experience volumetric strains known as 'contraction' (negative volume change) or 'dilation' (positive volume change, getting bigger). Sands that are not already in a dense state will first contract at low shear strains, become denser, and then begin to dilate at higher shear strains (Marshall et al., 2012). Therefore, different zones of ground around a tunnel will contract or dilate, as shown in Figure 1.5. This means that the ground displacement vectors will not all point to one focus.

Another difference between drained and undrained behaviour is that for undrained soils, the pattern of ground deformations is approximately the same, regardless of the magnitude of deformations. For drained soils, the pattern of ground movements, including their extent and the positions of zones of dilation and contraction, varies as the magnitude of deformations increases. We will examine this in more detail in the following sections on surface and subsurface ground movements.

1.3 MOVEMENTS OF THE GROUND SURFACE

The ground movements towards the face of the tunnel we saw in Section 1.2 cause settlements of the ground surface, as shown in Figure 1.6.

Transverse to the direction of tunnelling (in the y-axis direction), we can see a surface settlement trough forming and along the centreline (the x-axis direction), we can see an 'S'-shaped longitudinal surface settlement

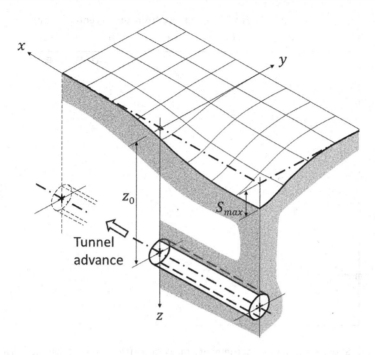

Figure 1.6 Movements of the ground surface (based on Attewell et al., 1986).

curve forming. The transverse surface settlement curve (or 'trough') has a shape that may be described by an inverted normal distribution curve, sometimes called a bell curve or a Gaussian curve, with the frequency of the mean representing the maximum settlement S_{max} over the centreline of the tunnel. The longitudinal surface settlement curve has a shape that may be described by an inverted cumulative normal distribution curve.

1.3.1 Transverse vertical settlements

An example of transverse surface settlements measured above a tunnel in London Clay (Standing et al., 1996; Nyren et al., 2001) is shown in Figure 1.7. This was a 4.85 m diameter open-face TBM with a precast concrete expanded wedgeblock segmental lining installed behind the shield. The depth to the tunnel axis was 31 m.

The settlements shown in Figure 1.7 were measured in the middle of St James's Park. These are referred to as 'greenfield' settlements because they are not affected by anything human-made, such as buildings, concrete slabs, piles or underground structures. Only measurements on one side of this particular tunnel are shown, because the settlement array only extended about 5 m in the other direction. Usually surface settlements are

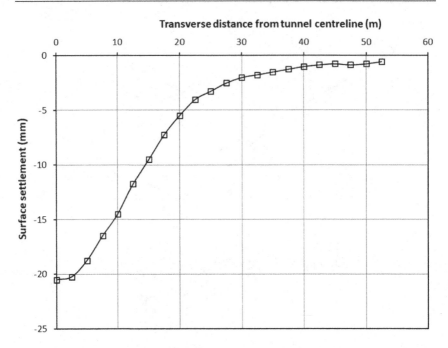

Figure 1.7 Short-term surface settlements measured after TBM passage at St James's Park, London (data from Grammatikopoulou et al., 2008).

symmetrical about the tunnel centreline, as long as the geology, surface levels and excavation process are symmetrical.

Settlements in the middle of a park are unlikely to cause problems, but greenfield settlements are very useful because our design models often assume the tunnel is built in a greenfield situation, and we can use this data to calibrate them.

1.3.2 Transient settlements

As a tunnel approaches and then passes a transverse settlement monitoring array, the settlements gradually increase, as illustrated in Figure 1.6. A real example from Crossrail on the North side of Hyde Park is shown in Figure 1.8 (Wan et al., 2017). The depth to the tunnel axis was 34.5 m. The TBM was 7.1 m diameter and used 'EPB' technology to apply a support pressure to the face, so even though this tunnel was much larger, the settlements were smaller in magnitude than at St James's Park. Face pressures were approximately 200 kPa, which corresponds to about 30% of the 'full overburden pressure', i.e. the vertical total stress at tunnel axis level (Wan et al., 2019). Tailskin grouting pressures were approximately 100 kPa (Wan et al., 2019). The first set of measurements shown are from when the

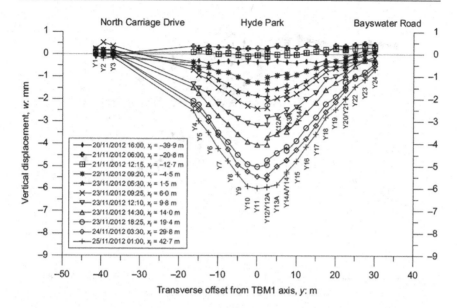

Figure 1.8 Surface settlements as an EPB approaches and passes a settlement monitoring array on the North side of Hyde Park (Wan et al., 2017).

face of the TBM was at −39.9 m approaching the monitoring array and the last set are from when the face was at +42.7 m beyond it.

Figure 1.8 shows that settlement initially increases slowly, then as the face gets closer to the array, the settlement increases more quickly. The rate then decreases again as the TBM moves further away. Note that the settlements were not monitored at regular TBM distance intervals, but increase in frequency as the tunnel passed directly underneath. The development of settlements as the tunnel advances can be demonstrated more clearly by plotting the settlements against face position, as shown in Figure 1.9. The largest magnitude settlements are the ones that are closest to the centreline, i.e. YSMP10 and YSMP12 (labelled as 'Y10' and 'Y12') in Figure 1.8.

In this case, when the face was under the array, the settlement was only about 30% of the final short-term settlement. This was because more ground movement could occur behind the face around the shield and during the grouting process than in front of the face, because the face pressure was higher than the grouting pressure. Based on several case histories in firm to stiff clays, Attewell & Woodman (1982) found that when tunnelling with no support pressure, between 30% and 50% of the final short-term settlement occurs ahead of the face with an average value at about 40%.

In other types of soil, Craig (1975) suggested the following percentages of final short-term settlement would occur, shown in Table 1.1.

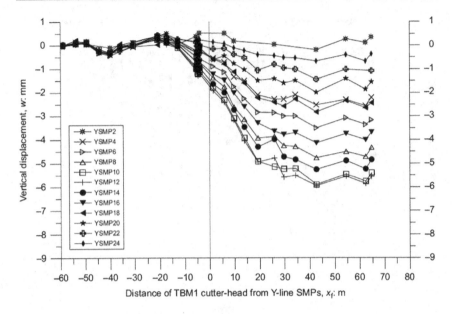

Figure 1.9 Longitudinal settlements on the Y-line array plotted against distance of TBM cutterhead from the array (Wan et al., 2017).

Another example, from the Second Heinenoord Tunnel in the Netherlands, is given in Figure 1.10 (van Jaarsveld et al., 2006). The ground was loose and medium dense sand and soft clay, and two bores were excavated by a large slurry TBM of 8.3 m diameter. Due to the soft soils and the high groundwater pressure, face pressures closer to the in situ stress were applied compared to the Crossrail EPB TBM that was shown in Figure 1.9, resulting in very little settlement ahead of the face. The reason that the second passage of the TBM resulted in less settlement than the first was that more grout was injected, presumably at a higher pressure. Figure 1.10 shows the same 'S'-shaped longitudinal settlement curve, illustrating that this general pattern of behaviour occurs in all types of soils and groundwater conditions.

Table 1.1 Percentages of final short-term settlement above the face and the tail of the shield (Craig, 1975).

Type of ground	Face	Tail of shield
Sand above water table	30–50%	60–80%
Sand below water table	0–25%	50–75%
Silts and soft clays	0–25%	30–50%

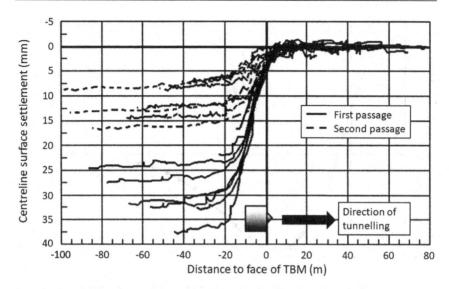

Figure 1.10 Longitudinal settlement profiles at various locations over the first and second TBM passages at the Second Heinenoord Tunnel in the Netherlands (van Jaarsveld et al., 2006).

1.3.3 Horizontal surface movements

Horizontal ground surface movements are much more difficult to measure. This is because using a precise level to obtain vertical displacements is more precise and accurate than using a total station to obtain 3D displacements. There are more accurate methods to measure horizontal displacements between fixed points, for example using micrometer sticks (Nyren et al., 2001), but these are relatively cumbersome and require monitoring points that are above ground, which cannot be easily achieved in public areas. For these reasons, measurements of horizontal ground surface movements are much less common than vertical movements.

Nyren et al. (2001) measured horizontal transverse and horizontal longitudinal movements as the tunnel face passed the monitoring array at St James's Park, shown in Figures 1.11 and 1.12, respectively.

In Figure 1.11, positive displacements are in the positive offset direction. Therefore, horizontal movements were mostly towards the tunnel. The point at which there was zero horizontal displacement should be directly above the tunnel centreline, but in this case it was 2.3 m to the right of the centreline. This asymmetry may have been due to a number of factors – geology, excavation method or measurement accuracy (the quoted accuracy of the micrometer stick measurements was ±1 mm).

Figure 1.12 shows the longitudinal horizontal displacements as the TBM passed the array. These were measured using a total station. They are baselined to measurements made before the TBM was in the zone of influence.

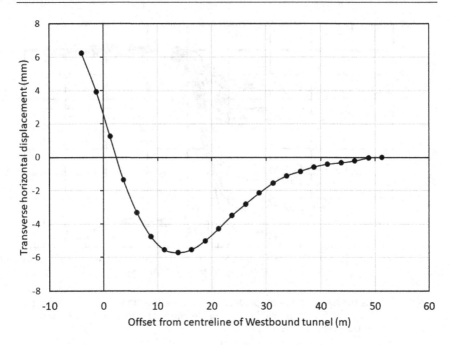

Figure 1.11 Transverse horizontal displacements after passage of the Westbound tunnel at St James's Park (Nyren et al., 2001).

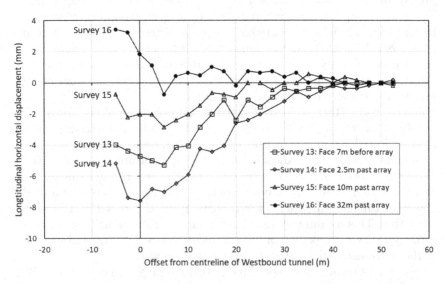

Figure 1.12 Longitudinal horizontal displacements during passage of the Westbound TBM at St James's Park (Nyren et al., 2001).

When the face was 7 m before the array, the ground surface had already moved towards the face, with a maximum of around 5 mm of movement above the tunnel centreline. The next survey was made when the face was 2.5 m past the array and the ground had moved even more. This was then reversed as the face moved past the array, because increments of ground movement were still orientated towards the face and shield of the TBM. By the time the face was 32 m past the array, longitudinal movements had returned to approximately zero, with some positive values close to the centreline.

1.3.4 Long-term settlements

Long-term settlements are caused by pore pressure changes in the soil. For drained soils, there will be no long-term pore pressure changes because these will have all occurred in the short-term (this is, after all, the definition of drained behaviour – that pore pressures will find a steady-state or equilibrium during the timescale of construction). An exception to this rule would be if pore pressures were expected to change after construction, perhaps due to switching off dewatering pumps, or any other foreseeable change in the groundwater table. If the pore pressure increases, then effective stress will decrease, causing heave. If the pore pressure decreases, then effective stress will increase, causing settlement.

For undrained soils, there will always be long-term settlements caused by the dissipation of excess pore pressures over time. This dissipation is driven by pore water flow away from areas where pore pressures are higher than they should be (due to an increase in mean total stress during construction) and into areas where pore pressures are lower than they should be (due to a decrease in mean total stress during construction). This dissipation continues until pore pressures stabilise at a steady-state or hydrostatic equilibrium. In a soil with low permeability, this can take many years. As the excess pore pressures dissipate, the effective stresses in the soil will also change, and the soil will contract or dilate.

Referring back to Figure 1.4, if the piston is loaded and this causes an increase in water pressure in the short-term, then over the long-term water will be squeezed out of the tube until hydrostatic equilibrium is attained and the load has been transferred to the spring. The spring has a much lower stiffness than the water (which is virtually incompressible) and so the piston will move downwards as the spring takes the load. This is analogous to consolidation settlement. If unloading has occurred, then the opposite will occur and over the long-term water will be drawn down the tube until hydrostatic equilibrium has been attained. In this case as the load is transferred to the spring, the piston will be pushed upwards, which is analogous to heave.

During this period of long-term settlement due to dissipation of excess pore pressures, ground behaviour will not be constant volume and

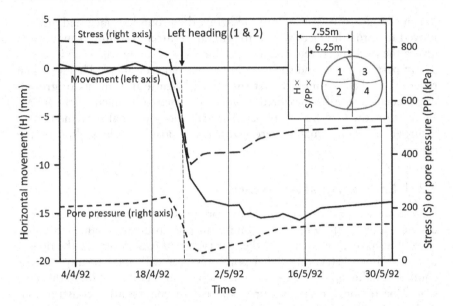

Figure 1.13 Total stress (S) and pore pressure (PP) measured by a spade cell and horizontal movement (H) measured by an inclinometer, as the Heathrow Trial Tunnel passes (redrawn from New & Bowers, 1994).

displacement vectors will not be directed to a single point. Wherever there has been a decrease in mean total stress, which in the short-term will cause a decrease in pore pressure, there will be swelling in the long-term as water flows from surrounding areas to equilibrate the difference. Wherever there has been an increase in mean total stress, which in the short-term causes an increase in pore pressure, there will be consolidation in the long-term as water is squeezed out and flows away. If you are unclear about this, refer to Section 1.2.1 again where undrained behaviour is explained.

Figure 1.13 shows an example of undrained behaviour close to an advancing shotcrete tunnel at Heathrow, UK (New & Bowers, 1994). As the side drift and then enlargement passes, the ground moves towards the excavation, causing a drop in total stress and pore pressure.

Pore pressures in undrained soils are affected by changes in mean total stress. In addition, tunnels in soft normally consolidated clay may generate positive excess pore pressures, as soft clays will try to contract when sheared, whereas tunnels in stiff overconsolidated clay will predominantly generate negative excess pore pressures in the surrounding ground as they try to dilate when sheared.

Therefore, as a tunnel advances, different zones of the ground are loaded or unloaded, principal stresses rotate (see Figure 1.3), and pore pressures change due to changes in both mean total stress and shear. This occurs in a complex 3D manner that will depend strongly on the construction method, the stratigraphy and the soil properties.

To minimise these long-term movements, the construction process should minimise changes in the stress state of the ground. However, even with a closed-face TBM applying support pressure very close to the in situ stress, there will always be changes to the stress state of the ground, and so some long-term movements are inevitable.

Long-term settlements due to the construction of the two running tunnels for the Jubilee Line Extension under St James's Park in London Clay are shown in Figure 1.14. Both tunnels were excavated in 1995/1996 by a 4.85 m diameter open-face TBM with a backhoe. These values were obtained from extensometers 5 m below the ground surface, since surface settlement levelling points were removed before 1999 (Avgerinos et al., 2016). The first tunnel constructed was the Westbound (WB), passing this location from 27th to 28th April 1995 (Nyren et al., 2001), and a typical transverse settlement trough was created. The Eastbound (EB) tunnel passed through 8 months later, from 8th to 10th January 1996. During this 8 month period (compare 'After WB' with 'Before EB' in Figure 1.14), the settlement increased, and also appeared to spread over a wider area. The next set of readings, on 1st March 1997, 14 months on from the 'After EB' data set, the settlement increased even more. The next readings, on 1st August 2006 more than 10 years after the Eastbound tunnel was constructed, and on 6th August 2011 more than 15 years after the Eastbound

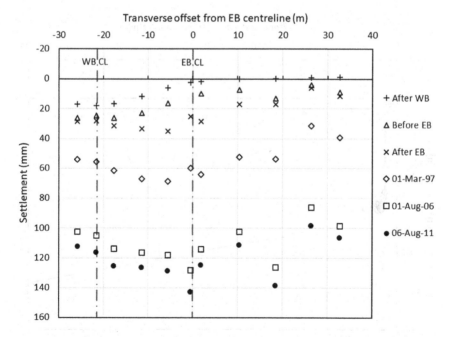

Figure 1.14 Long-term settlements due to the construction of Westbound (WB) and Eastbound (EB) Jubilee Line Extension running tunnels under St James's Park (data from Avgerinos et al., 2016).

tunnel was constructed, show the settlements continue to increase to quite large values. There is evidence in these last three readings that the trend has been slowing to a long-term equilibrium or steady-state value.

For tunnels in stiff overconsolidated clay, the predominantly negative excess pore pressures generated during tunnel construction should cause heave in the long term as the clay draws in water from surrounding areas and swells. So what is causing the long-term settlements we see in stiff clays? Numerical modelling by Shin et al. (2002) shows that if the tunnel is acting as a drain, then there will be long-term settlements. This is because a permanent reduction in pore pressure caused by the tunnel draining the ground around it will cause consolidation, in the same way as tree root suctions will cause settlement of a building. If the tunnel lining is impermeable, then there will be a slight heave of the ground in the long term.

In soft, normally consolidated or lightly consolidated clays, we can expect the magnitude of settlements to be greater than for stiff over-consolidated clays in both the short and long term. Belshaw & Palmer (1978) presented results from extensive monitoring of the Thunder Bay Tunnel in Ontario, Canada. The measurements at one settlement point above the tunnel centreline are shown in Figure 1.15. At the monitoring location, the stratigraphy consisted of about 3 m of silty sand underlain by 3.1 m of silt and then 17.7 m of silty clay and clay extending to bedrock. The consistency of the

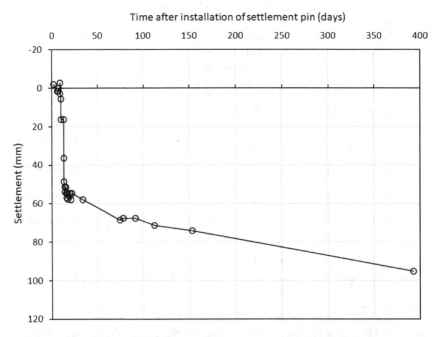

Figure 1.15 Long-term settlement monitoring of a surface settlement pin over the Thunder Bay Tunnel centreline (Belshaw & Palmer, 1978).

clay was soft above the tunnel, becoming firm at tunnel invert level. The tunnel was constructed by an open-face TBM with an excavated diameter of 2.47 m at a depth to axis of about 11 m. Although the area of the face was only about a quarter of the size of the St James's Park tunnels, the short-term settlements at Thunder Bay were nearly double the magnitude, due to the softer ground.

Figure 1.15 shows that over the first year after construction, settlements continued to increase and it is likely the trend would have continued had measurements been made over a longer period of time.

1.4 SUBSURFACE GROUND MOVEMENTS

Measuring subsurface ground movements is more difficult than measuring surface settlements, and so there are fewer data available. Usually, these movements have to be measured by installing instruments in boreholes to measure relative movements or rotations at different depths. Some examples of this can be found in Section 1.2.

Vertical subsurface ground movements can be measured by extensometers. A casing containing magnets is grouted into a borehole. An extensometer can then be lowered into the casing to measure the relative distances between the magnets. The level of the top of the casing must be measured as well at the same time as every set of extensometer readings, usually with a precise level. Therefore, to obtain a subsurface settlement trough, a number of boreholes are required. This is expensive, and difficult to achieve in an urban area, so most case studies of subsurface settlements tend to be in parks or other open areas and they are far less common than surface settlement monitoring.

Horizontal subsurface ground movements can be measured by inclinometers. A special casing is grouted into a borehole. The inclinometer is then lowered into the borehole, or it could be left in place. The inclinometer has several articulating sections and is capable of measuring changes in rotation between each section. The changes in rotation can be used to calculate horizontal displacements in two orthogonal directions.

It is possible to install a combined extensometer/inclinometer in a single borehole. An example of this is the proprietary system called 'shape accel array' (Lipscombe et al., 2015).

The vectors shown previously in Figures 1.1 and 1.2 were inferred from inclinometers and extensometers installed in boreholes.

1.5 STABILITY

Stability in geotechnical engineering is about avoiding failure or collapse. Instability is where dry soil, or soil and water fall or flow in an uncontrolled manner.

As we have seen in the preceding sections, as the tunnel is advanced, ground movements occur until the stresses in the ground and the support find a new equilibrium. If a new equilibrium cannot be found, because the destabilising forces are larger than the resistance provided by soil strength and support forces, then the ground movements will continue increasing until we have a collapse.

Chapters 2–5 will describe how to calculate and quantify stability as a factor of safety. When stability is quantified in this way, the factor of safety is a measure of how close we are to a global failure and we can expect that when the factor of safety is large, ground movements will be small, and when the factor of safety is approaching 1, ground movements will be large. This relationship between the magnitude of ground movements and stability will be further developed in Chapter 12.

1.5.1 The consequences of instability

In an open-face tunnel, instability may cause collapse, endangering the underground workforce and equipment, and surface infrastructure and people, who may fall into the hole or have a building collapse on top of them (Figure 1.16). In a tunnel being excavated by a closed-face machine,

Figure 1.16 Collapse of the Munich Metro (*Construction Today,* 1994).

workers may be protected, but instability of the face can cause overexcavation, excessive settlements of nearby structures and large holes opening up at the surface. These can all be very dangerous to the general public and are best avoided.

Cohesionless soils below the water table can be so unstable that they will flow through a small hole if even a small hydraulic gradient is present (Figure 1.17). In cases like these, the integrity of the lining and its gaskets is very important, as any flow of soil into the tunnel means a void forming outside the tunnel, which leads to unequal pressures acting on the lining and possibly structural failure of the ring. It has been known for tunnels to be flooded by silt and/or sand and water flowing through small gaps or holes in the lining, leading to total abandonment of the TBM and a complete restart (e.g. in Preston, UK: Thomas, 2011) or costly rescue works (e.g. Hull wastewater flow transfer tunnel, UK: Brown, 2004; Grose & Benton, 2005, 2006). Even in well-sealed tunnels being constructed using a closed-face TBM, massive overexcavation can occur if the support pressure is insufficient, as the ground will flow towards the cutterhead of the machine (e.g. Storebælt Eastern Railway Tunnel, Denmark: Biggart & Sternath, 1996).

Stability can also be a problem if there is *too much* support pressure in the tunnel. For instance, if compressed air is used and the pressure is too high, it can blow out to the surface as it did on the Docklands Light Railway Lewisham Extension in 1998 (New Civil Engineer, 1998). A crater 22 m wide and 7 m deep was created, and all the windows of the adjacent school building were broken. Fortunately, it was the weekend and no one was hurt.

Heading stability is the single most important aspect of soft ground tunnelling. It is often what determines the choice of construction method, for both conventional tunnelling and mechanised tunnelling, as shown in the lists below. These decisions must be based on a thorough understanding of the geology and the geotechnical behaviour.

Figure 1.17 Sand and water flowing through an opened grout plug in an old cast iron lining.

Mechanised tunnelling (i.e. TBM) decisions:

- choice of TBM type, i.e. open-face, slurry, EPB
- choice of segmental lining type, i.e. bolted with gaskets or expanded wedgeblock
- method for head interventions, i.e. atmospheric, compressed air or use of divers, or use of ground improvement such as grouting or dewatering

Conventional tunnelling (e.g. backactor or roadheader followed by shotcrete lining) decisions:

- choice of construction sequence, how face is divided, when invert is closed
- choice of available contingency measures
- choice of support types and toolbox measures, such as face shotcrete, temporary inverts, canopy tubes, spiles and face dowels
- whether ground improvement is needed, e.g. permeation grouting, jet grouting, ground freezing, dewatering

1.5.2 The causes of instability

Four main factors influence stability:

- *Cohesion* and *support pressure*, which improve stability
- *Gravity* and *seepage forces*, which worsen stability

Without either cohesion to hold the soil grains together or a support pressure applied to the face, a vertical face will fall down due to gravity or seepage forces. This is illustrated by Figure 1.18 which shows images from a 1 g model test (i.e. a scale model but not in a centrifuge) of a heading in dry sand by Messerli et al. (2010). As the plunger is retracted the sand falls into the tunnel ending up with an angle of repose approximately equal to its angle of friction.

Gravity is always present, but seepage forces only occur when there is a hydraulic gradient (also called a head difference) between the ground and the tunnel. As groundwater seeps through the ground towards the face it pushes the soil grains apart with a 'seepage force' proportional to the hydraulic gradient, in the direction of flow. This decreases stability. Conversely, if the head of a TBM were filled with water or slurry with a higher pressure than the groundwater pressure, it would flow into the ground, and this would aid stability.

A cohesionless soil needs only a very small hydraulic gradient for it to fail due to seepage, just a few centimetres of head is enough, as anyone who has tried to build a dam on a sandy beach will know. On the other hand, a

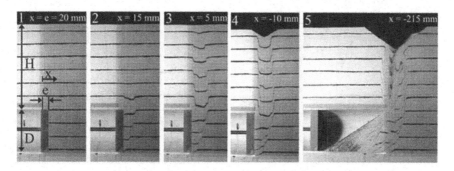

Figure 1.18 Gravity failure of dry sand in a 1 g model (from Messerli et al., 2010).

moist (the technical term is 'partially saturated') sand above the water table can have a small amount of apparent cohesion caused by capillary suction in the pores, as anyone who has built a sandcastle will also know. This may be just enough for a drained material to remain standing in small exposures for a short time. However, even a small amount of perched groundwater could cause local instability, so an open-face tunnel in these ground conditions needs to be planned and executed with great care.

Clays can often be observed standing in vertical faces, sometimes in large diameter open-face tunnels with no face support. This is because clays have cohesion. This cohesion is largely due to the clay's very low permeability. During the timescale of construction, as the soil is unloaded by removal of the soil that used to be next to it, the soil grains relax towards the face and this causes the pore water pressure between the soil grains to decrease. This drop in pore water pressure, known as 'suction' or 'negative excess pore pressure', holds the grains together. The low permeability of clay means that positive or negative excess pore pressures can exist for a long time because it takes so long for groundwater to flow to or from the surrounding ground to dissipate them. This is why it is called 'undrained' behaviour.

When these negative excess pore pressures are eventually dissipated, the clay will behave in a drained manner, i.e. more like a sand. Drained cohesion in clays is usually very small or zero. As shown above, *all* headings *will fail* without either cohesion or support pressure; it's just that in clays, it may take a very very long time.

For all these reasons, drained and undrained stability are quite different. The geometry of failure is different, and the calculations are done in a different way, so in this book they have been split into two separate chapters – Chapter 2 for undrained stability and Chapter 3 for drained stability.

A rule of thumb is that a soil with a permeability less than 10^{-7} to 10^{-8} m/s will behave in an undrained manner during the timescale of a typical tunnel construction, and a soil with higher permeability will behave in a drained manner (Anagnostou & Kovári, 1996). If it is unclear whether a soil will

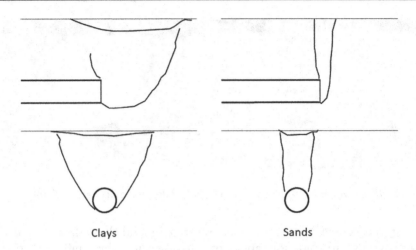

Clays   Sands

Figure 1.19 Geometry of stability failure in clay and sand (redrawn from Mair & Taylor, 1997), in centrifuge tests on clay by Mair (1979) and on sand by Chambon & Corté (1994).

behave in a drained or undrained manner, stability calculations need to be done for both cases and the worst case used for design. It is important to remember that real soils are variable, heterogeneous and rarely isotropic, and that ground mass behaviour can be different to the behaviour of small samples in geotechnical laboratory tests.

Interestingly, undrained stability depends on the depth of the tunnel, whereas drained stability is independent of depth and depends only on tunnel diameter. Another difference is the shape of the stability failure. As shown in Figure 1.19, in the drained case, the soil fails in a steep-sided chimney (see also Figure 1.18), which may or may not extend to the surface, whereas in the undrained case, the failure geometry has a much wider extent and is more of a cone shape.

1.6 TUNNEL LINING MOVEMENTS

After installing a tunnel lining, the lining's self-weight and the ground and groundwater pressures acting on the outside of the lining will cause it to deform. Since in soft ground the lining is installed either close to the face or at the back of the shield, but cannot be installed ahead of the face, it will not be directly affected by ground movements ahead of the face of the tunnel, but only those behind the face or shield. This means that the lining will deform much less than the ground around the tunnel and thus monitoring of lining movements tells us only part of the story of ground deformation, but rather more about the interaction between the lining and the ground.

Once installed, the lining will interact with the ground until an equilibrium state is found. This will depend on the stiffness and thickness of the lining and how many joints it has, as well as the ground behaviour. In soft ground, it is usual for inward movements of the tunnel lining, sometimes referred to as 'convergence', to be larger in the vertical direction than in the horizontal direction. This type of deformation is called 'squatting' and occurs because the vertical ground pressure applied to the lining is usually higher than the horizontal.

Even in overconsolidated soils with K_0 greater than 1, it is usual for tunnel linings to squat, as evidenced by large numbers of convergence measurements made in London Underground tunnels by Wright (2013), where average squat was between 0.5% and 1.0% of diameter. This may be due at least in part to the tunnel acting as a drain on the ground around it. The higher horizontal than vertical permeability in the London Clay (and indeed all horizontally bedded sedimentary soils, which is most of the soft ground on the planet) results in an asymmetric draw-down and hence larger consolidation strains in the vertical than the horizontal direction. This results in a higher vertical ground pressure applied to the lining than horizontal, causing squatting.

1.6.1 Case studies of tunnel lining movements

Usually, lining deformations in soft ground are small. If a total station is used, which is common practice because it is easy, convenient and already needed for setting out, then the accuracy of the measurements is approximately ±2–3 mm (Bock, 2003). For a concrete lining, either precast or sprayed, we expect strains at working loads to be less than 0.1% (e.g. Muir Wood, 2000). This strain corresponds to 1 mm change in diameter per metre diameter, thus 4 mm convergence of opposite points on the lining for a 4 m diameter tunnel, or 10 mm convergence for a 10 m diameter tunnel (Jones, 2007). Therefore, in most cases, at expected working loads, the magnitude of the lining displacements will not be much more than the accuracy of measurement with a total station.

The purpose of lining displacement monitoring in a soft ground tunnel is not to understand what is going on in any detail, but to give warning of something catastrophic occurring. Some typical data from lining displacement monitoring at Heathrow Terminal 4 station are given by Clayton et al. (2003) and shown in Figure 1.20.

In Figure 1.20, we can see that the vertical lining displacements were larger than the horizontal, despite this tunnel being constructed at 18 m depth in London Clay at Heathrow Terminal 4, where the site investigation indicated the K_0 value was 1.5. This was partly due to redistribution of stresses in the ground around the heading, and partly due to prior construction of adjacent platform tunnels, which would have reduced the horizontal stress in the ground.

An extreme example of lining displacements, from monitoring of the Heathrow Express Concourse Tunnel at the Central Terminal Area leading up to its collapse, is shown in Figure 1.21 (HSE, 2000).

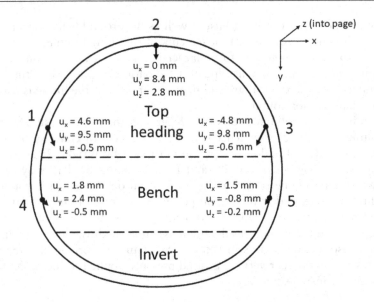

Figure 1.20 Typical lining displacements (exaggerated scale) for a sprayed concrete lining in soft ground (data from Table 2 in Clayton et al., 2003).

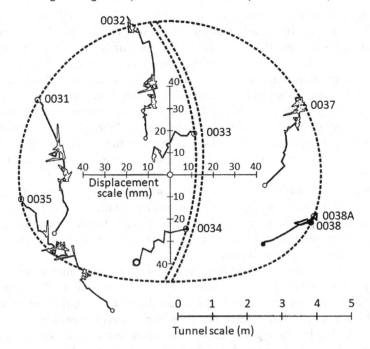

Figure 1.21 Lining displacements measured in Heathrow Express CTA concourse tunnel Ch.30 leading up to collapse (redrawn from HSE, 2000; contains public sector information published by the Health and Safety Executive and licensed under the Open Government Licence).

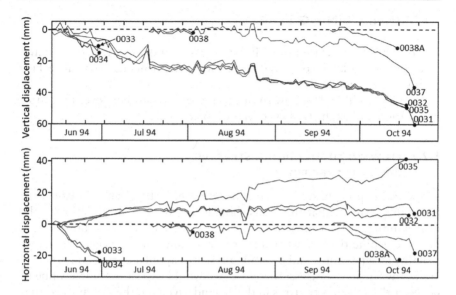

Figure 1.22 Vertical and horizontal lining displacements in the Heathrow Express CTA Station Concourse tunnel Ch.30 plotted against time leading up to collapse on 20th/21st October 1994 (redrawn from HSE, 2000; contains public sector information published by the Health and Safety Executive and licensed under the Open Government Licence).

The lining displacements shown as vector plots in Figure 1.21 can also be viewed in graph form against time in Figure 1.22. You can see that displacements never really stabilised and continued increasing over 4 months before collapse occurred. The magnitude of the lining displacements (compare to Figure 1.20) was much larger than should have been expected based on the trial tunnel constructed beforehand (Deane & Bassett, 1995).

There were many factors involved in the collapse, which were discussed at length in the Health and Safety Executive report (HSE, 1996), but even so, the lining displacements should have provided sufficient warning that the tunnel lining was suffering distress, but the monitoring data was ignored. The lining suffered from poor workmanship and quality control, but the beneficent ground conditions meant this did not result in immediate collapse. However, as well as time-dependent behaviour of the London Clay, there were construction operations that would have increased the load on the lining with time, such as subsequent construction of the adjacent platform tunnels, compensation grouting, and remedial works to the concourse tunnel invert. In less favourable ground conditions, fewer of these factors could have caused earlier and more sudden failure.

1.7 TUNNEL LINING STRESSES

It is difficult to measure tunnel lining stresses, and so there are not many case studies available compared to surface settlements or even subsurface ground movements.

Firstly, we will look at how tunnel lining stresses develop over time after installation and as the tunnel continues to advance.

1.7.1 How do tunnel lining stresses develop over time?

We would expect, based on the conceptual model we have developed so far in this chapter, that the stress should increase after installation of the tunnel lining, as the tunnel advances, until a stable value is obtained some distance behind the face when a short-term equilibrium has been achieved. Usually, face pressures and grouting pressures applied during construction are below the in situ stress; therefore, we would expect the stable value to be lower than the in situ stress in the ground. In effect, the in situ stress that was present before construction has been shared between the lining and arching in the ground around the tunnel.

How much of the ground load is taken by the lining depends on the timing of lining installation, the axial and flexural stiffness of the lining, and the soil behaviour. In soft ground, we would expect the lining to be significantly stiffer than the ground, and so it should attract more of the load. In hard rock, we would almost always expect the rock to be stiffer than the lining such that the rock itself is the main supporting element in the system.

In the long term, equilibrium of the lining-ground system may change due to a variety of factors, such as:

- the lining may creep and/or shrink
- the lining may deteriorate, perhaps due to chemical attack, corrosion, age or damage
- there will be temperature changes within the tunnel that cause the lining to expand or contract, increasing or decreasing the stress in the lining
- surface excavation, e.g. for a basement, may reduce the vertical ground load on the tunnel and reduce the lining stress, and may cause distortion of the tunnel
- surface loading, e.g. due to building construction, may have a similar but opposite effect on the lining stress
- nearby tunnel excavation may change the stress state in the ground and hence the lining stresses will change
- the pore pressures in the ground may change due to long-term changes to the water table or flow regime, changing in turn the water pressure applied to the lining

- in undrained soils, dissipation of excess pore pressures will cause consolidation or swelling of the soil, which may change the stresses applied to the lining

A comparison is made in Figures 1.23–1.25 of previous stress measurements in London Clay by Skempton (1943), Cooling & Ward (1953), Ward & Thomas (1965), Muir Wood (1969), Barratt et al. (1994) and Bowers & Redgers (1996). These were all made using load cells or strain gauges in precast concrete and cast iron segmental linings. These data could have

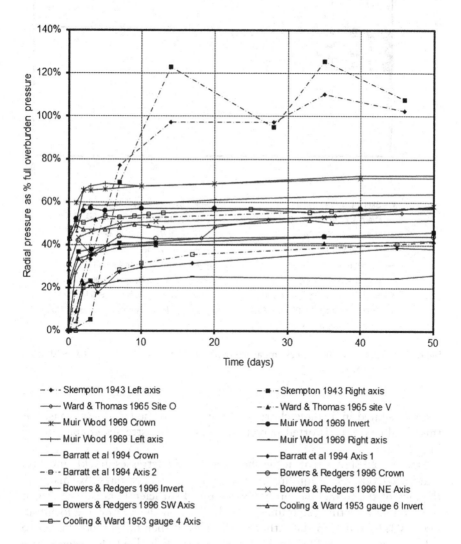

Figure 1.23 Tunnel lining stress measurements in London Clay, up to 50 days (Jones, 2007).

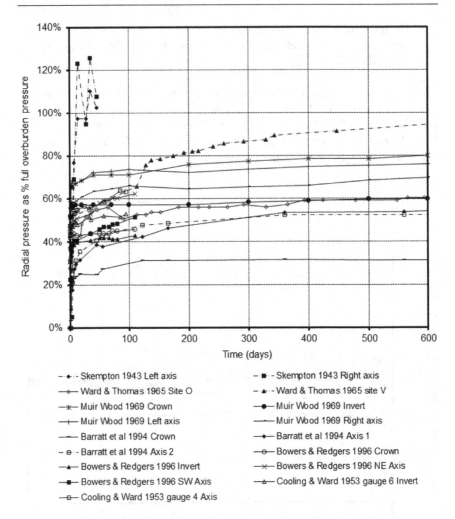

Figure 1.24 Tunnel lining stress measurements in London Clay, up to 600 days (Jones, 2007).

been presented in a single chart with log time on the horizontal axis, but this would give a distorted perception of how the stresses change in the long term.

Figure 1.23 shows that in the short term, lining stresses quickly increase and a relatively stable equilibrium is found within a few days. By this time, the face of the tunnel has probably advanced sufficiently far ahead that it is no longer affecting the lining-ground system at the monitoring location. Figures 1.24 and 1.25 show that in all cases, it then takes much longer to achieve a proper equilibrium, and this is most likely due to long-term pore pressure and temperature changes (Jones, 2007).

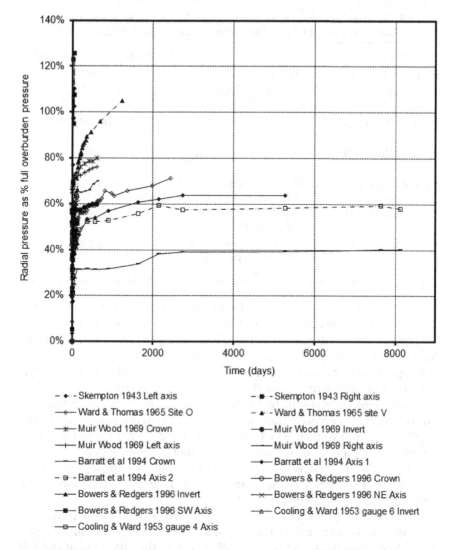

Figure 1.25 Tunnel lining stress measurements in London Clay, up to 19.5 years (Jones, 2007).

For segmental linings, the maximum load has always been found to occur in the long-term. Skempton (1943) found the maximum load to be approximately equal to that corresponding to the hydrostatic full overburden pressure (that is, the initial in situ stress with $K_0 = 1.0$). Ward & Thomas (1965) found that one of the tunnels they studied reached full overburden pressure, while the second one did not but was continuing to increase when measurements ceased. They therefore arrived at the conclusion that hydrostatic full overburden pressure would eventually act on the lining in the long-term. Since then, measurements by Muir Wood (1969), Barratt

Table 1.2 Maximum stresses in tunnels in London Clay, expressed as a percentage of the lining stress if full overburden pressure were applied (Jones, 2007).

Publication	Tunnel	Maximum load (% overburden)
Skempton (1943)	Unknown, London	102–108%
Cooling & Ward (1953)	9′ diameter water tunnel, London	53–64%
Ward & Thomas (1965)	'Site O', Victoria Line, at 3.5 years	71% (not stabilised)
Ward & Thomas (1965)	'Site V', Victoria Line, at 6.5 years	105%
Muir Wood (1969)	Heathrow Cargo Tunnel, Heathrow	60–80%
Barratt et al. (1994)	Jubilee Line, Regent's Park, at 19.5 years	40–64%
Bowers & Redgers (1996)	Jubilee Line Extension, St James's Park, at 4 months	43–62%

et al. (1994), Bowers & Redgers (1996) and Jones (2007) have all shown that load can stabilise at a value well below that corresponding to full overburden pressure. Values of maximum load as a percentage of hydrostatic full overburden pressure are shown in Table 1.2.

It is possible that the amount of stress in the tunnel lining is related to the amount of ground deformation that was allowed to occur during construction, which may explain the higher loads in the older tunnels that probably caused larger ground deformations. The exception to this rule would be the Heathrow Cargo tunnel, which was constructed with an unusually high degree of face support at shallow cover and had a volume loss of only 0.2% (this low value of 'volume loss' means that ground movements would have been very small – approximately an order of magnitude smaller than the other tunnels presented in Table 1.2).

Therefore, it seems reasonable to assume that so long as the tunnelling process is well-controlled and the ground movements are minimised, then in the long-term the tunnel lining will be supporting 40–70% of the full overburden load in London Clay. If, however, the ground movements are large and the tunnel lining is relatively permeable, there is a potential for up to 100% of full overburden to act on the tunnel lining in the long-term.

In other types of soil there are very few data, but long-term lining stresses in all kinds of soft ground tend to be in the range of 25–100% of full overburden. Some case studies are summarised in Table 1.3.

Suzuki et al.'s (1996) two tunnels were of similar construction and were in similar geology, but they were at different depths, and different grouting pressures were applied. At the Yodo River, grouting pressures were approximately 50% of overburden pressure, whereas at the Shigino Route, they were 75–105%. So at Shigino Route, where higher grouting pressures

Table 1.3 Case studies of lining stress measurements not in London Clay.

Publication	Tunnel	Maximum load (% overburden)
Belshaw & Palmer (1978)	Thunder Bay trunk sewer, Ontario, Canada, 2.38 m OD TBM tunnel, in soft to firm clay, at 1 year	25–57%
Palmer & Belshaw (1980)	Thunder Bay (as above but second array of instrumentation), in soft to firm clay, at 1 year	24–61%
Suzuki et al. (1996)	Yodo River cable tunnel, Osaka, Japan, 3.7 m OD TBM tunnel, in clay/silt/ sand/gravel, at 10 months	40–82%
	Shigino Route cable tunnel, Osaka, Japan, 3.55 m OD TBM tunnel, in clay/silt/ sand/gravel, at 10 months	35–60%
Sakurai & Izunami (1988)	Kobe Municipal Subway, Japan, approx. 10 m wide × 8 m high shotcrete tunnel with rockbolts, in hard clay and dense gravel layers, time of measurement unknown	25% average

were applied and ground deformation was minimised, the eventual ground pressure acting on the tunnel was lower. This is perhaps counterintuitive, but if there is any pattern in lining stress measurements in soft ground, it seems to be that the smaller the ground deformations, the lower the long-term lining stress. This runs counter to what we expect from simple models so it is important to remember this as you work through later chapters of this book.

1.7.2 Design based on precedent practice

If we had more measurements of lining stress, they could be very useful for dimensioning the lining at a preliminary design stage and for calibration of numerical models during detailed design. However, even in London Clay, which is probably the most measured and studied geomaterial, there is a large amount of variation in the measured stresses. This is partly because of the different construction methods employed but is also due to other factors. For instance, none of the published work presented so far in this chapter takes account of temperature, but it has been shown that temperature variations in the tunnel will not only affect the instrumentation but will also cause changes in the tunnel-ground system equilibrium (Jones, 2007). As the temperature in the tunnel increases, the lining will expand, and this will increase the ground pressure acting on it. Similarly, as the temperature in the tunnel decreases, the lining will contract away from the ground, reducing the ground pressure. This factor was identified and defined as 'ground reaction temperature sensitivity' in my thesis (Jones, 2007). It is likely that the long-term increases in lining stress in the London Underground tunnels measured by Ward & Thomas (1965) and Barratt et al. (1994) were caused, at least in part, by increasing tunnel temperatures over those years.

Usually, it is not the average load that is important to lining design, but how much the ground load varies around the tunnel lining. This is because in relatively shallow soft ground tunnels, we are not worried about hoop (axial) loads in the lining, which are usually well below the axial capacity, but bending moments. In London Clay, vertical ground stresses after construction (as measured in the lining at tunnel axis level in Figures 1.23–1.25) have always been found to be higher than the horizontal ground stresses (as measured at the tunnel crown or invert in Figures 1.23–1.25). This is despite the fact that in situ horizontal stress is higher than the in situ vertical stress prior to construction. However, this does agree with the lining displacements that are typically observed, which were presented in Section 1.6, where tunnel linings in soft ground invariably squat.

Another aspect that is particularly important for shotcrete lining design is the rate at which the lining stresses increase as the face continues to advance. This is because the shotcrete is gaining strength with time and we need to know that it has sufficient strength at early age to withstand these stresses. The published work we have looked at so far has not provided much detail in terms of relating lining stress to the position of the face. The following unique case study from the Heathrow Express Terminal 4 Station concourse tunnel will demonstrate the kind of detailed information that can be obtained with carefully installed and interpreted instrumentation.

1.7.3 Heathrow Express Terminal 4 Station concourse tunnel case study

The Heathrow Terminal 4 station concourse tunnel was constructed after the Heathrow Express collapse at the Central Terminal Area (HSE, 2000). A lot of instrumentation was installed to verify the design, some of which was presented earlier in this chapter in Figures 1.2 and 1.20. More details can be found in Clayton et al. (2002), van der Berg et al. (2003), Clayton et al. (2006) and Jones (2007). In this section, we will focus on the radial pressures applied by the ground onto the lining, as measured by radial pressure cells installed before spraying the shotcrete lining. These allowed the ground pressure to be measured from lining installation onwards.

The layout of the Heathrow Express Terminal 4 station is shown in Figure 1.26. It consists of two platform tunnels with a central concourse

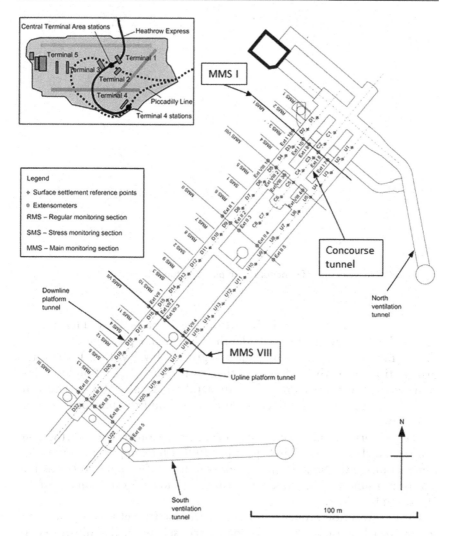

Figure 1.26 Plan of tunnels at Heathrow Express Terminal 4 station, showing location of concourse tunnel and layout of monitoring points and instruments (from van der Berg et al., 2003).

tunnel at the northeastern end. These three tunnels are connected by a series of cross-passages and to the north and south ventilation tunnels at either end, which were constructed last.

The platform tunnels are over 220 m long with a cross-sectional area of 62 m², and the concourse tunnel is 64 m long with a cross-sectional area of 49 m². A cross-section of the concourse and platform tunnels is shown in Figure 1.27, which also shows the surface level and geological strata. The concourse tunnel axis is at a depth of 17.2 m below ground

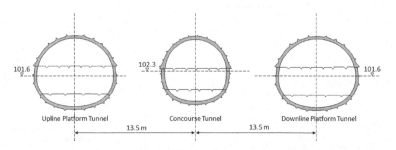

Figure 1.27 Cross-section of concourse and platform tunnels.

level and the tunnel is entirely within the London Clay. Piezometers across the site and at different depths indicated a piezometric level in the Terrace Gravels at approximately ground level with a hydrostatic distribution from there down to the basal beds of the London Clay, well below the tunnel horizon (van der Berg et al., 2003). The centreline spacing between the concourse tunnel and the platform tunnels is 13.5 m.

The platform tunnels were constructed first. Construction of the concourse tunnel commenced in September 1996 after completion of the adjacent sections of the permanent secondary lining in the upline and downline platform tunnels. The concourse tunnel headwall was completed on 7th November 1996.

The construction sequence for the concourse tunnel was top heading, bench, top heading, bench, double-invert, schematically illustrated in Figure 1.28. The invert was closed five rounds from the face. The advance length varied from 0.8 m to 1.2 m depending on ground conditions and design requirements, including the proximity of sensitive structures. The primary support consisted of 350 mm of sprayed concrete (shotcrete), reinforced with two layers of welded wire mesh (8 mm diameter at 150 mm centres) and full-section lattice girders 'Type 110 ROM E3'. The exposed ground was supported by a 50–100 mm shotcrete sealing layer applied immediately after each advance. The 350 mm total thickness included the sealing layer.

We will focus on the pressure cells installed in Main Monitoring Section VIII (MMS VIII) of the concourse tunnel. The location of MMS VIII is shown in Figure 1.29 and in the location plan (Figure 1.26).

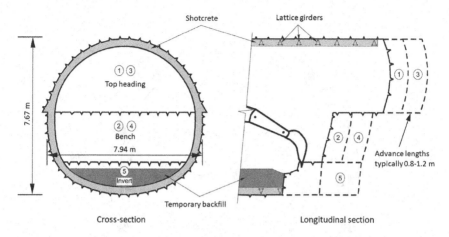

Figure 1.28 Concourse tunnel construction sequence (from van der Berg et al., 2003).

At each section, 12 tangential pressure cells and 12 radial pressure cells were installed. The locations are shown in Figure 1.30.

Table 1.4 lists construction events that may have affected stress measurements at MMS I and MMS VIII.

A selection of the recorded radial pressures for MMS VIII is shown in Figure 1.31. The positions of the radial pressure cells are marked by radial lines normal to the extrados of the lining. The outer perimeter represents the in situ stress normal to the lining calculated at the positions of the radial pressure cells, based on a bulk unit weight for the Made Ground, Terrace Gravel and London Clay of 19.5 kN/m^3 and a coefficient of earth pressure at rest (K_0) of 1.5 (Powell et al., 1997).

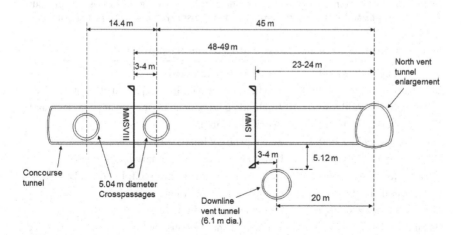

Figure 1.29 Long section of the concourse tunnel showing locations of MMS I and MMS VIII.

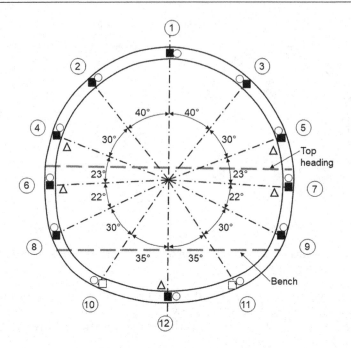

○	Pair of embedded vibrating wire strain gauges
□	Radial pressure cell
■	Tangential and radial pressure cells
△	Slot-cutting (displaced 1m longitudinally)
#	Location reference no. (e.g. radial pressure cell = PCR#)

Figure 1.30 Main monitoring section schematically showing locations of pressure cells and strain gauges embedded in the sprayed concrete primary lining.

Table 1.4 Construction events during concourse tunnel construction.

26th October 1996 12:00	Top heading excavated
26th October 1996 12:00	Top heading primary lining sprayed at positions 1–3
27th October 1996 05:30	Top heading primary lining sprayed at positions 4 and 5
27th October 1996 17:00	Bench excavated
28th October 1996 19:00	Bench primary lining sprayed
28th October 1996 15:30	Invert excavated and primary lining sprayed
7th–14th November 1996	Crosspassage construction (c.f. Figure 1.29)
18th November 1996	Compensation grouting
12th December 1996 (approx.)	Invert secondary lining cast
6th January 1997	Full round sheet waterproofing membrane installed, followed by casting of the secondary lining approximately 6 weeks later

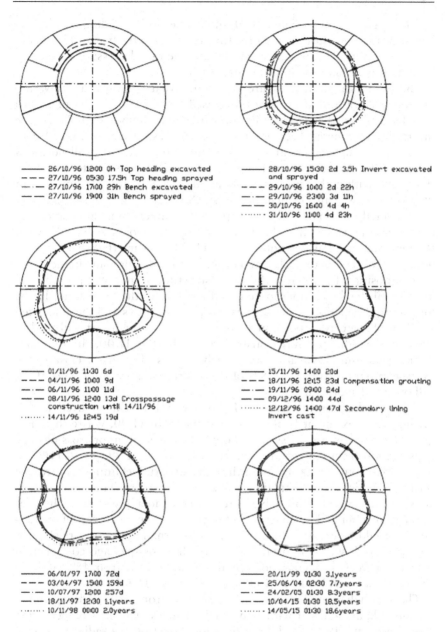

Figure 1.31 MMS VIII radial stresses (L-R, top-bottom: top heading and bench, invert, 6–19 days, 20–47 days, 72 days to 2 years, 3.1 years to 18.6 years).

The first diagram in Figure 1.31 shows the initial readings of the top heading radial pressure cells. At the time of spraying the bench, 13.5 hours after the top heading was sprayed, the average radial pressure on the top heading was 23% of the in situ radial stress.

The second diagram in Figure 1.31 shows that upon invert excavation the top heading radial pressures increased from 23% to 31% of in situ radial stress and the bench radial pressures had climbed to 16%. As the invert sprayed concrete became stiffer and gained strength, the lining began to act as a complete ring and pressure built up gradually at all positions except PCR1. A distinctive radial pressure distribution began to develop at the invert, with higher pressures measured at PCR10 and 11 compared to PCR12. This was also evident in the MMS I radial pressure cells and can be attributed to the non-circular shape of the tunnel lining; at position 12, the lower curvature makes the structural response more flexible relative to the high curvature at positions 10 and 11. This means that there is more unloading of the London Clay, and hence lower radial pressure in the short-term at position 12, whereas the stiffer structural response at positions 10 and 11 will tend to attract more radial pressure in the short term. In the long term, dissipation of negative excess pore pressures and the associated swelling of the clay at position 12 in the centre of the invert will tend to even out the radial pressure distribution. At 31/10/96 11:00, almost 5 days after top heading excavation and nearly 3 days after invert closure, the average radial pressure was 38% of the in situ radial stress or 43% of the full overburden pressure.

The next diagram in Figure 1.31 shows very little change over the following week, except at PCR8 and 10 at 06/11/96 11:00, where there is a noticeable increase in pressure. Although the dates given for cross-passage construction from the construction records were 07/11/96 to 14/11/96, it may be that some breaking out of the concourse tunnel lining occurred on 06/11/96, which would explain the change in pressure. The cross-passage construction caused a noticeable increase in the radial pressures at the sides of the tunnel but had much less effect near the crown and invert of the tunnel, as would be expected. By 14/11/96 12:45 the radial pressure changes had stabilised, with the overall effect of the cross-passage construction being an increase in average horizontal ground pressure from 45% of the in situ radial stress at 04/11/96 10:00 to 53% at 14/11/96 12:45.

The fourth diagram in Figure 1.31 shows that from 20 to 47 days, ground pressures had stabilised with only small changes evident. Compensation grouting on the 18/11/96 did not have any effect on the radial pressures. Casting the secondary lining invert on the 12/12/96 caused a significant increase at PCR12.

From 72 days to 18.6 years (last two diagrams in Figure 1.31), there were only gradual changes in radial pressure, tending to increase and even out the pressure distribution. The higher level of unloading of the ground in the area around PCR12 evident soon after closure of the invert resulted in negative excess pore pressures, which over time dissipated due to ground-water flow, leading to gradually increasing radial pressure due to swelling of the London Clay.

In conclusion, although the ground this tunnel was excavated through would already have been disturbed by construction of the parallel platform tunnels and therefore this was not really a 'greenfield' situation, the trend in the long term showed very little increase in load after the construction period was over. This may have been due to the control of ground deformation during construction, reducing the magnitude and extent of excess pore pressure, or due to the waterproof membrane making the tunnel impermeable so it did not act as a drain on the ground around it. Or indeed both of these effects may have been important.

1.8 SUMMARY

Using case studies, this chapter has described the evolution of ground movements, pore pressures, effective stresses and lining stresses as a tunnel advances, and what happens to them in the long-term. Drained and undrained behaviour has been explained and the factors affecting stability have been described. You should now have a conceptual model in your head of ground and tunnel lining behaviour and how they change with face position and time.

REFERENCES

Anagnostou, G. & Kovári, K. (1996). Face stability conditions with earth-pressure-balanced shields. *Tunn. Undergr. Space Technol.* 11, No. 2, 165–173.

Attewell, P. B. & Woodman, J. P. (1982). Predicting the dynamics of ground settlement and its derivatives caused by tunnelling in soil. *Ground Eng.* 15, No. 8, pp. 13–18, 22, 36.

Attewell, P. B., Yeates, J. & Selby, A. R. (1986). *Soil movements induced by tunnelling and their effects on pipelines and structures.* Glasgow: Blackie.

Avgerinos, V., Potts, D. M. & Standing, J. R. (2016). The use of kinematic hardening models for predicting tunnelling-induced ground movements in London Clay. *Géotechnique* 66, No. 2, 106–120.

Barratt, D. A., O'Reilly, M. P. & Temporal, J. (1994). Long-term measurements of loads on tunnel linings in overconsolidated clay. *Proc. Tunnelling '94*, pp. 469–481. London: IMM.

Belshaw, D. J. & Palmer, J. H. L. (1978). Results of a program of instrumentation involving a precast segmented concrete-lined tunnel in clay. *Can. Geotech. J.* 15, 573–583.

Biggart, A. R. & Sternath, R. (1996). Storebaelt eastern railway tunnel: construction. *Proc. Inst. Civ. Engrs – Civ. Engrg* 114, Special issue 1, 20–39.

Bock, H. (2003). Geotechnical instrumentation of tunnels. *Summerschool on rational tunnelling* (ed Kolymbas, D.), Innsbruck, Austria, pp. 187–224. Berlin: Logos Verlag.

Bowers, K. H. & Redgers, J. D. (1996). Discussion: observations of lining load in a London Clay tunnel. *Proc. Int. Symp. on Geotechnical Aspects of Underground Construction in Soft Ground* (eds Mair, R. J. & Taylor, R. N.), London, UK. Rotterdam: Balkema.

Brown, D. A. (2004). Hull wastewater flow transfer tunnel: recovery of tunnel collapse by ground freezing. *Proc. Inst. Civ. Engrs – Geotech. Engrg* **157**, April, issue GE2, 77–83.

Chambon, J. F. & Corté, J. F. (1994). Shallow tunnels in cohesionless soil: stability of tunnel face. *J. Geotech. Eng. ASCE* **120**, No. 7, 1150–1163.

Clayton, C. R. I., Thomas, A. H. & van der Berg, J. P. (2003). SCL tunnel design in soft ground – insights from monitoring and numerical modelling. *Summerschool on rational tunnelling* (ed Kolymbas, D.), Innsbruck, Austria, pp. 61–92. Berlin: Logos Verlag.

Clayton, C. R. I., van der Berg, J. P., Heymann, G., Bica, A. V. D. & Hope, V. S. (2002). The performance of pressure cells for sprayed concrete tunnel linings. *Géotechnique* **52**, No. 2, 107–115.

Clayton, C. R. I., van der Berg, J. P. & Thomas, A. H. (2006). Monitoring and displacements at Heathrow Express Terminal 4 station tunnels. *Géotechnique* **56**, No. 5, 323–334.

Construction Today (1994). Police probe repeat Munich tunnel breach. *Construction Today*, October issue, pp. 4–5.

Cooling, L. F. & Ward, W. H. (1953). Measurements of loads and strains in earth supporting structures. *Proc. 3rd ICSMFE*, Session 7/3, Vol. 2, pp. 162–166.

Craig, R. (1975). Discussion at meeting of British Tunnelling Society. *Tunnels and Tunnelling* **7**, 61–65.

Deane, A. P. & Bassett, R. H. (1995). The Heathrow Express trial tunnel. *Proc. Inst. Civ. Engrs – Civ. Engrg* **113**, July, 144–156.

Grammatikopoulou, A., Zdravkovic, L. & Potts, D. M. (2008). The influence of previous stress history and stress path direction on the surface settlement trough induced by tunnelling. *Géotechnique* **58**, No. 4, 269–281.

Grose, B. & Benton, L. (2005). Hull wastewater flow transfer tunnel: tunnel collapse and causation investigation. *Proc. Inst. Civ. Engrs – Geotech. Engrg* **158**, October, Issue GE4, 179–185.

Grose, B. & Benton, L. (2006). Discussion: Hull wastewater flow transfer tunnel: tunnel collapse and causation investigation. *Proc. Inst. Civ. Engrs – Geotech. Engrg* **159**, April, Issue GE2, 125–128.

HSE (1996). *Safety of New Austrian Tunnelling Method (NATM) tunnels*, Health and Safety Executive, p. 86. London: HSE Books, HMSO.

HSE (2000). *The collapse of NATM tunnels at Heathrow Airport – A report on the investigation by the Health and Safety Executive into the collapse of New Austrian Tunnelling Method (NATM) tunnels at the Central Terminal Area (CTA) of Heathrow Airport on 20/21 October 1994*. Norwich: Her Majesty's Stationery Office/HSE Books.

Jones, B. D. (2007). *Stresses in sprayed concrete tunnel junctions*. EngD thesis, University of Southampton.

Lipscombe, R., Carter, C., Perkins, O., Thurlow, P. & Guerrero, S. (2015). The use of Shape Accel Arrays (SAA) for measuring retaining wall deflection. *Crossrail Project – Infrastructure design and construction*, Vol. 1. London: ICE Publishing. Also available at: https://learninglegacy.crossrail.co.uk/documents/use-shape-accel-arrays-saa-measuring-retaining-wall-deflection/ [last accessed: 23rd June 2020].

Mair, R. J. (1979). *Centrifugal modelling of tunnel construction in soft clay*. PhD thesis, University of Cambridge.

Mair, R. J. & Taylor, R. N. (1997). Bored tunnelling in the urban environment. Theme Lecture, Plenary Session 4. *Proc. 14th Int. Conf. Soil Mechanics and Foundation Engineering*, Hamburg, Vol. 4, pp. 2353–2385.

Marshall, A. M. (2009). *Tunnelling in sand and its effect on pipelines and piles*. PhD thesis, University of Cambridge.

Marshall, A. M., Farrell, R., Klar, A. & Mair, R. (2012). Tunnels in sands: the effect of size, depth and volume loss on greenfield displacements. *Géotechnique* 62, No. 5, 385–399.

Messerli, J., Pimentel, E. & Anagnostou, G. (2010). Experimental study into tunnel face collapse in sand. *Physical Modelling in Geotechnics* (eds Springman, S., Laue, J. & Seward, L.), pp. 575–580. London: Taylor & Francis.

Muir Wood, A. M. (1969). Written contribution, plenary session 4. *Proc. 7th Int. Conf. Soil Mechanics and Foundation Engineering*, Vol. 3, Mexico, pp. 363–365.

Muir Wood, A. M. (2000). *Tunnelling: management by design*. London: E & FN Spon.

New, B. M. & Bowers, K. H. (1994). Ground movement model validation at the Heathrow Express trial tunnel. *Tunnelling '94, Proc. 7th Int. Symp. IMM and BTS*, London, UK, pp. 310–329. London: Chapman and Hall.

New Civil Engineer (1998). Docklands Light Railway Lewisham Extension blowout. February 1998.

Nyren, R. J., Standing, J. R. & Burland, J. B. (2001). Surface displacements at St James's Park greenfield reference site above twin tunnels through the London Clay. *Building Response to Tunnelling: Case Studies from Construction of the Jubilee Line Extension, London – Vol. 2: Case Studies* (eds Burland, J. B., Standing, J. R. & Jardine, F. M.), CIRIA Special Publication 200, Chapter 25, pp. 387–400. London: CIRIA.

Palmer, J. H. L. & Belshaw, D. J. (1980). Deformations and pore pressures in the vicinity of a precast, segmented, concrete-lined tunnel in clay. *Can. Geot. J.* 17, 174–184.

Powell, D. B., Sigl, O. & Beveridge, J. P. (1997). Heathrow Express – design and performance of platform tunnels at Terminal 4. *Proc. Tunnelling '97*, London, pp. 565–593. London: IMM.

Sakurai, S. & Izunami, R. (1988). Field measurements of the Kobe Municipal Subway Tunnel excavated in soil ground by NATM. *Proc. 2nd Int. Symp. on Field Measurements in Geomechanics* (ed. Sakurai, S.), Kobe, Japan, 6th–9th April 1987, pp. 861–869. Rotterdam: Balkema.

Shin, J. H., Addenbrooke, T. I. & Potts, D. M. (2002). A numerical study of the effect of groundwater movement on long-term tunnel behaviour. *Géotechnique* 52, No. 6, 391–403.

Skempton, A. W. (1943). Discussion: Tunnel linings with special reference to a new form of reinforced concrete lining, by G. L. Groves. *J. Inst. Civ. Eng.* 20, No. 5, March, 53–56.

Standing, J. R., Nyren, R. J., Longworth, T. I. & Burland, J. B. (1996). The measurement of ground movements due to tunnelling at two control sites along the Jubilee Line Extension. *Proc. Int. Symp. Geotechnical Aspects of Underground Construction in Soft Ground* (eds Mair, R. J. & Taylor, R. N.), London, pp. 659–664. Rotterdam: Balkema.

Suzuki, M., Kamada, T., Nakagawa, H., Hashimoto, T. & Satsukawa, Y. (1996). Measurement of earth and water pressures acting on the great depth shield segments. *Proc. Int. Symp. Geotechnical Aspects of Underground Construction in Soft Ground* (eds Mair, R. J. & Taylor, R. N.), London, pp. 613–619. Rotterdam: Balkema.

Terzaghi, K. (1936). The shearing resistance of saturated soils. *Proc. 1st Int. Conf. Soil Mechanics*, Vol. 1, pp. 54–56.

Thomas, A. H. (2003). *Numerical modelling of sprayed concrete lined (SCL) tunnels*. Ph.D. thesis, University of Southampton.

Thomas, T. (2011). Silt inflow stops Preston EPBM. *Tunnelling Journal*, March issue, p. 6. Also available at: http://tunnellingjournal.com/news/silt-stops-epbm-in-preston/?doing_wp_cron [last accessed 27th March 2017].

van der Berg, J. P., Clayton, C. R. I. & Powell, D. B. (2003). Displacements ahead of an advancing NATM tunnel in the London Clay. *Géotechnique* 53, No. 9, 767–784.

van Jaarsveld, E. P., Plekkenpol, J. W. & Messemaeckers van de Graaf, C. A. (2006). Ground deformations due to the boring of the Second Heinenoord Tunnel. *Tunnelling. A Decade of Progress – GeoDelft 1995–2005*, pp. 11–17. London: Taylor & Francis Group.

Wan, M. S. P., Standing, J. R., Potts, D. M. & Burland, J. B. (2017). Measured short-term ground surface response to EPBM tunnelling in London Clay. *Géotechnique* 67, No. 5, 420–445.

Wan, M. S. P., Standing, J. R., Potts, D. M. & Burland, J. B. (2019). Pore water pressure and total horizontal stress response to EPBM tunnelling in London Clay. *Géotechnique* 69, No. 5, 434–457.

Ward, W. H. & Thomas, H. S. H. (1965). The development of earth loading and deformation in tunnel linings in London Clay. *Proc. 6th ICSMFE*, Vol. 2, Divisions 3–6, Montreal, Canada, 8th–15th September, pp. 432–436.

Wright, P. J. (2013). Validation of soil parameters for deep tube tunnel assessment. *Proc. Inst. Civ. Engrs – Geotech. Engrg* 166, GE1, February, 18–30.

Chapter 2

Undrained stability

As explained in Section 1.5, drained and undrained stability are quite different. The geometry of failure is different, and the calculations are done in a different way. Therefore, they have been split into two separate chapters – this chapter covers undrained stability and Chapter 3 will cover drained stability.

A rule of thumb is that a soil with a permeability less than 10^{-7} to 10^{-8} m/s will behave in an undrained manner during the timescale of a typical tunnel construction, and a soil with higher permeability will behave in a drained manner (Anagnostou & Kovári, 1996). If it is unclear whether a soil will behave in a drained or undrained manner, stability calculations need to be done for both cases and the worst case used for design.

It is important to remember that real soils are variable, heterogeneous and rarely isotropic, and that ground mass behaviour can be different to the behaviour of small samples in geotechnical laboratory tests. Many of the calculations presented here are simple models and will assume the ground is homogeneous and isotropic and follows simple behavioural rules, so it is important to think about how the true ground behaviour may affect the potential failure mechanisms. For instance, a permeable layer of soil within an open face with groundwater pressure present could cause an uncontrolled failure, even if 90% of the ground is clay.

After working through this chapter you will understand:

- how tunnel headings collapse or blow-out
- how upper and lower bound limit states, limit equilibrium, numerical modelling and empirical data help us to estimate the true collapse load
- undrained stability ratio and critical stability number

After working through this chapter, you will be able to:

- calculate the factor of safety for undrained stability of a heading in a variety of situations
- calculate blow-out pressures in clay for hydraulic fracturing and passive failure

DOI: 10.1201/9780429470387-2

2.1 OVERVIEW OF STABILITY THEORY

Stability is an ultimate limit state. This means that we are designing and planning the construction to avoid a catastrophic failure. We want to try to predict when or how that failure will occur, and then ensure that we have a factor of safety that makes it very unlikely.

Since at failure the strength of the ground will be fully mobilised, heading stability lends itself well to plasticity solutions. A plasticity solution ignores what happens before failure and looks only at the fully mobilised failure geometry to calculate the load required to get there.

A heading may be geometrically simplified as shown in Figure 2.1, where:

D is the excavated diameter in m
C is the cover from the crown of the excavation to the ground surface in m
P is the unsupported length in m

In the case of a closed-face tunnel boring machine (TBM), P may be equal to zero. For a sequentially excavated tunnel lined with shotcrete, some assumptions about the values of P, C and D may need to be made.

For non-circular tunnels, Pound (2005) used numerical analysis to show that even for elliptical or rectangular tunnels, with a width three times the height or a height three times the width, or for semi-circular top headings, stability may be approximated by an equivalent circular tunnel with the same face area, where the cover C is taken as the distance from the actual crown to the surface and the depth to axis is taken as the depth to the centroid of the face. Errors were no more than 1.5%. So the precise geometry is not important, as long as a value of D is chosen that gives the same face area. Pound's results were for undrained cohesive soils, and as we will see

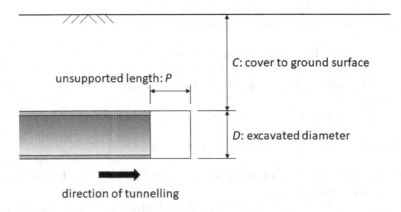

Figure 2.1 Simplified geometry of a tunnel heading.

they may not apply to drained non-cohesive soils – no one has done that research yet.

There are several ways we can determine stability:

1. Assuming a 'kinematically admissible mechanism', such that if a structure is loaded to this value it *must* collapse. This usually involves assuming the ground is made up of several large blocks that slide into the face. It demonstrates that failure must occur at this load, but there may be situations (e.g. other geometries or configurations of blocks) where failure may occur at a lower load. For this reason this is also known as an 'upper bound solution'.

2. Assuming a 'statically admissible stress field', such that if a structure is loaded to this value it *cannot* collapse. This is done by determining a set of stresses in the ground that are in equilibrium with the external loads and do not exceed the strength of the ground (Atkinson, 2007), thus demonstrating that failure cannot occur in this set of circumstances. The true failure load must therefore be higher, so this may be overconservative. This is known as a 'lower bound solution'.

3. These upper and lower bounds bracket the true collapse load. Combining them to find a solution that is kinematically and statically admissible is known as a 'limit equilibrium solution'. However, a limit equilibrium solution may not represent the true collapse load, as it relies on assumptions of geometry of failure, ignores stress–strain behaviour and ignores compatibility of strains.

4. Empirical data from heading stability failures in the field and in centrifuge tests can be used to develop relationships that may help predict the true collapse load.

5. Numerical modelling.

2.2 UNDRAINED STABILITY

The study of undrained stability began because of a collapse near Stockholm in 1964, recounted by Broms & Bennermark (1967):

> On Friday, November 20, 1964, a slide took place at Edsådalen, near Stockholm. The slide occurred in a soft clay when a 6.5 ft diameter hole was cut at the base of a vertical sheet pile wall that supported a 35 ft deep excavation. The slide took place approximately 1-1/2 hr after the clay behind the sheet pile wall had been exposed. The clay surface had been inspected approximately 1 hr before the slide occurred, revealing that the exposed surface was "hard and dry". The first indication that failure was imminent was the observation (made a few seconds before the actual slide by one of the three men present in the excavation) that the exposed clay surface started to move. Before any warning could be

given, all three men were buried under about 1,500 cu ft of clay, and one of the men lost his life.

In order to try to predict failures of this kind and prevent similar disasters, Broms and Bennermark used laboratory tests where clay was extruded under pressure through vertical circular openings, and also field observations of both stable and collapsed openings, to try to characterise stability, and they published the results in a seminal paper in 1967. They defined a stability ratio N, which is the ratio of full overburden pressure at the axis of the opening minus any support pressure divided by the undrained shear strength of the soil c_u, given by:

$$N = \frac{\gamma(C + D/2) + \sigma_s - \sigma_t}{c_u} \tag{2.1}$$

N is the stability ratio and is dimensionless
γ is the bulk unit weight of the soil in kN/m³
C is the cover in m
D is the diameter of the opening as defined in Figure 2.1 in m
σ_s is the effect at tunnel level of a surcharge at the surface (for instance consisting of a stockpile of bulk materials, traffic loads, a flexible raft foundation or a body of water) in kPa
σ_t is an internal face pressure provided by compressed air, slurry in the head of a slurry TBM or earth pressure in an earth pressure balance (EPB) machine in kPa
c_u is the undrained shear strength of the soil in kPa

For headings where the unsupported length $P = 0$, as defined in Figure 2.1, Broms & Bennermark (1967) found that the face should be stable if N is less than 6 and may collapse if N is greater than between 6 and 8, for C/D ratios greater than 3.5.

The soft normally consolidated clays Broms and Bennermark tested tended to generate positive excess pore pressures due to shearing of the soil near the hole. This reduced the effective stress and hence the strength of the soil, resulting in a short-term failure. As the pore water migrated away from the hole, consolidation increased the strength of the clay somewhat, perhaps giving the hard and dry surface appearance observed before the failure at Edsådalen.

For overconsolidated clays, there is a tendency for negative excess pore pressures to be generated during shearing, which increases the effective stress in the short-term. As pore water moves towards these areas over time, the clay will swell and the strength of the clay will decrease. Therefore, it is likely that the strength of a clay, its overconsolidation ratio and its mass permeability will have an effect on the true value of the stability ratio, and will also determine the stand-up time.

Further examination of Equation 2.1, and remembering that a low value of N is more stable and a higher value of N is less stable, can lead us to the following conclusions:

- Increasing the depth of the tunnel will make the heading less stable.
- Increasing the surcharge pressure will make the heading less stable.
- Decreasing the support pressure will make the heading less stable.
- Decreasing the undrained shear strength will make the heading less stable.

2.2.1 Heading stability in homogeneous clay

Davis et al. (1980) published both upper and lower bound solutions for an undrained soil. They are reasonably close together and so the true collapse load may be determined with reasonable accuracy.

Mair (1979, cited in Kimura & Mair, 1981) used centrifuge tests and case histories of tunnel heading failures to develop relationships to help predict the true collapse load. The data lay between the upper and lower bound solutions of Davis et al. (1980). The most common method used in practice for undrained stability calculations are Mair's design charts based on centrifuge testing (Kimura & Mair, 1981).

Mair (1979) performed a suite of centrifuge tests of a tunnel heading in clay at different C/D and P/D ratios, reported in Kimura & Mair (1981). The design curves based on these centrifuge tests are shown in Figure 2.2, as well as some of the centrifuge test data. P/D ratios between zero and infinity were also tested and used to interpolate the design curves for $P/D =$ 0.5, 1 and 2. These are not shown in Figure 2.2, because the values of P/D did not correspond to any of the curves.

A 'critical stability ratio' N_c is defined, which is the value of stability ratio at which failure occurs. Another way of saying this is if $N > N_c$, failure occurs. The ratio N_c/N can also be thought of as the factor of safety for heading stability.

Figure 2.2 shows that as C/D increases, i.e. as the depth of the tunnel increases or the diameter decreases, the heading becomes more stable. This is because there is more space for arching in the ground around the tunnel as it gets deeper (drained stability, on the other hand, is independent of depth, as we will see in the following chapter). But remember that as the depth of the tunnel increases, the value of N in Equation 2.1 also increases. Therefore, moving the tunnel deeper will not necessarily improve stability – this will depend on the diameter of the tunnel and whether the undrained shear strength increases with depth.

Another interesting attribute of this chart is that as P/D increases, the heading becomes less stable. Therefore, reducing the unsupported length will improve stability.

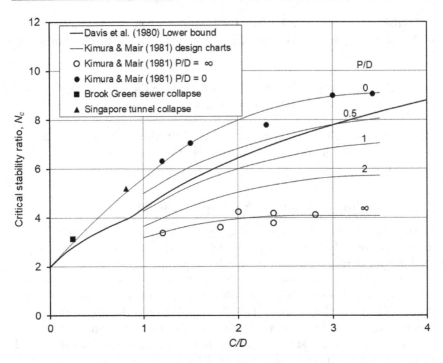

Figure 2.2 Design chart based on Mair's (1979) centrifuge tests (from Kimura & Mair, 1981), and Davis et al. (1980) lower bound envelope for *P/D* = 0. Values for Brook Green sewer collapse and Singapore tunnel collapse from Mair & Taylor (1997).

Also shown on the chart is the lower-bound plasticity solution provided by Davis et al. (1980). It has a kink in it because the critical stress field that causes collapse is different below and above *C/D* = 0.86. If the stability ratio is below the lower bound, then the heading cannot fail; therefore, it is reassuring that the *P/D* = 0 line from Mair's centrifuge tests lies above it. The undrained plasticity solutions will not be presented in detail in this book, as the centrifuge tests give a better estimate of the true collapse load.

WORKED EXAMPLE 2.1   OPEN-FACE TBM TUNNEL IN HOMOGENEOUS CLAY

An open-face shield mounted with a roadheader is to be used to excavate a 7.5 m diameter tunnel in stiff, overconsolidated clay that can safely be assumed to behave in an undrained manner during the timescale of construction. The roadheader can only reach a maximum of 1.5 m ahead of the shield, the shield is 5 m long, the tailskin is 2 m long and grouting is done through grout ports in the tailskin as the machine advances. The tunnel axis is 18 m below the ground surface, the bulk unit weight of the clay is 20 kN/m³ and the undrained shear strength is 100 kPa.

The geometry may be idealised as shown in Figure 2.3.

a. With no surcharge or internal tunnel pressure, calculate the factor of safety on undrained stability.
b. If a surcharge were applied, what value would cause collapse?
c. With no surcharge applied, at what value of undrained shear strength would collapse occur?
d. What is the factor of safety on undrained shear strength, i.e. the undrained shear strength divided by the undrained shear strength at which collapse would occur? Compare this to the value calculated in part a.

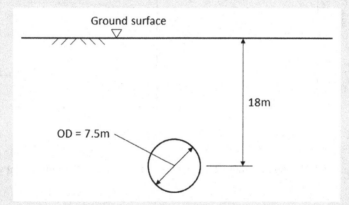

Figure 2.3 Worked Example 2.1 cross-section.

a. We first need to calculate the critical stability number N_c at the ultimate limit state. We can assume that deformations are large enough at the ultimate limit state to cause closure of the ground around the shield and tailskin; therefore, the unsupported length P is 1.5 m.

Using Mair's design chart based on centrifuge tests with the following parameters:

$$P/D = 1.5/7.5 = 0.2$$

$$C/D = (18 - 7.5/2)/7.5 = 14.25/7.5 = 1.9$$

A line is drawn vertically on the design chart in Figure 2.4 at $C/D = 1.9$. Then, the value of N_c at $P/D = 0.2$ may be found by interpolating between the N_c values at $P/D = 0$ (7.8) and $P/D = 0.5$ (6.8), giving $N_c = 7.4$.

If you are using a ruler to interpolate, watch the y-axis gridlines, which go up in steps of 2.

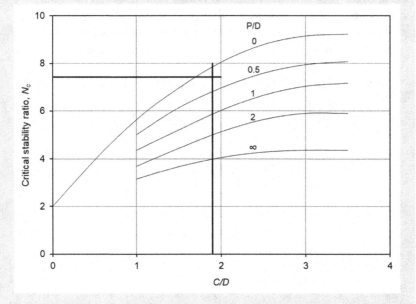

Figure 2.4 Worked Example 2.1 using the design chart to find the critical stability number.

Now the value of stability ratio N needs to be calculated. This is given by:

$$N = \frac{\gamma(C + D/2) + \sigma_s - \sigma_t}{c_u} = \frac{20(18) + 0 - 0}{100} = \frac{360}{100} = 3.6$$

The factor of safety may be given by:

$$\frac{N_c}{N} = \frac{7.4}{3.6} = 2.06$$

b. Since collapse occurs when $N_c = N$, we can use the stability ratio equation to find the value of surcharge σ_s that would cause collapse:

$$N = \frac{360 + \sigma_s}{100} = N_c = 7.4$$

Rearranging to solve for σ_s:

$$\sigma_s = 7.4 \times 100 - 360 = 380 \text{ kPa}$$

c. The stability ratio equation can be used in a similar way to calculate the value of undrained shear strength that would cause collapse with no surcharge:

$$N = \frac{360}{c_u} = N_c = 7.4$$

Therefore:

$$c_u = \frac{360}{7.4} = 48.6 \text{ kPa}$$

d. The factor of safety on undrained shear strength (i.e. the design value divided by the collapse value) is 100/48.6 = 2.06. The previously calculated factor of safety was exactly the same. Therefore the factors of safety on actions and on undrained shear strength are the same, which will become important later when we look at Eurocode 7 design approaches.

WORKED EXAMPLE 2.2   SPRAYED CONCRETE TUNNEL IN HOMOGENEOUS CLAY

A tunnel with an 85 m² cross-sectional area is to be constructed in clay in an urban area using a sprayed concrete lining, at a depth to the centroid of the cross-sectional area of 15 m. A cross-section is shown in Figure 2.5.

The ground is silty clay, with a permeability $k = 1 \times 10^{-9}$ m/s. The ground has a bulk unit weight of 19 kN/m³, an undrained shear strength of 60 kPa, a drained cohesion of 5 kPa and a drained angle of friction of 22°. The water table is at the surface. Assume there is no surcharge.

a. Assuming the tunnel is excavated full-face in 1 m advances, calculate the factor of safety on stability. Is this sufficient for Eurocode 7 or GEO Report 249?

b. Describe ways of improving the factor of safety by inspection of the stability ratio equation and the critical stability ratio design chart. Try some of them and see what happens.

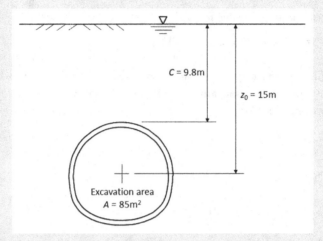

Figure 2.5 Worked Example 2.2 cross-section.

a. The permeability indicates this is an undrained stability problem. First calculate stability ratio N:

$$N = \frac{\gamma z_0 + \sigma_s - \sigma_t}{c_u}$$

Now, $\sigma_t = 0$ because there is no support pressure and $\sigma_s = 0$ because the question says to assume there is no surcharge. So, N is given by:

$$N = \frac{\gamma z_0}{c_u} = \frac{19 \times 15}{60} = \frac{285}{60} = 4.75$$

We need to calculate an equivalent diameter D and calculate C:

$$D = \sqrt{\frac{4A}{\pi}} = \sqrt{\frac{4 \times 85}{\pi}} = 10.40 \text{ m}$$

The cover value C should be the actual depth to the crown, not the value calculated using the depth to axis and the equivalent radius. For design, a cautious value of P should be used to take account of possible overexcavation of the face. It is also common to dome the face rather than leaving it flat. In this case we will assume $P = 1.5$ m.

Now calculate C/D and P/D:

$$\frac{C}{D} = \frac{9.8}{10.4} = 0.94$$

$$\frac{P}{D} = \frac{1.5}{10.4} = 0.14$$

These values are used in the chart in Figure 2.6. Note that the curve for $P/D = 0.5$ does not plot below $C/D = 1$, but it is reasonable to extrapolate such a small distance, especially as all the curves appear to be converging on $N_c = 2$ at $C/D = 0$.

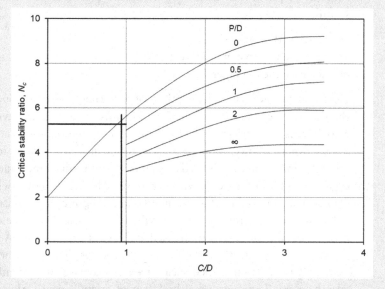

Figure 2.6 Worked Example 2.2 using the design chart to find the critical stability number.

Reading from the chart, $N_c = 5.28$. Therefore the factor of safety may be given by:

$$\frac{N_c}{N} = \frac{5.28}{4.75} = 1.11$$

Eurocode 7 design approach 1 (EN 1997-1: 2004, 2009) would require a factor of safety greater than 1.4. GEO Report 249 (Golder Associates, 2009) recommends a factor of safety greater than 1.5.

b. In order to improve the factor of safety on stability, we can either try to increase the critical stability number N_c or decrease the stability ratio N. By inspection of the stability ratio equation, several possible strategies are available to minimise N:

$$N = \frac{\gamma z_0 + \sigma_s - \sigma_t}{c_u}$$

- increase the undrained shear strength c_u of the soil
- decrease the depth z_0 of the tunnel

- decrease surcharge σ_s
- increase internal support pressure σ_t

By looking at the critical stability number chart, we can identify some more possible strategies to maximise the value of critical stability number N_c:

- increase the cover C
- decrease the unsupported length P
- decrease the diameter D

Note that by decreasing the depth z_0 of the tunnel, we can reduce the stability ratio N in the equation (which is good), but this will also decrease the cover C (which is bad – refer to the stability chart x-axis, where a lower value of C will result in a lower value for critical stability number N_c). Also, in most natural soils, undrained shear strength will increase with depth, so reducing the cover will also decrease the value of undrained shear strength c_u used to calculate the stability ratio N, which will result in a higher value of N (again bad). Therefore, undrained stability is usually improved by increasing the depth of the tunnel, at least for shallow tunnels where $C/D < 1$. If undrained shear strength increases with depth, even by a small amount, then the factor of safety is nearly always improved by increasing the depth of the tunnel.

For example, if the tunnel were raised by 4.6 m, this would make $C = 5.2$ m and $C/D = 0.5$. From the undrained stability chart $N_c = 3.80$, which is a lower (worse) value than before. The depth to axis is now $z_0 = 10.4$, so assuming undrained shear strength is the same as before (60 kPa), then stability ratio $N = 3.29$, which is a lower (better) value than before. Now, the factor of safety $N_c/N = 3.80/3.29 = 1.15$, which is slightly better overall than the previous value of 1.11. However, one would expect the undrained shear strength to be lower nearer to the surface, so on balance the factor of safety will probably be worse. In fact, the value of undrained shear strength would only need to be less than 57.7 kPa for the overall factor of safety to be worse.

Alternatively, if the tunnel were lowered 5.8 m making $C = 15.6$ m, then $C/D = 1.5$, and from the stability chart $N_c = 6.77$. The depth to axis is now $z_0 = 20.8$ m, so assuming undrained shear strength is constant, then the stability ratio $N = 6.59$. The factor of safety given by $N_c/N = 6.77/6.59 = 1.03$, which is slightly worse than before. However, if the undrained shear strength increased with depth at a rate of 8 kPa/m (a typical value) at 5.8 m deeper it would increase from 60 kPa to 106.4 kPa, and the stability ratio would now be $N = 3.71$, making the factor of safety $N_c/N = 6.77/3.71 = 1.82$.

In a clay, ground treatment options to increase the undrained shear strength are limited and probably only jet grouting will work, but in some other types of soils compaction or permeation grouting or ground freezing may improve stability by increasing the undrained shear strength.

For a sprayed concrete tunnel in clay, the most common methods of improving stability are to reduce the unsupported length or to divide the face. For a tunnel of this size (85 m², or 10.4 m equivalent diameter), there would be logistical benefits to dividing the face as well, for instance allowing the excavator to reach the crown by constructing a top heading first. Although beyond the scope of this book, designers need to understand the interaction between size of plant, construction logistics and stability, and how to optimise production while mitigating the risk of instability.

One possible method of dividing the face is shown in Figure 2.7:

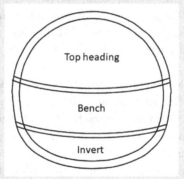

Figure 2.7 Worked Example 2.2 – Cross-section showing possible method of face division using a top heading–bench–invert sequence.

This is a top heading–bench–invert sequence. It is staggered such that although the equivalent diameter is reduced, the unsupported length P is increased, as shown in Figure 2.8.

It is often assumed that the face area is reduced to the size of the partial heading. This is difficult to justify but often the reason given is because the weight of the bench and invert yet to be excavated provides a kind of berm to support the lower part of the face, and because the failure mechanism, at least for shallow tunnels, is soil coming down from above and in front, not from below.

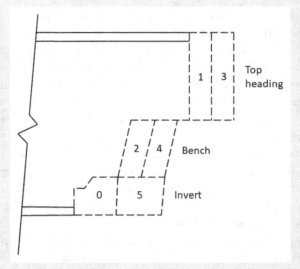

Figure 2.8 Worked Example 2.2 – long section showing top heading–bench–
invert face division.

Ascertaining a value of unsupported length is not straightforward either. Usually, the maximum distance from the excavated
face to full ring closure is taken as the unsupported length, because
in soft ground effective support is only really achieved when the
ring is closed.

In Figure 2.8, the previous invert '0' is excavated and about to be
sprayed, then stage 1 will begin. This represents the maximum unsupported length, which for 1 m top heading and bench advances and a
2 m invert advance will be between 4 and 6 m, depending on whether
it can be assumed that the recently sprayed invert is strong enough
to provide effective support when top heading 1 is excavated, and
what allowance for overexcavation ahead of the leading edge of the
top heading is assumed. A reasonable assumption could be that the
unsupported length is 5 m. If a temporary invert is sprayed in the top
heading, we could assume the unsupported length is 1.5 m.

For this example, we will assume P is 1.5 m, and that the face
area of the top heading is 35 m².

We need to calculate a new equivalent diameter D:

$$D = \sqrt{\frac{4A}{\pi}} = \sqrt{\frac{4 \times 35}{\pi}} = 6.68 \text{ m}$$

The cover C has not changed.

The depth to axis z_0 has reduced slightly and could be found by assuming it is at the centre of the equivalent circle.

$$z_0 = C + \frac{D}{2} = 9.8 + \frac{6.68}{2} = 13.14 \text{ m}$$

Using $P/D = 1.5/6.68 = 0.22$, and $C/D = 9.8/6.68 = 1.47$, the critical stability ratio N_c from the design chart in Figure 2.2 is now approximately 6.5. The decrease in D caused by subdividing the face led to an increase in C/D and a significant increase in N_c.

Since z_0 has changed, N is now given by:

$$N = \frac{\gamma z_0}{c_u} = \frac{19 \times 13.14}{60} = \frac{250}{60} = 4.16$$

Now, the factor of safety $N_c/N = 6.5/4.16 = 1.56$, which is a significant improvement on 1.11 and would be considered sufficient according to Eurocode 7 (EN 1997-1: 2004, 2009) and GEO Report 249 (Golder Associates, 2009).

2.2.2 Heading stability in clay with undrained shear strength increasing with depth

When calculating stability ratio N to compare with design charts for critical stability number N_c, a single value of undrained shear strength must be used, but it is unclear what this value should be if, as is often the case, the undrained shear strength is not constant, but increases with depth. Mair & Taylor (1997) recommended using the value of undrained shear strength at tunnel axis level, but this value may underestimate stability ratio N and hence be unsafe. Atkinson & Mair (1981) recommended using the average value of undrained shear strength between the surface and the tunnel axis, but this may overestimate N and be too conservative.

Pound (2005) ran 3D numerical analyses with a linear increase of undrained shear strength with depth, for $P/D = 0$, and compared the results to models with a constant undrained shear strength. The variation of undrained shear strength with depth was defined by the following equation:

$$c_u = A(B + z) \tag{2.2}$$

c_u is the undrained shear strength in kPa
A and B are constants
z is the depth in m

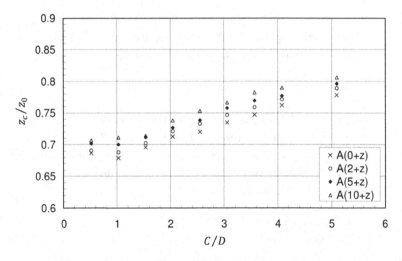

Figure 2.9 Relative depth below ground level at which undrained shear strength is equal to the critical value in constant undrained shear strength analysis (from Pound, 2005).

To encompass the range of soil strengths from soft normally consolidated clays to heavily overconsolidated clays, B was given values of 0, 2, 5 and 10. A was reduced in the model until failure occurred.

Pound (2005) defined an equivalent depth z_c at which the value of undrained shear strength was equal to the value in the constant strength analysis with the same C/D ratio. Figure 2.9 shows this equivalent depth z_c divided by depth to axis z_0 plotted against C/D ratio.

Figure 2.9 shows that for lower values of B, i.e. when the value of undrained shear strength at the ground surface was lower, the equivalent depth was nearer the surface (z_c/z_0 had a lower value). For higher values of B, the equivalent depth was nearer the tunnel (z_c/z_0 had a higher value). But for any particular value of C/D, the relative depth did not vary that much.

It also shows that the relative depth increases for higher values of C/D, meaning that failure is governed by a value of undrained shear strength located closer to the tunnel face for deeper tunnels. This may be because the failure mechanism is more localised at depth.

Overall, within a range of C/D from 0.5 to 6, the relative depth z_c/z_0 varies from 0.675 to 0.825.

Müller (2015) found that for $P/D = \infty$, the equivalent undrained shear strength at failure was about 0.8 times the value at axis level, when it increases with depth using the equation $c_u = 30 + 4.5z$. This would result in z_c/z_0 values between approximately 0.6 and 0.7.

Therefore, contrary to the recommendation of Mair & Taylor (1997), the value at tunnel axis level should not be used. Rather, for a linearly increasing undrained shear strength, the value at 0.6–0.8 times the depth to axis should be used as a first estimate. If you have a nonlinear undrained shear strength variation with depth, or layered soils, you will need to use conservative assumptions or numerical modelling.

2.2.3 Heading stability in clay with overlying coarse-grained soils

A special case, but one that is very common, is when the tunnel is in clay, but the overlying deposits consist of coarse-grained cohesionless soils.

Grant & Taylor (2000) conducted 'plane strain' centrifuge tests to investigate this; the tunnel extends from one end of the model to the other and contains a latex bag filled with compressed air that is slowly decompressed, simulating gradual reduction of support pressure over an infinitely long tunnel ($P/D = \infty$).

They found that applying a surcharge to the top of the model simulated by water in a latex membrane or loose sand resulted in critical stability number values that plotted very close to the design line for $P/D = \infty$ in Figure 2.2 if the cover C was assumed to include the clay only. Therefore, the loose sand appeared to act as though it were an applied surcharge just like the water.

On the other hand, when dense sand was overlying the clay there was a noticeable improvement in critical stability number for tunnels with small depths of clay cover above the crown of $C_{clay}/D < 2$, enhancing the stability. Grant & Taylor (2000) recommended that "if a significant layer of relatively dense coarse grained material is present above a tunnel in clay, an N_c value of 4 (for $P/D = \infty$) would seem to be appropriate regardless of the depth of clay cover above the tunnel crown". This is of limited use in practice, since P/D in soft ground is always a finite distance.

More research is required to explore the effect of the density of the coarse-grained material and further testing at shallow cover, as well as testing at other values of P/D. At present, therefore, it is recommended to include overlying coarse-grained soils in the value for surcharge and to take the cover C as the distance from the crown to the top of the clay.

WORKED EXAMPLE 2.3   A TUNNEL IN CLAY WITH OVERLYING COARSE-GRAINED SOILS

A 6 m diameter tunnel is to be built in soft clay with overlying sand using an EPB TBM as shown in Figure 2.10. What is the minimum support pressure needed to prevent collapse with a factor of safety of 1.5?

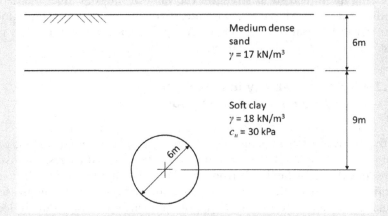

Figure 2.10 Worked Example 2.3 cross-section.

Assume that $P/D = 0$ because it is an EPB machine and unsupported length P is zero.

Calculate C/D based on the cover of clay:

$$\frac{C}{D} = \frac{6}{6} = 1$$

Now look up the critical stability number N_c in Mair's design chart (c.f. Figure 2.2; also see Figure 2.11):

$$N_c = 5.64.$$

Rearrange the stability ratio from Equation 2.1:

$$N = \frac{\gamma(C + D/2) + \sigma_s - \sigma_t}{c_u}$$

$$\sigma_t = \gamma(C + D/2) + \sigma_s - Nc_u$$

To avoid collapse with a factor of safety of 1.5, $N \leq N_c/1.5$, therefore:

$$\sigma_t = \gamma(C + D/2) + \sigma_s - \frac{N_c c_u}{1.5}$$

Now taking the sand as surcharge, and assuming no other surcharge is applied at the surface:

$$\sigma_s = \gamma_{sand} z_{sand} = 17 \times 6 = 102 \text{ kPa}$$

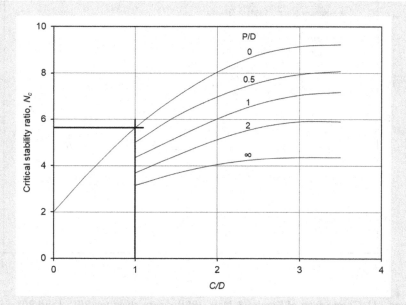

Figure 2.11 Worked Example 2.3 using the design chart to find the critical stability number.

Therefore the required minimum support pressure is:

$$\sigma_t = 18 \times 9 + 102 - \frac{5.64 \times 30}{1.5} = 264 - 112.8 = 151.2 \text{ kPa}$$

2.2.4 Numerical modelling of heading stability in clay

Except for the special case where $P/D = \infty$, which can be modelled as a 2D plane strain problem, heading stability must be modelled in 3D. There are a number of ways to bring a tunnel heading to failure in a numerical model, which are all based on the stability ratio equation (Equation 2.1):

- Reduce the undrained shear strength until failure occurs
- Increase the surcharge until failure occurs
- Decrease the tunnel support pressure until failure occurs.

Increasing the surcharge or decreasing the support pressure are the two ways in which 1 g and centrifuge models of heading failure are controlled (Casarin & Mair, 1981; Kimura & Mair, 1981).

Müller (2015) found that each of these three methods gave similar results and are therefore interchangeable (Figure 2.12). Bradley (2013) also got

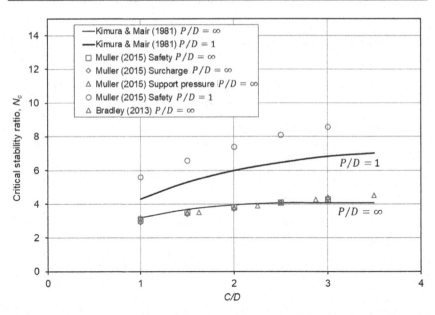

Figure 2.12 Comparison of Kimura & Mair (1981) design chart for $P/D = 1$ and $P/D = \infty$ with numerical modelling by Müller (2015) and Bradley (2013). 'Safety' is reducing undrained shear strength to failure, 'Surcharge' is increasing surcharge to failure, and 'Support pressure' is reducing support pressure to failure.

very similar results at $P/D = \infty$ using the reduction of support pressure method (and applying surcharge if failure was still not achieved). Müller's results for $P/D = 1$, however, did not agree with the design chart.

Pound (2005), Müller (2015) and Bradley (2013) all modelled a tunnel heading with $P/D = 0$, and their results are shown in Figure 2.13. Again, similar to $P/D = 1$ in Figure 2.12, the numerical model appeared to consistently overestimate the value of critical stability ratio N_c at failure compared to the design charts based on centrifuge tests and case histories produced by Kimura & Mair (1981) and Mair & Taylor (1997). This may be because for the 2D plane strain situation ($P/D = \infty$) the failure is well-defined in the numerical model, with a sharp increase in displacement, whereas for other values of P/D Bradley (2013) found that the displacement increased more gradually, making it difficult to define a precise point where failure occurred. Pound (2005) defined failure as an increase of maximum face displacement of more than $0.1D$ for a 1% decrease in undrained shear strength (or approximately a 1% increase in N_c). If this were applied to Bradley's models then the N_c values at higher C/D ratios would be higher than shown in Figure 2.13. It should also be noted that Kimura & Mair (1981) did not specify how failure was defined in their centrifuge tests, so this may also be a source of discrepancy.

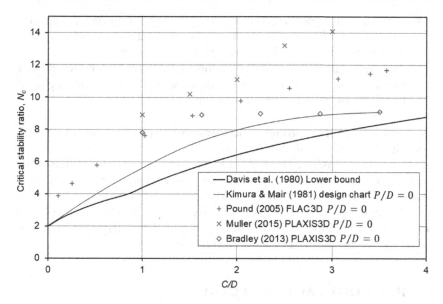

Figure 2.13 Comparison of Kimura & Mair (1981) design chart and Davis et al. (1980) lower bound with numerical modelling by Pound (2005), Müller (2015) and Bradley (2013), all for P/D = 0.

Therefore, numerical modelling of typical headings in clay will not be straightforward. However, in most practical situations, maximum face areas, maximum unsupported lengths, or minimum support pressures will be limited to a significantly lower value by allowable deformations.

2.2.5 Summary of undrained stability

Undrained stability may be assessed by calculating the stability ratio N, and comparing it to the critical stability number N_c in Mair's design chart in Figure 2.2. The factor of safety is given by N_c/N.

Undrained stability depends on the depth and diameter of the tunnel, the value of undrained shear strength, the unsupported length, support pressure applied from within the tunnel and surcharge pressure applied at the surface. Stability can be improved either by increasing N_c or by reducing N, which can be achieved by:

- reducing the diameter (or reducing the equivalent diameter by reducing the face area)
- increasing the cover
- decreasing the unsupported length
- decreasing the depth or surcharge pressure
- increasing the support pressure applied within the heading

- increasing the undrained shear strength, either by ground improvement or by relocation of the tunnel (usually if the tunnel is moved deeper the undrained shear strength will be higher)
- presupport measures such as canopy tubes, spiling or face dowels (these can be thought of as effectively either reducing the unsupported length or applying a support pressure)

Where undrained shear strength increases with depth the value at 0.6–0.8 times the depth to axis should be used.

Where a tunnel is in clay overlain by coarse-grained cohesionless soils, treat the cohesionless soils as a surcharge and assume the cover C used in the design chart is the cover of clay only.

Collapse of a tunnel heading in clay does not usually occur suddenly but is a progressive failure. Therefore, it can be difficult to define when failure occurs in a laboratory test or numerical model.

2.3 BLOW-OUT FAILURE IN CLAY

Blow-out failure only occurs in tunnels with a pressurised face, which can be compressed air, slurry in a slurry TBM or earth pressure in an EPB TBM. There are several ways in which a blow-out can occur:

- softening and erosion
- hydraulic fracturing
- passive failure

Passive failure is like a stability failure in reverse: a large cone of soil is forced upwards, failing along large shear surfaces. The main difference to stability failure is that the weight of the soil is a favourable rather than an unfavourable load, and the volume of soil involved is usually larger. Therefore, passive failure requires high face pressures.

2.3.1 Softening and erosion

Softening and erosion appears to be most common for pressurised water conveyance tunnels during operation where the head of water is greater than the distance to the top of the clay or the surface. A leak out of the lining can, over time, gradually soften and erode the clay, eventually creating a path out of the clay to the surface or into a more permeable soil, resulting in flooding and/or excessive loss of water from the tunnel. This happened to a raw water tunnel near the village of Datchet in 2006, where water eroded a path through the London Clay and suddenly burst out, spouting 15 ft into the air and producing flood waters 3 ft deep (BBC, 2006). 14 properties were flooded, most only in their gardens, before the tunnel could be isolated. This was an unbolted concrete wedgeblock tunnel commissioned in 1976 (Pawsey & Humphrey, 1976), so it would have relied on the ground pressure from the clay, and the impermeability of the clay itself, to remain watertight (Wood, 2008). The tunnel was at approximately 30 m depth below the ground surface, but the reservoir level was approximately 18 m higher than the ground surface level above the tunnel (Pawsey & Humphrey, 1976). Therefore, even though the overburden pressure at the tunnel depth, about 600 kPa, was significantly higher than the internal water pressure of about 480 kPa, a blow-out still occurred.

2.3.2 Hydraulic fracturing in clay

Hydraulic fracturing is a localised effect that can happen in the short term, or may occur after some softening and erosion has reduced the effective cover. This is where the clay is fractured by a high localised fluid pressure that exceeds the tensile or shear strength of the clay (Marchi et al., 2014), creating a path for the escape of support fluid. This will usually occur at the crown of the tunnel. In a tunnel with compressed air, the loss of air pressure could be followed by flooding of the tunnel or collapse of the face. In a slurry TBM, a sudden loss of slurry can be followed by overexcavation due to failure of the face caused by loss of support pressure. Also, release of the slurry at the surface or into an overlying lake or river could have environmental consequences.

Looking back at the Docklands Light Railway tunnel blow-out described in Section 1.5.1 and shown in Figure 2.14, compressed air was needed for construction of a crosspassage between the two running tunnels near the deepest point under the River Thames, where the water pressure was about 2.7 bar at axis level (270 kPa). One bulkhead was 60 m beyond the crosspassage and the other one was much further outbye towards the portal, where the cover was only 8 m. The blow-out occurred near the outbye bulkhead through the tunnel lining as the bulkheads were being tested up to 3 bar and pressure had reached only 2.1 bar (210 kPa). Eight of the

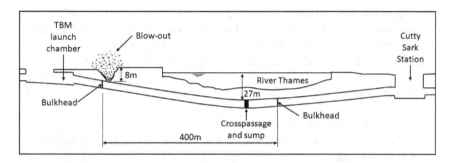

Figure 2.14 Details of the Docklands Light Railway Extension blow-out – long section of the South drive (based on details in *New Civil Engineer*, 1998).

tunnel's 1.2 m long 5.2 m diameter precast concrete rings were blown apart by the blast (New Civil Engineer, 1998). The full overburden pressure at the crown at this location was less than 160 kPa. The large crater caused by the blast was probably due to the large volume of air that was released.

For hydraulic fracturing to occur, simple models assume the fluid pressure usually has to exceed the full overburden pressure of soil and water (Holzhäuser, 2003), and generally accepted practice is to limit support fluid pressures to this value (e.g. Guglielmetti et al., 2008). This is usually significantly lower than the pressure needed for passive failure to occur.

Bezuijen & Brassinga (2006) show from field data and centrifuge tests that bentonite slurry blow-outs caused by hydraulic fracturing can occur at much lower face pressures than those needed for passive failure predicted by finite element or kinematic analysis methods. This is because these methods do not take account of the fact that slurry is a fluid. Bezuijen & Brassinga found this limit to be approximately the pore pressure plus 2 to 3 times the effective stress for their case.

Marchi et al. (2014) show that fracture initiation in clay may be caused by either tension or shear. The most important factor is the confining pressure, otherwise known as the minor principal stress. As fluid pressure increases, the radial stress in the surrounding clay increases but the circumferential stress decreases. Failure occurs either when the circumferential stress reaches the tensile strength or when the difference between the radial and circumferential stresses causes shear failure.

The pressure at tensile fracture is given by (Mitchell & Soga, 2005):

$$P_f = 2\sigma_0 - u_0 + \sigma'_t \qquad (2.3)$$

P_f is the fracture pressure
σ_0 is the initial confining pressure (the minor principal total stress)
u_0 is the initial pore pressure
σ'_t is the effective tensile strength (tension positive)

This equation means that fracturing pressure increases with initial confining pressure with a gradient of 2. However, this assumes the soil is linear elastic, and in reality the gradient may be less than 2 (Marchi et al., 2014).

The pressure at shear fracture is given by (Soga et al., 2005):

$$P_f = \sigma_0 + nc_u \qquad (2.4)$$

P_f is the fracture pressure
σ_0 is the initial confining pressure (the minor principal total stress)
n is a constant
c_u is the undrained shear strength

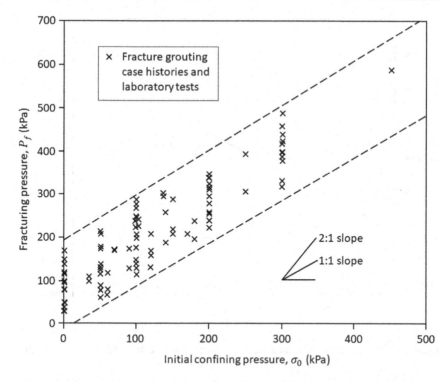

Figure 2.15 Fracturing pressure as a function of initial confining pressure of different clay soils, from Marchi et al. (2014).

For a clay with a positive liquidity index (i.e. the water content is above the plastic limit), $n = 1$, but for a clay with a negative liquidity index $n = 1.5$ to 2.

From a large number of experimental tests and case histories of fracture grouting, Marchi et al. (2014) showed that the lower bound to all the measured fracture pressures was approximately the initial confining pressure, as shown in Figure 2.15. Therefore, although the undrained shear strength or effective tensile strength of the soil may result in a higher value of fracturing pressure, it seems the fracture pressure cannot be lower than the minor principal total stress.

2.3.3 Passive failure in clay

Since hydraulic fracturing is the critical mechanism when support fluid is used, passive failure in clay is only likely to be the limiting case for EPB TBMs. Most studies of passive failure in clay, involving numerical models or kinematic limit state analysis, have assumed a uniform face pressure (Mollon et al., 2013), whereas in an EPB machine we would expect a face pressure that increases with depth. Despite this shortcoming, a design

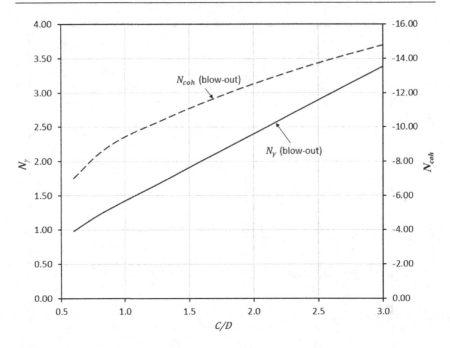

Figure 2.16 Design chart for critical passive failure blow-out (from Mollon et al., 2013).

chart by Mollon et al. (2013) based on their asymmetric 'M2' velocity field method is shown in Figure 2.16, with values given in Table 2.1 to aid interpolation for design purposes. The values of N_γ and N_{coh} from the design chart should be used in the following equation:

$$\sigma_b = \gamma D N_\gamma - c_u N_{coh} + \sigma_s \qquad (2.5)$$

σ_b is the critical face pressure for a passive failure blow-out
γ is the bulk unit weight of the clay
D is the tunnel diameter
N_γ is a stability number taking account of soil weight
N_{coh} is a stability number taking account of soil cohesion
σ_s is the surcharge pressure in kPa

This equation is derived from the more general stability equation, which is:

$$\sigma_t = \gamma D N_\gamma - c N_{coh} + \sigma_s N_s \qquad (2.6)$$

N_γ, N_{coh} and N_s are stability numbers for the effect of soil weight, cohesion and other effects, respectively, which are dimensionless
σ_t is the required support pressure in kPa
c is Mohr–Coulomb cohesion in kPa

Table 2.1 Values used to produce Figure 2.16 for critical passive failure blow-out (from Mollon et al., 2013).

C/D	N_γ (passive failure blow-out)	N_{coh} (passive failure blow-out)
0.6	0.98	−7.02
0.8	1.22	−8.47
1.0	1.42	−9.43
1.3	1.71	−10.44
1.6	2.01	−11.40
2.0	2.40	−12.53
2.5	2.90	−13.75
3.0	3.39	−14.80

For undrained constant volume behaviour, $N_s = 1$ and $c = c_u$. In order to make Equation 2.1 and Equation 2.6 equivalent, all that is needed is to substitute $N_\gamma = C/D + 0.5$ and $N_{coh} = N$. But in Mollon et al.'s 'M2' velocity field method, $N_\gamma \neq C/D + 0.5$, because the velocity field is asymmetric (the maximum velocity is set $0.4D$ above the centre of the face), and therefore the results cannot be used with Equation 2.1, but have to be used with the more general Equation 2.6.

Unfortunately, no centrifuge modelling of passive failure in clay has been published, but Mollon et al.'s M2 method is corroborated to some degree by their numerical modelling of a heading using FLAC3D, for which a comparison is shown in Figure 2.17 for $D = 10$ m, $\gamma = 18$ kN/m³ and $c_u = 20$ kPa.

Figure 2.17 shows that the kinematic analysis overestimates the critical blow-out pressures compared to the numerical modelling results. This is unsurprising because it is an upper bound and defines when the ground *must* fail, and therefore will always be on the unsafe side. However, it must be close to the true collapse geometry and conditions, as it is not too far above the FLAC3D results.

Figure 2.17 also gives us a feel for the magnitude of face pressure required to cause passive failure. Even for this low value of undrained shear strength (20 kPa), the critical face pressure is about 1.5 times the full overburden pressure at axis level. At an undrained shear strength of 30 kPa the critical face pressure is double the full overburden pressure (Mollon et al., 2013).

Also shown on Figure 2.17 is the minimum face pressure required to avoid collapse, from the design charts based on centrifuge testing produced by Kimura & Mair (1981), for the same geometry $D = 10$ m and $P = 0$, soil bulk unit weight $\gamma = 18$ kN/m³ and undrained shear strength $c_u = 20$ kPa. Also shown, for comparison, is the full overburden pressure at axis level. The aim of design would be to find a safe zone of face pressures between collapse and blow-out, allowing for factors of safety and for variability of the applied pressure.

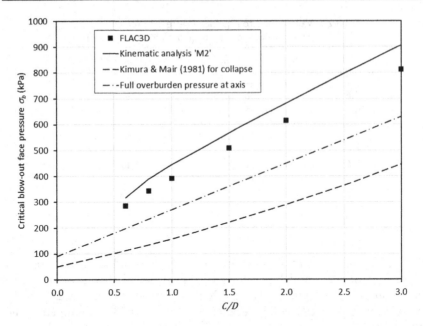

Figure 2.17 Comparison of critical blow-out pressures using the design chart based on 'M2' velocity field kinematic analysis by Mollon et al. (2013) and a 3D finite difference model in FLAC3D, also by Mollon et al. (2013), for $D = 10$ m, $\gamma = 18$ kN/m³ and $c_u = 20$ kPa.

In summary, to estimate the critical face pressure that would cause a passive failure blow-out, the design chart in Figure 2.16 and Equation 2.5 can be used to provide an initial estimate, and if accuracy is of critical importance, a 3D numerical model can be used. Where a fluid such as slurry or compressed air is used for face support, a hydraulic fracturing blow-out is more likely, and face pressures should be limited to below the minor principal total stress, calculated at the crown of the tunnel, or to a limit based on the worst case of shear or tensile hydraulic fracturing using Equation 2.3 and Equation 2.4.

WORKED EXAMPLE 2.4   CRITICAL BLOW-OUT PRESSURE FOR A TUNNEL IN CLAY

A 5 m diameter tunnel is to be built in clay with shallow cover of only 3 m. The clay has bulk unit weight $\gamma = 18$ kN/m³, undrained shear strength $c_u = 25$ kPa, it has a positive liquidity index, and effective tensile strength $\sigma_t' = 0$ kPa. The coefficient of earth pressure at rest $K_0 = 0.6$ and the water table is at the surface. What is the critical blow-out pressure for an EPB machine?

For the face pressure, hydraulic fracturing should not normally be possible in an EPB machine unless a large amount of soil conditioning is pumped into the ground. Therefore we will do a calculation for passive failure.

$$C/D = 3/5 = 0.6$$

From Table 2.1, $N_\gamma = 0.98$ and $N_{coh} = -7.02$ for passive failure.
Using Equation 2.5:

$$\sigma_b = \gamma D N_\gamma - c_u N_{coh} + \sigma_s = 18 \times 5 \times 0.98 - 25 \times (-7.02) + 0$$
$$= 88.2 + 175.5 = 263.7 \text{ kPa}$$

While grouting the annulus between the segmental lining and the ground behind the shield and tailskin, hydraulic fracturing may cause grout to escape to the surface. The limiting pressure may be given by either a tensile (Equation 2.3) or shear (Equation 2.4) mechanism, and the critical location will be at the crown of the tunnel.

The initial confining pressure σ_0 is the minor principal total stress, which in this case is the horizontal total stress, which is given by:

$$\sigma_0 = K_0 (\sigma_v - u_0) + u_0$$

because K_0 is the ratio of horizontal to vertical effective stress.

The initial pore pressure u_0 at the crown is 30 kPa and the vertical total stress is given by:

$$\sigma_v = \gamma z_{crown} = 18 \times 3 = 54 \text{ kPa}$$

Therefore, the initial confining pressure is:

$$\sigma_0 = K_0 (\sigma_v - u_0) + u_0 = 0.6(54 - 30) + 30 = 44.4 \text{ kPa}$$

Assuming the effective tensile strength σ_t' is zero, the pressure at tensile fracture is therefore:

$$P_{f,tension} = 2\sigma_0 - u_0 + \sigma_t' = 2 \times 44.4 - 30 + 0 = 58.8 \text{ kPa}$$

and the pressure at shear fracture is:

$$P_{f,shear} = \sigma_0 + n c_u = 44.4 + 1 \times 25 = 69.4 \text{ kPa}$$

The maximum grouting pressure at the crown will not cause hydraulic fracturing if it is kept below the lower value of 58.8 kPa, or more conservatively the initial confining pressure (or in situ minor principal total stress) of 44.4 kPa could be used. The grouting pressure needs to be higher than the pore pressure of 30 kPa at the crown, giving a fairly narrow band of allowable pressures, especially if factors of safety and variability are accounted for.

The passing of the TBM will of course change the stress state in the ground, so the minor principal total stress will not be this initial undisturbed value. Stresses calculated by 3D numerical modelling could be used but there will still be considerable uncertainty in the results of this calculation.

WORKED EXAMPLE 2.5   CRITICAL COLLAPSE AND BLOW-OUT PRESSURES FOR A TUNNEL UNDER A RIVER

A 12 m diameter bored tunnel is to be constructed using a closed-face TBM (either a slurry or EPB) beneath a river with clay cover of 6 m, as shown in the section in Figure 2.18. The highest credible river level during construction is 12 m and the lowest is 6 m.

a. Calculate the face pressure required to avoid collapse, with factor of safety of 1.5 and allowance for variability of ±25 kPa.
b. Calculate the hydraulic fracturing pressure for slurry in the face of a slurry TBM.
c. Calculate the blow-out pressure for an EPB TBM.

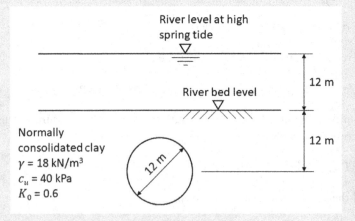

Figure 2.18 Worked Example 2.5 cross-section.

a. First calculate the face pressure required to avoid collapse, with factor of safety of 1.5 and allowance for variability of ±25 kPa:

$$C/D = 6/12 = 0.5$$

$$P/D = 0$$

From Mair's design chart for undrained stability, $N_c = 3.94$. To achieve a factor of safety (N_c/N) of 1.5, requires:

$$\frac{N_c}{N} = 1.5 = \frac{N_c}{\left(\dfrac{\gamma(C+D/2)+\sigma_s -\sigma_t}{c_u}\right)} = \frac{3.94}{\left(\dfrac{18\times12+120-\sigma_t}{40}\right)}$$

Rearranging for σ_t gives:

$$\sigma_t = \gamma(C+D/2)+\sigma_s - \frac{N_c c_u}{1.5} = 18\times12+120-\frac{3.94\times40}{1.5} = 231 \text{ kPa}$$

Allowing for variability, the target minimum support pressure is given by:

$$P_{st,collapse} = 231+25 = 256 \text{ kPa}$$

b. If a slurry TBM were used, the fluid pressure that would cause hydraulic fracture would be above the minor principal total stress at the crown, which for $K_0 < 1$ is the horizontal total stress σ_h. In this case, the critical case is when the river level is at its lowest and initial pore pressure at the crown $u_0 = 120$ kPa (ignoring the effect of the TBM on pore pressure, and assuming the ground is affected by tide level).

$$\sigma_v = (6\times10+6\times18) = 168 \text{ kPa}$$

$$\sigma'_v = \sigma_v - u_0 = 168-(12\times10) = 48 \text{ kPa}$$

$$\sigma'_h = K_0\sigma'_v = 0.6\times48 = 28.8 \text{ kPa}$$

$$\sigma_h = \sigma'_h + u_0 = 28.8+120 = 148.8 \text{ kPa}$$

Note that no factor of safety or allowance for variation has yet been applied, but the pressure at which hydraulic fracturing may occur is much lower than the target minimum support pressure required to prevent collapse. Therefore, it is not possible to use a slurry TBM safely in this situation without risk of escape of slurry into the river, a decrease of support pressure to the hydrostatic value and continuous failure of the face resulting in overexcavation.

c. If an EPB machine were used, hydraulic fracturing could be avoided and the upper limit on face pressure would be defined by passive failure.

Extrapolating from the curves in Figure 2.16, $N_\gamma = 0.86$ and $N_{cob} = -6.3$.

$$\sigma_b = \gamma D N_\gamma - c_u N_{cob} + \sigma_s = 18 \times 12 \times 0.86 - 40 \times (-6.3) + 6 \times 10$$

$$= 185.76 + 252 + 60 = 498 \text{ kPa}$$

If a factor of safety of 1.5 were placed on the undrained shear strength, the passive failure blow out pressure would decrease to 414 kPa. Also taking away the 25 kPa allowance for variability, this would give a target maximum face pressure of 389 kPa.

Therefore, there is a safe range of operating pressures between 256 kPa and 389 kPa.

2.3.4 Uplift failure of a tunnel heading invert in clay

Very little is known about this, and there are no records of it ever occurring in practice. However, it is theoretically possible that a tunnel in clay, or with clay in the invert, could fail if there is a permeable layer beneath the clay with a high enough water pressure to cause uplift failure of the clay. A similar failure mechanism is known to occur in shafts or other deep excavations. As excavation unloads the ground, failure is only resisted by the weight of the soil and its shear resistance. Either passive failure or hydraulic fracturing of the clay could occur, resulting in inundation of the heading by water and soil.

For more background information on this type of failure, refer to Section 4.2 on shaft uplift failure.

2.3.5 Summary of undrained blow-out failure

Blow-out failure only occurs in tunnels with a pressurised face, which can be compressed air, slurry in a slurry TBM or earth pressure in an EPB TBM. There are several ways in which a blow-out can occur:

- softening and erosion
- hydraulic fracturing
- passive failure

Passive failure is like a stability failure in reverse: a large cone of soil is forced upwards, failing along large shear surfaces. However, passive failure is different in that the weight of the soil is a favourable rather than an unfavourable load, and the volume of soil involved is usually larger. Therefore, passive failure requires high face pressures.

Softening and erosion appears to be most common for pressurised water conveyance tunnels during operation where the head of water is greater than the distance to the top of the clay or the surface.

Hydraulic fracturing is a localised effect that can happen in the short-term, or may occur after some softening and erosion has reduced the effective cover. This is where the clay is fractured by a high localised fluid pressure (the fluid can be air, water, soil conditioning, slurry or grout) that exceeds the tensile or shear strength of the clay, creating a path for the escape of support fluid. Hydraulic fracturing can occur at much lower face pressures than those needed for passive failure.

Uplift failure may also be a risk where a permeable layer with high pore water pressure lies beneath a tunnel in clay, causing failure of the clay upwards into the heading, potentially followed by flooding of water and soil into the tunnel.

2.4 PROBLEMS

The following questions should each take about 45 minutes.

Q2.1. A 6.3 m excavated diameter tunnel is to be driven using an EPB TBM through soft clay, at a depth to axis of 27.4 m (see Figure 2.19). The tunnel is below a lake with maximum water depth of 7.5 m. Assume that the soft clay has a bulk unit weight of 18.5 kN/m³ and a characteristic value of undrained shear strength of 50 kPa.

   i. Calculate the factor of safety on undrained stability (N_c/N) for an EPB face pressure of 0 kPa. State whether the tunnel face is stable.

   ii. Calculate the target minimum EPB face pressure required to provide a factor of safety >1.5, with an allowance for variability of ±25 kPa.

   iii. Calculate the value of undrained shear strength at which the face would be stable during a head intervention (i.e. with no EPB face pressure applied) with a factor of safety of 1.5.

   iv. Assuming that the undrained shear strength is 50 kPa again and does not vary with depth, at what depth to axis would the tunnel not require application of an EPB face pressure?

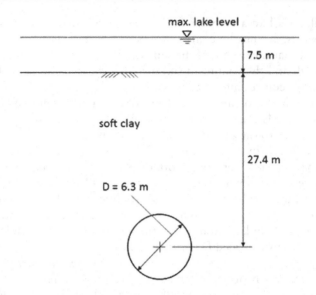

Figure 2.19 Cross-section of EPB TBM tunnel drive beneath lake.

Q2.2. A 9.5 m excavated diameter tunnel is to be driven using an open-face TBM through stiff clay, at a depth to axis of 12 m, as shown in Figure 2.20. Assume that the bucket excavators in the shield can reach a maximum of 1.5 m ahead of the leading edge of the shield. The stiff clay has a bulk unit weight of 20 kN/m³ and is overlain by 3 m of sandy gravel with a bulk unit weight of 18 kN/m³. The undrained shear strength of the stiff clay is 50 kPa at the interface with the sandy gravel and increases with depth by 8 kPa per metre depth.

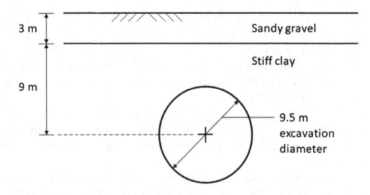

Figure 2.20 Cross-section of open-face TBM tunnel in stiff clay.

i. Using the undrained shear strength at the tunnel axis level, calculate the factor of safety on undrained stability (N_c/N). Assume there is no surcharge.

ii. Now use the undrained shear strength at 0.6 times the depth to axis (refer to Section 2.2.2) to calculate the factor of safety on undrained stability (N_c/N). Assume there is no surcharge.

iii. Now repeating the same calculation as in (ii), but with a surcharge of 75 kPa, calculate the factor of safety on undrained stability (N_c/N).

iv. What can you conclude from these calculations? If a factor of safety of 1.4 were required for undrained stability, is it stable in situations (ii) and (iii)? If not, what can be done about it?

Q2.3. The same tunnel in Q2.2 may encounter silt and sand layers within the stiff clay stratum that have pore pressures within them, in the bottom half of the face. At this location the tunnel is 1.5 m deeper, as shown in Figure 2.21. Assume the pore pressure is hydrostatic with a piezometric level at the ground surface, and that seepage in the bottom 1 m of the face is acceptable. Assume that $K_0 = 1.0$.

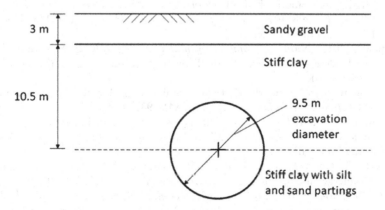

Figure 2.21 Cross-section of open-face TBM tunnel in stiff clay with silt and sand partings below axis level.

i. To find the compressed air pressure that could cause hydraulic fracturing of the stiff clay and air loss, calculate the horizontal stress at the crown of the tunnel, assuming this could reduce during excavation to a value equal to either twice the undrained shear strength (since the vertical total stress will be

zero, Mohr's circle for undrained behaviour sets this condition) or the initial horizontal total stress, whichever is lower.

ii. Calculate the pore pressure 1 m above the invert level. Is the
allowable compressed air pressure calculated in (i) sufficient
to counteract this?

iii. Calculate the compressed air pressure required to cause passive failure. Which mode of blow-out – hydraulic fracturing or
passive failure, is more likely to occur?

REFERENCES

Anagnostou, G. & Kovári, K. (1996). Face stability conditions with earth-
pressure-balanced shields. *Tunn. Undergr. Space Technol.* **11**, No. 2,
165–173.

Atkinson, J. H. (2007). *The mechanics of soils and foundations*, 2nd Edition.
London: Taylor & Francis.

Atkinson, J. H. & Mair, R. J. (1981). Soil mechanics aspects of soft ground tunnelling. *Ground Eng.*, July, 20–38.

BBC (2006). *Homes flooded by reservoir leak*. BBC News, Saturday 8th April
2006. Available at: http://news.bbc.co.uk/1/hi/england/berkshire/4892522.
stm [last accessed 9th March 2017].

Bezuijen, A. & Brassinga, H. E. (2006). Blow-out pressures measured in a centrifuge model and in the field. *Tunnelling: A decade of progress – GeoDelft
1995–2005* (eds Bezuijen, A. & van Lottum, H.), pp. 143–148. Leiden:
Taylor & Francis/Balkema.

Bradley, J. (2013). *Numerical analysis of tunnel heading stability*. MSc Tunnelling
and Underground Space, Finite Element Methods Project, University of
Warwick, UK.

Broms, B. B. & Bennermark, H. (1967). Stability of clay at vertical openings.
ASCE J. Soil Mech. Found. Eng. Div. SM1 **93**, 71–94.

Casarin, C. & Mair, R. J. (1981). The assessment of tunnel stability in clay by
model tests. *Soft ground tunnelling: Failures and displacements*, pp. 33–44.
Rotterdam: Balkema.

Davis, E. H., Gunn, M. J., Mair, R. J. & Seneviratne, H. N. (1980). The stability
of underground openings in cohesive materials. *Géotechnique* **30**, 397–416.

EN 1997-1:2004 (2009). *Eurocode 7: Geotechnical design – Part 1: General rules*,
incorporating corrigendum February 2009. Brussels: European Committee
for Standardization.

Golder Associates (2009). *Ground control for slurry TBM tunnelling*. GEO Report
249. Hong Kong: The Government of the Hong Kong Special Administrative
Region.

Grant, R. J. & Taylor, R. N. (2000). Stability of tunnels in clay with overlying
layers of coarse grained soil. *Proc. ISRM Int. Symp.*, 19th–24th November,
Melbourne, Australia. Published by CRC Press/ISRM.

Guglielmetti, V., Grasso, P., Mahtab, A. & Xu, S. (2008). *Mechanized tunnelling
in urban areas*. London: Taylor & Francis Group.

Holzhäuser, J. (2003). Geotechnical aspects of compressed air support on TBM tunnelling. *Engineering and health in compressed air working*, pp. 359–371. London: Thomas Telford.

Kimura, T. & Mair, R. J. (1981). Centrifugal testing of model tunnels in soft clay. *Proc. 10th Int. Conf. Soil Mech. & Found. Engrg*, Stockholm, Vol. 1, pp. 319–322.

Mair, R. J. (1979). *Centrifugal modelling of tunnel construction in soft clay*. PhD thesis, University of Cambridge.

Mair, R. J. & Taylor, R. N. (1997). Bored tunnelling in the urban environment. Theme Lecture, Plenary Session 4. *Proc. 14th Int. Conf. Soil Mechanics and Foundation Engineering*, Hamburg, Vol. 4.

Marchi, M., Gottardi, G. & Soga, K. (2014). Fracturing pressure in clay. *ASCE J. Geotech. Geoenviron. Eng.* **140** (2). https://doi.org/10.1061/(ASCE)GT.1943-5606.0001019

Mitchell, J. K. & Soga, K. (2005). *Fundamentals of soil behavior*, 3rd Edition. London: Wiley.

Mollon, G., Dias, D. & Soubra, A.-H. (2013). Continuous velocity fields for collapse and blowout of a pressurized tunnel face in purely cohesive soil. *Int. J. Numer. Anal. Meth. Geomech.* 37, 2061–2083.

Müller, R. (2015). *Design Study Report*. MSc Tunnelling and Underground Space, Finite Element Methods Project, University of Warwick, UK.

New Civil Engineer (1998). Docklands Light Railway Lewisham Extension blowout. February 1998.

Pawsey, D. B. H. & Humphrey, A. W. (1976). The Queen Mother Reservoir, Datchet – some aspects of its design and construction. *Ground Eng.*, October, 27–30.

Pound, C. (2005). Improved solutions for the stability of a tunnel heading in cohesive soil. *Proc. Underground Construction 2005*. London: Brintex.

Soga, K., Ng, M. Y. A. & Gafar, K. (2005). Soil fractures in grouting. *Proc. 11th Int. Congress on Computer Methods and Advances in Geomechanics*, Turin, Italy, pp. 397–406.

Wood, R. (2008). Wraysbury Reservoir Inlet Tunnel – in situ structural secondary lining of existing tunnel. *Water Treatment and Supply*, 159–161. Available at: http://waterprojectsonline.com/case_studies/2008/Thames%20Wraysbury%20Res%20Inlet%20tunnel%202008.pdf [last accessed 9th March 2017].

Chapter 3

Drained stability

The most common methods used in practice for stability calculations are Mair's design charts based on centrifuge testing for the undrained case (Kimura & Mair, 1981), and Anagnostou & Kovári's method for the drained case (Anagnostou & Kovári, 1994, 1996a and 1996b). This chapter will first provide some background and will then describe how to use Anagnostou & Kovári's method to solve drained stability problems.

After working through this chapter, you will understand:

- how tunnel headings collapse or blow-out in the drained case
- how upper bound and lower bound limit states, limit equilibrium solutions, numerical modelling and empirical data help us to estimate the required support pressure
- the role of groundwater pressure and seepage forces in drained stability
- how slurry and earth pressure balance (EPB) tunnel boring machines (TBMs) support the face
- slurry penetration and stand-up times

After working through this chapter, you will be able to:

- calculate the minimum and maximum support pressures for tunnelling in drained soils to avoid collapse, overmucking or blow-out failures

3.1 DRAINED STABILITY WITHOUT SEEPAGE

As discussed in the previous chapters, four main factors influence stability:

- *Cohesion* and *support pressure*, which improve stability
- *Gravity* and *seepage forces*, which worsen stability

DOI: 10.1201/9780429470387-3

Without either cohesion to hold the soil grains together or a support pressure applied to the face, a vertical face will fall down due to gravity. In the case of a tunnel below the water table, seepage forces will also contribute to instability. Seepage forces occur when there is a hydraulic gradient between the ground and the tunnel. As groundwater seeps through the ground towards the face, it pushes the soil grains apart with a 'seepage force' proportional to the hydraulic gradient, in the direction of flow.

'Drained' soils are permeable enough that excess pore pressures are dissipated during the timescale of construction. For tunnel heading stability, this means that any negative excess pore pressures that may have helped hold the soil grains together cannot be counted on, as water will have time to flow in from the surrounding ground. Therefore, the failure of drained soils depends on the drained cohesion c' and the internal angle of friction ϕ, rather than the undrained shear strength.

Note that in drained stability it has become common practice to refer to the cover using the letter 'H' rather than the letter 'C', although some papers still use 'C'. These will be used interchangeably in this chapter.

3.1.1 Dry cohesionless soils

The simplest situation to consider is that of dry, cohesionless soils, i.e. where cohesion $c' = 0$. Failure is governed by the geometry of the heading, the unit weight of the soil and the angle of friction ϕ. It is certain that a support pressure is required. This may be provided by slurry pressure in a slurry machine, compressed air in an open face, or by effective stress and fluid pressure in an EPB machine.

Atkinson & Potts (1977) performed 1 g (i.e. scale models on the laboratory floor) and centrifuge tests (scale models accelerated in a centrifuge) on dry cohesionless Leighton Buzzard sand. They also developed upper bound and lower bound limit state solutions for a plane strain tunnel, i.e. where the unsupported length $P = \infty$.

If you remember from Chapter 2, the upper bound is any kinematically admissible mechanism, and at this load, the heading *must fail*. The lower bound is a statically admissible stress field, which nowhere violates the failure criterion for the soil, and at this load, the heading *cannot fail*. The upper and lower bounds bracket the true collapse load, and ideally they are close enough together that a target support pressure may be set.

Figure 3.1 shows the upper and lower bound solutions for a dry cohesionless soil from Atkinson & Potts's paper (1977). The vertical axis is the support pressure σ_t, normalised by the unit weight of the soil γ and the diameter D. The horizontal axis is the ratio of cover H to diameter D, and it can be seen that the depth of the tunnel does not affect either the upper or lower bound limit state solutions. Remember this is different to an undrained stability problem, where depth is a factor (c.f. Figure 2.2).

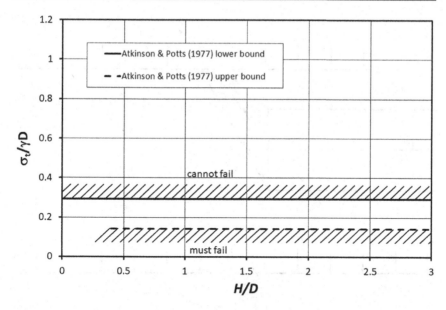

Figure 3.1 Upper and lower bound solutions for a dry cohesionless soil (Atkinson & Potts, 1977) with angle of friction $\phi = 35°$.

The upper bound is based on an infinitely long triangular wedge of soil falling from the crown of the tunnel. The geometry of this wedge depends on the angle of friction ϕ. At low values of H/D, the wedge intersects the surface, so no values can be determined.

Although intended for dry cohesionless soils, it may be possible to use this solution for cohesionless soils below the water table if it can be assumed that there is no flow of groundwater and hence no seepage forces. In this case one would use the submerged unit weight, which is given by:

$$\gamma' = \gamma - \gamma_w \tag{3.1}$$

γ' is the submerged unit weight in kN/m³
γ is the bulk unit weight in kN/m³
γ_w is the unit weight of water, usually assumed to be 10 kN/m³

3.1.2 Dry drained soils with cohesion

Leca & Dormieux (1990) found upper and lower bound limit state solutions for $c' - \phi$ soils, that is, drained soils with cohesion. This was a great step forward because many natural soils have some cohesion. The main difference compared to Atkinson & Potts (1977) was that the kinematic mechanisms (which provided the upper bound) were in three dimensions and P/D was zero rather than infinity, which was more realistic. Also, the

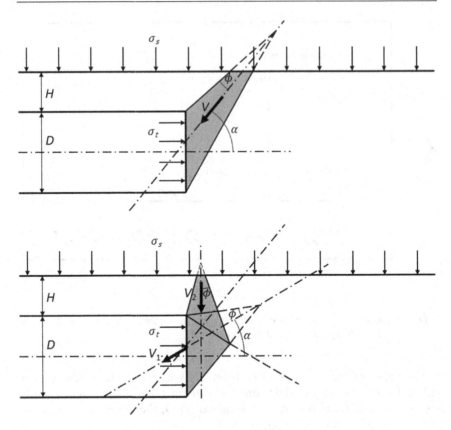

Figure 3.2 Upper bound mechanisms from Leca & Dormieux (1990).

lower bound included three different stress states, whereas Atkinson & Potts assumed one plane strain stress state. The geometry of the upper bound is either one or two truncated cones, as shown in Figure 3.2. The value of the angle α is varied to find the worst case.

Figure 3.3 shows the effect of increasing the value of cohesion from 0 to 5 to 10 kPa for a soil with $\phi = 35°$. Both the upper and lower bounds move downwards, meaning that the required support pressure to prevent instability decreases as cohesion increases. At $c' = 10$ kPa, Leca & Dormieux's upper bound is at zero. If the upper bound represented the true collapse mechanism, which we don't yet know for sure, then no support pressure would be required to maintain stability. This shows that tunnel headings in drained soil above the water table may be stable at quite small values of cohesion. This can be provided by moisture in the soil (like a sandcastle), a low clay content or cementing of the soil grains. However, driving a tunnel in such a situation must be done carefully, because even a small amount of perched water can cause instability due to seepage forces, and low values of cohesion cannot always be relied on given the inherent variability of geological materials.

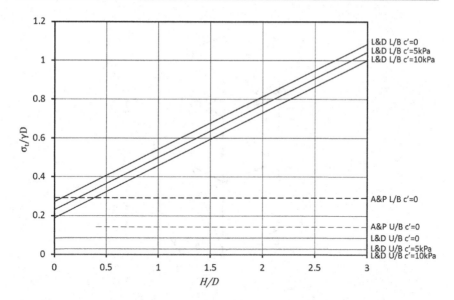

Figure 3.3 Effect of drained cohesion c' on required support pressure σ_t in a dry $c' - \phi$ soil with $\phi = 35°$. ('A&P' = Atkinson & Potts, 'L&D' = Leca & Dormieux, 'L/B' = lower bound, 'U/B' = upper bound).

Unfortunately, although the geometry of Leca & Dormieux's model is closer to reality than Atkinson & Potts's, the upper and lower bounds are quite far apart and this makes it difficult to use these solutions with any certainty about the value of the true collapse load. Atkinson & Potts assumed a plane strain tunnel at both the upper and lower bounds (because this was what they were modelling in the 1 g and centrifuge tests), essentially assuming an unlined infinitely long tunnel, whereas Leca & Dormieux assumed a three-dimensional failure at the face with unsupported length $P = 0$ for the upper bound.

Anagnostou & Kovári (1994, 1996a, 1996b) used a limit equilibrium solution they attributed to Horn (1961) and developed it to allow direct calculation of the required support pressure in any situation above or below the water table, with or without seepage, for tunnels where $P = 0$. The geometry is shown in Figure 3.4. This solution is potentially much more useful for practical situations involving closed-face tunnelling machines, although it cannot be guaranteed to represent the true collapse load because the geometry is a simplification. Messerli et al. (2010) extended the equations to allow for short unsupported lengths and Anagnostou & Perazzelli (2013) investigated the effect of non-uniform support pressures.

Figure 3.5 shows a comparison of Anagnostou & Kovári's limit equilibrium solution with the limit state solutions, for a dry cohesionless soil with $\phi = 35°$. Those of you familiar with Mair & Taylor's excellent 'Theme Lecture' paper at the 14th ICSMFE in 1997 will recognise Figure 3.5 as

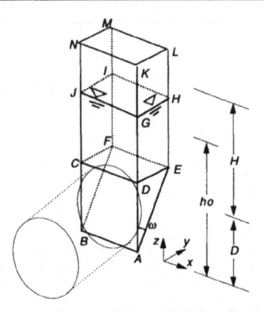

Figure 3.4 Wedge and prism model (from Anagnostou & Kovári, 1996b, after Horn, 1961).

very similar to their Figure 9 (Mair & Taylor, 1997). Interestingly, the support pressure calculated by the limit equilibrium solution is not completely constant and as well as being lower for $H/D < 0.5$, it also does increase very, very slightly with increasing H/D after that. The limit equilibrium solution is consistent with the limit state solutions in that the upper bounds

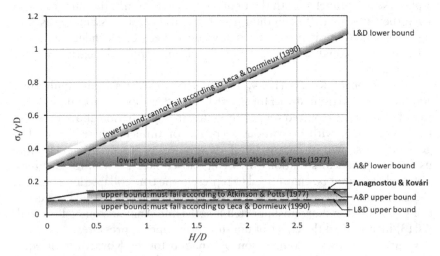

Figure 3.5 Comparison of Anagnostou & Kovári's (1994) limit equilibrium solution with limit state solutions by Leca & Dormieux (1990) ('L&D') and Atkinson & Potts (1977) ('A&P') for a dry cohesionless soil with $\phi = 35°$.

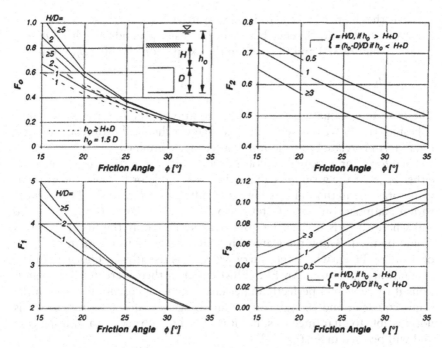

Figure 3.6 Dimensionless coefficients F_0, F_1, F_2 and F_3 based on the wedge-prism model shown in Figure 3.4 for use in Equation 3.2 (Anagnostou & Kovári, 1996b). Note that the labels on the F_0 nomogram have been switched around compared to those incorrectly printed in Anagnostou & Kovári (1996a & 1996b), as explained in Appendix A.

(where the heading must fail) are below it, and the lower bounds (where the heading cannot fail) are above it.

A derivation of the wedge and prism limit equilibrium solution used by Anagnostou & Kovári is given in Appendix A. Based on these equations, Anagnostou & Kovári (1996a, 1996b) defined four dimensionless coefficients F_0, F_1, F_2 and F_3, which may be read from nomograms shown in Figure 3.6 and used in Equation 3.2. The derivation is not included here as it is more important that you use Figure 3.6 and Equation 3.2 and understand their implications first. Once you understand the principles, feel free to go to Appendix A for the full derivation.

$$s' = F_0\gamma'D - F_1c' + F_2\gamma'\,\Delta h - F_3c\frac{\Delta h}{D} \tag{3.2}$$

s' is the effective support pressure required to prevent collapse
F_0, F_1, F_2 and F_3 are dimensionless coefficients
γ' is the submerged unit weight, which is the bulk unit weight minus the unit weight of water
c' is the drained cohesion

D is the diameter of the tunnel excavation

Δh is the head difference between the in situ groundwater head h_0 (c.f. Figure 3.4) and the head in the excavation chamber h_f, given by:

$$\Delta h = h_0 - h_f \tag{3.3}$$

The F_0 term in Equation 3.2 is related to gravity, the F_1 term is related to cohesion, the F_2 term is related to seepage via the head difference Δh, and the F_3 term is related to cohesion and seepage combined.

Therefore, for a tunnel in cohesionless soil where there is no seepage, either because the ground is above the water table or because the groundwater pressure in the soil is balanced by a fluid pressure in the TBM's excavation chamber, only the F_0 term would be used. For a tunnel in $c' - \phi$ soil where there is no seepage towards the face, only the first two terms would be used. Below the water table, these conditions occur in the front of a slurry TBM, where the slurry pressure is usually set higher than the groundwater pressure. An EPB TBM could in theory create these conditions if there were a perfectly impermeable plug of conditioned soil in the screw conveyor and enough soil conditioning were injected. In reality it is not possible to guarantee this, but it is useful to consider as a limiting case and will be used in Section 3.2.4.

For a tunnel in cohesionless soil where there is seepage, only the F_0 and F_2 terms are used. For a tunnel in $c' - \phi$ soil where there is seepage, all the terms are used.

3.1.3 Comparison of analytical methods with centrifuge tests and finite element models

As mentioned previously, the wedge and prism limit equilibrium method makes assumptions about the geometry of failure, so it would give us confidence if it could be validated by field measurements, laboratory testing or numerical analysis. Figures 3.7 and 3.8 show comparisons between 1 g model tests by Messerli et al. (2010) and Chambon & Corté (1994) with the limit state upper bound and the wedge-prism limit equilibrium solution, and to finite element models by Vermeer et al. (2002).

Messerli et al.'s tests were on a uniform fine sand with angle of friction $\phi = 33°$ and unit weight $\gamma = 17$ kN/m³. Analytical solutions are presented for comparison, with $\phi = 33°$ and $\gamma = 17$ kN/m³ used as input parameters.

Vermeer et al. (2002) performed 25 finite element calculations and determined that the relationship between the soil weight stability number N_γ (= $\sigma_t/\gamma D$) and angle of friction ϕ is given by:

$$N_\gamma = \frac{1}{9\tan\phi} - 0.05 = \frac{\sigma_t}{\gamma D} \tag{3.4}$$

This relationship was found to hold true as long as $H/D > 1$.

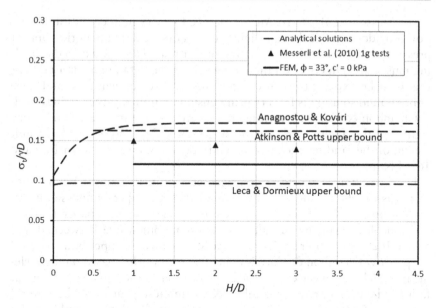

Figure 3.7 Comparison of centrifuge tests by Messerli et al. (2010) on dry cohesionless sand with limit state solutions (Atkinson & Potts (1977) upper bound and Leca & Dormieux (1990) upper bound), a limit equilibrium solution (Anagnostou & Kovári, 1996a) and finite element models ('FEM' = FE models by Vermeer et al., 2002).

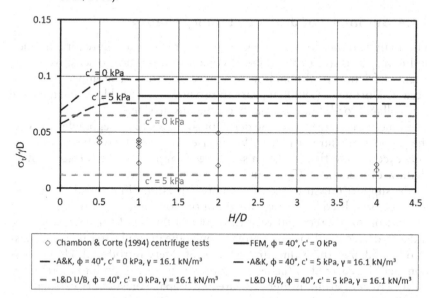

Figure 3.8 Comparison of centrifuge tests by Chambon & Corté (1994) on dry cohesionless sand with Leca & Dormieux's (1990) upper bound ('L&D U/B'), Anagnostou & Kovári's (1994) limit equilibrium solution ('A&K') and finite element models ('FEM' = FE models by Vermeer et al., 2002).

For Messerli et al.'s tests, Anagnostou & Kovári's limit equilibrium solution provides a good, though slightly conservative, estimate of the support pressure at failure. Atkinson & Potts's upper bound (below which the heading must fail) is in effect on the wrong side of the data points, but perhaps this is to be expected because the failure is three-dimensional and their upper bound solution is for a two-dimensional unlined plane strain tunnel.

Chambon & Corté's centrifuge tests were on Fontainebleau sand. They estimated the angle of friction ϕ was between 38° and 42° and the unit weight of the sand for the tests shown in Figure 3.8 was between 16.0 and 16.2 kN/m³.

For Chambon & Corté's tests, Leca & Dormieux's upper bound with $c' = 0$ is on the wrong side. Remember that an upper bound should be on the unsafe side. Chambon & Corté suggest in their paper that the Fontainebleau sand has a value of cohesion somewhere between 0 and 5 kPa. If c' were closer to 5 kPa, Leca & Dormieux's upper bound would drop below Chambon & Corté's data. The relationship based on finite element models (Vermeer et al., 2002) overestimates the support pressure at failure when compared to Chambon & Corté's tests, but would be closer if a value of $c' \neq 0$ were used. Anagnostou & Kovári's solution would require a higher minimum support pressure than Chambon & Corté's tests indicate is necessary, even if $c' = 5$ kPa is assumed, so it is on the safe side.

3.1.4 Summary of drained stability theory

The analytical solutions show that support pressure at failure for a drained soil is very sensitive to the value of cohesion c', but not so sensitive to the angle of friction ϕ, although it does have an effect. The upper bound of Leca & Dormieux (1990) appears to provide a reasonable estimate of the true collapse load, but it is usually on the unsafe side.

All methods suggest that the depth of the tunnel has negligible effect on the support pressure at failure. Vermeer et al. (2002) have found that the only exception to this is when a surcharge is applied to the surface above a shallow tunnel with $C/D < 2$ and $\phi < 25°$.

The support pressures required to prevent failure in drained $c' - \phi$ soils may be normalised by the unit weight of the soil γ and the diameter D. The proportion of γD required is of the order of 0.05–0.2 for most sands. In order to decide what face pressure to apply, the groundwater pressure must also be added to this value, and a factor of safety introduced. How this is done will be covered in the next section when we will go into more detail on the effect of groundwater pressure and seepage on stability calculations for closed-face TBMs, and what the results tell us about how to operate these machines.

Both the relationships based on parametric finite element studies by Vermeer et al. (2002) and the limit equilibrium solution proposed by

Anagnostou & Kovári (1994, 1996a & 1996b) seem to provide a reasonable, and slightly conservative, estimate of the collapse load when compared to centrifuge and 1 g laboratory tests.

3.2 APPLICATION OF DRAINED STABILITY TO CLOSED-FACE TBMS

As demonstrated in the previous sections, drained soils with little or no cohesion require a support pressure to be applied to maintain stability. Closed-face tunnelling machines are often used in these soils, but it is difficult to know what support pressure to specify. If there is instability, the uncontrolled flow of ground into the face will lead to overexcavation, which may not be easily detected, and this in turn may cause excessive surface settlements or collapse. This has occurred all over the world, for example the 'Lavender Street incident' during EPB tunnelling for the CTRL in London (Lovelace, 2003), or the 37 incidents, some of which reached the surface, during construction of the SMART tunnel in Malaysia using a slurry TBM (CEDD, 2012).

3.2.1 Application to slurry TBMs

A slurry TBM supports the ground by filling the head with bentonite slurry under pressure. The slurry pressure must be greater than the groundwater pressure because otherwise it would not be possible to pump it in, but also because we need the excess slurry pressure to apply a support pressure to the soil grains and to include an allowance for variability. This is shown in Figure 3.9.

All slurry machines will experience fluctuations in the slurry pressure, due to dynamic effects such as rotation of the cutterhead, excavation and pumping, and due to the difficulty in balancing the slurry feed in and the slurry extraction rate out. Since a drop in the slurry pressure could result in instability and hence overexcavation, an allowance for variability is added to the target minimum face pressure. Sometimes a factor of safety of some kind is also included, as well as an allowance for the effects of surcharge on the surface (Golder Associates, 2009), as shown in the following equation:

$$P_{st} = s' + u + v + q \tag{3.5}$$

P_{st} is the slurry pressure in the excavation chamber in kPa
s' is the effective support pressure in kPa
u is the pore pressure in the ground in kPa
v is the allowance for variability of slurry pressure (a function of the TBM and how it controls slurry pressure) in kPa
q is an average surcharge pressure applied to the surface in kPa

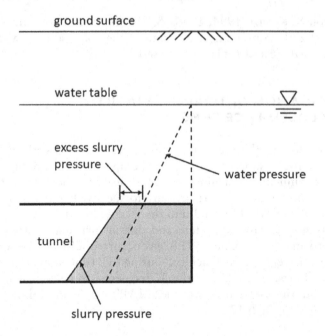

ground surface

water table

excess slurry
pressure

water pressure

tunnel

slurry pressure

Figure 3.9 Slurry pressure diagram for a slurry TBM (after Anagnostou & Kovári, 1996b).

In Figure 3.9, the slurry pressure increases at a faster rate with depth than the groundwater pressure. This is because slurry, when mixed with excavated soil, has a higher unit weight, typically assumed to be around 12 kN/m³, whereas water is just under 10 kN/m³. This means that the minimum excess slurry pressure is always at the crown. Therefore, stability calculations for slurry TBMs are always performed at the crown level as this is the worst case (Golder Associates, 2009).

Because the slurry pressure is always greater than the groundwater pressure, the slurry will flow into the ground. As it displaces the groundwater in the soil's pore spaces, the slurry flow is slowed down and soil particles suspended in the slurry are filtered out and block the pores. Bentonite slurry, at the right concentrations, also has thixotropic properties – this means that it forms a gel when not agitated. Therefore, a 'filter cake' is formed, which acts as a kind of membrane that allows the excess slurry pressure to be applied to the soil grains, counteracting instability.

The ideal situation is where the filter cake forms close to the excavation surface, and Anagnostou & Kovári refer to this as the 'membrane model'. As mentioned earlier in this chapter, drained stability can be approximated using a wedge and prism model (Anagnostou & Kovári, 1994, who attribute the model to Horn, 1961). The geometry of the model was shown in Figure 3.4. As the slurry penetrates into the soil, the excess slurry pressure acting to support the soil grains is spread over the penetration distance, so as the slurry penetrates through the 'wedge', the resultant support force it applies to the wedge and hence the prisms above is gradually compromised. For higher permeability soils, penetration occurs at a faster rate.

One way to reduce the slurry penetration rate is to use a higher viscosity slurry, but this increases slurry pumping and treatment costs. For continuous TBM tunnelling, usually penetration rate is not a problem unless the soil permeability is very high (such as in open gravels), because the TBM is constantly advancing and excavating the ground. It is usually only when there is a standstill of the TBM that penetration is a problem, and often this can be mitigated by temporarily increasing the slurry viscosity. Anagnostou & Kovári (1994) found that increasing the slurry pressure also helps.

WORKED EXAMPLE 3.1 CALCULATING THE TARGET MINIMUM SLURRY PRESSURE

A 9 m diameter tunnel is to be built in sand with cover $H = 15$ m. The sand has bulk unit weight $\gamma = 18$ kN/m³, angle of friction $\phi = 30°$ and effective cohesion $c' = 0$ kPa. The variability of slurry pressure achievable by this machine is ±25 kPa. Surcharge q may be up to 75 kPa. The groundwater table is at the ground surface.

Assume the membrane model applies, i.e. a filter cake forms and the slurry does not penetrate into the soil a significant distance.

Calculate the pore pressure u:

$$u = H\gamma_w = 15 \times 10 = 150 \text{ kPa}$$

Find value of F_0 from nomograms, for $H/D = 1.67$ (need to interpolate between $H/D = 1$ and $H/D = 2$), and $h_0 = H + D$ (i.e. use the dashed lines). This gives $F_0 = 0.22$.

The other coefficients F_1, F_2 and F_3 are not required because $c' = 0$, and because P_{st} will be higher than the pore pressure in the ground, so it is a safe assumption that there will be no seepage towards the face.

The submerged unit weight is given by:

$$\gamma' = \gamma - \gamma_w = 18 - 10 = 8 \text{ kN/m}^3$$

Therefore, the effective support pressure is given by:

$$s' = F_0\gamma'D = 0.22 \times 8 \times 9 = 15.84 \text{ kPa}$$

The target minimum slurry pressure at the crown should therefore be:

$$P_{st,crown} = s' + u + v + q = 15.84 + 150 + 25 + 75 = 265.84 \text{ kPa}$$

Note that the effective support pressure, which is the force applied by the slurry to the soil grains needed to maintain stability of the face, is very small relative to the target minimum slurry pressure – it only makes up 6% of the total in this case.

Note also that the effect of surcharge q has been added to the target minimum slurry pressure without any reduction. This assumption is very conservative and could be refined by use of a numerical model or by using the extension to the wedge-prism method described in Appendix A.

No factor of safety has been applied to this calculation as yet. GEO Report 249 (Golder Associates, 2009) recommends applying a partial factor of 1.2 to c' and tan ϕ. Eurocode 7 (EN 1997-1:2004, 2009) would require a partial factor of 1.25 on these strength parameters.

3.2.2 Slurry infiltration during TBM standstills

As the worked example shows, the target minimum slurry pressure will be higher than the pore pressure in a cohesionless soil. Therefore, during a standstill of the TBM, the slurry will flow into the ground. The support pressure is no longer applied to a thin membrane at the tunnel face, but is spread over the depth of infiltration.

Slurry will penetrate the ground to a maximum distance e_{max}, which depends on the permeability of the ground, the viscosity of the slurry and the excess pressure. Another way of thinking about this is that for a given ground permeability and slurry viscosity, there will be a 'stagnation gradient' at which flow will stop. The stagnation gradient is the excess pressure Δp divided by the maximum penetration e_{max}.

The maximum penetration of slurry e_{max} into a soil can be given by:

$$e_{max} = \frac{\Delta p_s d_{10}}{\alpha \tau_f} \tag{3.6}$$

Δp_s is the total pressure difference across the filter cake (the difference between the slurry pressure and the pore pressure in the ground)

d_{10} is the characteristic grain diameter at which 10% of the soil passes through a sieve (in other words, 90% of the soil grains have a larger characteristic diameter)

τ_f is the shear strength of the slurry

α is a parameter that describes the relationship between grain size and the effective radius of a flow channel, and normally takes a value between 2 and 4 (see Krause, 1987, and Kilchert & Karstedt, 1984). Anagnostou & Kovári (1994) take $\alpha = 2$, following the German standard for diaphragm walling DIN 4126 (1986).

Anagnostou & Kovári (1994) modelled a TBM standstill by integrating the support pressure over the maximum slurry penetration distance in the wedge-silo model. Figure 3.10 shows that face stability decreases with increasing characteristic grain size. Curve B shows the effect of increasing the excess slurry pressure compared to curve A. Stability is improved, but the face quickly becomes unstable at around $d_{10} = 2$ mm. For finer soils the excess pressure applied in curve A would have been sufficient so there is no real benefit gained. Curve C shows the effect of increasing the bentonite concentration from 4% to 7% (increasing the slurry's yield strength from 15 Pa to 80 Pa). This allows a face to be stable even in very coarse-grained soils.

Since we may want to know how the penetration of the slurry increases with time, the following hyperbolic equation proposed by Krause (1987) may be used:

$$\frac{e_t}{e_{max}} = \frac{t}{a+t} \qquad\qquad (3.7)$$

t is the time
e_t is the penetration distance at time t

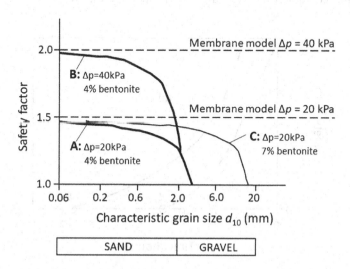

Figure 3.10 Reduction of safety factor with increasing characteristic grain size d_{10}, due to slurry penetration in a soil with $c' = 0$ and $\phi = 37.5°$ (redrawn from Anagnostou & Kovári, 1996b).

a is the time at which half the penetration distance has been reached, which can be determined by laboratory testing using an infiltration column

Anagnostou & Kovári (1994) use a more complicated model for slurry penetration than shown in Equation 3.7, but the results were similar. Again, by integrating the support pressure over the penetration distance in the wedge-silo model, they showed that stability decreases over time as slurry penetrates into the ground. This gives a 'stand-up time' after which the face will be unstable, shown as t_{cr} in Figure 3.11. Stand-up time can be increased by increasing the slurry pressure, as demonstrated by the curves for Δp = 40 kPa and Δp = 80 kPa.

Stand-up time has a near-reciprocal relationship with permeability, so if the permeability is reduced by an order of magnitude, then the stand-up time is increased by an order of magnitude.

3.2.3 Slurry infiltration during excavation

Anagnostou & Kovári (1994) gave a rule of thumb that if the soil d_{10} value is smaller than 0.6 mm, then infiltration will be small and the membrane model can be assumed to apply. However, they only looked into the effect of slurry penetration on the effective zone of application of support pressure. Broere & van Tol (2000) identified an additional effect which is the generation of positive excess pore pressures in the ground ahead of the face, which has a tendency to occur when fine or medium sands are being excavated by a TBM in a confined aquifer (i.e. with an impermeable layer above).

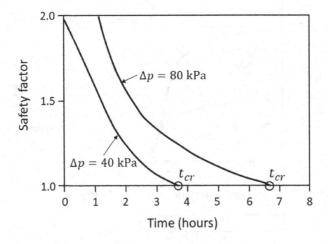

Figure 3.11 Safety factor as a function of time for a 4% bentonite slurry with yield strength 15 Pa, in a coarse-grained soil with d_{10} = 6 mm, k = 10^{-4} m/s, c' = 0 and ϕ = 37.5° (from Anagnostou & Kovári, 1996b).

Positive excess pore pressures can be generated in the sand due to repeated excavation of the filter cake by the cutter bits followed by infiltration of the slurry, driving filtrate water into the ground and elevating the pore pressure. These positive excess pore pressures cause a reduction in effective stress and hence a reduction in the shear resistance of the soil. Also, higher pore pressures in the soil effectively cancel out some of the effective support pressure that can be used to support the soil grains.

Using a modified wedge-silo limit equilibrium model developed by Broere (1998), Broere & van Tol (2000) modelled this effect and found that the required minimum support pressure was significantly higher, at 80 kPa above the in situ pore pressure compared to only 16 kPa above the in situ pore pressure when using Anagnostou & Kovári's (1994) full membrane model. This was corroborated by measurements made by piezometers installed in the alignment of the Second Heinenoord Tunnel. A different tunnel in Rotterdam in somewhat coarser sand and with no clearly definable impermeable overlying stratum had a maximum positive excess pore pressure of only 5 kPa, demonstrating that the effect depends on the presence of sand lenses or confined sand layers.

A graph from Broere & van Tol (2000) is shown in Figure 3.12. This shows measurements of pore pressure as the cutterhead approaches the piezometer. During excavation, significant positive excess pore pressures are generated ahead of the face. During ringbuilding every 1.5 m these positive excess pore pressures dissipate and the pore pressure returns

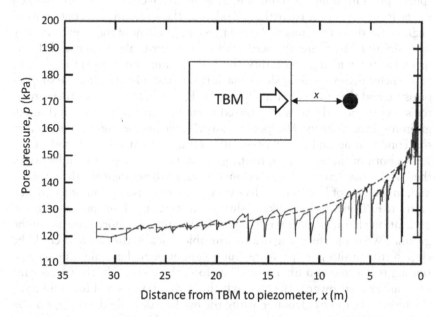

Figure 3.12 Piezometer measurements of pore pressure ahead of a slurry TBM at the Second Heinenoord Tunnel (Broere & van Tol, 2000).

to the initial in situ value of approximately 120 kPa (with some tidal variation evident).

A further paper by Broere & van Tol (2001) says that if the soil permeability is between 10^{-3} and 10^{-5} m/s, then the effect of positive excess pore
pressures ahead of the face will be significant. This paper also extends the
model to less permeable silty sands by including a transient groundwater flow
model to replace the steady-state model of the previous paper. This allowed
for a slower dissipation of excess pore pressures during the ringbuilding cycle.

3.2.4 Application to earth pressure balance TBMs

Face stability calculations for EPB machines are more complicated. The cutterhead breaks up the soil and the muck moves through openings in the cutterhead into the 'excavation chamber'. The excavation chamber is kept filled
with muck under pressure ('EPB pressure'), and it is this pressure that provides the face stability. Muck is removed in a controlled manner from the
excavation chamber using an Archimedes screw, and the EPB pressure is controlled by balancing the screw speed with the penetration rate of the TBM
into the ground. It is important to note that EPB pressure is usually measured
by load cells in the excavation chamber, which actually measure total stress,
i.e. the total of both the effective support pressure and the pore pressure.

At the back end of the screw, the muck falls out onto a belt conveyor at
atmospheric pressure, so there is a gradient from the EPB pressure to atmospheric pressure along the screw. The gradient in effective stress is maintained
by the tortuosity of the route the soil takes up the screw and the cohesion and
angle of friction of the muck. A pore pressure gradient in the screw can only
be maintained by filling the screw with a low permeability material. If the
muck has too high a permeability, soil conditioning can be used to reduce it.

It is helpful, as we did for slurry machines, to consider the effective support
pressure and the pore pressure separately. The 'effective support pressure' is
the stress in the soil grains transferred from the machine to the soil through
grain-to-grain contacts. The 'pore pressure' is the pressure of the fluid (water,
soil conditioning and air) between the soil grains. At the back end of the
screw, both of these are zero. In the ground, far away from the influence of
the TBM, they have undisturbed in situ values dependent on their weight
and, in the case of horizontal effective stress, stress history and other effects.

In an EPB machine, unlike a slurry machine, the fluid pressure in the
excavation chamber cannot reliably be higher than the pore pressure in the
ground. With a perfect plug of impermeable muck in the screw, it could be
close to the in situ pore pressure, but it cannot be higher unless soil conditioning is injected at a high rate. Therefore, there is nearly always seepage
of groundwater towards the face, which, as we know, is bad for stability.

The effect can be calculated using the method described in Anagnostou
& Kovári (1996a). It can be illuminating to calculate the two extremes:
in the first case (A) with zero pore pressure in the excavation chamber,

giving the maximum destabilising hydraulic gradient in the ground, and in the second case (B) with the pore pressure in the excavation chamber equal to the in situ pore pressure in the ground. In the first case A, the destabilising seepage force needs to be counteracted by an increased effective support pressure, but the effective support pressure is equal to the total EPB pressure because there is no pore pressure to drive against. In the second case B, there is no destabilising seepage pressure, so the effective support pressure is only what is required to maintain stability due to gravity, but we have to add the pore pressure to get the EPB pressure. It turns out, if you run the numbers, that the EPB pressure required for Case A is always less than in Case B. In other words, the extra effective support pressure required to resist the destabilising seepage forces is less than the in situ pore pressure.

WORKED EXAMPLE 3.2   TARGET MINIMUM EPB PRESSURE

An 8 m diameter tunnel is to be built in sandy silt with cover $H = 16$ m and the water table located at the ground surface, i.e. $H_w = 16$ m and $h_0 = 24$ m. The sandy silt has saturated unit weight $\gamma = 18$ kN/m^3, angle of friction $\phi = 20°$ and effective cohesion $c' = 0$ kPa.

The variability of EPB pressure achievable by this machine is ±25 kPa. Surcharge q may be up to 25 kPa.

We want to use Equation 3.2:

$$s' = F_0\gamma'D - F_1c + F_2\gamma'\Delta h - F_3c\frac{\Delta h}{D}$$

Since $c = 0$ kPa, we need to find F_0 and F_2 only.

Find value of F_0 and F_2 from the nomograms, for $H/D = 2$, and $h_0 = H + D$ (i.e. use the dashed lines). As shown in Figure 3.13, this gives $F_0 = 0.50$ and $F_2 = 0.60$.

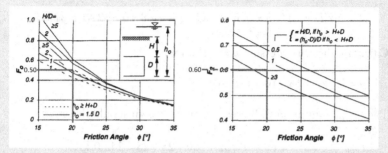

Figure 3.13 Worked Example 3.2 – reading nomograms to find values of F_0 and F_2 (corrected versions of nomograms from Anagnostou & Kovári, 1996b).

The submerged unit weight is given by:

$$\gamma' = \gamma - \gamma_w = 18 - 10 = 8 \text{ kN/m}^3$$

Case A: maximum seepage

For Case A, where pore pressure in the chamber $h_f = 0$, the head difference Δh between the ground and the TBM that is driving seepage is given by:

$$\Delta h = h_0 - h_f = 24 - 0 = 24 \text{ m}$$

The effective support pressure required to maintain stability is given by:

$$s' = F_0 \gamma' D + F_2 \gamma' \Delta h = 0.50 \times 8 \times 8 + 0.60 \times 8 \times 24 = 32 + 115.2 = 147.2 \text{ kPa}$$

The pore pressure the TBM is pushing against is zero.

The target minimum EPB pressure at the crown should therefore be:

$$P_{EPB,crown} = s' + u + v + q = 147.2 + 0 + 25 + 25 = 197.2 \text{ kPa}$$

No factor of safety has been applied to this calculation as yet. GEO Report 249 (Golder Associates, 2009) recommends applying a partial factor of 1.2 to c' and tan ϕ. Eurocode 7 (EN 1997-1:2004, 2009) requires a partial factor of 1.25.

Case B: no seepage

For Case B, a perfect impermeable plug is maintained in the screw and pore pressure in the chamber is equal to the in situ pore pressure in the ground. Therefore, there is no seepage towards the TBM, and $\Delta h = 0$.

The effective support pressure required to maintain stability is given by:

$$s' = F_0 \gamma' D + F_2 \gamma' \Delta h = 0.50 \times 8 \times 8 + 0.60 \times 8 \times 0 = 32 + 0 = 32.0 \text{ kPa}$$

Calculate the pore pressure u at the crown:

$$u = H_w \gamma_w = 16 \times 10 = 160 \text{ kPa}$$

The target minimum EPB pressure at the crown should therefore be:

$$P_{EPB,crown} = s' + u + v + q = 32 + 160 + 25 + 25 = 242.0 \text{ kPa}$$

No factor of safety has been applied to this calculation.

In Case A the effective support pressure is much higher than in Case B, at 147.2 kPa versus 32.0 kPa, but in Case B, a higher EPB pressure is required to maintain stability. Therefore, one would think, surely it is better to drain the ground and run the machine with a lower EPB pressure? Why do we go to great lengths to get the soil conditioning right so we can 'maintain a plug in the screw'? One reason is that, particularly in high permeability soils, uncontrolled water flow through the screw could flood the tunnel. But there is another very good reason: the higher the effective support pressure, the higher the effective stress in the soil's grain to grain contacts. The shear strength of a soil is proportional to the effective stress, so this means the soil will behave like a stronger material. This is undesirable because we want it to flow nicely through openings in the cutterhead, fill the excavation chamber and move along the screw, not arch around openings and block the screw. The increased friction between soil grains will also generate a lot of heat and increase wear of metal parts. This means an increase in maintenance stops and slower over-all progress. In some soils, the machine cannot be advanced at all.

The same principles govern the stability calculations for both slurry and EPB machines, even though they work in different ways. The main differ-ence is that in a slurry machine, the slurry pressure is always greater than the groundwater pressure, but we need a filter cake to form to also apply an effective support pressure to support the soil grains. In an EPB machine, the machine is applying an effective support pressure to the soil grains, but we need to maintain a plug in the screw to counteract the groundwater pressure and minimise the hydraulic gradient.

Soil conditioning is critical to making the muck in an EPB machine behave the way we need it to. Ideally we turn it into a low-permeability, cohesive and plastic material to minimise the work required to get it to flow through the machine, minimise friction and wear of metal parts, and create a good plug in the screw so that the pore pressure in the excavation chamber is close to the in situ pore pressure. Often, the spoil needs to be suitable for disposal, so we compromise between creating the perfect EPB material and producing spoil that can be reused as fill or transported and disposed of safely.

3.3 BLOW-OUT FAILURE IN DRAINED SOILS

For drained soils, as for undrained, there can be two types of blow-out: a large-scale passive failure of a block of soil moving upwards towards the surface (as described in Leca & Dormieux, 1990), or an escape of fluid (slurry, grout or compressed air) through a fracture.

3.3.1 Passive failure in drained soils

To understand the shape of a passive failure in drained soils, we have to look at numerical or physical models, because real-world failures are so

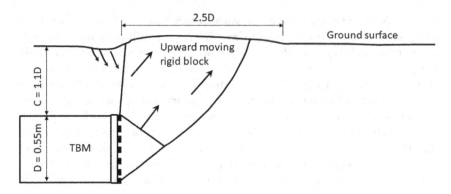

Figure 3.14 Blow-out passive failure mechanism in a reduced-scale model of an earth pressure balance TBM (redrawn from Berthoz et al., 2012).

hard to observe. Berthoz et al. (2012) performed a series of reduced-scale 1 g laboratory tests in moist sand with bulk unit weight $\gamma = 13$ kN/m^3, drained cohesion $c' = 0.5$ kPa and angle of friction $\phi = 36°$. They used a scale model 0.55 m diameter EPB TBM. It advanced 0.95 m into the ground and then the face pressure was increased by reducing the extraction rate of the Archimedes screw while continuing to advance the machine. By measuring displacements at the surface and within the ground, they found that the passive failure geometry consisted of rigid blocks as shown in Figure 3.14. Displacements within the blocks were similar, while displacements outside the blocks were small.

It is clear from Figure 3.14 that we cannot simply adapt Anagnostou & Kovári's (1994) wedge and prism model to passive failure as we no longer have a vertical prism.

Wong et al. (2010, 2012) performed centrifuge tests of a tunnel in saturated sand. The tunnel was impermeable, therefore there was no groundwater flow. A rigid piston simulated the increase of face pressure. This is unrealistic in some respects as the face displacement caused by a closed-face tunnelling machine or an open face with compressed air applied will always be pressure-controlled, not displacement-controlled. Figure 3.15 shows displacement vectors from their tests.

Figure 3.15 shows very different displacement vectors at $C/D = 2.2$ and $C/D = 4.3$. A very similar pattern of vectors was found in numerical models of the centrifuge tests (Wong et al., 2012). At $C/D = 2.2$, the failure is directed upwards towards the nearby ground surface, whereas at $C/D = 4.3$, the failure is localised in front of the tunnel face, with very little effect on the more distant ground surface except some slight settlement. Berthoz et al. (2012) found that at $C/D = 2$, when surcharge of 50 kPa was added to the ground surface, there was no heave of the surface at all during passive failure, and in fact there was settlement. In this case, the passive failure was localised to horizontal displacements away from the tunnel face in the

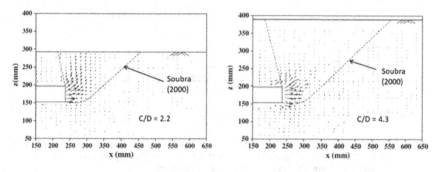

Figure 3.15 Displacement vectors at face displacement of 0.5D in a centrifuge test for
C/D = 2.2 and C/D = 4.3, from Wong et al. (2012).

direction of tunnelling, the same pattern found by Wong et al. (2012) for
C/D = 4.3 on the right-hand side of Figure 3.15. One could consider the
surcharge as, in effect, making the tunnel seem to be deeper (as though C
were increased). In terms of vertical stress, 50 kPa surcharge in Berthoz
et al.'s reduced-scale model is the equivalent of increasing the cover from
1.1D to approximately 8D. This indicates that passive failure can have different forms with shallow and deep cover or with varying levels of surcharge pressure.

Figure 3.15 also shows an upper-bound kinematic mechanism proposed
by Soubra (2000). The experimental and numerical studies of Berthoz et al.
(2012) and Wong et al. (2010, 2012) both show that the geometry of failure
found in kinematic analyses of passive failure often includes a much greater
volume of soil than that found in practice. Berthoz et al. (2012) argue that
this may be because the kinematic analysis assumes a failure at much higher
displacements than is practical. Alternatively, Wong et al. (2012) suggest
that the assumption of associative plasticity (sometimes called the 'normality condition', where the angle of dilation is equal to the critical state angle
of friction ϕ_{cs}) in the kinematic analysis may be the cause of the difference.

As described earlier in this chapter in Section 3.1.1, there are two plasticity limit states, the upper bound and the lower bound. The lower bound
is based on a statically admissible stress state and gives a limiting value
of face pressure at which the face cannot fail, that is definitely safe. The
upper bound is based on a kinematic mechanism, and gives a value of face
pressure at which the face will definitely fail, but it is an unsafe prediction because it could fail at a lower value. A handy diagram is provided by
Berthoz et al. (2012), reproduced in Figure 3.16.

Kinematic analysis usually requires partial definition of the geometry
of failure, usually with one or two parameters that need to be optimised.
Many researchers have attempted to provide realistic kinematic mechanisms for collapse and blow-out of tunnels in drained soils, though there
has been less success with blow-out than with collapse. The aim is to make

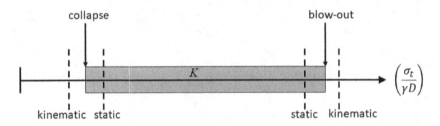

Figure 3.16 Diagram explaining limit states for a domain of safe face pressures K (redrawn from Berthoz et al., 2012). σ_t is the tunnel support pressure, γ is the unit weight of the soil and D is the tunnel diameter.

the geometry of the kinematic analysis as realistic as possible, so it gives a close approximation of the tunnel support pressure needed to prevent collapse or blow-out. Then it can become a useful tool for determining the safe range of face pressures between collapse and blow-out.

Leca & Dormieux (1990) proposed a 3D upper-bound mechanism based on a single truncated cone, shown in long section in Figure 3.17.

Kinematic analysis assumes the presence of a 'velocity discontinuity surface'. Within the surface a rigid block is moving and outside of the surface there is no movement. This seems a reasonable assumption for drained soils given the displacements presented by Berthoz et al. (2012), shown earlier in Figure 3.14. The velocity discontinuity surface is at an angle ϕ to the direction of motion of the block, because this is the angle that offers the lowest shear resistance, and this is why the shape is conical. The direction of motion has to be along the axis of the cone. The angle α is then found at which blow-out occurs most easily, i.e. with the lowest tunnel support pressure.

The problem with modelling the velocity discontinuity surface as a cone of circular section is that when it intersects the tunnel face at an angle, the intersection is an ellipse. This means that there are zones on either side of

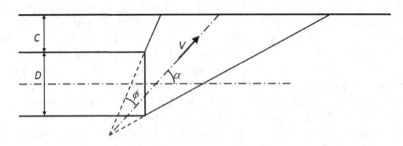

Figure 3.17 Long section through single cone kinematic mechanism for blow-out (redrawn from Leca & Dormieux, 1990). V is the resultant blow-out force acting on the cone, ϕ is the angle of friction and α is the cone projection angle.

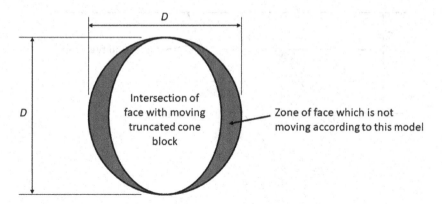

Figure 3.18 Intersection of a circular cone at an angle with a circular tunnel face (redrawn from Leca & Dormieux, 1990).

the face which do not move at all, as shown in Figure 3.18. It also may not reflect the true shape of the failure.

A principle of kinematic analysis is that if the geometry can be changed to show that failure can occur at a lower tunnel support pressure, then that new geometry must be more realistic, because we assume that in the real world failure occurs in the easiest way possible.

In order to better approximate the true geometry of failure, and hence to move the upper bound closer to the true failure load, several researchers have added complexity to Leca & Dormieux's model. Soubra (2000) tried two truncated cones with a log spiral in between, as shown by the dashed lines in Figure 3.15. Mollon et al. (2009) used multiple truncated cones, as shown schematically in Figure 3.19. Each time, results were improved, i.e. the tunnel support pressure at failure was reduced and could therefore be assumed to be closer to the true blow-out pressure that might occur in a real situation (have another look at Figure 3.16). This may be, at least in part, because the volume of the failure has been reduced (compare the volume within the velocity discontinuity surface in Figure 3.19 with that in Figure 3.17), but also because the support pressure is applied to the whole face area and therefore will provide the same force (load V in Figures 3.17 and 3.19) at a lower tunnel support pressure.

To get over the problem illustrated by Figure 3.18, Mollon et al. (2010) improved on the multiblock mechanism by applying a spatial discretisation technique to ensure the intersection of the moving blocks with the tunnel face matched the whole circular face. Mollon et al. (2011) then abandoned multiple blocks in favour of a log spiral shape, again using spatial discretisation to ensure the failure intersected with the whole face. Each time, an incremental improvement in the upper bound was achieved, indicating that the failure mechanisms were becoming more realistic. Looking back at the vectors in Figure 3.15 or the sketch in Figure 3.14, and comparing the

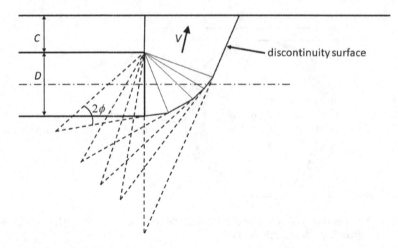

Figure 3.19 Multiple truncated cone mechanism for passive failure (redrawn from Mollon et al., 2009 with permission from ASCE).

shape of the failure with Figure 3.20, we can see that the log spiral seems to fit the shape far better than the earlier mechanisms proposed by Leca & Dormieux (1990) or Soubra (2000) (see Figures 3.17 and 3.15 respectively).

A design chart for critical blow-out pressures based on the work of Mollon et al. (2011) is shown in Figure 3.21 for a cohesionless soil with $\phi = 20°$ and $\phi = 40°$. As C/D increases, the critical blow-out pressure increases rapidly, indicating that blow-out to the surface may be a risk for shallow

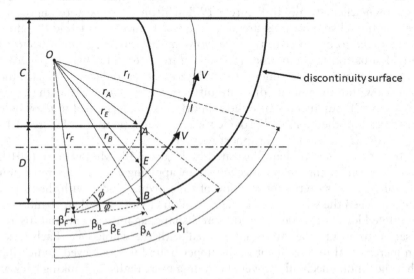

Figure 3.20 Log spiral mechanism for passive failure (redrawn from Mollon et al., 2011).

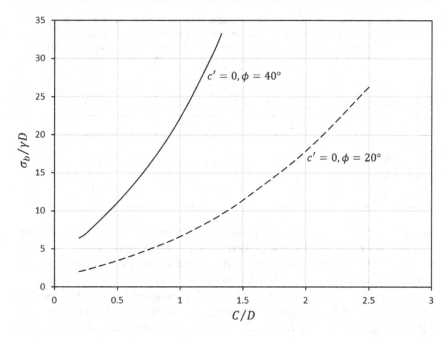

Figure 3.21 Design chart based on the log-spiral passive failure geometry for purely frictional cohesionless soils (redrawn from Mollon et al., 2011). σ_b is the blow-out pressure.

tunnels only. This should be compared to collapse of a heading in frictional soil, which does not depend on C/D.

There does not seem to be a consensus on whether kinematic analysis gives the right answers. Wong et al. (2010) for some reason do not achieve passive failure in their centrifuge tests, as a continuous increase of displacement at constant face pressure is not found despite attaining very large displacements, nor is it found in their numerical model (Wong et al., 2012), so no firm conclusions can be drawn from their work.

Berthoz et al. (2012) reached passive failure in their experiments at face pressures 1–2 orders of magnitude lower than predicted using a 2-cone kinematic mechanism proposed by Subrin (2002). They ascribe this to the smaller volume of failure and the fact that the very large deformations needed to mobilise yield on all surfaces could not be achieved in the experiment. There also must be an error in their kinematic analysis because Mollon et al.'s (2011) design charts give much lower values than the Subrin (2002) values, but still higher than Berthoz et al.'s, as shown in Table 3.1.

Using numerical models in FLAC3D, Dias et al. (2008) found that a blow-out failure occurred only in the upper part of the face (Figure 3.22) for purely frictional soils, and at a face pressure 40% lower than predicted using kinematic analysis methods of Leca & Dormieux (1990) and Soubra (2000). They hypothesised that this mode of failure can occur when the

Table 3.1 Comparison of critical blow-out pressures σ_b from Berthoz et al. (2012) experiments and kinematic analysis using the method of Subrin (2002) with values from design charts by Mollon et al. (2011).

Experiment ID	Berthoz et al. (2012)		Mollon et al. (2011) σ_b (kPa)
	Experiment σ_b (kPa)	Subrin (2002) σ_b (kPa)	
MC1-B1	34	612	168
MC1-B2	10		
MC3-B2	21	612	174
MC5-B1	47	515	115
MC5-B3	21		

face pressure is uniform, and presumably therefore would be less likely in an EPB or slurry machine, where face pressure increases from crown to invert. This would explain the full-face failure found in Berthoz et al.'s (2012) experiments with a model EPB machine in purely frictional soil.

On the other hand, Mollon et al. (2013b) compare FLAC3D numerical analyses with their log spiral kinematic mechanism, and find very close agreement, as shown in Table 3.2, for cohesive-frictional soils. They state that similar agreement was found for "several cases of frictional soils, with or without cohesion (not shown in this paper)". It is unfortunate that these are not shown in their paper as it would have been interesting to see if the failure for the purely frictional soils was only in the top half of the face as found by Dias et al. (2008), and what impact this may have had on the critical blow-out pressure.

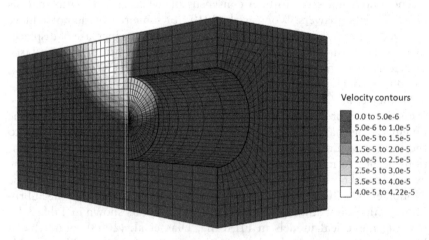

Velocity contours

■	0.0 to 5.0e-6
	5.0e-6 to 1.0e-5
	1.0e-5 to 1.5e-5
	1.5e-5 to 2.0e-5
	2.0e-5 to 2.5e-5
	2.5e-5 to 3.0e-5
	3.5e-5 to 4.0e-5
□	4.0e-5 to 4.22e-5

Figure 3.22 Displacement velocity field from FLAC3D model of passive failure, for $C/D = 0.5$ in Fontainebleau sand, drained cohesion $c' = 0$ kPa, angle of friction $\phi = 42°$, bulk unit weight $\gamma = 15.7$ kN/m³ (from Dias et al., 2008 with permission from ASCE).

Table 3.2 Critical blow-out pressures as found by the M1 kinematic analyses and the FLAC3D numerical models for $D = 10$ m and $\gamma = 18$ kN/m^3, from Mollon et al. (2013b) with permission from ASCE.

	$\phi = 17°, c' = 7\ kPa$			$\phi = 25°, c' = 10\ kPa$		
	Blow-out pressure (kPa)		*Difference*	*Blow-out pressure (kPa)*		*Difference*
C/D	M1	FLAC3D	(M1-FLAC3D)/ FLAC3D	M1	FLAC3D	(M1-FLAC3D)/ FLAC3D
0.6	682.4	635	7.46%	1112.2	1091	1.94%
0.8	878.6	864	1.69%	1487.3	1521	−2.22%
1	1096.6	1113	−1.47%	1903.5	2004	−5.01%
1.5	1777.1	1842	−3.52%	3301.2	3488	−5.36%
2	2637.7	2740	−3.73%	5213.3	5337	−2.32%
3	5243.1	5253	−0.19%	–	–	–

3.3.2 Blow-outs caused by hydraulic fracturing

Bezuijen & Brassinga (2006) show from field data and centrifuge tests that bentonite slurry blow-outs can occur at much lower face pressures than those predicted by finite element or kinematic analysis methods. This is because these methods do not take account of the fact that slurry is a fluid. The limit then becomes the pressure at which hydraulic fracturing of the ground can occur. Generally accepted practice is to limit support fluid pressures to the pore pressure plus the vertical effective stress at the crown of the tunnel (e.g. Guglielmetti et al., 2008). Bezuijen & Brassinga found the limit to be approximately the pore pressure plus two to three times the effective stress for their case.

Broere (2015) also highlights the importance of pore pressures. If the slurry is infiltrating the pores of the ground, then pore pressures will be elevated in front of or above the face. This causes a drop in the effective stress and hence the shear resistance of the ground. Although the infiltration will apply stabilising seepage forces to the soil grains, and in theory a vertical face of cohesionless soil can be stable as long as the hydraulic gradient is greater than 2 (van Rhee & Bezuijen, 1992), a continuous infiltration cannot be relied on in practice.

Holzhäuser (2003) discusses several types of compressed air blow-outs. The air pressure is almost always set higher than the pore pressure in the ground, at least at the crown level. The difference between air pressure and slurry pressure is that air pressure does not vary with height, because its density is only 1.225 kg/m^3. Therefore, wherever the air can flow to, it applies approximately the same pressure. The air tends to displace the pore water, which can dry out and erode flow channels, leading to a sudden blow-out and loss of air pressure, followed by collapse.

If a less permeable layer is some distance above the crown, the pore pressure in the ground in between can become equal to that in the working chamber. Thus, the compressed air pressure may act at a level where the overburden is insufficient to contain it. In this case, blow-out can occur due to hydraulic fracturing even though the applied compressed air pressure is lower than the full overburden pressure at the crown level. This is referred to as a 'gasometer' blow-out (Holzhäuser, 2003). This happened on the Blackwall Tunnel construction under the Thames, reported by Moir (1897), resulting in flooding of the tunnel by river water. So calculations of maximum allowable air pressure need to take account of the geology of the site.

As a minimum, compressed air pressure should be kept below the overburden pressure. The British Standard for health and safety in tunnelling BS 6164:2019 is even more conservative and states that only the dry density of the soil should be used in the calculation of limiting air pressure, as the compressed air tends to drive away groundwater and dry out the soil.

Slurry blow-outs can also occur, for example during construction of the Second Heinenoord Tunnel where large quantities of bentonite escaped into the Old Meuse River above the tunnel. Bezuijen & Brassinga (2006) investigated this event and using TBM records, finite element modelling and centrifuge modelling they established that it was caused by hydraulic fracturing of the sand above the crown of the tunnel, possibly exacerbated by slurry penetration causing positive excess pore pressures and hence reducing the effective stress in the soil. Bezuijen & Brassinga came to the conclusion that a fluid escaping through a fracture caused by a strain localisation needs a smaller face pressure than a passive failure. They found that the fluid pressure required for a blow-out minus the pore pressure is two to three times the effective in situ stress in the soil.

Although bentonite slurry produces a filter cake, this is destroyed when the TBM starts advancing, and then positive pressures can be generated in the soil in front of and just above the face, reducing the soil's effective stress and making a blow-out more likely.

There is also the possibility that a slurry TBM could meet an abandoned well, a poorly backfilled borehole, or some other underground void, or the permeability of the ground could suddenly increase. This could cause a sudden loss of pressure, leading to collapse and/or overexcavation. Also, if there is a route to the surface, the slurry only needs enough head to reach the surface and it will flood the local area. After many incidents of local streets being flooded with bentonite slurry and sinkholes appearing along the route of the Singapore Circle Line and the SMART tunnel in Kuala Lumpur, Malaysia (CEDD, 2012), 'variable density' TBMs have been developed, which allow a higher density slurry to be used in the head during excavation (Bäppler et al., 2017). It is then diluted to allow it to be pumped out of the tunnel for treatment. The higher density means that a

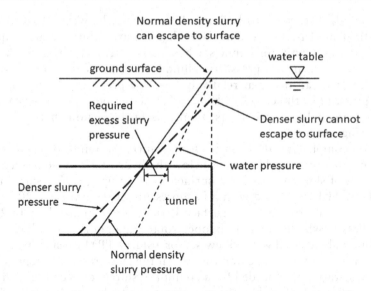

Figure 3.23 Normal density and higher density slurry pressure diagrams.

higher slurry pressure can be used at shallow cover without risk of slurry reaching the surface, as illustrated in Figure 3.23.

3.3.3 Summary of blow-outs in drained soils

For all types of blow-outs, increasing cover will generally reduce the risk. An understanding of the ground and the groundwater, and how they interact with the support method, will help predict blow-out scenarios.

Numerical and physical modelling techniques appear to be fairly reliable methods for analysing collapse or passive failure of a heading. However, the focus has been on collapse and more research is needed to improve our knowledge of passive failure and blow-outs in general.

The kinematic analysis methods that have been applied to passive failure of drained soils have not taken account of groundwater flow, and neither has much of the numerical or physical modelling. Most theoretical approaches suffer from the inability to model the actual interaction between the support fluid and the soil grains. Hydraulic fracturing, and soil layers, lenses or channels with varying permeability, are all difficult to model, and all may result in blow-outs at lower face pressures than predicted using currently available simple models. More research is needed to investigate all these scenarios and further improve our predictions. Pan & Dias (2016a, 2016b) developed methods of calculating pore pressures due to steady-state seepage in a finite element model and used them in a kinematic analysis of collapse. They also looked at the effect of anisotropic permeability. If they or others apply this method to blow-out in future the results will be interesting.

Although there seem to be large discrepancies between physical and analytical models, the results show that passive failure requires quite large face pressures for tunnels where $C/D > 1$. On the other hand, hydraulic fracturing, or pressurised fluid finding an escape route by some other means, can occur at relatively low face pressures. If the support fluid pressure at the crown is kept to below the pore pressure plus the minor principal effective stress, then it is fairly certain that hydraulic fracturing cannot occur.

To be completely safe from bentonite slurry escaping to the surface, the face pressure should be kept below the hydrostatic value needed for a column of slurry to reach the surface. For example, with a slurry unit weight of 11 kN/m³, and with a face pressure at the crown of 200 kPa, the slurry will not reach the surface if the cover is more than 18.2 m, regardless of whether there is an open route via a poorly backfilled borehole or an abandoned well. Likewise, for a slurry TBM passing beneath a river, slurry cannot escape into the river if the face pressure is kept below the hydrostatic value needed for a column of slurry to reach the riverbed with a pressure higher than the water pressure at the riverbed level. This is hard, if not impossible, to achieve, since the slurry pressure usually needs to be set higher than the pore pressure in the ground. In this case, the properties of the ground between the crown and the riverbed become of crucial importance as it is the only thing stopping a pollution incident similar to the one we looked at from the Second Heinenoord Tunnel (Bezuijen & Brassinga, 2006).

3.4 PIPING

In cohesionless soils below the water table, it is possible for soil and water to flow into the tunnel due to destabilising seepage forces. This was discussed in Chapter 1 and is known as 'piping'. Tunnels have been flooded by silt and/or sand and water flowing through small gaps or holes in the lining, leading to total abandonment of the TBM and a complete restart (e.g. in Preston, UK: Thomas, 2011) or costly rescue works (e.g. Hull wastewater flow transfer tunnel, UK: Grose & Benton, 2005, 2006; Brown, 2004).

The only way to avoid this risk is to pay close attention to the quality of gaskets, bolts and grommets in segments, and ensure that rings are built to the specified tolerances and grouted through the tailskin immediately behind the brush seals. Any evidence of water inflow bringing soil with it should be dealt with immediately by grouting or sealing. At Hull, joints between segments may have been opened up by movements of the tunnel rings relative to the shaft, so any post-construction deformations, caused by adjacent tunnel construction for example, should be assessed and monitored carefully.

3.5 PROBLEMS

Q3.1. The following question is about slurry TBM face stability.

    i. According to the method described in GEO Report 249, what four values need to be calculated and added together to give the target minimum slurry pressure at the face? Write down the equation and explain each term.

    ii. Describe qualitatively what the 'effective support pressure' is and explain how it is applied to the face of a slurry TBM during the excavation cycle and, with reference to the 'wedge and prism model', explain how its effect may change as time elapses during stoppages.

    iii. A slurry TBM is to be driven through sandy silt below a lake as shown in Figure 3.24. Calculate the effective support pressure required if the saturated unit weight $\gamma = 19$ kN/m³, drained cohesion $c' = 5$ kPa and angle of friction $\phi = 25°$.

    iv. Calculate the target minimum slurry pressure at the crown, assuming the slurry pressure can be controlled to ±20 kPa and ignoring the effect of surcharge.

    v. From your answer to (iv), what parameters are the most important to know accurately before and during the tunnel drive?

    vi. If this were a mixshield TBM, what air pressure would be needed if the slurry level behind the submerged wall were 2.5 m below the crown? Assume the slurry has a unit weight of 12 kN/m³.

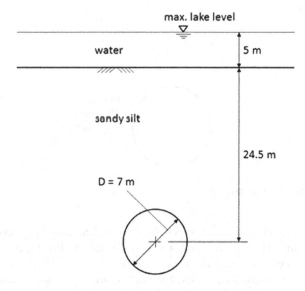

max. lake level

water   5 m

sandy silt

24.5 m

D = 7 m

Figure 3.24 Q3.1 Cross-section of slurry TBM tunnel below lake.

Q3.2. A contractor has proposed using an EPB TBM for a 2 km section of a 5 m ID metro tunnel between two construction shafts. The bolted precast concrete segmental lining is 250 mm thick. The overcut diameter of the TBM is 5.7 m. The soil has been characterised as varying from sandy silt to silty clay with an angle of friction between $\phi_{min} = 20°$ and $\phi_{max} = 30°$, and drained cohesion $c'_{min} = 0$ kPa and $c'_{max} = 10$ kPa. The depth to axis is 20 m and the water table is at a depth of 5 m below ground level (Figure 3.25). The saturated unit weight of the soil $\gamma = 20$ kN/m³.

  i. Calculate the minimum face pressure required to maintain stability in the drained case, using the method of Anagnostou & Kovári. Ignore all factors of safety, any allowance for variability of face pressure and any surface surcharge.

 ii. If it were possible to identify an area with the most favourable ground conditions to be used for a head intervention, show that it would still not be possible to enter the head without compressed air. Demonstrate this using calculations for the drained case.

iii. If dewatering were used to avoid the need for compressed air, calculate the level the water table needs to be drawn down to.

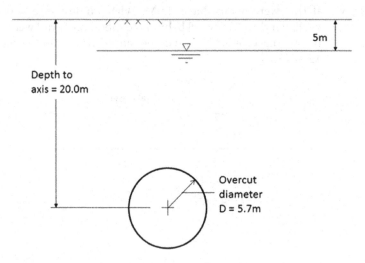

Figure 3.25 Q3.2 Section through metro tunnel.

Q3.3. A tunnel with a 70 m² cross-sectional area is to be constructed in an urban area using a sprayed concrete lining, at a depth to the centroid of the cross-sectional area of 19 m as shown in Figure 3.26. The ground is silty sand with some gravel, with a bulk unit weight of 18 kN/m³ and a permeability of 10^{-7} m/s.

The ground has an angle of friction of 25°. The tunnel face is well above the water table. Assume there is no surcharge.

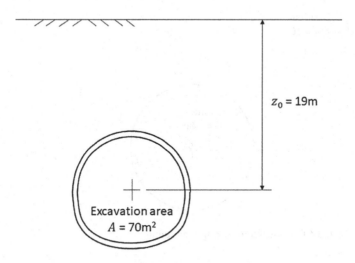

$z_0 = 19m$

Excavation area
$A = 70m^2$

Figure 3.26 Q3.3 Cross-section of sprayed concrete lined tunnel.

   i. If the tunnel is excavated full-face, what value of drained cohesion is required to provide a factor of safety of 1.0? Use Anagnostou & Kovári's nomograms, but explain any assumption you had to make that was unrealistic compared to the real situation described in the question.

  ii. What other analytical (i.e. not numerical) method could be used to obtain a better answer to (i)?

 iii. Describe, with the aid of sketches, two different sprayed concrete construction sequences that could be used instead of full-face, and say how they improve face stability.

 iv. Describe, with the aid of sketches, two other toolbox measures that could be used to improve the face stability, and explain how they work.

Q3.4. A contractor has proposed using an EPB TBM for a section of a 7 m ID rail tunnel between two shafts. The bolted segmental lining is 300 mm thick. The overcut diameter of the TBM is 7.8 m. The soil has been characterised as varying from sandy silt to silty clay with an angle of friction $\phi = 25°$ and drained cohesion $c' = 0$ kPa, and an undrained shear strength of 50 kPa. The depth to axis is 16 m and the water table is at a depth of 3 m below ground level (Figure 3.27). The bulk unit weight of the soil $\gamma = 20$ kN/m³ both above and below the water table.

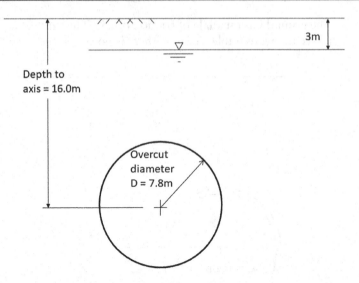

Figure 3.27 Q3.4 Cross-section through rail tunnel.

    i. Calculate the theoretical minimum face pressure required to maintain face stability in the drained case, using the method of Anagnostou & Kovári. Assume that the piezometric head of water pressure in the chamber (h_f) is equal to the pre-existing piezometric head in the ground (h_0). Ignore all factors of safety, any allowance for variability of face pressure and any surface surcharge.

   ii. Demonstrate, with the aid of calculations, the effect of draining the ground such that the piezometric head of water pressure in the chamber (h_f) is equal to zero. Why might it be desirable to maintain h_f as close to h_0 as possible, and how might this be achieved?

  iii. What value of cohesion would the soil need for the face to be stable without applying any support pressure?

  iv. Calculate the theoretical minimum face pressure required to maintain stability in the undrained case, using Mair's heading stability chart. Ignore all factors of safety and ignore any allowance for variability of face pressure and any surface surcharge.

   v. How are factors of safety introduced in GEO Report 249 (Golder Associates, 2009) for drained soils at the ultimate limit state? How would this approach affect the target minimum face pressures calculated in (i) and (iv)? Which is the critical case for design – drained or undrained?

Q3.5. It is proposed to use an EPB TBM for an 8 m ID flood alleviation sewer between two shafts. The bolted segmental lining is 300 mm thick. The overcut diameter of the TBM is 8.7 m. The soil has been characterised as sandy silt with an angle of friction $\phi = 23°$ and drained cohesion $c' = 0$ kPa. The depth to axis is 16 m and the water table is at a depth of 2 m below ground level (Figure 3.28). Assume the bulk unit weight of the soil $\gamma = 20$ kN/m³ both above and below the water table.

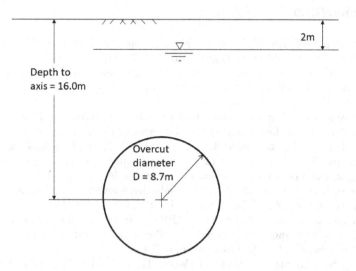

Figure 3.28 Q3.5 Cross-section through proposed flood alleviation sewer.

i. Calculate the theoretical minimum face pressure required to maintain face stability. Assume that the piezometric head of water pressure in the chamber (h_f) is equal to the pre-existing piezometric head in the ground (h_0). Ignore all factors of safety and ignore any allowance for variability of face pressure and any surface surcharge.

ii. Demonstrate, with the aid of calculations, the effect of draining the ground such that the piezometric head of water pressure in the chamber (h_f) is equal to zero. Why might it be desirable to maintain h_f as close to h_0 as possible, and how might this be achieved?

iii. What value of cohesion would the soil need for the face to be stable without applying any support pressure?

iv. Assuming that the piezometric head of water pressure in the chamber (h_f) is equal to the pre-existing piezometric head in the ground (h_0), as calculated in part (i) for sandy silt with an angle of friction $\phi = 23°$ and drained cohesion $c' = 0$ kPa, calculate the target minimum face pressure required for a factored surcharge of 75 kPa and a maximum variability of face pressure of ±40 kPa.

   v. How are factors of safety introduced in GEO Report 249 for drained soils at the ultimate limit state? How would this approach affect the target minimum face pressure calculated in (iv)?

  vi. How could a head intervention be achieved in this tunnel?

REFERENCES

Anagnostou, G. & Kovári, K. (1994). The face stability of slurry shield-driven tunnels. *Tunnels and Deep Space* **9**, No. 2, 165–174.

Anagnostou, G. & Kovári, K. (1996a). Face stability conditions with earth-pressure-balanced shields. *Tunn. Undergr. Space Technol.* **11**, No. 2, 165–173.

Anagnostou, G. & Kovári, K. (1996b). Face stability in slurry and EPB shield tunnelling. *Geotechnical Aspects of Underground Construction in Soft Ground* (eds Mair, R. J. & Taylor, R. N.), pp. 453–458. Rotterdam: Balkema.

Anagnostou, G. & Perazzelli, P. (2013). The stability of a tunnel face with a free span and a non-uniform support. *Geotechnik* **36**, Heft 1, 40–50.

Atkinson, J. H. & Potts, D. M. (1977). Subsidence above shallow tunnels in soft ground. *Proc. ASCE Geot. Engrg Div.* **103**, GT4, 307–325.

Bäppler, K., Battistoni, F. & Burger, W. (2017). Variable density TBM – combining two soft ground TBM technologies. *Proc. AFTES International Congress*, Paris, France, 13th–15th November 2017, Paper C3-2.

Berthoz, N., Branque, D., Subrin, D., Wong, H. & Humbert, E. (2012). Face failure in homogeneous and stratified soft ground: theoretical and experimental approaches on 1g EPBS reduced scale model. *Tunn. Undergr. Space Technol.* **30**, 25–37. Figures 3.14 and 3.16 reprinted with permission from Elsevier.

Bezuijen, A. & Brassinga, H. E. (2006). Blow-out pressures measured in a centrifuge model and in the field. *Tunnelling: A decade of progress – GeoDelft 1995–2005* (eds Bezuijen, A. & van Lottum, H.), pp. 143–148. Leiden: Taylor & Francis/Balkema.

Broere, W. (1998). Face stability calculation for a slurry shield in heterogeneous soft soils. *Proc. Tunnels and Metropolises* (eds Negro Jr., A. & Ferreira, A. A.), Sao Paulo, Brazil, pp. 215–218. Rotterdam: Balkema.

Broere, W. (2015). On the face support of microtunnelling TBMs. *Tunn. Undergr. Space Technol.* **46**, 12–17.

Broere, W. & van Tol, A. F. (2000). Influence of infiltration and groundwater flow on tunnel face stability. *Proc. Geotechnical Aspects of Underground Construction in Soft Ground* (eds Kusakabe, O., Fujita, K. & Miyazaki, Y.), Tokyo, Japan, pp. 339–344. Rotterdam: Balkema.

Broere, W. & van Tol, A. F. (2001). Time-dependant infiltration and groundwater flow in a face stability analysis. *Proc. Modern Tunneling Science and Technology* (eds Adachi, T., Tateyama, K. & Kimura, M.), Kyoto, Japan, pp. 629–634. Lisse: Swets & Zeitlinger.

BS 6164:2019. *Health and safety in tunnelling in the construction industry – code of practice*. London, UK: British Standards Institution.

CEDD (2012). *Catalogue of notable tunnel failure case histories (up to November 2012)*. Prepared by the Mainland East Division Geotechnical Engineering Office, Civil Engineering and Development Department, The Government of the Hong Kong Special Administrative Region. Available at: http://www.cedd.gov.hk/eng/publications/geo/doc/HK%20NotableTunnel%20Cat.pdf [last accessed 3rd June 2014].

Chambon, J. F. & Corté, J. F. (1994). Shallow tunnels in cohesionless soil: stability of tunnel face. *J. Geotech. Eng. ASCE* 120, No. 7, 1150–1163.

Dias, D., Janin, J.-P., Soubra, A.-H. & Kastner, R. (2008). Three-dimensional face stability analysis of circular tunnels by numerical simulations. *Proc. ASCE Geo Congress 2008: Characterization, Monitoring and Modeling of Geosystems*, pp. 886–893.

DIN 4126 (1986). *Ortbeton-Schlitzwände Konstruktion und Ausführung* [in German – now withdrawn].

EN 1997-1:2004 (2009). *Eurocode 7: Geotechnical design – Part 1: General rules*, incorporating corrigendum February 2009. Brussels: European Committee for Standardization.

Golder Associates (2009). *Ground control for slurry TBM tunnelling*. GEO Report 249. Hong Kong: The Government of the Hong Kong Special Administrative Region.

Guglielmetti, V., Grasso, P., Mahtab, A. & Xu, S. (2008). *Mechanized tunnelling in urban areas*. London: Taylor & Francis Group.

Holzhäuser, J. (2003). Geotechnical aspects of compressed air support on TBM tunnelling. *Engineering and health in compressed air working*, pp. 359–371. London: Thomas Telford.

Horn, M. (1961). Horizontaler Erddruck auf senkrechte Abschlussflächen von Tunnelröhren. In *Landeskonferenz der Ungarischen Tiefbauindustrie*, Horizontal earth pressure on perpendicular tunnel face. *Proceedings of the Hungarian National Conference of the Foundation Engineer Industry*, Budapest, Hungary, pp. 7–16.

Kilchert, M. & J. Karstedt (1984). *Schlitzwände als Tragund Dichtungwände, Band 2, Standsicherheitberechnung von Schlitzwänden*, pp. 28–34. Berlin: DIN.

Kimura, T. & Mair, R. J. (1981). Centrifugal testing of model tunnels in soft clay. *Proc. 10th Int. Conf. Soil Mech. & Found. Engrg*, Stockholm, Vol. 1, pp. 319–322.

Krause, T. (1987). *Schildvortrieb mit flüssigkeits- und erdgestutzter Ortsbrust*. PhD thesis, Technischen Universität Carolo-Wilhelmina, Braunschweig.

Leca, E. & Dormieux, L. (1990). Upper and lower bound solutions for the face stability of shallow circular tunnels in frictional material. *Géotechnique* 40, No. 4, 581–605.

Lovelace, N. (2003). LUL fears for Central Line tunnels after CTRL ground collapse. *New Civil Engineer*, 13th February 2003. Available at: http://www.nce.co.uk/lul-fears-for-central-line-tunnels-after-ctrl-ground-collapse/792589.article [last accessed 3rd June 2014].

Mair, R. J. & Taylor, R. N. (1997). Bored tunnelling in the urban environment. Theme Lecture, Plenary Session 4. *Proc. 14th Int. Conf. Soil Mechanics and Foundation Engineering*, Hamburg, Vol. 4.

Messerli, J., Pimentel, E. & Anagnostou, G. (2010). Experimental study into tunnel face collapse in sand. *Physical Modelling in Geotechnics* (eds Springman, S., Laue, J. & Seward, L.), pp. 575–580. London: Taylor & Francis.

Moir, E. W. (1897). Contribution to discussion on the Blackwall Tunnel. *Minutes of the Proceedings of the Institution of Civil Engineers* vol. CXXX, pp. 80–96.

Mollon, G., Dias, D. & Soubra, A.-H. (2009). Probabilistic analysis and design of circular tunnels against face stability. *Int. J. Geomech.* **9**, No. 6, November, 237–249.

Mollon, G., Dias, D. & Soubra, A.-H. (2010). Face stability analysis of circular tunnels driven by a pressurized shield. *J. Geotech. Geoenviron. Eng.* **136**, No. 1, 215–229.

Mollon, G., Dias, D. & Soubra, A.-H. (2011). Rotational failure mechanisms for the face stability analysis of tunnels driven by a pressurized shield. *Int. J. Numer. Anal. Meth. Geomech.* **35**, 1363–1388.

Mollon, G., Dias, D. & Soubra, A.-H. (2013b). Range of the safe retaining pressures of a pressurized tunnel face by a probabilistic approach. *J. Geotech. Geoenviron. Eng.* **139**, No. 11, 1954–1967.

Pan, Q. & Dias, D. (2016a). Face stability analysis for a shield-driven tunnel in anisotropic and nonhomogeneous soils by the kinematical approach. *Int. J. Geomech.* **16**, No. 3.

Pan, Q. & Dias, D. (2016b). The effect of pore water pressure on tunnel face stability. *Int. J. Numer. Anal. Meth. Geomech.* **40**, 2123–2136.

Soubra, A.-H. (2000). Three-dimensional face stability analysis of shallow circular tunnels. *Proc. Int. Conf. on Geotechnical and Geological Engineering*, Melbourne, Australia, November 19–24, pp. 1–6.

Subrin, D. (2002). *Études théoriques sur la stabilité et le comportement des tunnels renforcés par boulonnage.* thèse de doctorat soutenue à l'Institut National des Sciences Appliquées de Lyon, France.

van Rhee, C. & Bezuijen, A. (1992). Influence of seepage on stability of sandy slope. *ASCE J. Geotech. Eng.* **8**, 1236–1240.

Vermeer, P. A., Ruse, N. & Marcher, T. (2002). Tunnel heading stability in drained ground. *Felsbau* **20**, No. 6, 8–18.

Wong, K. S., Ng, C. W. W., Chen, Y. M. & Bian, X. C. (2010). Centrifuge modelling of passive failure of tunnel face in saturated sand. *Proc. 7th Int. Conf. Physical Modelling in Geotechnics* (eds Springman, S., Laue, J. & Seward, L.), pp. 599–604. London: Taylor & Francis Group.

Wong, K. S., Ng, C. W. W., Chen, Y. M. & Bian, X. C. (2012). Centrifuge and numerical investigation of passive failure of tunnel face in sand. *Tunn. Undergr. Space Technol.* **28**, 297–303. Figure 3.15 Reprinted with permission from Elsevier.

Chapter 4

Stability of shafts

Unlike a tunnel heading, at the base of a shaft gravity is on our side. However, seepage forces or constrained water pressures can still cause instability, as can heave pressures due to time-dependent swelling of clay.

Shafts can experience instability in four main ways. The first is hydraulic failure of the soil in the base of the shaft due to an upwards seepage of water pushing the soil grains apart. The second is a base heave failure. The third is a geotechnical uplift failure of the base of the shaft during excavation caused by water pressure in a confined aquifer acting on the underside of an impermeable layer of soil. Finally, the fourth is a buoyancy uplift failure of a shaft after base slab construction, where the whole shaft is pushed out of the ground by water pressure. These will be described in the following sections, along with worked examples.

After working through this chapter, you will understand:

- the types of stability failures that can occur in a shaft and what causes them
- the importance of groundwater and permeability to shaft stability

After working through this chapter, you will be able to:

- calculate the ultimate limit state for each type of shaft stability failure according to Eurocode 7

4.1 HYDRAULIC FAILURE IN A SHAFT DURING EXCAVATION

This is a well-known geotechnical problem for excavations in cohesionless soils. If the hydraulic gradient is too large at the base of the shaft, then the upwards seepage of water will cause hydraulic failure to occur, colloquially known as 'boiling' or 'quicksand'. Hydraulic failure occurs when the seepage forces reduce the effective stresses to zero and the soil grains begin to

DOI: 10.1201/9780429470387-4

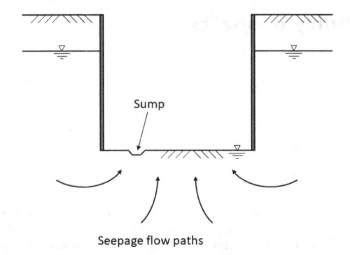

Seepage flow paths

Figure 4.1 Section through a circular dry caisson or underpinned shaft showing ground-
    water flow.

move apart. The result is as horrifyingly dangerous as it sounds: a complete
loss of bearing capacity at the base of the shaft.

For a dry caisson or underpinned shaft, there will be one or more sumps
in the base of the excavation to pump out the water, maintaining the
groundwater level just below the excavation level, as shown in Figure 4.1.
This will cause a groundwater flow from the surrounding ground, shown
by the arrows, generating large upwards seepage forces through the soil in
the base of the shaft.

To verify whether hydraulic failure will be an issue, a flow calculation
needs to be made to determine the pore pressures in the ground and/or the
hydraulic gradients. This cannot be done on paper by sketching a flownet
as we might do for a trench supported by sheet piles or an embedded retaining wall, because a circular shaft is not a 2D plane strain problem, but is
either axisymmetric for a circular shaft, or a 3D problem if the shaft is
non-circular (Polubarinova-Kochina, 1962) and requires calculation by a
numerical method.

Consider a unit volume of soil in the base of a shaft excavation, shown
in section in Figure 4.2.

Soil is a porous medium made up of soil particles with water filling the
voids in between. The porosity n is given by:

$$n = \frac{V_v}{V} \tag{4.1}$$

n is the porosity, the fraction of the total volume that is made up of
    voids. Therefore, $(1 - n)$ is the fraction of the total volume that is
    made up of solid particles.

V_v is the volume of voids
V is the total volume

There are three forces acting on the soil particles in the unit volume (Polubarinova-Kochina, 1962), all in units of kN/m³ because they are a force per unit volume. The first is the weight of the soil particles, acting downwards. Taking the downwards direction as positive, we get:

$$F_1 = (1-n)\gamma_s \, \mathbf{j} \qquad (4.2)$$

F_1 is the weight of soil particles in kN/m³
γ_s is the unit weight of the soil particles in kN/m³
$\mathbf{j}$ is the unit vector $(0, 0, 1)$

The second force is the effect of water pressure on the solid particles:

$$F_2 = (1-n) \, \text{grad } u \qquad (4.3)$$

F_2 is the water pressure force on the soil particles in kN/m³
u is the pore pressure
grad u is the pore pressure gradient, where 'grad' is the gradient of a
scalar field. It is itself a vector, given by: grad $u = \left(\frac{\partial u}{\partial x}, \frac{\partial u}{\partial y}, \frac{\partial u}{\partial z}\right)$

For vertical flow, we could replace grad u with $(u_2 - u_1)/\Delta z$, where Δz is the vertical distance across the unit volume and u_1 and u_2 are the pore pressures, as shown in Figure 4.2.

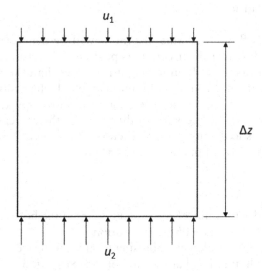

Figure 4.2 A volume of soil experiencing an upwards vertical seepage of water.

The third force is the seepage force, the drag effect of the water moving between the soil particles. It can also be thought of as the resistance of the soil to the passage of water, and it is this resistance that gives the value of permeability. It is given by:

$$\mathbf{F}_3 = n(\text{grad } u + \gamma_w \mathbf{j}) \tag{4.4}$$

$\mathbf{F}_3$ is the seepage force in kN/m³
γ_w is the unit weight of water in kN/m³

Summing the three forces gives us the resultant seepage force $\mathbf{S}_z$. If positive, then the soil is stable, if negative then it is boiling.

$$\mathbf{S}_z = \mathbf{F}_1 + \mathbf{F}_2 + \mathbf{F}_3 = (1-n)\gamma_s \mathbf{j} + (1-n) \text{ grad } u + n \text{ grad } u + n\gamma_w \mathbf{j} \tag{4.5}$$

$\mathbf{S}_z$ is the resultant seepage force in kN/m³

Rearranging this gives:

$$\mathbf{S}_z = (1-n)\gamma_s \mathbf{j} + \text{grad } u + n\gamma_w \mathbf{j} \tag{4.6}$$

Now the bulk unit weight can be expressed as:

$$\gamma_b = (1-n)\gamma_s + n\gamma_w \tag{4.7}$$

Therefore:

$$\mathbf{S}_z = \gamma_b \mathbf{j} + \text{grad } u \tag{4.8}$$

For an upwards flow, the value of grad u will be negative, because we are taking the downwards direction as positive. Therefore, Equation 4.8 says that as long as the bulk unit weight is greater than the upwards pore pressure gradient, the ground will remain stable. If the resultant force on the soil particles $\mathbf{S}_z$ becomes negative, they will move upwards and 'boil'. Therefore, it is the pore pressure gradient that is the critical factor.

For design we should apply partial factors according to Eurocode 7 (EN 1997-1:2004+A1:2013) Equation (2.9a), where:

$$u_{dst;d} \leq \sigma_{stb;d} \tag{4.9}$$

$u_{dst;d}$ is the design value of the pore pressure at the bottom of a column of soil at the base of the excavation
$\sigma_{stb;d}$ is the design value of stabilising total vertical stress at the bottom of a column of soil at the base of the excavation
The 'design value' means a characteristic value modified by a partial factor

The Eurocode 7 approach is simplified, in that it assumes vertical flow of groundwater and that the pore pressure at the base of the excavation is zero. Since grad u is the gradient of pore pressure and γ_b is effectively the gradient of total stress, then we can see that Equations 4.8 and 4.9 are similar. The difference is that Equation 4.8 is expressed in terms of gradients and Equation 4.9 is expressed in terms of values of pore pressure and total stress an arbitrary distance below the base of the excavation, and assuming they are zero at the base of the excavation.

In Eurocode 7, partial factors need to be applied according to Table A.17 to obtain the design values of actions and resistance. These may be altered by the National Annex. In the case of the UK they are the same as the recommended values, as shown in Table 4.1.

Therefore, we should multiply the pore pressure gradient by 1.35 and multiply the bulk unit weight by 0.9 to obtain the design values. If the design value of pore pressure gradient is lower than the design value of the bulk unit weight, then there is sufficient factor of safety against the ultimate limit state. This can be expressed as the following inequality:

$$1.35 \times \text{grad } u \leq 0.9 \times \gamma_b \mathbf{j} \qquad (4.10)$$

It may in some cases be easier to deal with hydraulic gradient rather than pore pressure gradient. Hydraulic gradient is related to pore pressure gradient by the equation:

$$\frac{1}{\gamma_w}\text{grad } u = \text{grad } h - \mathbf{j} \qquad (4.11)$$

grad h is the hydraulic gradient, where h is the groundwater head. For vertical flow parallel to the z-axis this would be equal to $\Delta h / \Delta z$.

Table 4.1 Partial factors for use in the hydraulic heave ultimate limit state (HYD) from Table A.17 of Eurocode 7 (EN 1997-1:2004+A1:2013) and from Table A.NA.17 of the UK National Annex to Eurocode 7 (NA+A1:2014 to BS EN 1997-1:2004+A1:2013).

	Symbol	Table A.17	Table A.NA.17
Permanent Unfavourable	$\gamma_{G,dst}$	1.35	1.35
Permanent Favourable	$\gamma_{G,stb}$	0.9	0.9
Variable Unfavourable	$\gamma_{Q,dst}$	1.5	1.5
Variable Favourable	$\gamma_{Q,stb}$	0	0

The seepage force F_3, expressed in terms of hydraulic gradient, is now given by:

$$F_3 = n\gamma_w \text{grad } h \tag{4.12}$$

The pore pressure force F_2 may be given by:

$$F_2 = (1-n)[\gamma_w \text{ grad } h - \gamma_w \mathbf{j}] \tag{4.13}$$

The F_1 force remains the same as in Equation 4.2. Using Equations 4.12 and 4.13 instead of Equations 4.3 and 4.4, the expression for resultant seepage force is:

$$S_z = F_1 + F_2 + F_3 = (1-n)\gamma_s \mathbf{j} + (1-n)[\gamma_w \text{ grad } h - \gamma_w \mathbf{j}] + n\gamma_w \text{grad } h \tag{4.14}$$

Using Equation 4.7 to simplify gives:

$$S_z = (\gamma_b - \gamma_w)\mathbf{j} + \gamma_w \text{ grad } h \tag{4.15}$$

Equation 4.15 says that as long as the upwards hydraulic gradient multiplied by the unit weight of water is less than the submerged unit weight $(\gamma_b - \gamma_w)$, then the ground will remain stable.

For design we should apply partial factors according to Eurocode 7 (EN 1997-1:2004, 2009) Equation (2.9b), where:

$$S_{dst;d} \leq G'_{stb;d} \tag{4.16}$$

$S_{dst;d}$ is the design value of seepage force, which is the second term on the right-hand side of Equation 4.15

$G'_{stb;d}$ is the design value of submerged weight, which is the first term on the right-hand side of Equation 4.15

The 'design value' means a characteristic value modified by a partial factor

The same partial factors in Table 4.1 should be used. Thus, the ultimate limit state is defined by the following inequality:

$$1.35 \times \gamma_w \text{ grad } h \leq 0.9 \times (\gamma_b - \gamma_w)\mathbf{j} \tag{4.17}$$

For soils with cohesion, the design value of cohesion (the characteristic value reduced by a partial factor) may be added as a stabilising action to the right-hand side of Equation 4.17.

The risk of hydraulic failure can be mitigated in many different ways, for example:

- Lengthening the seepage path; this may be achieved by constructing the shaft walls using secant piles or diaphragm walls that extend a significant distance below the formation level of the excavation. Increasing the distance over which the head difference acts will reduce the hydraulic gradient.

- Alternatively, these walls could extend into an impermeable stratum, providing a complete cut-off. Then the interior of the shaft can be dewatered prior to or during excavation such that there is no groundwater flow.
- Dewater the ground in and around the shaft using deep wells. This will reduce or remove any upwards hydraulic gradient within the shaft.
- Use grouting or ground freezing to increase the cohesion and decrease the permeability of the soil.
- Keep the shaft flooded so that there is no hydraulic gradient and use the caisson-sinking method along with a long-reach excavator or grab from the surface to excavate. This is known as a 'wet caisson'. The water can be pumped out once the caisson is well embedded into an impermeable soil, providing cut-off. If the soil is cohesionless all the way down, the base slab can be poured underwater using a tremie and guided by divers, then the water pumped out when it has gained sufficient strength. Note that using divers introduces new risks and it is difficult to assure quality of the base slab.

WORKED EXAMPLE 4.1 HYDRAULIC FAILURE IN A SHAFT

A 12 m diameter shaft is to be built using a dry caisson in sand. The sand has bulk unit weight $\gamma_b = 18$ kN/m^3, angle of friction $\phi = 30°$ and effective cohesion $c' = 0$ kPa. A numerical seepage analysis finds that the maximum hydraulic gradient near the surface of the excavation in the shaft is 1.5, in the upwards vertical direction.

Calculate the design value of both destabilising and stabilising forces and say whether the shaft excavation is within the ultimate limit state for hydraulic failure according to Eurocode 7.

The design value of the upwards destabilising seepage force is given by:

$$S_{dst;d} = \gamma_{G,dst} \times \gamma_w \text{ grad } h = 1.35 \times 10 \times 1.5 = 20.25 \text{ kN/m}^3$$

The design value of the downwards stabilising force due to submerged unit weight is given by:

$$G'_{stb;d} = \gamma_{G,stb} \times (\gamma_b - \gamma_w) \text{ j} = 0.9 \times (18 - 10) \times 1 = 7.2 \text{ kN/m}^3$$

Since the bulk unit weight γ_b and the unit weight of water γ_w both act vertically downwards, then they are multiplied by the resultant of the unit vector j in the vertical downwards direction, which is equal to 1.

The design value of the destabilising seepage force is greater than the design value of the stabilising force. Therefore, the ultimate limit state is exceeded and the shaft cannot be built this way. Either a wet caisson, embedded walls, dewatering or ground improvement is needed.

WORKED EXAMPLE 4.2   HYDRAULIC FAILURE IN A DIAPHRAGM WALL BOX

A 10 m deep excavation is made in sand within two parallel diaphragm walls 20 m deep. The sand has bulk unit weight $\gamma_b = 17$ kN/m³, angle of friction $\phi = 32°$ and effective cohesion $c' = 0$ kPa. The diaphragm walls are assumed impermeable and the flownet shown in Figure 4.3 has been created.

Calculate the maximum hydraulic gradient. Then calculate the design values of destabilising and stabilising forces. Does the design pass the ultimate limit state for hydraulic failure according to Eurocode 7?

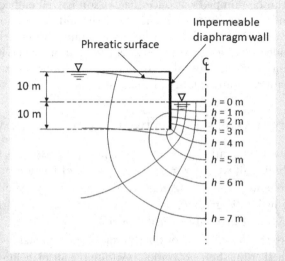

Figure 4.3 Worked Example 4.2 diaphragm wall flownet (based on an example in Powrie, 1997).

From Figure 4.3, the maximum hydraulic gradient will be close to the diaphragm wall and is a drop in head of 4 m over a vertical distance of 10 m. Therefore, the hydraulic gradient is 0.4.

The design value of the upwards destabilising seepage force is given by:

$$S_{dst;d} = \gamma_{G,dst} \times \gamma_w \text{ grad } h = 1.35 \times 10 \times 0.4 = 5.4 \text{ kN/m}^3$$

The design value of the downwards stabilising force due to submerged unit weight is given by:

$$G'_{stb;d} = \gamma_{G,stb} \times (\gamma_b - \gamma_w) \, j = 0.9 \times (17 - 10) \times 1 = 6.3 \text{ kN/m}^3$$

Since $G'_{stb;d} \geq S_{dst;d}$, it is within the ultimate limit state and the excavation will not fail.

4.2 BASE HEAVE FAILURE OF A SHAFT IN CLAY

In a similar manner to tunnel heading stability in clay, covered in Chapter 2, the base of a shaft may fail if the shear stresses along failure planes exceed the undrained shear strength. The ground around the shaft moves downwards and the ground within the shaft moves upwards, with a slip circle around the bottom of the shaft lining. This failure is resisted by the undrained shear strength of the ground. Any surcharge on the ground surface around the shaft will be unfavourable, and any surcharge within the shaft, for example provided by compressed air, will be favourable.

Terzaghi (1943) considered the base heave stability of excavations, but only in plane strain, i.e. assuming an infinitely long excavation. Bjerrum & Eide (1956) were the first to consider rectangular, square and circular excavations. Although not presented in this way in their paper, we can define stability ratio in the same way as for heading stability:

$$N = \frac{\gamma H + q - \sigma_{ca}}{c_u} \tag{4.18}$$

N is the stability ratio, which is dimensionless
γ is the bulk unit weight of the clay in kN/m^3
H is the depth of the shaft in m
q is a surcharge on the ground surface around the outside of the shaft in kPa
σ_{ca} is compressed air pressure within the shaft, if provided, or any permanent surcharge within the shaft, in kPa. This is favourable and so should only be counted if permanently applied.
c_u is the undrained shear strength in kPa

For a given ratio of shaft depth H to shaft diameter D, there is a critical stability number N_c at which the base will fail. Note that Bjerrum & Eide (1956) also gave N_c values for strutted excavations that are rectangular in plan, in which case N_c also depends on the ratio of width to length. In this section we will only consider circular shafts.

Values of N_c may be read from a design chart (Figure 4.4) and are identical to those provided by Skempton (1951) for the bearing capacity of footings, because the situations are analogous – the deep excavation is considered to be like a negatively loaded perfectly smooth footing (Bolton et al., 2008). Bjerrum & Eide (1956) proved this method to be sufficiently accurate for design purposes by reference to 14 case studies of failures or partial failures. For the seven complete failures they studied, the factor of safety was estimated to be between 0.82 and 1.16.

In exactly the same way as for tunnel heading stability, the factor of safety is the ratio N_c/N. For design according to Eurocode 7 (EN 1997-1:2004+A1:2013), shaft base stability is a 'GEO' limit state. Therefore, the factor of safety needs to be greater than 1.4.

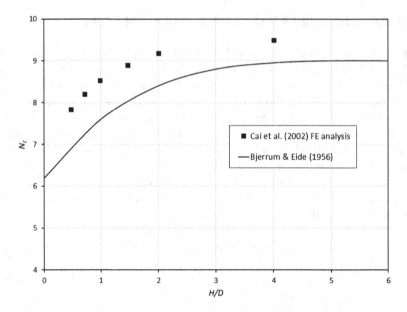

Figure 4.4 Critical stability number for shaft base stability in clay.

WORKED EXAMPLE 4.3 – BASE HEAVE
FAILURE OF A SHAFT IN CLAY

A 10 m deep 6 m diameter shaft is to be excavated in soft clay with an undrained shear strength of 30 kPa and a bulk unit weight of 19 kN/m³. What is the factor of safety?

What is the maximum depth the shaft can be excavated with a factor of safety of 1.4?

First, we need to calculate the stability ratio:

$$N = \frac{\gamma H + q}{c_u} = \frac{19 \times 10 + 0}{30} = \frac{190}{30} = 6.33$$

The ratio of depth to diameter, $H/D = 10/6 = 1.67$, therefore from the design chart in Figure 4.4, $N_c = 8.2$.

Therefore, the factor of safety is given by:

$$\frac{N_c}{N} = \frac{8.2}{6.33} = 1.30$$

This is insufficient for a design to Eurocode 7, which requires a factor of safety of 1.4. The maximum depth may only be found by trial and error, because each time the depth is changed, both N and N_c change.

For a depth of 9 m:

$$N = \frac{19 \times 9}{30} = 5.7$$

And for $H/D = 9/6 = 1.5$, $N_c = 8.1$. Therefore, the factor of safety is given by:

$$\frac{N_c}{N} = \frac{8.1}{5.7} = 1.42$$

Thus, a depth of 9 m is an acceptable design.

Cai et al. (2002) used axisymmetric finite element modelling to investigate base heave failure of circular shafts in soft clays. They found that results were not affected by the thickness of the clay below the excavation, as long as the thickness was greater than $\sqrt{2}R$, where R is the radius of the shaft. If a hard stratum were closer to the base of the excavation than this, stability would be improved.

Cai et al.'s finite element analysis results are shown in Figure 4.4 for stiff walls with no embedment below excavation level, in a homogeneous clay layer with constant undrained shear strength. We can see that Bjerrum & Eide's method gives more conservative, i.e. lower, values of N_c.

Cai et al. (2002) also looked at embedment of diaphragm walls or secant piles below the excavation level. As the ratio of embedment depth to excavation depth increased, there was a linear increase in critical stability ratio N_c. If we define a modification factor λ_w based on their results to modify N_c to account for wall embedment, it can be given by:

$$\lambda_w = 1 + 0.435 \frac{h_w}{H} \tag{4.19}$$

λ_w is the modification factor for wall embedment below excavation level
h_w is the wall embedment depth below excavation level in m
H is the excavation depth in m

WORKED EXAMPLE 4.4 – BASE HEAVE FAILURE OF A DIAPHRAGM WALL SHAFT IN CLAY

A 20 m deep 20 m diameter shaft is to be excavated in clay with an undrained shear strength of 70 kPa and a bulk unit weight of 20 kN/m³ using diaphragm walls with an embedment depth of 10 m below the excavation level. An allowance for crane loading of 10 kPa surcharge should be made. What is the factor of safety?

First we need to calculate the stability ratio:

$$N = \frac{\gamma H + q}{c_u} = \frac{20 \times 20 + 10}{70} = \frac{410}{70} = 5.86$$

The ratio of depth to diameter, $H/D = 20/20 = 1.0$, therefore from the design chart in Figure 4.4, $N_c = 7.6$.

If the shaft walls had zero embedment, the factor of safety would be $7.6/5.86 = 1.30$. This would not be sufficient if we were designing to Eurocode 7.

Using Equation 4.19 to calculate the modification factor for wall embedment depth:

$$\lambda_w = 1 + 0.435 \frac{h_w}{H} = 1 + 0.435 \frac{10}{20} = 1.22$$

Therefore, the modified critical stability ratio is:

$$N_c^* = \lambda_w N_c = 1.22 \times 7.6 = 9.3$$

Thus the factor of safety against base heave failure is:

$$\frac{N_c^*}{N} = \frac{9.3}{5.86} = 1.59$$

4.3 UPLIFT FAILURE IN A SHAFT DURING EXCAVATION

If there is an impermeable layer of soil overlying a permeable layer with pore pressure present, it is possible that as excavation progresses, there may come a point where the weight of the overlying impermeable layer is less than the pore pressure in the permeable layer beneath. If this happens, then the water will cause an uplift failure, pushing the impermeable soil upwards.

The risk of uplift failure may be mitigated by reducing the pore pressure in the underlying permeable soil by pumping from wells drilled into it. Sometimes pumping may not be necessary and passive relief wells may be used with water channelled to the excavation sump.

4.3.1 Verification of the uplift ultimate limit state using Eurocode 7

Eurocode 7 (EN 1997-1:2004+A1:2013) provides partial factors for uplift failure, which are given in Table 4.2 in the third column headed 'Table A.15'. National Annexes to Eurocodes may give different guidance, and the partial factors given by the UK National Annex (NA+A1:2014 to

Table 4.2 Partial factors for use in the uplift ultimate limit state (UPL) from Table A.15 of Eurocode 7 (EN 1997-1:2004+A1:2013) and from Table A.NA.15 of the UK National Annex to Eurocode 7 (NA+A1:2014 to BS EN 1997-1:2004+A1:2013).

	Symbol	*Table A.15*	*Table A.NA.15*
Permanent Unfavourable	$\gamma_{G,dst}$	1.0	1.1
Permanent Favourable	$\gamma_{G,stb}$	0.9	0.9
Variable Unfavourable	$\gamma_{Q,dst}$	1.5	1.5
Variable Favourable	$\gamma_{Q,stb}$	0	0

BS EN 1997-1:2004+A1:2013) are given in the fourth column of Table 4.2, headed 'Table A.NA.15'. Note that for this limit state, the UK National Annex requires a different value for the permanent unfavourable actions.

The ultimate limit state is verified by comparing design values of stabilising and destabilising forces, using Equation (2.8) in Eurocode 7. The design value of stabilising forces needs to be equal to or higher than the design value of the destabilising forces, as follows:

$$V_{dst;d} \leq G_{stb;d} + R_d \tag{4.20}$$

$V_{dst;d}$ is the design value of the destabilising vertical force
$G_{stb;d}$ is the design value of the stabilising vertical force
R_d is the design value of any additional resistance to uplift

The stabilising force in this case is the weight of the lower permeability layer plus any surcharge acting at the base of the excavation. The design value is calculated by multiplying the total vertical stress at the base of the lower permeability layer by the partial factor for permanent favourable actions in Table 4.2.

The destabilising force is the groundwater pressure at the base of the lower permeability layer. The design value is calculated by multiplying the groundwater pressure by the partial factor for permanent unfavourable actions in Table 4.2.

WORKED EXAMPLE 4.5   UPLIFT FAILURE DURING SHAFT EXCAVATION

A 20 m deep 10 m ID shaft is excavated using the underpinning method, entirely in clay, as shown in Figure 4.5. The shaft is to be built in the UK according to the UK National Annex to Eurocode 7. The clay stratum extends 5 m below the formation level of the excavation and below that the ground is sandy gravel. The groundwater pressure is hydrostatic with a piezometric level 10 m below the ground surface. The clay has a bulk unit weight of 20 kN/m³ and a characteristic value of undrained shear strength of 60 kPa.

Calculate the pore pressure at the base of the clay. Then calculate the design values of destabilising and stabilising forces. Does the design pass the ultimate limit state for uplift failure according to Eurocode 7?

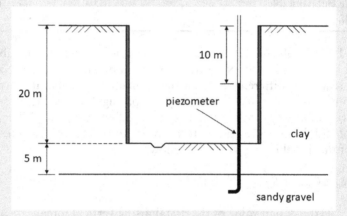

Figure 4.5 Worked Example 4.5 underpinned shaft at risk of uplift failure during excavation.

The destabilising force is the groundwater pressure at the base of the clay. From Figure 4.5, the groundwater head at the base of the clay is 15 m. Taking the unit weight of water γ_w = 10 kN/m³, the groundwater pressure is given by:

$$u = \gamma_w h = 10 \times 15 = 150 \text{ kPa}$$

The design value of the destabilising pressure is given by:

$$V_{dst;d} = \gamma_{G,dst} \times u = 1.1 \times 150 = 165 \text{ kPa}$$

The stabilising force is the weight of the clay. The total vertical stress at the base of the clay below the shaft excavation is given by:

$$\sigma_v = \gamma_{b,clay} \times t_{clay} = 20 \times 5 = 100 \text{ kPa}$$

The design value of the stabilising stress is given by:

$$G_{stb;d} = \gamma_{G,stb} \times \sigma_v = 0.9 \times 100 = 90 \text{ kPa}$$

Since $V_{dst;d} > G_{stb;d}$ the shaft fails in uplift.

Further question: At what excavation depth will the shaft fail the uplift ultimate limit state?

Let z be the depth of excavation. The thickness of the clay will be given by:

$$t_{clay} = 25 - z$$

The total vertical stress will therefore be given by:

$$\sigma_v = \gamma_{b,clay} \times t_{clay} = 20 \times (25 - z) = 500 - 20z$$

The groundwater pressure at the base of the clay is unchanged and so the destabilising force is also unchanged. The ultimate limit state is reached when the design value of the stabilising force equals the design value of the destabilising force:

$$V_{dst;d} = G_{stb;d} = 165 \text{ kPa}$$

Therefore:

$$165 = 0.9 \times (500 - 20z)$$

And:

$$z = 15.83 \text{ m}$$

So the shaft will fail the ultimate limit state for uplift when the excavation has reached 15.83 m depth.

It seems intuitively unlikely that the shaft would fail in uplift with a 9 m thickness of reasonably strong clay below the base of the excavation. This is because we have neglected the fact that we would need to mobilise quite large shear surfaces in the clay to make this failure actually happen. Eurocode 7 allows for including this by using the term R_d in Equation 4.20.

We do not know the geometry of a large-scale uplift failure. A numerical or physical model could be used to find the most likely failure geometry, but to my knowledge nothing has been published on this subject. To start with we could assume a cylindrical failure with vertical sides equal to the thickness of clay and a diameter equal to the internal diameter of the shaft.

The value of shear resistance on this surface would be equal to the characteristic value of undrained shear strength multiplied by the surface area of the cylinder. This could be added to the weight of the cylinder to give the total stabilising force, as shown in Figure 4.6.

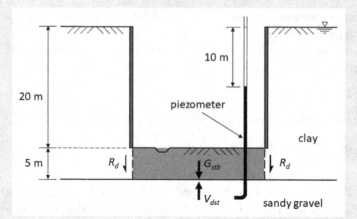

Figure 4.6 Worked Example 4.5 Shaft section showing forces for uplift failure.

The additional resistance due to shear is given by:

$$R_d = \pi D t_{clay} \frac{c_u}{\gamma_{cu}} = \pi \times 10 \times 5 \times \frac{60}{1.4} = 6732 \text{ kN}$$

where γ_{cu} is the partial factor for undrained shear strength resistance from Table A.16 of Eurocode 7, and is equal to 1.4.

In this case we need to calculate $V_{dst;d}$ and $G_{stb;d}$ as forces rather than pressures.

$$V_{dst;d} = \gamma_{G,dst} \times u\left(\frac{\pi D^2}{4}\right) = 1.1 \times 150 \times 78.54 = 12959 \text{ kN}$$

$$G_{stb;d} = \gamma_{G,stb} \times \sigma_v\left(\frac{\pi D^2}{4}\right) = 0.9 \times 100 \times 78.54 = 7069 \text{ kN}$$

The combined value of $R_d + G_{stb;d}$ is now greater than $V_{dst;d}$, so the design passes the ultimate limit state and is acceptable.

4.3.2 Geometry of uplift failure during excavation

As was mentioned in Worked Example 4.5, the geometry of uplift failure is unknown. The simplest shape would be a cylinder, and this would be consistent with the 'normality condition', where for an undrained soil with $\phi = 0$, the failure surfaces should be parallel to the direction of motion. However, it is known that for tunnel heading collapse, the failure does not involve the motion of rigid blocks bounded by a velocity discontinuity surface, but is best described as a continuous deformation of the soil. Mollon et al. (2013a) modelled collapse and blow-out of a tunnel and found that the geometry of the plastic zone was not cylindrical, but torus shaped. Therefore, the assumption of a cylindrical block failure is probably not accurate. More research is needed to investigate this.

4.4 UPLIFT FAILURE OF A SHAFT AFTER BASE SLAB CONSTRUCTION

Sometimes groundwater is drained to a sump and pumped away during construction, or the groundwater level is lowered by deep wells in or around the shaft. Then a watertight reinforced concrete base slab is cast, and once the concrete has gained sufficient strength, the pumping is stopped or the pipes are grouted and water pressure will build up under the base slab. If the upwards force due to the water pressure is larger than the weight of the shaft structure plus the friction between the shaft lining and the ground, then the entire shaft could quite simply be pushed out of the ground like a cork out of a bottle.

In this section we are assuming that the base slab has sufficient structural capacity to withstand the shear forces and bending moments induced by the water pressure acting upwards, and that there is a shear connection between the base slab and the shaft lining.

Similar to the mechanism described in Section 4.2, this is also an uplift limit state according to Eurocode 7 (EN 1997-1:2004+A1:2013). Partial factors will be applied as listed in Table 4.2.

The ultimate limit state is verified by comparing design values of stabilising and destabilising forces, using Equation (2.8) in Eurocode 7. The design value of stabilising forces needs to be equal to or higher than the design value of the destabilising forces, as follows:

$$V_{dst;d} \leq G_{stb;d} + R_d \tag{4.21}$$

$V_{dst;d}$ is the design value of the destabilising vertical force
$G_{stb;d}$ is the design value of the stabilising vertical force
R_d is the design value of any additional resistance to uplift

Remember that 'design values' are determined by applying the partial factors from Table 4.2 to characteristic values of the forces.

The stabilising force in this case is the weight of the shaft lining and base slab plus any additional permanent weight within the shaft already installed at the time of base slab construction, such as intermediate slabs, roof slab or secondary lining. The design value is calculated by multiplying the total weight by the partial factor for permanent favourable actions in Table 4.2.

The destabilising force is the groundwater pressure under the base slab. The design value is calculated by multiplying the groundwater pressure by the partial factor for permanent unfavourable actions in Table 4.2.

The additional resistance to uplift is the friction between the shaft lining and the surrounding ground. This friction is similar to shaft friction of a bored pile, and so a similar methodology can be used.

Eurocode 7 (EN 1997-1:2004+A1:2013) only allows pile design based on static pile tests, or on the results of static pile tests 'in comparable situations'. Unless we have pile test data available, the best we can do for most shaft designs is to make a cautious estimate of the friction based on empirical knowledge of bored pile shaft friction in tension in similar ground conditions and depths, and the soil parameters obtained from site investigation. This is a large subject area, and depends greatly on the type and quality of site investigation data available and local experience of pile testing. You will need to do some research. The simplest method is described in this chapter, based on soil shear strength data from laboratory tests, but this should be used only for preliminary design.

Depending on the construction method of the shaft, the interface between the lining and the ground will be disturbed or softened to a greater or lesser degree. A review of shaft sinking methods can be found in Allenby & Kilburn (2015). Underpinned shafts may cause more unloading of the ground, but are usually grouted after installation of each ring, providing a good and rough contact between the cut ground and the grout. Shafts underpinned using cast concrete or shotcrete will have a similar interface. Caisson-sinking usually entails keeping an annulus filled with bentonite slurry between the lining and the ground until the required depth is reached, at which point the annulus is grouted. Therefore, the ground at the perimeter may have been remoulded by the cutting edge and softened by the slurry. Note that shafts are larger and shaft sinking invariably takes longer than installation of a single pile, and so soil softening and disturbance will likely be more significant.

In clay soils in the short-term, the shaft friction for bored piles can be estimated from the following equation:

$$f_s = \alpha c_u \qquad\qquad (4.22)$$

f_s is the shaft friction in kPa
α is a coefficient that depends on the properties of the shaft lining-ground interface
c_u is the undrained shear strength in kPa

The value of the interface coefficient α is empirical and can vary from around 0.3 to 1.0 (Craig, 1997). A typical value for a bored pile in stiff fissured clays is 0.45 (Skempton, 1959; Fleming, 1997).

An effective stress approach can also be used in clay, assuming that excess pore pressures in the ground close to the shaft lining may dissipate in the timescale of construction, and is always used in drained soils such as silts, sands and gravels.

$$f_s = K_s \sigma_v' \tan\delta \qquad (4.23)$$

K_s is the coefficient of earth pressure at the shaft-ground interface, i.e.
$\quad K_s \sigma_v'$ is the effective horizontal stress
σ_v' is the effective vertical stress
δ is the angle of friction of the shaft-ground interface

The effective vertical stress is usually assumed to be the in situ value prior to construction. If the groundwater level has changed, this should be taken into account.

The coefficient of earth pressure at the shaft-ground interface can be calculated using Jaky's formula for normally consolidated soils:

$$K_s = 1 - \sin\phi' \qquad (4.24)$$

ϕ' is the soil's drained angle of friction

For overconsolidated soils, a modified version of this formula may be used:

$$K_s = (1 - \sin\phi')OCR^{\sin\phi'} \qquad (4.25)$$

OCR is the soil's overconsolidation ratio, which is the maximum value of effective stress experienced by the soil in the past divided by its present value

Sometimes pile designers use the normally consolidated value even for over-consolidated soils (Powrie, 1997). This provides a conservative value for overconsolidated soils and would allow for stress relief and remoulding of the soil at the pile perimeter.

The angle of friction of the shaft-ground interface δ will depend on the properties of the interface, and some judgement may be required to take into account the method of shaft construction. For most bored piles in cohesionless soil, δ can be assumed to be approximately equal to the angle of friction ϕ' of the soil. Powrie (1997) recommends that this should be the critical state value rather than the peak value. In clays, δ may be lower, and has been taken as 12° to 16° for London Clay, for example (Fleming, 1997).

WORKED EXAMPLE 4.6   UPLIFT FAILURE OF A SHAFT AFTER BASE SLAB CONSTRUCTION

A 12 m deep 15 m ID shaft is excavated using the caisson-sinking method in sandy gravel, as shown in Figure 4.7. The shaft is to be built in the UK according to the UK National Annex to Eurocode 7. The shaft lining is 350 mm thick and the base slab is 2 m thick and both are reinforced concrete. The unit weight of the reinforced concrete is 24 kN/m³.

The groundwater pressure is hydrostatic with a long-term maximum expected groundwater level at 2 m below the ground surface. During caisson sinking, the sandy gravel is dewatered using deep wells around the shaft perimeter, to a level below the base slab formation level. A 2 m thick reinforced concrete base slab is then cast and allowed to gain sufficient strength before the dewatering wells are turned off.

The sandy gravel is normally consolidated, has a bulk unit weight of 19 kN/m³ and effective shear strength parameters $c' = 0$ and $\phi' = 35°$. These are the characteristic mean values determined from a series of tests from 0 to 12 m depth, in three boreholes outside the shaft perimeter.

Calculate the pore pressure under the base slab. Then calculate the design values of destabilising forces, stabilising forces and additional resistance to uplift due to friction. Does the design pass the ultimate limit state for uplift failure according to Eurocode 7?

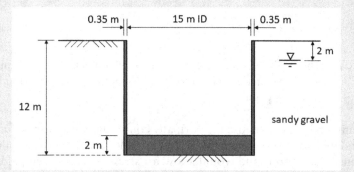

Figure 4.7 Worked Example 4.6 shaft at risk of uplift failure after base slab construction.

Note that for this design to work, the base slab needs sufficient bending moment and shear capacity to withstand the groundwater pressure, and there needs to be shear transfer between the base slab and the shaft lining, provided by shear keys, dowels, or by excavating the base slab under the last ring.

The destabilising force is the groundwater pressure at the formation level of the base slab. From Figure 4.7, the groundwater head at this level is 10 m. Taking the unit weight of water $\gamma_w = 10$ kN/m³, the groundwater pressure is given by:

$$u = \gamma_w h = 10 \times 10 = 100 \text{ kPa}$$

This time we will use forces rather than stresses. Let A be the area of the shaft base. The design value of the destabilising force is given by:

$$V_{dst;d} = \gamma_{G,dst} \times u \times A = 1.1 \times 100 \times \pi \left(\frac{15.7}{2} \right)^2 = 21295.2 \text{ kN}$$

The stabilising force is the weight of the shaft lining and base slab. This is given by:

$$G_{stb} = \gamma_{conc} \times \left(V_{base} + V_{lining} \right) = 24 \times \left(\pi \left(\frac{15}{2} \right)^2 \times 2 + \pi \times 15.35 \times 0.35 \times 12 \right)$$

$$= 24 \times (353.4 + 202.5) = 13343 \text{ kN}$$

The design value of the stabilising force is given by:

$$G_{stb;d} = \gamma_{G,stb} \times G_{stb} = 0.9 \times 13341.6 = 12009 \text{ kN}$$

Since $V_{dst;d} > G_{stb;d}$, we need to consider the additional resistance provided by friction between the lining and the ground.

It is usual practice to fill an annulus between the shaft lining and the ground with bentonite slurry during caisson sinking. Once the required depth has been reached, the bentonite is replaced by cementitious grout. Therefore, the friction between the grouted lining and the ground should be similar to a bored pile.

Eurocode 7 (EN 1997-1:2004+A1:2013) only allows pile design based on static pile tests, or on the results of static pile tests 'in comparable situations' (Frank et al., 2005). To estimate the friction between the shaft lining and the ground, the best we can do for most shaft designs is to make a cautious estimate of the friction based on empirical knowledge of bored pile shaft friction in tension in similar soil, and the soil parameters obtained from site investigation.

Table A.NA.16 of the UK National Annex to Eurocode 7 (NA+A1:2014 to BS EN 1997-1:2004+A1:2013) specifies partial factors for uplift on soil parameters and resistances.

According to this table, the tangent of the soil's angle of friction should be divided by 1.25.

Also, the rules of A.3.3.2 and A.3.3.3 relating to piles should be followed, where correlation factors are specified to account for the number of tests. For mean characteristic strengths based on three profiles, the correlation factor $\xi_3 = 1.42$. This is the UK value, and you should look up the National Annex of the country the project is in for the appropriate value.

A.3.3.2 of the UK National Annex also says that where ground test results are used to calculate characteristic resistances, a model factor should be applied, equal to 1.4.

The interface angle of friction, for sand and gravel, we will assume is equal to the soil's internal angle of friction. Therefore:

$$\tan\delta = \frac{\tan\phi'}{\xi_3 \gamma_{s;t} \gamma_{\phi'}} = \frac{\tan 35°}{1.42 \times 1.0 \times 1.25} = 0.394$$

The characteristic value of shaft friction $q_{s;i;k}$ is given by:

$$q_{s;i;k} = K_s \sigma_v' \tan\delta = (1 - \sin\phi')\sigma_v' \tan\delta = (1 - \sin 35°) \times (19 - 10) \times 6 \times 0.394$$

$$= 9.08 \text{ kPa}$$

The effective stress σ_v' is taken as the average value over the depth of the shaft, i.e. at 6 m depth. In a more complex geological situation you may need to calculate values of shaft friction for slices of the shaft and add them together.

The design value for stabilising resistance due to shaft friction R_d is therefore:

$$R_d = \frac{A_s q_{s;i;k}}{1.4} = \frac{\pi \times 15.7 \times 12 \times 9.08}{1.4} = 3840.3 \text{ kN}$$

The total of design stabilising force and resistance is now:

$$G_{stb;d} + R_d = 12009 + 3840.3 = 15849 \text{ kN}$$

This is still insufficient to prevent floatation of the shaft, because the design destabilising force is 21295.2 kN. The weight of the shaft could be increased by increasing the thickness of the base slab. Alternatively, the design could be refined by increasing the number of soil tests, and thereby increasing the characteristic strength and reducing the correlation factor, or by performing tension pile tests to determine shaft resistance.

If the base slab thickness were increased to 4 m thick, then the ground-water pressure acting on the underside of the base slab increases to 120 kPa and the design value of destabilising force becomes:

$$V_{dst;d} = \gamma_{G,dst} \times u \times A = 1.1 \times 120 \times \pi \left(\frac{15.7}{2}\right)^2 = 25554 \text{ kN}$$

The stabilising force is increased as the weight of the base slab is larger, and is now:

$$G_{stb} = \gamma_{conc} \times \left(V_{base} + V_{lining}\right) = 24 \times \left(\pi \left(\frac{15}{2}\right)^2 \times 4 + \pi \times 15.35 \times 0.35 \times 14\right)$$

$$= 24 \times (706.9 + 236.3) = 22636 \text{ kN}$$

The design value of stabilising force is now:

$$G_{stb;d} = \gamma_{G,stb} \times G_{stb} = 0.9 \times 22636 = 20372 \text{ kN}$$

The shaft is deeper, so the average vertical effective stress is larger, which increases the shaft friction:

$$q_{s;i;k} = K_s \sigma_v' \tan\delta = (1 - \sin\phi') \sigma_v' \tan\delta = (1 - \sin 35°) \times (19 - 10) \times 7 \times 0.394$$

$$= 10.60 \text{ kPa}$$

The design value for stabilising resistance due to shaft friction R_d is also increased as the shaft depth is increased to 14 m:

$$R_d = \frac{A_s q_{s;i;k}}{1.4} = \frac{\pi \times 15.7 \times 14 \times 10.60}{1.4} = 5227.1 \text{ kN}$$

So, now the total of design stabilising force and resistance is:

$$G_{stb;d} + R_d = 20372 + 5227 = 25599 \text{ kN}$$

This is now greater than the design destabilising force of 25554 kN and so the design is now adequate.

4.5 LONG-TERM HEAVE UNDER A SHAFT BASE SLAB

During excavation, clay soils in the base of a shaft will be unloaded, i.e. there will be a decrease in total stress. This unloading will be experienced in the short-term as a decrease in pore pressure. In the long-term, pore water will gradually flow from surrounding areas until these negative excess pore pressures are dissipated. When this happens, the clay will increase in volume. Under the base of a shaft, this swelling will generate heave pressures, trying to push the base of the shaft upwards.

There is more than one failure scenario to consider. The base slab may fail in bending or shear due to the heave pressure. If the base slab does not fail, it is also possible that the whole shaft could be pushed upwards out of the ground, which could cause serviceability or structural problems for utility connections or tunnel junctions, or indeed for any headhouse or other structure constructed on top of the shaft.

The risk of heave may be mitigated by laying compressible void formers on the ground before casting the base slab. Then the clay can swell without applying pressure to the base slab. In large shafts, tension piles may be used to anchor the base slab.

4.6 SUMMARY OF SHAFT STABILITY

During construction, shafts can experience instability in several ways:

- In drained soils, upwards seepage forces can push the soil particles apart causing hydraulic failure (also known as 'boiling' or 'quicksand' or 'hydraulic heave'), resulting in a complete loss of bearing capacity.
- In undrained soils, the base can fail if the shear stresses exceed the undrained shear strength along failure surfaces, in a similar manner to undrained stability failure of a tunnel heading.
- As excavation progresses, a low permeability soil overlying a higher permeability soil can be pushed upwards due to trapped groundwater pressure, causing uplift failure.
- If groundwater pressure can build up under a shaft base slab, a shaft may be pushed up out of the ground. This is another form of uplift failure.
- Long-term heave pressures may build up beneath a base slab due to swelling of the clay, which may push the shaft upwards. It is likely that this will be a serviceability issue, but it could in some cases cause a structural failure.

4.7 PROBLEMS

Q4.1. A sheet pile circular shaft is to be excavated in cohesionless silt and sand with a bulk unit weight of 17 kN/m³. A seepage analysis determines that the maximum hydraulic gradient will be 0.45.
    i. Calculate the upwards destabilising seepage force.
    ii. Calculate the downwards stabilising force.

iii. Apply the appropriate partial factors to the stabilising and destabilising forces according to Eurocode 7. Is the design within the ultimate state?

Q4.2. A secant pile wall shaft is to be excavated to 15 m depth, below the water table, in sand with no cohesion. The sand has a bulk unit weight of 18 kN/m³.

i. Ignoring the need for factors of safety, what is the maximum hydraulic gradient that can exist at the base of the shaft before hydraulic failure (i.e. boiling or piping) will occur?

ii. Including factors of safety for a HYD limit state according to Eurocode 7, what is the maximum hydraulic gradient allowed?

iii. Describe ways in which the design of the shaft or the construction method could be altered to reduce the hydraulic gradient.

Q4.3. A 10 m deep 5 m diameter shaft is to be excavated in soft clay with an undrained shear strength of 35 kPa and a bulk unit weight of 18 kN/m³.

i. Calculate the undrained stability ratio at the maximum depth of 10 m.

ii. From the design chart in Figure 4.4, find the value of critical stability number N_c.

iii. Calculate the factor of safety. Does the shaft design meet the requirements of Eurocode 7 for base heave stability?

iv. Can a 30 kPa surcharge be safely applied around the outside of the shaft?

Q4.4. A 15 m deep 7.5 m diameter shaft is to be excavated in soft clay with an undrained shear strength of 30 kPa and a bulk unit weight of 17 kN/m³.

i. Calculate the factor of safety. Does the shaft design meet the requirements of Eurocode 7 for base heave stability?

ii. At what depth is a factor of safety of 1.4 exceeded?

iii. The design is changed to a circular sheet pile shaft. What embedment of the sheet piles is required to give a factor of safety of 1.4?

Q4.5. A 12 m diameter 32 m deep underpinned shaft is to be excavated in stiff clay with an undrained shear strength of 150 kPa and a bulk unit weight of 20 kN/m³. At 36 m below ground level, the geology changes from stiff clay to sandy gravel. The pore pressure distribution is hydrostatic, with a piezometric level 2 m below ground level.

i. Calculate the pore pressure at the base of the clay.

ii. Calculate the design values of destabilising and stabilising forces, ignoring the resistance provided by the undrained shear strength of the clay. Show that the design does not pass the ultimate limit state for uplift failure according to Eurocode 7.

    iii. Calculate the additional resistance to uplift provided by the undrained shear strength of the clay. Show that the design still does not pass the ultimate limit state for uplift failure according to Eurocode 7. Remember that for this calculation R_d, $V_{dst;d}$ and $G_{stb;d}$ need to be calculated as forces rather than pressures.

    iv. Since the shaft is to be used as a storage tank, it can be made shallower and with a larger diameter, as long as the volume remains the same. Find the depth at which the design will pass the UPL ultimate limit state.

Q4.6. A 10 m OD shaft is to be excavated to 40 m depth in stiff fissured clay, at which point a 2 m thick reinforced concrete base slab will be cast. The ground is dewatered during construction using deep wells to avoid uplift failure during construction caused by a permeable stratum below formation level, but the pumps will be switched off once the base slab concrete has gained its design strength. The long-term maximum water table will be at 5 m below ground level with hydrostatic distribution to well below the shaft bottom. The shaft lining is 400 mm thick. The clay has an undrained shear strength of 200 kPa and drained parameters are $c' = 0$ and $\phi' = 22°$, obtained from three boreholes. The overconsolidation ratio, $OCR = 2.0$. The bulk unit weight is 21 kN/m³.

    i. Show that the shaft fails the uplift ultimate limit state in the long-term, if friction between the shaft lining and the ground is ignored.

    ii. Estimate the shaft friction that is available in the short-term (using undrained shear strength) and in the long-term (using drained parameters) and show that the shaft now passes the uplift ultimate limit state.

REFERENCES

Allenby, D. & Kilburn, D. (2015). Overview of underpinning and caisson shaft-sinking techniques. *Proc. ICE Geotech. Engrg* **168**, issue GE1, 3–15.

Bjerrum, L. & Eide, O. (1956). Stability of strutted excavations in clay. *Géotechnique* **6**, No. 1, 32–47.

Bolton, M. D., Lam, S. Y. & Osman, A. S. (2008). Keynote Lecture: Supporting excavations in clay – from analysis to decision-making. *Proc. 6th International Symposium on Geotechnical Aspects of Underground Construction in Soft Ground*, IS-Shanghai, Vol. 1, pp. 12–25. London: CRC Press.

Cai, F., Ugai, K. & Hagiwara, T. (2002). Base stability of circular excavations in soft clay. *ASCE J. Geotech. Geoenviron. Eng.* **128**, No. 8, 702–706.

Craig, R. F. (1997). *Soil mechanics*, 6th edition. London: Chapman & Hall.

EN 1997-1:2004+A1:2013. *Eurocode 7: Geotechnical design – Part 1: General rules*, incorporating corrigendum February 2009. Brussels: European Committee for Standardization.

Fleming, W. G. K., Weltman, A. J., Randolph, M. F. & Elson, W. K. (1985). *Piling engineering*. Glasgow and London: Surrey University Press.

Fleming, W. G. K. (1997). *Pile design and practice notes*. Kvaerner Cementation Foundations Limited, March 1997.

Frank, R., Bauduin, C., Driscoll, R., Kavvadas, M., Krebs Ovesen, N., Orr, T. & Schuppener, B. (2005). *Designers' guide to Eurocode 7: geotechnical design*. London: Thomas Telford Limited.

Mollon, G., Dias, D. & Soubra, A.-H. (2013a). Continuous velocity fields for collapse and blowout of a pressurized tunnel face in purely cohesive soil. *Int. J. Numer. Anal. Meth. Geomech.* 37, 2061–2083.

NA+A1:2014 to BS EN 1997-1:2004+A1:2013. *UK National Annex to Eurocode 7: Geotechnical design – part 1: general rules*, incorporating corrigendum February 2009. British Standards Institution.

Polubarinova-Kochina, P. Ya. (1962). *Theory of ground water movement*, translated from the 1952 Russian edition by J. M. R. De Wiest. Princeton, NJ: Princeton University Press.

Powrie, W. (1997). *Soil mechanics – concepts and applications*. London: E & FN Spon.

Skempton, A. W. (1951). The bearing capacity of clays. *Proc. Building Research Congress*, London, Division 1, pp. 180–189.

Skempton, A. W. (1959). Cast-in-situ bored piles in London Clay. *Géotechnique* 9, No. 4, 153–173.

Terzaghi, K. (1943). *Theoretical soil mechanics*. New York, London: John Wiley & Sons.

Chapter 5

Stability and Eurocode 7

Eurocode 7 (EN 1997-1:2004+A1:2013) does not have a specific section on the design of tunnels, though there are sections on retaining structures, spread foundations, pile foundations, anchorages and embankments. Perhaps a section on tunnels will be added in a future revision.

Face stability is a 'GEO' ultimate limit state according to Clause 2.4.7.1 (1)P of Eurocode 7 (EN 1997-1:2004+A1:2013), defined as "failure or excessive deformation of the ground, in which the strength of soil or rock is significant in providing resistance". As we saw in Chapter 4, there are other forms of instability, such as uplift, which is the 'UPL' limit state, and hydraulic failure, which is the 'HYD' limit state, both of which can apply to shafts or to tunnels. In this chapter we will focus on large-scale face stability to pick up on issues relating to the selection of characteristic values and the selection of partial factors.

The first thing we need to do is decide on the 'characteristic value' of each soil parameter we are using, based on the results of laboratory or in situ tests on a number of soil samples. There isn't just one characteristic value of each geotechnical parameter, each limit state will have its own. So the characteristic value of a soil parameter selected for design of the tunnel lining may be different to the characteristic value used for assessing stability.

After working through this chapter, you will understand:

- how Eurocode 7 ensures stability through the use of characteristic values and partial factors
- how to select characteristic values of parameters
- the use of statistics and probability theory in geotechnical design

After working through this chapter, you will be able to:

- select an appropriate characteristic value of a geotechnical parameter for use in design
- calculate the ultimate limit state for each type of stability failure

DOI: 10.1201/9780429470387-5

5.1 SIZE OF THE ZONE OF GROUND GOVERNING THE OCCURRENCE OF THE LIMIT STATE

Characteristic values are defined by Eurocode 0 (EN 1990:2002 +A1:2005) as the 5th percentile value. This means that the probability of the true value being lower than the characteristic value is 5% or 0.05. This applies to all kinds of construction materials, such as concrete or timber. However, in Eurocode 7, a characteristic value (clause 2.4.5.2, EN 1997-1:2004+A1:2013) for a geotechnical parameter is defined as a cautious estimate of the true mean value over the zone of ground governing the behaviour at the limit state. If statistical methods are used, this means we need to have a probability of less than 5% that the true mean value over the zone of ground governing the occurrence of the limit state is lower than the characteristic value. Therefore, we need to know something about the probability distribution of the geotechnical parameter, and how it may vary over the zone of ground we are interested in.

If the limit state involves hydraulic fracturing, or failure of a discrete block in a fissured clay, then the characteristic value needs to be much more cautious because failure may depend on a small zone of ground. In effect, failure may depend on a single value. In this case, the characteristic value should be the 5th percentile value of the probability distribution for that parameter. Note that the probability distribution for a soil parameter is not just a normal distribution defined by the mean and standard deviation of the test data, it needs to be adjusted to take account of the level of confidence, which is based on the number of data points; this will be covered in the next section.

The size of the shear surfaces that need to be mobilised at failure due to undrained or drained stability are usually large and therefore the value chosen should be somewhere between the 5th percentile value and the mean of the individual test data, because failure is not governed by a single value of undrained shear strength or drained shear strength parameters. If the zone of ground governing the occurrence of the limit state can be estimated to be very large compared to the 'scale of fluctuation', then the mean value with 95% confidence level can be used (see next section), and we are in effect assuming that the variability within that zone of ground is similar to the variability in the site investigation data (Hicks & Nuttall, 2012).

The 'scale of fluctuation' may be interpreted as the distance within which soil properties are largely correlated (Schneider & Fitze, 2011). The determination of the scale of fluctuation is complex and outside the scope of routine site investigation campaigns, but there is some research available. For instance, Phoon & Kulhawy (1999a & 1999b) quote a range of values of vertical scale of fluctuation between 1 m and 6 m and horizontal scale of fluctuation between 3 m and 80 m for strength parameters in a variety of different natural soil types.

5.2 CORRECTING FOR CONFIDENCE IN THE SITE INVESTIGATION

In order to determine the characteristic value of a geotechnical parameter for a particular limit state, first we need to calculate the mean and standard deviation of the site investigation test data. Since the site investigation test data may in some cases be based on a small number of samples, we need to make allowances for the degree of confidence we have that our sample size reflects the true variability of the ground we are tunnelling through. Statistical methods exist to quantify this uncertainty and can make a powerful case for spending more money on site investigation to reduce uncertainty.

In the following, we will assume that the geotechnical parameters follow a normal distribution. However, the normal distribution is not always the best fit to geotechnical data and you should consider whether other distributions may be more appropriate, for example a Weibull or Lognormal distribution (Masoudian et al., 2019).

Eurocode 7 (EN 1997-1:2004+A1:2013, clause 2.4.5.2 (4)P) requires the designer to take account of the type and number of samples used in the site investigation. The number of samples will affect the level of confidence in the estimate of the mean and standard deviation. With a small number of samples of a given soil stratum, we have less confidence in the calculated mean and standard deviation, and so we need to correct for this lack of confidence. The 'Student' or 't' distribution allows us to do this. Tables of 't' distribution values may be found quite easily on the World Wide Web.

For hydraulic fracturing or the case of a localised block failure, we want to find the 5th percentile value, i.e. we want to know that 95% of the ground is stronger than the characteristic value. If there were an infinite number of samples taken in the site investigation (i.e. perfect knowledge of the probability distribution), the 't' distribution becomes the same as the normal distribution, and 95% of values will be higher than $X_m - 1.645S$, where X_m is the mean and S is the standard deviation of the test data. However, if the number of samples was 5, effectively the 't' distribution becomes flatter and wider than the normal distribution and 95% of values will be higher than $X_m - 2.015S$. Therefore, with a small number of samples, the characteristic value of a soil strength parameter will be further away from the mean of the test data, and the design will therefore be less efficient.

For the case of a large-scale stability failure, we need a 95% confidence that the true mean value will be higher than the characteristic mean value. This characteristic mean value may be calculated using the following expression from Schneider (1999):

$$X_k = X_m - \frac{f}{\sqrt{n}} S \qquad (5.1)$$

X_k is the characteristic mean value
X_m is the mean of the test data
f is a coefficient related to the type of distribution, confidence limits
    and number of test values
n is the number of test values (also called 'sample size')
S is the standard deviation of the test data

The coefficient f may be obtained from statistical tables of the 'Student' or
't' distribution using a one-tail confidence limit of 5%. f has a higher value
when n is small, and decreases exponentially as n increases. Figure 5.1
shows how the value of $f/\sqrt{n}$ decreases with the number of test values n.

Schneider (1999) explains how this calculation of the true mean value
can be difficult to achieve using a statistical calculation in practice. If the
number of samples is small, then the high value of f means that the characteristic value is very low – generally much lower than prior knowledge
and experience would suggest is reasonable, and sometimes lower than is
theoretically possible.

Schneider (1999) proposes a simple method, which is widely used in
practice, where the characteristic value is equal to the mean minus half a
standard deviation (this is the grey horizontal line in Figure 5.1):

$$X_k = X_m - 0.5S \tag{5.2}$$

with notation as for Equation 5.1

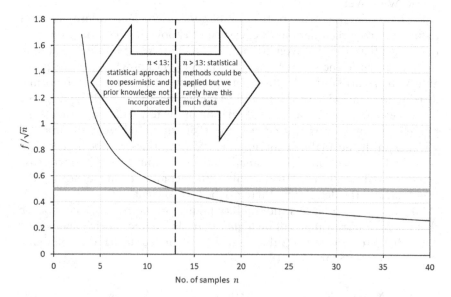

Figure 5.1 Improvement in confidence with increasing no. of samples (Schneider, 1999).

For parameters that vary with depth, multivariate statistics should be used. This is explained by Bond & Harris (2008), where worked examples are also provided.

There is often a theoretical lower limit to soil strength parameters, which can be calculated. The shear strength cannot be exceeded by the shear stress present in the in situ stress state, usually defined by the coefficient of earth pressure at rest, K_0. If the characteristic value calculated using statistical methods were found to be lower than the theoretical minimum, then the theoretical minimum should be used instead. This anomaly is often caused by assuming that the soil parameter follows a normal distribution, when in fact it is bounded by a lower limit and would probably be better modelled by a Weibull or Lognormal distribution. An example of undrained shear strength test data from a real project is shown in Figure 5.2, where values below the theoretical minimum can be seen.

Another anomaly caused by the normal distribution assumption is that as sample size increases, the probability of a rare very high value being found increases, but because there is a lower limit (e.g. the laboratory or in situ test cannot provide negative strength values), this causes a trend where as the sample size increases, the mean also increases (see Figure 5.2 where rare high values can be seen that if mirrored about the mean would probably be negative). The normal distribution is symmetric and its standard deviation and mean are calculated based on all the data, regardless of which side of the mean they are on. The lognormal distribution, as a counter-example, is not symmetric. The 'geometric' mean and standard deviation provided by the lognormal distribution will not, on average, vary with sample size. Therefore, where the site investigation is extensive and there is a relatively large amount of data, consideration should be given to what probability distribution functions are used.

5.3 MODELLING SPATIAL VARIABILITY OF SOIL PARAMETERS EXPLICITLY

Where the zone of ground governing the occurrence of the limit state is more than 20 times larger than the scale of fluctuation, then a cautious estimate of the true mean value can be used. Where the zone of ground governing the occurrence of the limit state is smaller or of similar magnitude to the scale of fluctuation, then a cautious estimate of the 5th percentile value can be used. If we are between these two cases, or where we suspect that failure may follow a path of least resistance, then we may wish to model spatial variation of parameters explicitly.

Spatial variation can be modelled explicitly by using a Monte Carlo method to assign random values of the parameters to different zones within a model. The model is run many times with different randomly generated variations of parameters. This is sometimes called the 'random

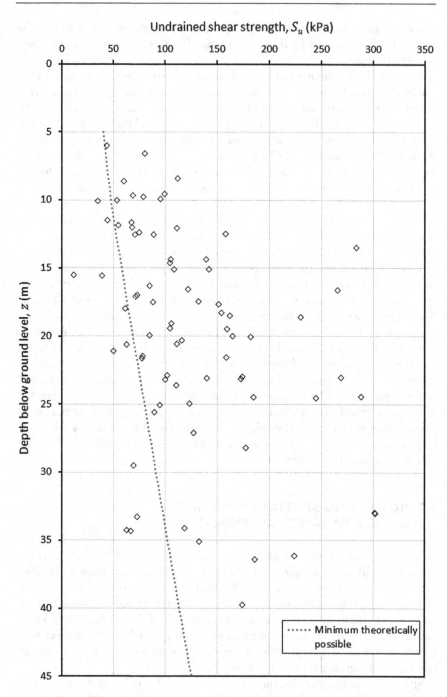

Figure 5.2 An example of real undrained shear strength test data, sample size *n* = 66.

field method'. The size of the zones is key – an 'auto-correlation distance' is defined, which is related to the scale of fluctuation, over which parameters can be considered to remain constant. For slope stability, Hicks & Nuttall (2012) and Griffiths & Fenton (2007) found that the mean factor of safety was lower when spatial variation was modelled, as failure surfaces followed a path of least resistance. This may be due to the different scales of problem and different scales of fluctuation modelled, and also how the areas modelled lined up with failure surfaces. They used the random field method to apply values of parameters to a finite element mesh that were spatially correlated, that is, where the distance between midpoints of the finite elements determined how much the parameters could vary. This is sometimes referred to as the 'random finite element method' or 'RFEM' (Fenton & Griffiths, 2007). Papaioannou et al. (2009) also found that ignoring spatial variation led to an overestimate of probability of failure, this time for a rock tunnel with 1100 m of overburden.

If you are interested in finding out more about probabilistic methods, start with the overview in Jones (2015). It is inevitable that as the cost of computing continues to decrease, these methods will become part of routine design practice.

5.4 APPLYING PARTIAL FACTORS

The design approach and partial factors used depend on the country of application, but generally speaking, partial factors are applied either to the strength parameters or to the effect of actions. In the United Kingdom, Design Approach 1 must be used, as required by the UK National Annex to Eurocode 7 (NA+A1:2014 to BS EN 1997-1:2004+A1:2013). Design Approach 1 has two combinations as described in Eurocode 7 clause 2.4.7.3.4.2 (EN 1997-1:2004+A1:2013).

Applying Design Approach 1 to undrained tunnel stability means that the factor of safety N_c/N must be greater than 1.35 (Combination 1) and the tunnel face must be stable when the characteristic value of undrained shear strength is divided by 1.4 (Combination 2). By inspection of Equation 2.1 in Chapter 2, it is clear that both these combinations will have the same effect, except Combination 1 requires a factor of safety of 1.35 and Combination 2 requires 1.4. Therefore, Combination 2 will always be the critical case.

Applying Design Approach 1 to drained stability is a little more tricky. The in situ pore pressure should be well known from the ground investigation, and as long as a cautious estimate of the piezometric level is made based on seasonal and, where appropriate, tidal variations relevant to the construction period, no safety factor is required. Therefore, only the drained shear strength parameters c and ϕ should be factored by dividing c and $\tan\phi$ by 1.25 according to Design Approach 1 Combination 2 (EN 1997-1:2004+A1:2013).

REFERENCES

Bond, A. & Harris, A. (2008). *Decoding Eurocode 7*. Abingdon: Taylor & Francis.

EN 1990:2002 +A1:2005 (2010). *Eurocode – Basis of structural design*, incorporating corrigenda December 2008 and April 2010. Brussels: European Committee for Standardization.

EN 1997-1:2004+A1:2013. *Eurocode 7: Geotechnical design – part 1: general rules*, incorporating corrigendum February 2009. Brussels: European Committee for Standardization.

Fenton, G. A. & Griffiths, D. V. (2007). Review of probability theory, random variables and random fields. *Probabilistic methods in geotechnical engineering* (eds Griffiths, D. V. & Fenton, G. A.), CISM Courses and Lectures No. 491, pp. 1–69. New York/Wien: Springer.

Griffiths, D. V. & Fenton, G. A. (2007). The Random Finite Element Method (RFEM) in slope stability. *Probabilistic methods in geotechnical engineering* (eds Griffiths, D. V. & Fenton, G. A.), CISM Courses and Lectures No. 491, pp. 317–346. New York/Wien: Springer.

Hicks, M. A. & Nuttall, J. D. (2012). Influence of soil heterogeneity on geotechnical performance and uncertainty: a stochastic view on EC7. *Proc. 10th Int. Probabilistic Workshop*, Stuttgart, Germany (eds Moorman, C., Huber, M. & Proske, D.), Mitteilung 67, pp. 215–228. Stuttgart: Institut für Geotechnik der Universität Stuttgart.

Jones, B. D. (2015). Probabilistic methods. *Tunnelling Journal*, October/November issue, 26–29.

Masoudian, M. S., Afrapoli, M. A. H., Tasalloti, A. & Marshall, A. M. (2019). A general framework for coupled hydro-mechanical modelling of rainfall-induced instability in unsaturated slopes with multivariate random fields. *Comput. Geotech.* **115**, paper 103162, 1–12.

NA+A1:2014 to BS EN 1997-1:2004+A1:2013. *UK National Annex to Eurocode 7: Geotechnical design – Part 1: General rules*, incorporating corrigendum February 2009. British Standards Institution.

Papaioannou, I., Heidkamp, H., Düster, A., Rank, E. & Katz, C. (2009). Random field reliability analysis as a means for risk assessment in tunnelling. *Proc. 2nd Int. Conf. on Computational Methods in Tunnelling*, Ruhr Universität Bochum, Germany. Aedificatio Publishers.

Phoon, K. -K. & Kulhawy, F. H. (1999a). Characterization of geotechnical variability. *Can. Geot. J.* **36**, 612–624.

Phoon, K. -K. & Kulhawy, F. H. (1999b). Evaluation of geotechnical property variability. *Can. Geot. J.* **36**, 625–639.

Schneider, H. R. (1999). Determination of characteristic soil properties. *Proc. XII European Conf. on Soil Mech. & Geot. Engrg – Geotechnical Engineering for Transportation Infrastructure: Theory and Practice, Planning and Design, Construction and Maintenance* (eds Barends, F. B. J. et al.), Amsterdam, The Netherlands, pp. 273–281. Rotterdam: Balkema.

Schneider, H. R. & Fitze, P. (2011). Characteristic shear strength values for EC7: guidelines based on a statistical framework. *Proc. XV European Conf. on Soil Mech. & Geot. Engrg*, Athens, Greece, Vol. 4.

Chapter 6

Global design using analytical solutions

This chapter describes the global design of tunnels and shafts, taking account of soil-structure interaction, to give estimates of forces in the tunnel lining. Details, particularly pertaining to segmental linings, such as joints, bolts, gaskets and reinforcement, will be dealt with in Chapters 9 and 10, and Chapter 11 will focus on sprayed concrete. You should be aware though, that details such as joints can have an important effect on the flexibility of the lining, so this current chapter should not be used in isolation.

A robust, rational and efficient design process should start with very simple calculations and gradually introduce layers of complexity. In this way, the sensitivity of the design to each layer of complexity may be assessed, errors or bugs in spreadsheets or numerical models easily found and gross errors avoided.

For example, a simple wished-in-place equilibrium calculation allows an upper bound to the hoop force to be obtained very quickly. This may be followed by an analytical solution that includes soil-structure interaction and will calculate bending moments, and a slightly lower hoop force. The next stage may be to introduce 3D effects via the convergence-confinement method. All these calculations can be easily done by hand or in a spreadsheet. Only after all these simple methods have been exhausted and you have a very good feel for the problem should you start using numerical analysis in 2D and 3D to introduce further layers of complexity, such as nonlinear soil and lining behaviour, non-circular tunnel shapes and interaction with other tunnels or structures.

After working through this chapter, you will understand:

- the limitations of analytical solutions for tunnel design
- the relative importance of soil stiffness, lining stiffness and lining flexibility to the axial forces and bending moments
- the principles of continuum and bedded beam analytical solutions

DOI: 10.1201/9780429470387-6

After working through this chapter, you will be able to:

- calculate simple wished-in-place equilibrium for a circular tunnel
- use the Curtis-Muir Wood solution to calculate axial forces and bending moments in circular tunnel and shaft linings

6.1 SIMPLE WISHED-IN-PLACE EQUILIBRIUM

As a real tunnel is constructed, there is redistribution of stresses in the ground around and ahead of the face, and then interaction between the lining and the ground, such that the stresses transmitted to the lining are not equal to the initial in situ stresses. If we assume in a design model that the tunnel lining is 'wished-in-place', we ignore this redistribution and interaction and just assume that we can magically materialise a perfectly stiff and infinitely long tunnel lining into existence without disturbing the soil. This will always provide an upper bound to hoop forces in the tunnel lining compared to more sophisticated analytical or numerical methods, so it can provide a useful 'reality check'.

Hoop forces (sometimes referred to as 'tangential' or 'axial' forces) at any position in the tunnel lining may be found by (theoretically) cutting the ring and resolving the internal and external forces in one direction, as shown in Figure 6.1, using the following notation:

N is the hoop force in kN/m (kN per metre length of tunnel)
$\sigma_{v,axis}$ is the initial in situ vertical total stress at axis level in kPa
$\sigma_{h,axis}$ is the initial in situ horizontal total stress at axis level in kPa

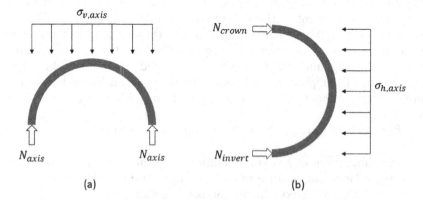

(a)   (b)

Figure 6.1 Wished-in-place equilibrium to calculate (a) hoop force at axis or (b) hoop force at crown and invert.

To calculate the hoop force at axis in Figure 6.1(a), the vertical total stress *at axis level* should be used, because although the initial in situ stress at the invert will be higher than at the crown, the tunnel must be in equilibrium, so the tunnel will be pushed upwards until the ground pressure on the crown equals the ground pressure on the invert. This is a simplification and not strictly true, because shear stresses between the lining and the ground can also be generated. Therefore, the vertical stress on the top of the tunnel can be different to the vertical stress on the bottom of the tunnel. It is also simple to add in the effect of the weight of the tunnel lining if you wish.

Figure 6.1(b) could also be made slightly more sophisticated by making the horizontal total stress increase with depth and applying the average value above axis level to calculate the crown hoop force and the average value below axis level to calculate the invert hoop force. Although it might be tempting to make things more complicated at this point, the whole rationale for this kind of basic initial analysis is that it may easily be done on the back of an envelope, or even in one's head, so it is better to keep things simple, unless the tunnel is particularly large and shallow. The equilibrium equations are given by:

$$N_{axis} = \sigma_{v,axis} r_0 \tag{6.1}$$

$$N_{crown} = N_{invert} = \sigma_{h,axis} r_0 \tag{6.2}$$

N_{axis} is the hoop force at axis in kN/m

$\sigma_{v,axis}$ is the vertical in situ stress in the ground at axis level in kPa (note that $1 \text{ kPa} = 1 \text{ kN/m}^2$)

N_{crown} is the hoop force at the crown (the highest point in the tunnel lining) in kN/m

N_{invert} is the hoop force at the invert (the lowest point in the tunnel lining) in kN/m

$\sigma_{h,axis}$ is the horizontal in situ stress in the ground at axis level in kPa

r_0 is the external radius of the tunnel (also referred to as the 'radius to extrados') in m

To calculate the hoop stress, the hoop force, which is in kN/m, need only be divided by the thickness of the lining.

Thus, a very simple calculation provides us with a conservative estimate of the hoop force in the tunnel lining. Bearing in mind of course that we have not yet introduced the relaxation of stress in the ground prior to installation of the lining, or the interaction between the lining and the ground once the lining is installed, and we have no way of calculating bending moments in the lining.

WORKED EXAMPLE 6.1   SIMPLE
WISHED-IN-PLACE EQUILIBRIUM, $K_0 = 1.0$

From the simplified geometry of a tunnel in Figure 6.2, calculate the vertical and horizontal total ground pressure acting on the tunnel lining, assuming $K_0 = 1.0$. Note that OD means 'outside diameter', i.e. the diameter to the extrados or outside surface of the lining, and ID means 'inside diameter', i.e. the diameter to the intrados or inside face of the lining.

γ_b is the bulk unit weight of the soil, i.e. of the soil particles and pore water combined.

Assume the tunnel is 'wished-in-place', i.e. there is no relaxation of stress during construction, there is a perfectly stiff lining so the initial stresses are applied directly to the tunnel lining, and assume the tunnel lining is weightless. Calculate the hoop force in the lining.

If the tunnel lining thickness $t = 0.3$ m, what is the hoop stress per metre length of tunnel?

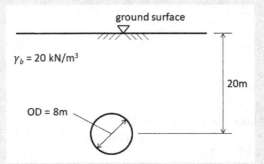

Figure 6.2 Worked Example 6.1 cross-section.

Vertical in situ total stress at axis level $\sigma_{v0} = \gamma_b z_0 = 20 \times 20 = 400$ kPa.

For $K_0 = 1.0$, horizontal in situ total stress at axis level $\sigma_{h0} = \sigma_{v0} = 400$ kPa.

To calculate the thrust in the lining at axis, assume buoyancy means that vertical ground pressures above and below the tunnel equilibrate so the average is the vertical stress at axis level σ_{v0}.

Therefore, for static equilibrium, the hoop force at axis is given by:

$$N_{axis} = \sigma_{v0} r_0 = 400 \times 4 = 1600 \text{ kN/m}$$

where r_0 is the radius to the extrados of the tunnel.

In the horizontal direction, since $\sigma_{h0} = \sigma_{v0} = 400$ kPa at axis level, $N_{crown} = N_{invert} = 1600$ kN/m. Alternatively, you could calculate the average horizontal ground pressure on the top half as 360 kPa and on the bottom half as 440 kPa, resulting in a hoop force at the crown of 1440 kN/m and at the invert of 1760 kN/m.

At axis, hoop stress is given by the hoop thrust divided by the cross-sectional area. Since the hoop thrust is already expressed in units per metre length of tunnel, this is given by:

$$\sigma_{axis} = \frac{N_{axis}}{t} = \frac{1600}{0.3} = 5333 \text{ kPa} = 5.33 \text{ MPa}$$

For varying horizontal in situ stress with depth, at the crown the hoop stress would be:

$$\sigma_{crown} = \frac{N_{crown}}{t} = \frac{1440}{0.3} = 4800 \text{ kPa} = 4.80 \text{ MPa}$$

and at the invert it would be:

$$\sigma_{invert} = \frac{N_{invert}}{t} = \frac{1760 \text{ kN/m}}{0.3 \text{ m}} = 5867 \text{ kPa} = 5.87 \text{MPa}$$

EXTENSION TO WORKED EXAMPLE 6.1 INCLUDING LINING WEIGHT

If the unit weight of the concrete lining $\gamma_c = 24$ kN/m^3, what is the radial ground pressure acting on the crown and invert?

Total weight of concrete ring is given by:

$$W = 2\pi r_m t \gamma_c = 2 \times \pi \times 3.85 \times 0.3 \times 24 = 174.2 \text{ kN/m}$$

where r_m is the radius to the centroid of the lining = 3.85 m.

For equilibrium in the vertical direction, the force due to vertical ground pressure on the top half of the ring, $2r_0\sigma_{v(crown)}$, plus the weight of the ring W is equal to the force due to vertical ground pressure on the bottom half of the ring, $2r_0\sigma_{v(invert)}$:

$$2r_0\sigma_{v(crown)} + W = 2r_0\sigma_{v(invert)}$$

Therefore the vertical component of ground pressure on the ring will always be larger at the invert by the weight of the ring. Assuming the average σ_{v0} is the same, then:

$$\sigma_{v(crown)} = \sigma_{v0} - \frac{W}{4r_0} = 400 - \frac{174.2}{16} = 389.1 \text{ kPa}$$

$$\sigma_{v(invert)} = \sigma_{v0} + \frac{W}{4r_0} = 400 + \frac{174.2}{16} = 410.9 \text{ kPa}$$

So, the effect of the ring weight is quite small, and generally, if one assumes it is weightless (as many more sophisticated methods such as the Curtis-Muir Wood equations do), the error is not large unless the tunnel is very large and shallow.

WORKED EXAMPLE 6.2 – SIMPLE WISHED-IN-PLACE EQUILIBRIUM, $K_0 \neq 1.0$

From the simplified geometry of a tunnel in Figure 6.3, calculate the vertical and horizontal total stress at tunnel axis level, assuming $K_0 = 0.7$. Remember that K_0 is the ratio of horizontal to vertical effective stress. The unit weight of the soil above the water table is denoted by γ_{unsat} and below the water table is denoted by γ_{sat}.

Assuming the tunnel is wished-in-place and perfectly stiff, i.e. the full overburden load is acting on the tunnel lining, calculate the hoop force in the lining at crown and invert level.

The tunnel lining is 200 mm thick concrete. Calculate the hoop stress at crown/invert and axis.

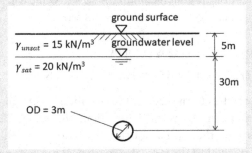

Figure 6.3 Worked Example 6.2 cross-section.

Vertical total stress at axis level σ_{v0} is given by:

$$\sigma_{v0} = 5 \times 15 + 30 \times 20 = 675 \text{ kPa}$$

The pore pressure at axis level u_0 is given by:

$$u_0 = 30 \times 10 = 300 \text{ kPa}$$

The vertical effective stress at axis level σ'_{v0} is therefore:

$$\sigma'_{v0} = \sigma_{v0} - u_0 = 675 - 300 = 375 \text{ kPa}$$

The horizontal effective stress at axis level σ'_{h0} is:

$$\sigma'_{h0} = K_0 \sigma'_{v0} = 0.7 \times 375 = 262.5 \text{ kPa}$$

Therefore the horizontal total stress at axis level σ_{h0} is:

$$\sigma_{h0} = \sigma'_{h0} + u_0 = 262.5 + 300 = 562.5 \text{ kPa}$$

Remember that $K_0 \neq \sigma_{h0}/\sigma_{v0}$. It is always $K_0 = \sigma'_{h0}/\sigma'_{v0}$.

Using the horizontal total stress at axis level for both the crown and invert hoop thrust calculation, because r_0 is small and z_0 is large:

$$N_{cr,inv} = \sigma_{h0}r_0 = 562.5 \times 1.5 = 843.75 \text{ kN/m}$$

Now calculate the hoop force at axis level:

$$N_{axis} = \sigma_{v0}r_0 = 675 \times 1.5 = 1012.5 \text{ kN/m}$$

At crown/invert the hoop stress is given by:

$$\frac{N_{cr,inv}}{0.2} = \frac{843.75}{0.2} = 4.22 \text{ N/mm}^2$$

At axis level the hoop stress is given by:

$$\frac{N_{axis}}{0.2} = \frac{1012.5}{0.2} = 5.06 \text{ N/mm}^2$$

Note that when $K_0 < 1.0$, the hoop force at axis is larger than the hoop force at the crown or invert. When $K_0 > 1.0$, the hoop force at axis is smaller than the hoop force at the crown or invert.

6.2 EMPIRICAL METHODS

A first estimate of the expected load in a tunnel lining may be obtained by considering case histories of lining forces where similar construction methods have been used in similar geology. Some data were provided in Section 1.7. Often these data are normalised by dividing by the full overburden pressure (the vertical total stress at the tunnel axis level). Even so, there are large differences between lining forces measured on different projects, even in similar geology, and this is because they are heavily dependent on the construction method.

It may be possible in the future to develop empirical relationships to allow preliminary estimates of lining forces to be made, but at present there is insufficient data because lining forces are so rarely measured.

6.3 SOIL-STRUCTURE INTERACTION USING THE CURTIS-MUIR WOOD SOLUTION

In order to make the calculation more realistic, soil-structure interaction may be introduced by using an analytical solution. These solutions allow the tunnel lining to deform elastically due to axial compression, flexure and shear stresses between the lining and the ground. This means that some load sharing with the ground is allowed and bending moments in the lining may be calculated. However, relaxation of the ground stresses prior to installation of the lining is not taken into account, unless of course we were to make some assumptions and reduce the in situ stresses we input to the model.

An 'analytical solution', or 'closed-form solution', is a calculation that may be performed by hand (or in a simple spreadsheet) without iteration. These methods are usually simple and based on elastic constitutive models, homogeneous materials and simple boundary conditions. Even if numerical modelling is going to be employed, an analytical solution may be used to verify that the numerical modelling program is functioning correctly (see Section 7.7 in the following chapter).

In order to reduce the number of variables, several assumptions are made. The tunnel is circular. The ground is homogeneous and isotropic, meaning that layered soils cannot be considered, and the elastic soil parameters must be the same in the horizontal and vertical directions.

In this book, we are going to look at the Curtis-Muir Wood solution (Curtis, 1974; Muir Wood, 1975; Curtis, 1975). Other solutions exist, most of which vary only slightly in their composition or notation, and either give exactly the same results or a negligible difference. A summary of solutions available at the time was provided by Duddeck & Erdmann (1985), noting their differences. Einstein & Schwartz (1979) later provided a 'simplified analysis' to achieve the same result as the Curtis-Muir Wood solution, which is not really any simpler (if you look this one up, be sure to also read the lively discussion!). Ahrens et al. (1982) did a thorough study of both the continuum solution and a bedded beam solution, and these methods were adopted in the German recommendations at the time (Duddeck, 1980).

The original analytical solution published by Sir Alan Muir Wood in 1975 (Muir Wood, 1975) contained some errors and an unnecessary simplification, which were corrected by John Curtis soon after (Curtis, 1975), who also extended the method to include time-dependent deformation of the ground and lining (i.e. viscoelasticity; Curtis, 1974). Hence the corrected analytical solution is commonly referred to as the 'Curtis-Muir Wood' solution.

6.3.1 Notation used in the Curtis-Muir Wood solution

Table 6.1 lists the notation used in the Curtis-Muir Wood solution. I have mostly used the same notation as Curtis (1974), but please be aware that it differs in some ways from notation used in the rest of this book. It is intended to be optimal for those of you who wish to refer to the original papers.

Table 6.1 Notation used in the Curtis-Muir Wood analytical solution.

c	As suffix to denote 'of the ground'
d	As suffix to denote 'due to distortional loading'
E	Young's modulus for the lining
E_c	Young's modulus for the ground
I	Second moment of area of lining per unit length of tunnel
I_e	Effective value of I for a jointed ring (i.e. a segmental lining)
l	As suffix to denote 'of the lining'
M	Bending moment in lining per unit length of tunnel
N	Hoop thrust in lining per unit length of tunnel
n	Number of segments in a ring of lining (also as suffix to denote 'normal')
P_0	Distortional stress (see Equation 6.3)
P_u	Uniform stress (see Equation 6.60)
r	Radius
r_m	Radius to the centroid of the tunnel lining
r_0	Radius to extrados of tunnel lining
R_c or Q_1	Compressibility factor (see Equation 6.61)
S_n, S_t	Normal and tangential (shear) stresses
t	Effective thickness of lining (also as suffix to denote 'tangential')
u	Radial movement of the ground (also as suffix to denote 'due to uniform loading')
u_0	Deformation of the ground in the absence of a lining at $r = r_0$
$\bar{u}$	Deformation of the ground inhibited by the presence of the lining
u_l	Deformation of the lining
$u_{0\theta}$	Circumferential movement of ground at $r = r_0$

(Continued)

Table 6.1 Notation used in the Curtis-Muir Wood analytical solution (*Continued*).

η	Ratio of radius of lining centroid to that of extrados
θ	Angle from crown (also as suffix to denote circumferential direction)
v	Poisson's ratio for the ground
v_l	Poisson's ratio for the lining
σ_v, σ_h	Pre-existing vertical and horizontal stress in undisturbed ground
$\sigma_r, \sigma_\theta, \sigma_z, \tau_{r\theta}$	Radial, tangential and out-of-plane normal stress and shear stress in the ground using polar coordinates
$\varepsilon_r, \varepsilon_\theta, \varepsilon_z$	Radial, tangential and out-of-plane strain in the ground using polar coordinates

6.3.2 Boundary conditions and ground stresses

The method is based on assuming the ground extends infinitely around the tunnel. The surface is not modelled, so this method is not appropriate for very shallow tunnels where arching around the crown will be limited by the cover available.

The method assumes the ground is homogeneous – that it has the same properties everywhere. This assumption means that layered soils or soils whose properties vary with depth or from one side of the tunnel to the other cannot be modelled.

It also assumes 'plane strain'. This means that the tunnel and the ground extend infinitely into the page, and there is zero strain in that direction – hence all strains occur within the 2D plane.

Pre-existing vertical and horizontal total stresses in the ground σ_v and σ_h are allowed to be different, but they do not vary with depth, i.e. we use the values at tunnel axis level.

The tunnel-ground boundary is at the radius to the extrados, r_0, and this is where ground stresses are calculated and interact with the tunnel lining. Where the axial or bending stiffness of the tunnel lining is being calculated, the radius to the centroid of the tunnel lining r_m is used.

Figure 6.4 shows the boundary conditions and ground stresses used in the Curtis-Muir Wood solution.

The ground and the lining are assumed elastic, therefore there is no non-linearity, no plasticity, no failure criterion, no groundwater flow and no soil consolidation. Groundwater pressure is taken into account by the use of total stress rather than effective stress plus pore pressure. An undrained soil response may be modelled by using the undrained Young's modulus and a Poisson's ratio of 0.5.

If $\sigma_v \neq \sigma_h$, then one can imagine separating the stress into two components, one which is uniform in all directions and one which is exactly opposite in the vertical and horizontal directions.

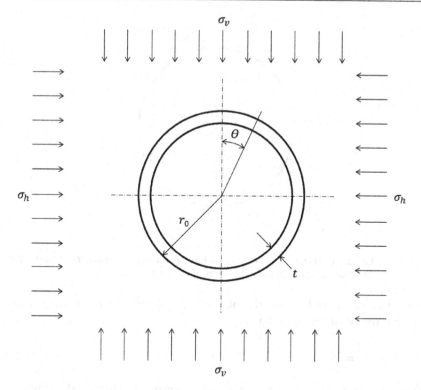

Figure 6.4 Boundary conditions and ground stresses in the Curtis-Muir Wood solution.

The distortional stress P_0 is defined, which is given by:

$$P_0 = \sigma_v - \sigma_h \tag{6.3}$$

Therefore, the distortional unloading of the ground is:

$$\frac{P_0}{2} \cos 2\theta \tag{6.4}$$

which varies with angle θ from the crown as shown in Figure 6.5.

So at the crown, where $\theta = 0°$, the distortional stress is $(\sigma_v - \sigma_h)/2$, and at axis level, where $\theta = 90°$, the distortional stress is $-(\sigma_v - \sigma_h)/2$. Try it for different values of θ.

It follows from the definition of distortional stress that the uniform stress P_u is the average of the vertical and horizontal stresses, i.e. $(\sigma_v + \sigma_h)/2$.

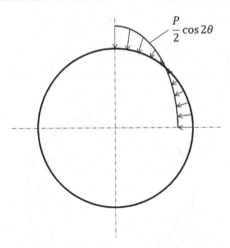

Figure 6.5 Distortional stress in the Curtis-Muir Wood solution from $\theta = 0$ to $\theta = 90°$. Only distortional stress in the first quadrant is shown.

If stresses S_{nc} and S_{tc} are the normal and shear stresses in the ground around the opening, such that:

$$S_{nc} = S_{tc} = \frac{P_0}{2} \tag{6.5}$$

then the normal stress $S_{nc}\cos2\theta$ acts normally around the opening and $-S_{tc}\sin2\theta$ acts tangentially around the opening. The minus sign is because for $\theta = 0$ at the crown and positive clockwise, Mohr's circle gives negative shear acting on the lining for $0 < \theta < \pi/2$. The arrow in Figure 6.6 shows the magnitude and direction of the shear.

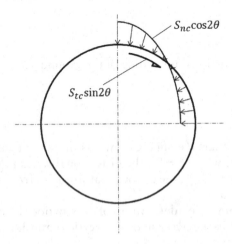

Figure 6.6 Magnitude and direction of shear in the Curtis-Muir Wood solution.

6.3.3 Elliptical deformation of a circular opening

First, elliptical deformation due to distortional stress of a circular opening in a homogeneous elastic material is considered, i.e. deformation of the ground with no tunnel lining. Airy's stress function is used for this purpose (Timoshenko & Goodier, 1970), with the form:

$$\phi = \left(ar^2 + br^4 + cr^{-2} + d\right)\cos 2\theta \qquad (6.6)$$

ϕ is a stress field
a, b, c and d are constants

We hypothesise that the solution has the form of Airy's stress function and we calculate the values of the constants a, b, c and d using equations derived from the boundary conditions and Hooke's Law. This will give us expressions for ground displacements around a circular opening.

In polar coordinates the radial, tangential and shear stresses in the ground are given by:

$$\sigma_r = \frac{1}{r}\frac{\partial \phi}{\partial r} + \frac{1}{r^2}\frac{\partial^2 \phi}{\partial \theta^2} \qquad (6.7)$$

$$\sigma_\theta = \frac{\partial^2 \phi}{\partial r^2} \qquad (6.8)$$

$$\tau_{r\theta} = -\frac{\partial}{\partial r}\left(\frac{1}{r}\frac{\partial \phi}{\partial \theta}\right) \qquad (6.9)$$

Substituting Equation 6.6 into Equations 6.7–6.9 gives:

$$\sigma_r = \left(-2a - 6cr^{-4} - 4dr^{-2}\right)\cos 2\theta \qquad (6.10)$$

$$\sigma_\theta = \left(2a + 12br^2 + 6cr^{-4}\right)\cos 2\theta \qquad (6.11)$$

$$\tau_{r\theta} = \left(2a + 6br^2 - 6cr^{-4} - 2dr^{-2}\right)\sin 2\theta \qquad (6.12)$$

At the tunnel boundary $(r = r_0)$, we have $\sigma_r = \tau_{r\theta} = 0$, so from Equations 6.10 and 6.12, we get:

$$a + 3cr_0^{-4} + 2dr_0^{-2} = -\frac{S_n}{2} \qquad (6.13)$$

$$a + 3br_0^2 - 3cr_0^{-4} - dr_0^{-2} = -\frac{S_t}{2} \qquad (6.14)$$

At the infinite boundary $(r = \infty)$, we have $\sigma_r = S_n \cos 2\theta$ and $\tau_{r\theta} = -S_t \sin 2\theta$, giving:

$$a + 3cr_\infty^{-4} + 2dr_\infty^{-2} = 0 \tag{6.15}$$

$$a + 3br_\infty^2 - 3cr_\infty^{-4} - dr_\infty^{-2} = 0 \tag{6.16}$$

Since r_∞ is very large, Equation 6.15 tells us that $a = 0$.

For the same reason, and because $a = 0$, Equation 6.16 tells us that $b = 0$. Rearranging Equation 6.13, and using $a = 0$, gives:

$$c = \frac{-S_n r_0^4}{6} - \frac{2dr_0^2}{3} \tag{6.17}$$

Inserting Equation 6.17 into 6.14 gives:

$$d = -\left(\frac{S_n + S_t}{2}\right)r_0^2 \tag{6.18}$$

Substituting for d in Equation 6.17 gives:

$$c = \frac{(S_n + 2S_t)r_0^4}{6} \tag{6.19}$$

Substituting for $a = 0$, c and d in Equation 6.10 gives:

$$\sigma_r = \left\{-(2S_t + S_n)\frac{r_0^4}{r^4} + 2(S_n + S_t)\frac{r_0^2}{r^2}\right\}\cos 2\theta \tag{6.20}$$

Substituting for $a = 0$, $b = 0$, and c in Equation 6.11 gives:

$$\sigma_\theta = \left\{(S_n + 2S_t)\frac{r_0^4}{r^4}\right\}\cos 2\theta \tag{6.21}$$

And substituting for $a = 0$, $b = 0$, c and d in Equation 6.12 gives:

$$\tau_{r\theta} = \left\{-(S_n + 2S_t)\frac{r_0^4}{r^4} + (S_n + S_t)\frac{r_0^2}{r^2}\right\}\sin 2\theta \tag{6.22}$$

Now, for plane strain, where the strain in the longitudinal tunnel axis direction (here denoted with subscript z) is assumed to be zero, Hooke's Law in polar cylindrical coordinates gives us the following three equations:

$$\varepsilon_r = \frac{1}{E}\left[\sigma_r - v(\sigma_\theta + \sigma_z)\right] \tag{6.23}$$

$$\varepsilon_\theta = \frac{1}{E}\left[\sigma_\theta - v(\sigma_r + \sigma_z)\right]$$

(6.24)

$$\varepsilon_z = \frac{1}{E}\left[\sigma_z - v(\sigma_r + \sigma_\theta)\right] = 0$$

(6.25)

Rearranging Equation 6.25 for σ_z gives:

$$\sigma_z = v(\sigma_r + \sigma_\theta)$$

(6.26)

Substituting for σ_z in Equations 6.23 and 6.24 gives:

$$\varepsilon_r = \frac{(1+v)}{E}\left[(1-v)\sigma_r - v\sigma_\theta\right]$$

(6.27)

$$\varepsilon_\theta = \frac{(1+v)}{E}\left[(1-v)\sigma_\theta - v\sigma_r\right]$$

(6.28)

Now, substituting values of σ_θ and σ_r from Equations 6.20 and 6.21 into Equations 6.27 and 6.28, and then simplifying the expressions, gives:

$$\varepsilon_r = \frac{(1+v)}{E}\left[(1-v)\left\{-(2S_t + S_n)\frac{r_0^4}{r^4} + 2(S_n + S_t)\frac{r_0^2}{r^2}\right\} - v\left\{(S_n + 2S_t)\frac{r_0^4}{r^4}\right\}\cos2\theta\right]$$

$$= \frac{(1+v)}{E}\left[2(1-v)(S_n + S_t)\frac{r_0^2}{r^2} - (2S_t + S_n)\frac{r_0^4}{r^4}\right]\cos2\theta$$

(6.29)

$$\varepsilon_\theta = \frac{(1+v)}{E}\left[(1-v)\left\{(S_n + 2S_t)\frac{r_0^4}{r^4}\right\} - v\left\{-(2S_t + S_n)\frac{r_0^4}{r^4} + 2(S_n + S_t)\frac{r_0^2}{r^2}\right\}\right]\cos2\theta$$

$$= \frac{(1+v)}{E}\left[(S_n + 2S_t)\frac{r_0^4}{r^4} - 2v(S_n + S_t)\frac{r_0^2}{r^2}\right]\cos2\theta$$

(6.30)

For a polar cylindrical coordinate system, the following relationships between strains and displacements apply:

$$\varepsilon_r = \frac{\partial u_r}{\partial r}$$

(6.31)

$$\varepsilon_\theta = \frac{1}{r}\left(\frac{\partial u_\theta}{\partial \theta} + u_r\right)$$

(6.32)

Therefore:

$$u_r = \int \varepsilon_r \; dr = \frac{(1+v)}{E}\left[\frac{(2S_t + S_n)}{3}\frac{r_0^4}{r^3} - 2(1-v)(S_n + S_t)\frac{r_0^2}{r}\right]\cos2\theta \quad (6.33)$$

And:

$$u_\theta = \int (r\varepsilon_\theta - u_r)d\theta = \int \left\{\frac{(1+v)}{E}\left[(S_n + 2S_t)\frac{r_0^4}{r^3} - 2v(S_n + S_t)\frac{r_0^2}{r}\right]\cos2\theta\right.$$

$$\left. - \frac{(1+v)}{E}\left[\frac{(2S_t + S_n)}{3}\frac{r_0^4}{r^3} - 2(1-v)(S_n + S_t)\frac{r_0^2}{r}\right]\cos2\theta\right\}d\theta$$

$$= \frac{(1+v)}{2E}\left[\frac{(2S_n + 4S_t)}{3}\frac{r_0^4}{r^3} + (2-4v)(S_n + S_t)\frac{r_0^2}{r}\right]\sin2\theta$$

$$= \frac{(1+v)}{E}\left[\frac{(S_n + 2S_t)}{3}\frac{r_0^4}{r^3} + (1-2v)(S_n + S_t)\frac{r_0^2}{r}\right]\sin2\theta \quad (6.34)$$

Equations 6.33 and 6.34, when combined with the displacements due to uniform compression (coming up later), are identical to the 'Kirsch problem' for an infinite elastic plate with a hole, adapted to plane strain (see e.g. Jaeger & Cook, 1976 or Hoek & Brown, 1980).

When $r = r_0$, i.e. at the periphery of the opening or the extrados of the lining, the deformations are obtained.

Radial deformation at the periphery of the opening, when $r = r_0$:

$$u_{0r} = \frac{-r_0(1+v)}{3E_c}\left[(5-6v)S_{nc} + (4-6v)S_{tc}\right]\cos2\theta \quad (6.35)$$

Tangential deformation at the periphery of the opening, when $r = r_0$:

$$u_{0\theta} = \frac{r_0(1+v)}{3E_c}\left[(4-6v)S_{nc} + (5-6v)S_{tc}\right]\sin2\theta \quad (6.36)$$

By substituting Equation 6.5 into 6.35 and 6.36, we get Equations 6.37 and 6.38, which describe the elastic elliptical deformation of the ground due to distortional stress:

$$u_{0rc} = \frac{-r_0(1+v)}{E_c}(3-4v)\frac{P_0}{2}\cos2\theta \quad (6.37)$$

$$u_{0\theta c} = \frac{r_0(1+v)}{E_c}(3-4v)\frac{P_0}{2}\sin2\theta \quad (6.38)$$

6.3.4 Elliptical deformation of a thin inextensible lining

Since this analytical solution considers elasticity, the principle of superposition holds. This means we can consider elliptical distortion and uniform compression of the lining separately, then add the displacements, strains or forces together at the end. In this section we are considering the effect of distortional load on a thin inextensible circular lining. We can then equate the displacements of the lining with the displacements of the ground found in the previous section, giving us the soil-structure interaction.

Looking back at Figure 6.6, the radial ground pressure is given by $S_n\cos2\theta$ and the tangential stress at the boundary is given by $S_t\sin2\theta$.

At the crown, where $\theta = 0$, the internal forces in a thin inextensible lining will be given by integration along the arc shown in Figure 6.7.

To find N at the crown, horizontal force equilibrium is used, integrating S_n and S_t over small arc lengths $r_0 d\theta$ from $\theta = 0$ to $\theta = \pi/2$.

To find M at the crown, moment equilibrium is used, where the stresses S_n and S_t over small arc lengths $r_0 d\theta$ are integrated from $\theta = 0$ to $\theta = \pi/4$ with lever arms $j = r_0\sin\theta$ and $k = r_0(1-\cos\theta)$, respectively, shown in Figure 6.7. The reason the integration is only from 0 to $\pi/4$ is because the deformation is elliptical and the moment M is equal to zero at $\theta = \pi/4$, and $M_{\theta=0} = -M_{\theta=90°}$. For this purpose we are assuming that the elliptical deformation will have a negligible effect on the geometry of the lever arms.

$$N_{\theta=0} = r_0 \int_0^{\frac{\pi}{2}} \left(S_n\cos2\theta\sin\theta - S_t\sin2\theta\cos\theta\right)d\theta \qquad (6.39)$$

$$M_{\theta=0} = r_0^2 \int_0^{\frac{\pi}{4}} \left(S_n\cos2\theta\sin\theta + S_t\sin2\theta(1-\cos\theta)\right)d\theta \qquad (6.40)$$

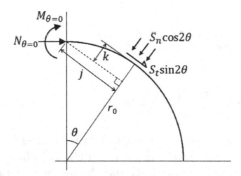

Figure 6.7 Resolving equilibrium to find internal forces at the crown of a thin inextensible lining subject to a distortional stress field.

Using the following equations (Equations 6.41–6.44 are found by 'integration by parts' – see Appendix B) helps us with the integration:

$$\int_0^{\frac{\pi}{2}} \cos2\theta \sin\theta \; d\theta = \left[\frac{\cos2\theta\cos\theta + 2\sin2\theta\sin\theta}{3}\right]_0^{\frac{\pi}{2}} = -\frac{1}{3} \tag{6.41}$$

$$\int_0^{\frac{\pi}{2}} \sin2\theta \cos\theta \; d\theta = \left[\frac{-(\sin2\theta \sin\theta + 2\cos2\theta \cos\theta)}{3}\right]_0^{\frac{\pi}{2}} = \frac{2}{3} \tag{6.42}$$

$$\int_0^{\frac{\pi}{4}} \cos2\theta \sin\theta \; d\theta = \left[\frac{\cos2\theta\cos\theta + 2\sin2\theta\sin\theta}{3}\right]_0^{\frac{\pi}{2}} = -\frac{1}{3} \tag{6.43}$$

$$\int_0^{\frac{\pi}{4}} \sin2\theta \cos\theta \; d\theta = \left[\frac{-(\sin2\theta \sin\theta + 2\cos2\theta \cos\theta)}{3}\right]_0^{\frac{\pi}{2}} = \frac{2}{3} \tag{6.44}$$

$$\int_0^{\frac{\pi}{4}} \sin2\theta \; d\theta = \frac{1}{2} \tag{6.45}$$

Therefore:

$$N_{\theta=0} = \frac{-r_0}{3}\left(S_n + 2S_t\right) \tag{6.46}$$

$$M_{\theta=0} = \frac{r_0^2}{6}\left(2S_n + S_t\right) \tag{6.47}$$

And it follows that:

$$N = N_{\theta=0}\cos2\theta = \frac{-r_0}{3}\left(S_n + 2S_t\right)\cos2\theta \tag{6.48}$$

$$M = M_{\theta=0}\cos2\theta = \frac{r_0^2}{6}\left(2S_n + S_t\right)\cos2\theta \tag{6.49}$$

Morgan (1961) showed that the moment M is induced by a change in the radius of curvature Δr from a circle to an ellipse, where $M = EI/\Delta r$. The ellipse major axis radius is $(r_0 + \delta)$ and the minor axis radius is $(r_0 - \delta)$. The largest change in curvature happens at these points, and hence the largest moments.

Morgan (1961) also showed that the radial deformation of a thin inextensible lining deforming elliptically is given by:

$$u_l = \frac{-Mr_0^2}{3EI} \tag{6.50}$$

Substituting for M in Equation 6.50 using the expression in Equation 6.49 gives radial deformation of the lining:

$$u_l = \frac{-r_0^4}{18EI}(2S_n + S_t)\cos2\theta \tag{6.51}$$

6.3.5 'Full slip' – no shear between lining and ground

In order to equate ground deformations and lining deformations, we need to first consider the interface between the lining and the ground. There are two limiting conditions, 'full slip' and 'no slip'.

'Full slip' means there is no shear between the lining and the ground and the interface is assumed to slip without resistance. 'No slip' means a perfect bond between the lining and the ground, and shear deformations are exactly equal. These limiting conditions may seem odd or unrealistic, but we can look at both 'full slip' and 'no slip', and the true situation is somewhere in between, so by using both we can bound the true solution.

'Full slip' means $S_t = 0$, and we may proceed as follows:

a. The ground will deform radially according to Equation 6.37.
b. The lining will deform radially under the load $S_n\cos2\theta$ in Equation 6.52:

$$u_l = r_0^4 2S_{nl}\cos2\theta/18EI \tag{6.52}$$

c. The ground will be restrained from deforming radially under the action of a load $S_{nc}\cos2\theta = -S_n\cos2\theta$ according to Equation 6.35:

$$\bar{u}_0 = \frac{-r_0(1+v)}{3E_c}(5-6v)S_n\cos2\theta \tag{6.53}$$

d. Since $u_0 + \bar{u}_0 = u_l$, then we combine Equations 6.35, 6.52 and 6.53 to find an expression for S_n:

$$S_n = \frac{3(3-4v)P_0/2}{5-6v+4Q_2} \tag{6.54}$$

where:

$$Q_2 = \frac{E_c}{E} \frac{1}{(1+v)} \frac{r_0^3}{12I} \tag{6.55}$$

Q_2 is just a shortcut to make the equation simpler; it doesn't appear to mean anything on its own. It appears in several of the following equations.

What we have done is calculate the radial deformation of an unlined tunnel due to distortional load, and how the lining will deform under such a load. Ensuring compatibility of radial deformations (by equating the two), we have used Equation 6.54 to calculate directly the load S_n that is required for equilibrium.

e. From Equations 6.48 and 6.49, we can calculate the moment and hoop force in the lining using the value of S_n calculated in Equation 6.54, or we can get there directly by substitution:

$$N_d = \frac{-P_0 r_0}{2} \frac{(3-4v)}{(5-6v+4Q_2)} \cos 2\theta \tag{6.56}$$

$$M = \frac{P_0 r_0^2}{2} \frac{(3-4v)}{(5-6v+4Q_2)} \cos 2\theta \tag{6.57}$$

6.3.6 'No slip' – full shear interaction between lining and ground

In this case we need to equate both radial and tangential deformations. Repeating a similar procedure to Section 6.3.5, we find:

$$S_n = \frac{(1-Q_2)P_0/2}{1+Q_2\left(\dfrac{3-2v}{3-4v}\right)} \tag{6.58}$$

$$S_t = \frac{(1+2Q_2)P_0/2}{1+Q_2\left(\dfrac{3-2v}{3-4v}\right)} \tag{6.59}$$

Putting S_n and S_t into Equations 6.48 and 6.49, we can calculate thrust and moment due to distortional load when there is no slip allowed between the lining and the ground.

6.3.7 Direct compression of the lining due to uniform load

In Sections 6.3.5 and 6.3.6 we calculated the effects of distortional load with either 'full slip' or 'no slip' while ignoring the uniform load. In this section we will calculate the effects of uniform load only. Since the analysis is elastic, we can simply superpose the effects of distortional load and uniform load to get the final result. The uniform load will only affect the axial thrust N. It will not affect the moment M, because a uniform compression of a circular lining will not generate any bending moments.

First, calculate the uniform loading from the ground:

$$P_u = \frac{\sigma_v + \sigma_h}{2} \tag{6.60}$$

There is an interaction, and in effect the ground will share some of this load through arching around the opening, so the thrust due to uniform loading is not equal to the uniform loading multiplied by the radius of the tunnel, but less than this.

The relative stiffness of the ground and lining determines how much of the load is taken by the lining. This was called the 'compressibility factor', R_c, by Muir Wood (1975) and called Q_1 by Curtis (1974), and is given by:

$$R_c = \frac{\eta r_0 E_c \left(1 - v_l^2\right)}{tE(1+v)} \tag{6.61}$$

v_l is the Poisson's ratio of the lining
η is the ratio of the radius to the lining centroid to that of the extrados

Note that there is an error in Muir Wood (1975: equation 36), which had η in the denominator (bottom part) of the quotient in Equation 6.61.

For the case of a rectangular section 'smoothbore' lining or a sprayed concrete lining, the centroid is positioned at the mid-thickness of the lining, denoted r_m, so Equation 6.61 can be rewritten:

$$R_c - \frac{E_c \left(1 - v_l^2\right)}{E(1+v)} \frac{r_m}{t} \tag{6.62}$$

The axial hoop force N_u due to uniform loading can be calculated using the following equation:

$$N_u = \frac{r_0 P_u}{1 + R_c} \tag{6.63}$$

The equations are now all derived. The following worked example will show how they are used to calculate the hoop force and bending moment in a tunnel lining.

WORKED EXAMPLE 6.3

The following example is for a rectangular section lining in undrained stiff clay with the parameters listed in Table 6.2.

A calculation of a wished-in-place, perfectly stiff lining, where the pre-existing ground stresses are applied directly to the lining with no interaction, would result in the following hoop forces:

630 kN/m at the crown and invert

900 kN/m at axis

765 kN/m average

Calculate expressions for hoop force and bending moment as a function of angle 2θ for both the full slip and no slip cases.

Then program the Curtis-Muir Wood solution into a spreadsheet. Calculate hoop forces and bending moments around the ring for both the full slip and no slip cases. Plot these on a graph.

Table 6.2 Soil and lining parameters for Worked Example 6.3.

Parameter	Value	Units	Description
E_c	80	MPa	Young's modulus of the soil
v	0.5	-	Poisson's ratio of the soil
σ_v	300	kPa	Vertical in situ total stress
σ_h	210	kPa	Horizontal in situ total stress
r_0	3	m	Radius to extrados
E	20	GPa	Young's modulus of the lining
t	0.3	m	Thickness of the lining
I	0.00225	m⁴	Second moment of area of the lining
v_l	0.2	-	Poisson's ratio of the lining

Uniform compression:

First, calculate the direct compression of the lining due to uniform load (see Section 6.3.7):

$$P_u = \frac{\sigma_v + \sigma_h}{2} = \frac{300 + 210}{2} = 255 \text{ kPa}$$

$$R_c = \frac{E_c\left(1 - v_l^2\right)}{E(1+v)} \frac{r_m}{t} = \frac{80000 \times \left(1 - 0.2^2\right)}{20 \times 10^6 \times (1+0.5)} \frac{2.85}{0.3} = 0.02432$$

$$N_u = \frac{r_0 P_u}{1 + R_c} = \frac{3 \times 255}{1 + 0.02432} = 746.84 \text{ kN/m}$$

Note that compared to the average hoop force calculated by simple wished-in-place equilibrium, N_u is lower, by a factor of $1/(1 + R_c)$. This occurs because the lining compresses and the resulting radial deformation of the ground results in the ground arching around the tunnel. Increasing R_c will increase load sharing with the ground and reduce the uniform direct compression of the lining N_u. This occurs when, for instance, the Young's modulus of the soil E_c is increased, or when the Young's modulus of the lining E is reduced.

Distortion – 'full slip' case:

Next we need to calculate the effect of distortional load S_n in the 'full slip' case:

$$P_0 = \sigma_v - \sigma_h = 300 - 210 = 90 \text{ kPa}$$

$$Q_2 = \frac{E_c}{E} \frac{1}{(1+v)} \frac{r_0^3}{12I} = \frac{80 \times 10^3}{20 \times 10^6} \frac{1}{(1+0.5)} \frac{3^3}{12 \times 0.00225} = 2.667 \text{ m}^{-1}$$

The hoop force due to distortional load is given by Equation 6.56:

$$N_d = \frac{-P_0 r_0}{2} \frac{(3-4v)}{5-6v+4Q_2} \cos 2\theta = \frac{-90 \times 3}{2} \frac{(3-4 \times 0.5)}{5-6 \times 0.5 + 4 \times 2.667} \cos 2\theta$$

$$= -10.658 \cos 2\theta$$

Now we can calculate the hoop force N, which is the sum of the distortional load (Equation 6.56) and the uniform load (Equation 6.63):

$$N = N_u + N_d = 746.84 - 10.658 \cos 2\theta$$

The bending moment M is given by Equation 6.57:

$$M = \frac{P_0 r_0^2}{2} \frac{(3-4v)}{5-6v+4Q_2} \cos 2\theta = \frac{90 \times 3^2}{2} \frac{(3-4 \times 0.5)}{5-6 \times 0.5 + 4 \times 2.667} \cos 2\theta = 31.974 \cos 2\theta$$

Distortion – 'no slip' case:

Using Equation 6.58:

$$S_n = \frac{(1-Q_2)P_0/2}{1+Q_2\left(\dfrac{3-2v}{3-4v}\right)} = \frac{(1-2.667) \times 90/2}{1+2.667 \times \left(\dfrac{3-2 \times 0.5}{3-4 \times 0.5}\right)} = -11.842 \text{ kN/m}$$

Using Equation 6.59:

$$S_t = \frac{(1+2Q_2)P_0/2}{1+Q_2\left(\dfrac{3-2v}{3-4v}\right)} = \frac{(1+2 \times 2.667) \times 90/2}{1+Q_2\left(\dfrac{3-2 \times 0.5}{3-4 \times 0.5}\right)} = 45 \text{ kN/m}$$

Inserting values for S_n and S_t into Equations 6.48 and 6.49 gives:

$$N_d = \frac{-r_0}{3}(S_n + 2S_t)\cos2\theta = \frac{-3}{3}(-11.842 + 2 \times 45)\cos2\theta = -78.158\cos2\theta \text{ kN/m}$$

$$M = \frac{r_0^2}{6}(2S_n + S_t)\cos2\theta = \frac{3^2}{6}(2 \times (-11.842) + 45)\cos2\theta = 31.974\cos2\theta \text{ kN/m}$$

The total hoop force N for the 'no slip' case is given by:

$$N = N_u + N_d = 746.84 - 78.158\cos2\theta \text{ kN/m}$$

Results from spreadsheet calculation varying θ:

The results of the Curtis-Muir Wood calculation are shown in Figure 6.8. Note that the hoop force is consistently lower than the average hoop force in the simple wished-in-place calculation (given in the question), and varies much less around the lining. This is because the lining is flexible and hence will deform elliptically with the ground, reducing radial stress where the in situ stress was highest to begin with (i.e. in the vertical direction) and increasing radial stress where the in situ stress was lowest (in the horizontal direction).

We can also see that the 'no slip' calculation gives a larger variation of hoop force around the tunnel than the 'full slip' calculation.

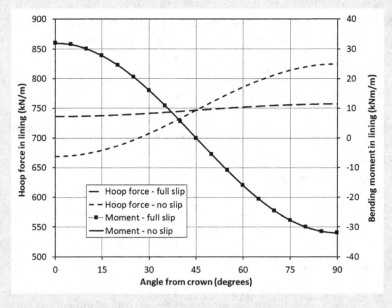

Figure 6.8 Worked Example 6.3 spreadsheet-generated graph showing hoop forces and bending moments in the lining.

Note that the bending moments are the same in both the 'full slip' and 'no slip' cases. This only happens when the Poisson's ratio of the ground is equal to 0.5.

The bending moment is low compared to the hoop force. This is nearly always the case and the section is likely to be mainly in compression in most cases, making design of tunnel linings in lightly reinforced or unreinforced concrete possible.

The variation of hoop force around the lining is higher when the no slip condition is in force. This is because the higher vertical stress is transferred to the lining via interface shear even where the angle of incidence is high, whereas in the full slip condition, the stress is redistributed more.

If the ground stiffness were increased, the effect would be to reduce both hoop forces and bending moments in the lining. Interestingly, if the ground stiffness were very low, and the no slip condition were in force, the solution would approach the wished-in-place perfectly stiff lining situation described in the question with 900 kN/m hoop force at axis and 630 kN/m hoop force at the crown and invert.

Figure 6.9 shows the effect of ground stiffness, which serves to redistribute loads and, with higher ground stiffness, to reduce them. A ground stiffness of 80 MPa would be typical of a stiff overconsolidated clay such as London Clay at strains of around 0.1%. A ground stiffness of 800 MPa would be typical for a soft rock such as chalk or weak sandstone.

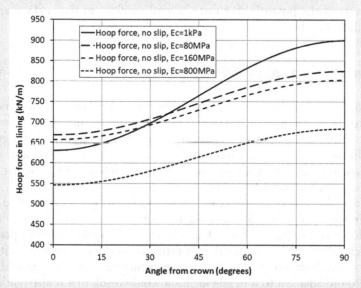

Figure 6.9 Worked Example 6.3 hoop forces at different values of ground stiffness E_c.

6.4 GLOBAL DESIGN OF SHAFTS

If we were to apply the Curtis-Muir Wood solution to a shaft, this would allow some soil-structure interaction to occur in our design, resulting in a more efficient lining design. However, on most soft ground tunnelling projects the horizontal principal stresses will be assumed equal, so no bending moments will be calculated. This would result in an unconservative design, because it is reasonable to expect some uneven loading to induce distortion and bending moments in the shaft lining. Therefore, it is recommended that the horizontal total stress in one of the principal directions is reduced by a factor of 0.8 to induce bending moments and ensure that the lining has sufficient structural capacity.

It will also become clear in later sections of this book that in many cases hoop forces are beneficial, because they increase the moment capacity of a reinforced concrete lining. With this in mind, careful attention needs to be paid to the shaft lining near to the surface where the in situ ground stresses are lowest, not just the deepest point where they are highest. It is also near to the surface that uneven loading is most likely to occur, perhaps unloading on one side due to adjacent excavation, or surcharges caused by structures, stockpiled materials, spoil heaps or construction plant such as a crane. Using the Curtis-Muir Wood solution with an unequal stress ratio of 0.8 makes the bending moments proportional to the ground stresses, and so close to the surface it may underestimate the true bending moments. Therefore, it is important to understand the potential for uneven loading close to the surface and to calculate its effect.

6.5 BEDDED BEAM MODELS

'Bedded beam' or 'beam-spring' models are analytical or numerical models consisting of beam elements representing the lining, and spring elements representing the interaction with the surrounding ground. Springs are usually in radial and tangential directions to represent radial and shear stresses, respectively. They are sometimes also referred to as 'hyperstatic reaction method' models or 'HRM' models.

Bedded beam models are generally thought of as less sophisticated than continuum analytical solutions such as the Curtis-Muir Wood solution (Duddeck & Erdmann, 1985). This is because the springs are independent of each other and hence they do not properly model soil-structure interaction, where it would be expected that deformation of the lining and hence the ground adjacent to it in one location would result in a redistribution of stress in the ground, affecting neighbouring areas (e.g. Tomlinson, 1995:

p. 199). This phenomenon was observed for a tunnel by Do et al. (2018), in a comparison of the HRM with a numerical continuum model in FLAC3D. For this reason, bedded beam models are not examined in much detail in this book.

A comparison of the boundary conditions and the resulting deformations and stresses for a bedded beam model (Schulze & Duddeck's method, 1964) and a continuum analytical solution (such as the Curtis-Muir Wood solution) is shown in Figure 6.10.

In Figure 6.10, the spring constant K_r is related to the ground stiffness and simulates load sharing between the ground and the tunnel lining. The force in the spring increases as the lining moves inwards, reducing the net load on the lining, and vice versa. It is often necessary to remove springs in locations where they would experience too much tension, and so springs

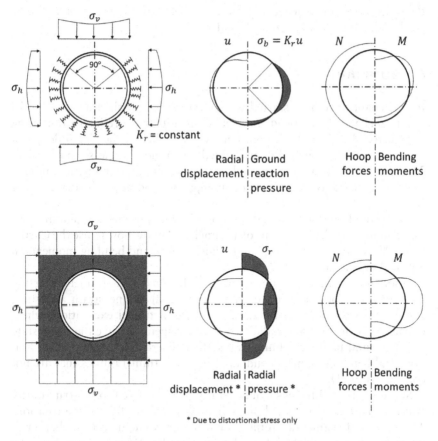

Figure 6.10 Comparison of bedded beam (top) vs continuum (bottom) analysis (redrawn from Duddeck & Erdmann, 1985).

may be omitted in the crown of the tunnel, as Schulze & Duddeck (1964) did, and as shown in Figure 6.10.

Bedded beam models may also be implemented in a structural frame analysis software package. One advantage of bedded beam models is that they do allow non-circular linings to be quickly and easily modelled and are therefore still used in preliminary design of sprayed concrete linings.

In 2D, these models can be used to simulate a solid monolithic ring with full rigidity or with a reduced moment of inertia to take account of joints in a segmental lining. Alternatively, joints can be modelled by introducing a rotational spring and tangential shear springs at joint locations. This will be discussed in Chapter 9.

In 3D bedded beam models, springs can also be added in the longitudinal direction to simulate longitudinal shear between the lining and the ground, and circumferential joints may also be added. It is usual to stagger the radial joints in adjacent rings in real tunnels, and so this should also be modelled. More details may be found in Chapter 9.

6.6 SUMMARY

In this chapter we have started with a very simple wished-in-place model of a tunnel and then looked at the effect of soil-structure interaction using the Curtis-Muir Wood solution. Finally, we looked at the global design of shafts and briefly touched upon bedded beam models and seismic design. This will be built upon in the following chapter where numerical modelling will be used to introduce more complexity and make the design more realistic.

You should now understand the boundary conditions and assumptions inherent in analytical solutions of a tunnel, and know how to use the Curtis-Muir Wood solution to estimate the hoop forces and bending moments in a tunnel or shaft lining.

Hoop forces and bending moments calculated in a tunnel using the Curtis-Muir Wood solution or other similar 2D plane strain analytical solutions do not account for the 3D nature of tunnel excavation, where extrusion of the face and radial ground movements change the stress state of the ground before the lining is installed. There are also important longitudinal bending moments and axial forces in tunnel linings that are not considered in 2D.

Shotcrete tunnel linings can generally be considered to be structurally continuous, but segmental linings have joints that allow some rotation. The design of segmental linings is covered in some detail in Chapter 9, where methods for calculating the rotational stiffness of joints will be explained.

6.7 PROBLEMS

Q6.1. A tunnel is to be constructed with the geometry and geology as shown in Figure 6.11. Assume the pore pressures below the water table are hydrostatic.

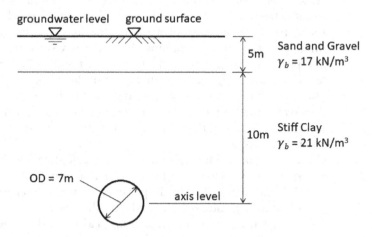

groundwater level ground surface

5m Sand and Gravel $\gamma_b = 17$ kN/m^3

10m Stiff Clay $\gamma_b = 21$ kN/m^3

OD = 7m

axis level

Figure 6.11 Q6.1 tunnel cross-section.

   i. What is the in situ vertical total stress at tunnel axis level?
  ii. Assuming $K_0 = 1.0$, what is the in situ horizontal total stress at tunnel axis level?
 iii. Using the basic method of applying the in situ total stress directly to the lining (assuming the lining is wished-in-place and perfectly stiff), what is the hoop thrust at axis in the tunnel lining, per metre length of tunnel? Ignore the self-weight of the lining.
  iv. Using the basic method of applying the in situ total stress directly to the lining, what is the hoop thrust at crown and invert per metre length of tunnel, using the horizontal in situ stress calculated at axis level?
   v. Using the Curtis-Muir Wood solution, calculate the hoop thrust in a smoothbore precast concrete segmental lining using the following parameters:

Radius to extrados of tunnel lining	$r_0 = 3.5$ m
Lining thickness	$t = 300$ mm
Young's modulus of the lining	$E = 30$ GPa
Poisson's ratio of the lining	$v_l = 0.2$
Undrained Young's modulus of the stiff clay	$E_c = 48$ MPa
Undrained Poisson's ratio of the stiff clay	$v = 0.5$
Number of segments	$n = 8$

Q6.2. Use the same tunnel geometry as in Question Q6.1, except with the groundwater level 3 m below the ground surface and $K_0 = 0.6$. Assume the bulk unit weight of the sand and gravel is unchanged. Calculate the following:

    i.   What is the in situ vertical effective stress at tunnel axis level?

    ii.   What is the in situ horizontal effective stress at tunnel axis level?

    iii. What is the in situ horizontal total stress at tunnel axis level?

    iv. Using the basic method of applying the in situ total stress directly to the lining (assuming the lining is wished-in-place and perfectly stiff), what is the hoop thrust at axis per metre length of tunnel? Ignore the self-weight of the lining.

    v.   Using the basic method of applying the in situ total stress directly to the lining, what is the hoop thrust at crown and invert per metre length of tunnel, using the in situ stress calculated at axis level?

    vi. Using the Curtis-Muir Wood solution, calculate the hoop thrust per metre length of lining as a function of the form $A + B\cos2\theta$, using the elastic parameters and tunnel lining described in Question 6.1(v), for the no slip case.

    vii. Using the Curtis-Muir Wood solution, calculate the bending moment per metre length of lining as a function of $\cos2\theta$, using the elastic parameters and tunnel lining described in Question 6.1(v), for the no slip case.

    viii. Sketch a graph plotting hoop thrust per metre length of lining on the y-axis and polar angle θ (clockwise from the crown of the tunnel) on the x-axis from 0 to 180° (crown to invert), marking the maximum and minimum values of hoop thrust. There should be no need to calculate intermediate points.

Q6.3. A 12 m internal diameter shaft 40 m deep is to be constructed using a shotcrete lining 0.35 m thick in stiff overconsolidated clay. Assume the Young's modulus of the shotcrete is 30 GPa and the Poisson's ratio is 0.2. The London Clay has a Young's modulus that increases with depth according to the relationship $E_c = 20 + 3.2z$ [GPa]. It has a unit weight of 20 kN/m³ and $K_0 = 0.8$. Pore pressures are hydrostatic with the water table 5 m below the surface.

    i.   Set up a spreadsheet to calculate hoop force and bending moment using the Curtis-Muir Wood solution.

    ii. Calculate the horizontal total stress at 10, 20, 30 and 40 m depth.

    iii. Assuming that the horizontal stress in the minor principal direction is 0.8 times the value in the major principal direction, calculate the hoop forces and bending moments at 10, 20, 30 and 40 m depth in the short-term undrained case. Remember that the soil stiffness is increasing with depth.

iv. To assess the sensitivity of the design to lower shotcrete stiffness at early age, calculate the hoop forces and bending moments at 40 m depth using a shotcrete Young's modulus of 5 and 15 GPa. Plot on a graph along with the 30 GPa values and describe the effect.

v. Using a shotcrete Young's modulus of 30 GPa, by how much would you need to reduce the initial in situ stresses in the ground to get the same hoop force you get with a shotcrete Young's modulus of 5 GPa? Is the difference related to the parameter R_c in Equation 6.62?

REFERENCES

Ahrens, H., Lindner, E. & Lux, K.-H. (1982). Zur Dimensionierung von Tunnelausbauten nach den 'Empfehlungen zur Berechnung von Tunneln im Lockergestein 1980' [Dimensioning of tunnel linings by applying the recommendations of 1980]. *Bautechnik* 59, No. 8, 260–273 and 303–311.

Curtis, D. J. (1974). Visco-elastic tunnel analysis. *Tunnels and Tunnelling*, November, 38–39.

Curtis, D. J. (1975). Discussion: The circular tunnel in elastic ground. *Géotechnique* 25, No. 1, 115–127.

Do, N. A., Dias, D. & Oreste, P. (2018). Numerical investigation of segmental tunnel linings-comparison between the hyperstatic reaction method and a 3D numerical model. *Geomech. Eng.* 14, No. 3, 293–299.

Duddeck, H. (1980). Empfehlungen zur Berechnung von Tunneln im Lockergestein, Deutsche Gesellschaft für Erd- und Grundbau. *Bautechnik* 57, No. 10, 349–356.

Duddeck, H. & Erdmann, J. (1985). On structural design models for tunnels in soft soil. *Undergr. Space* 9, 246–259.

Einstein, H. H. & Schwartz, C. W. (1979). Simplified analysis for tunnel supports. *J. Geotech. Engng Div. ASCE* 105, No.4, April, 499–518.

Hoek, E. & Brown, E. T. (1980). *Underground excavations in rock*. London: IMM.

Jaeger J. C. & Cook, N. G. W. (1976). *Fundamentals of rock mechanics*, 2nd Edition. London: Chapman & Hall.

Morgan, H. D. (1961). A contribution to the analysis of stress in a circular tunnel. *Géotechnique* 11, 37–46.

Muir Wood, A. M. (1975). The circular tunnel in elastic ground. *Géotechnique* 25, No. 1, 115–117.

Schulze, H. & Duddeck, H. (1964). Spannungen in Schildvorgetriebenen Tunneln. *Beton- und Stahlbeton* 59, 169–175.

Timoshenko, S. P. & Goodier, J. N. (1970). *Theory of elasticity*, 3rd Edition – International Student Edition. New York: McGraw-Hill Book Company.

Tomlinson, M. J. (1995). *Foundation design and construction*, 6th Edition. Harlow: Addison Wesley Longman.

Chapter 7

Global design using numerical modelling

Numerical models are similar to the analytical models we used in Chapter 6. Both are defined by geometry and boundary conditions and have material properties that govern the constitutive behaviour of the ground and the tunnel lining. They are just more complex and so special software is needed to build the model and to solve the equations.

Numerical models allow a problem to be discretised into a mesh of elements that are 'finite'. This is to differentiate them from the infinitesimal elements one might find in calculus. Across each element, field quantities, such as stress or displacement, are only allowed to vary in a simple manner, typically following a linear or quadratic equation. This allows a solution to be found by ensuring that the number of unknowns is matched by the number of equations. If the real variation of field quantities is more complex, which it often is, then this means that the outputs will be approximate. Accuracy may be improved by making the elements smaller or more complex. Smaller elements mean more elements, and hence more equations to solve. More complex elements (for instance, switching from linear to quadratic) will also mean more equations. Consequently, there is always a trade-off between accuracy and the amount of time the computer takes to run the model.

After working through this chapter, you will understand:

- the principles of numerical modelling in 2D and 3D
- the limitations of numerical modelling
- the effect of using different boundary conditions at the edges of the model and at the tunnel excavation boundary
- how groundwater is modelled in a soil in the short- and long-term

After working through this chapter, you will be able to:

- build a 2D or 3D numerical model of a tunnel
- set suitable boundary conditions and distances
- select suitable element types and mesh density
- debug your model

DOI: 10.1201/9780429470387-7

- validate your model
- interpret and present results

There are two main types of numerical analysis that are suitable for the routine design of tunnels in soft ground: the finite element method and the finite difference method. The finite element method, broadly speaking, creates a large number of equations from considerations of equilibrium, compatibility, constitutive behaviour and boundary conditions. It puts all the equations into a big matrix and inverts the matrix to find a solution to the unknowns, which are the stresses, strains and displacements. The finite difference method, on the other hand, finds equilibrium gradually by stepping towards the solution in increments of displacement or velocity. This is done on an element-by-element basis and there is no need to solve a large matrix of equations. The stepwise solution method means that the materials can follow almost any stress path to failure and makes it easy to incorporate complex constitutive behaviour.

Information on specific programs may be found online or in their manuals. The theoretical background to the finite element method may be found in textbooks such as the one by Cook et al. (2002), or more specifically, for geotechnical problems, see Potts & Zdravković (1999, 2001). For the finite difference method, the FLAC or FLAC3D theoretical background manuals (Itasca, 2006) are probably the best source of information.

Once the problem has been defined and a preliminary design obtained by using simple calculations and analytical solutions, the next stage is to begin introducing further layers of complexity to the model. This might involve:

- Geometry – non-circular tunnel lining, multiple soil layers, non-horizontal ground surface, other tunnels, basements, piles or surface structures, groundwater, asymmetric surface surcharge or excavation, tunnel excavation sequence
- Constitutive models – soil constitutive model (e.g. effective stress analysis, plasticity, nonlinearity, stress path direction, contraction or dilation under shear, consolidation and swelling), lining constitutive model (e.g. stress-strain nonlinearity, plasticity, shrinkage, creep, thermal effects, crack formation, joints, plastic hinges)
- Dynamic analysis for earthquake resistant design
- Long-term effects – degradation of temporary linings or sacrificial zones due to chemical attack or because they are not designed for long-term durability, long-term groundwater level changes or steady-state flow scenarios, consolidation or swelling of clays due to permanent changes in pore pressure

Most of these effects can only be included in a model by using numerical analysis, using either a finite element or a finite difference method, but would be difficult or impossible to achieve in an analytical model.

I am again going to emphasise how important it is to understand that the aim here is not to immediately produce a model that is as complex and as close to reality as the design budget allows. The aim is to understand the influence of each layer of complexity. This is a subtle but crucial difference. You must resist all temptation to launch straight into developing a single very complex model from the beginning, because this will make it difficult, if not impossible, to identify modelling errors and properly understand the relative importance of the different factors.

Another way to think of it is that the objective of any campaign of modelling is to better understand the problem by using a suite of numerical experiments. Each experiment is an idealised version of reality where some boundary conditions and parameters are simplified or controlled and a small number of parameters are varied, one at a time, within realistic limits, to isolate their effect and understand which parameters are important and which are not. The output will be a set of lining loads that represent the range of expected values in a range of possible scenarios.

This chapter will cover the following steps as part of a rational approach to numerical modelling:

7.1   Boundary conditions at the tunnel perimeter
7.2   Boundary conditions at the edges of the model
7.3   Boundary distances
7.4   Element types for the lining and the ground
7.5   Mesh density and refinement
7.6   Modelling groundwater
7.7   Validation and error checking
7.8   Constitutive models
7.9   Interpretation and presentation of results
7.10 3D numerical analysis

This will be followed by a summary and then some example problems.

7.1 BOUNDARY CONDITIONS AT THE TUNNEL PERIMETER

In a 2D plane strain model, it is not possible to model deformations in the ground ahead of the face due to excavation and prior to lining placement. There are a number of ways to simulate the excavation and installation of the tunnel lining in a 2D model:

7.1.1   Wished-in-place tunnel lining
7.1.2   The convergence-confinement method
7.1.3   Gap method
7.1.4   The grout pressure method

7.1.5   Surface contraction
7.1.6   Core softening

These will be described in the following sub-sections.

7.1.1 Wished-in-place tunnel lining

In a 2D plane strain numerical model, a rectangular block of soil is modelled, as shown in Figure 7.1.

In Step 1, the initial in situ stresses are imposed on the ground, and usually this is solved to equilibrium to set up the stress matrices needed for subsequent steps. In some programs this is called the 'Initial' or 'Geostatic' step. After this step has solved, the displacements and strains are usually reset to zero. In Step 2, the tunnel is excavated and lined and then solved to equilibrium.

In this case, the analysis is what is called 'wished-in-place'. This is because the ground only notices what has happened when the changes are calculated. So it is as though the whole length of the tunnel has been excavated and lined instantaneously, as if by magic. This is similar to what happens in the Curtis-Muir Wood solution, in that the initial in situ stresses

Step 1: Initial in situ stress conditions

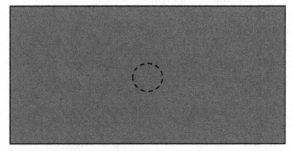

Step 2: Excavation and lining of the tunnel

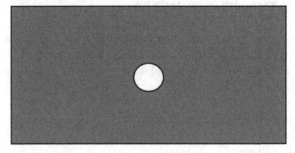

Figure 7.1 Geometry of a 2D plane strain numerical model of a tunnel.

in the ground are applied to a 'wished-in-place' tunnel, and only then does some deformation and redistribution of stress occur due to soil-structure interaction.

7.1.2 The convergence-confinement method

The main problem with analytical solutions, such as the Curtis-Muir Wood solution, is that they are 2D and thus cannot explicitly account for redistribution of stresses in the ground prior to lining installation. The same can be said of a wished-in-place 2D numerical analysis. One way to take account of this effect would be to use a 3D numerical analysis that models the advancing tunnel. Since 3D numerical models are complex and time-consuming, an approximate method was developed that allows some relaxation of the in situ stresses before inputting them to the 2D analytical solution or 2D numerical model. This is known as the 'convergence-confinement' method (Panet, 2001). It is based on the concept of a 'ground reaction curve' (Fenner, 1938; Pacher, 1964) and the definition of a 'confinement loss factor' by Panet & Guellec (1974).

A ground reaction curve is shown in Figure 7.2. Ground pressure curve 'I' shows what would happen if a tunnel were excavated in a very stiff and strong rock mass – the ground pressure can reduce from the in situ pressure p_0 to zero without any support installation. Ground pressure curve 'II' is a ground mass that undergoes plastic deformation – the ground pressure would reduce somewhat with increasing radial deformation, but at some point failure would occur. Theoretically, in an elastic-perfectly plastic model, the ground pressure should remain constant when plastic failure has occurred. But some softening may lead to an increase in ground pressure after failure begins, as indicated by the dashed line. If a tunnel lining were

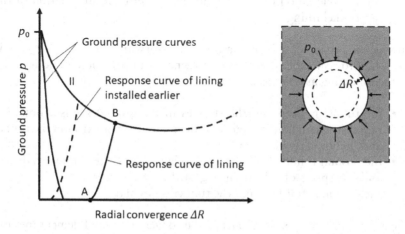

Figure 7.2 Conceptual model of the ground reaction curve.

installed at 'A', then it will be loaded and will undergo its own deformation until it meets the ground pressure curve 'II' at 'B'. A lining installed earlier, i.e. closer to the face, would have to support a higher load.

The convergence-confinement method assumes that a certain amount of convergence has occurred prior to installation of the tunnel lining, which has reduced the ground pressure that will act on the lining. Convergence depends on the distance of the section in question from the working face, on the unsupported length of the heading, and on the stiffness of the support. It may also be time-dependent as most soils and rocks exhibit time-dependent creep or consolidation behaviour.

It should be remembered that in using the convergence-confinement method we are ignoring stresses and strains in the longitudinal direction, which may be important. In fact, convergence occurs not only because there is an unsupported length of heading that allows radial movement but also because extrusion of the face into the tunnel allows radial movement to occur in the ground ahead of the face (Janin, 2017). Therefore, the convergence-confinement method is always inductive; we tell it how much we think the ground will converge based on empirical evidence, simple plasticity solutions or judgement. Convergence (or confinement loss factor) is always an input, not an output.

The convergence-confinement approach can be used with the Curtis-Muir Wood or any other 2D plane strain analytical method. It is also used with 2D plane strain numerical models. In principle, the stress σ applied to the tunnel lining is given by:

$$\sigma = (1-\lambda)\sigma_0 \tag{7.1}$$

σ_0 is the initial in situ stress value
λ is the 'confinement loss factor', between 0 and 1, which simulates the relaxation of the ground stresses prior to installation of the tunnel lining

This is, in principle, a simple and straightforward idea. However, it is important to remember that the convergence-confinement method is based on the following assumptions:

- the tunnel is deep and the stresses in the ground are isotropic, i.e. radial stress and radial deformation are uniform all around the tunnel and do not vary with depth
- the ground is homogeneous (e.g. no layering or variation with depth) and all aspects of behaviour (e.g. stiffness) are isotropic
- excavation is full-face and the tunnel is circular

The most common approach to using the convergence-confinement method in a 2D numerical model is shown in Figure 7.3. The initial stresses are set

Step 1: Initial in situ stress conditions

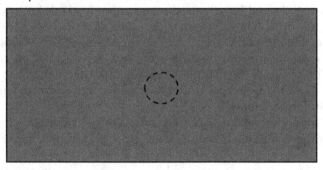

Step 2: Excavation of the tunnel and application of internal radial pressure at soil nodes

Step 3: Installation of lining and removal of internal radial pressure

Figure 7.3 The convergence-confinement method applied in a 2D numerical model.

up, then the elements of soil within the tunnel boundary are removed and an internal pressure is applied to the boundary equal to the initial stresses at each node multiplied by $(1 - \lambda)$. This is solved to equilibrium. In the final step, the internal pressure is removed, the tunnel lining is installed and the model is again solved to equilibrium.

The value of the confinement loss factor λ needs to be carefully calibrated. It will depend heavily on the excavation method: for example, an open-face tunnel will have a much higher value of λ than a closed-face tunnel boring machine (TBM). This λ value is usually either based on empirical data and judgement or on calibration to a 3D model of an advancing tunnel. The λ value is varied until the maximum surface settlement matches the value expected from case histories of similar tunnels or the maximum settlement in a 3D model.

Panet (1995) proposed equations based on an axisymmetric elastoplastic model for determining λ in a deep, homogeneous, isotropic medium:

$$\lambda = \alpha_0 + (1 - \alpha_0)\left[1 - \left(\frac{mr_0}{mr_0 + d} \right)^2 \right] \tag{7.2}$$

d is the distance between the point considered and the face
r_0 is the radius to the extrados
α_0 and m are dimensionless parameters where α_0 is the value of λ at $d = 0$. Panet (1995) recommends $\alpha_0 = 0.25$ and $m = 0.75$.

This equation is only appropriate for deep tunnels in homogeneous, isotropic soil. The convergence-confinement method is nevertheless often used for shallow tunnels, anisotropic ground, heterogeneous ground and for non-circular tunnels, often inappropriately according to Janin (2017). Gilleron et al. (2017) found that the value of λ found by calibration to a 3D model could be significantly different to the value of λ from Panet's equation for shallow tunnels in heterogeneous, anisotropic soil. Therefore, Panet's equation is of limited use for soft ground tunnelling, which nearly always occurs near to the surface in heterogeneous soil where K_0 is not equal to 1.0.

7.1.2.1 The β-factor method

An alternative to λ is to use what is known as the 'β-factor method' (e.g. Möller & Vermeer, 2008). β is defined differently to λ in that it represents the proportion of initial stresses that remain after relaxation, rather than the proportion lost during deconfinement (compare Equation 7.3 with Equation 7.1). It is also expressed in terms of effective stress rather than total stress, which makes sense because the pore pressure cannot arch around the tunnel.

$$\sigma' = \beta \sigma'_0 \tag{7.3}$$

σ' is the effective stress
β is the stress reduction factor
σ_0' is the initial in situ effective stress before excavation

7.1.2.2 The target volume loss method

Another variant of the convergence-confinement method of numerical analysis is where, based on field data from past projects, a target volume loss is set (e.g. Addenbrooke et al., 1997). The volume loss in the model can be measured at the tunnel boundary, or in the case of undrained analysis, it can be measured at the surface, which is a bit simpler. Using this method, the internal pressure is reduced and then the lining is installed in the same way as for the convergence-confinement method, but the aim is to achieve a target volume loss rather than a maximum settlement. This is some-times called the 'volume loss method'. When calibrating the convergence-confinement method to case histories, it makes more sense to use volume loss, which should be fairly constant for tunnels constructed using simi-lar methods in similar geology, rather than maximum settlement, which is strongly dependent on tunnel diameter.

An example of the volume loss method is shown in Figure 7.4. The aver-age stress in the tunnel lining, as a percentage of the full overburden pres-sure, is plotted on the y-axis and the volume loss is plotted on the x-axis. An undrained analysis was used with an elastoplastic soil model. The und-rained shear strength was set so that the stability number was equal to 2. In the 2D analysis, several models were run with different values of con-finement loss factor λ. In the 3D analysis, several models were run with dif-ferent values of unsupported length P. As P or λ increases, the volume loss

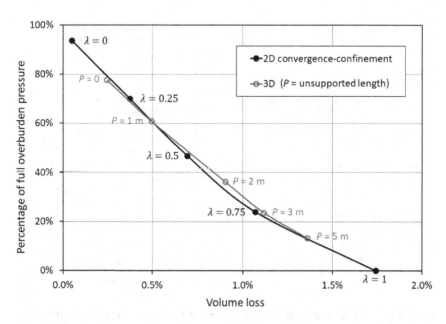

Figure 7.4 2D numerical analysis using the convergence-confinement method compared with 3D numerical analysis.

increases, and the trends are virtually collinear. Therefore, by using a 3D numerical model to simulate the construction sequence, a suitable value of confinement loss factor can be selected for use in 2D analysis. For example, if the unsupported length were 3 m, then an appropriate value of confinement loss factor λ would be 0.75.

There is evidence to suggest that using the convergence-confinement method to predict ground movements and lining stresses in a 2D analysis will be inaccurate. Gilleron et al. (2017) showed that stress paths and plastic zones in the ground around the tunnel were completely different in a 2D numerical model when compared to a 3D numerical model of an advancing tunnel, even when the confinement loss factor λ in the 2D model was calibrated to give the same maximum settlement as in the 3D model. Stress paths are shown in Figure 7.5. The difference in stress paths should be expected to result in very different patterns of ground movements, because the deformation behaviour of soils depends strongly on the values of deviatoric stress q and mean effective stress p'. This criticism probably applies to all methods of 2D tunnel analysis, not just the convergence-confinement method.

Mean effective stress p' is given by:

$$p' = \frac{\sigma_1' + \sigma_2' + \sigma_3'}{3} \tag{7.4}$$

σ_1', σ_2' and σ_3' are the principal effective stresses

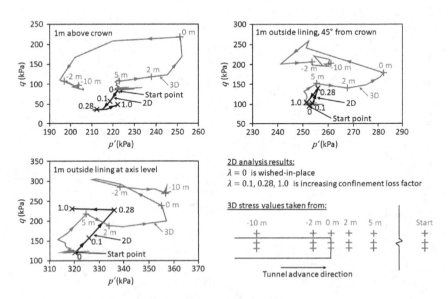

Figure 7.5 Stress paths in (p', q) space in the ground 1 m from the extrados of the tunnel lining versus confinement loss factor in 2D and longitudinal distance from the face in 3D (redrawn from Gilleron et al., 2017).

Deviatoric stress q is given by:

$$q = \sqrt{\frac{\left(\sigma_1 - \sigma_2\right)^2 + \left(\sigma_2 - \sigma_3\right)^2 + \left(\sigma_3 - \sigma_1\right)^2}{2}} \qquad (7.5)$$

σ_1, σ_2 and σ_3 are the principal total stresses

Jones (2012) found that the stress reduction required to give a specified value of volume loss depended on the constitutive model and parameters used for the ground. These different values of β and different constitutive models and parameters would result in a large variation of predicted lining forces at the same value of volume loss. Möller & Vermeer (2008) similarly found that a different value of β was needed to match surface settlements (typically 0.3–0.4) and lining forces (typically 0.5–0.7) in the same 2D model. Janin et al. (2015) also warned against calibrating the convergence-confinement method to a single type of measurement, and identified the application of a single value of confinement loss factor to all points on the tunnel perimeter as the cause of significant differences between 2D analysis and field measurements, whereas a 3D analysis gave much better agreement. Therefore, the sensitivity of the convergence-confinement method to its inputs should be carefully assessed, particularly if it is to be used for tunnel lining design. Just because a 2D model can match surface settlements does not mean that the lining forces it predicts will also be reliable, and vice versa.

The convergence-confinement method generally assumes that the ratio of horizontal to vertical stress that will be applied to the tunnel lining is the same as the initial in situ stress ratio. This will lead to strange results when $K_0 > 1.0$. After installation of the tunnel lining, it will converge more horizontally than vertically because the horizontal stress will be larger than the vertical. We know from large numbers of tunnel circularity measurements in London Clay by Wright (2013), that even when $K_0 > 1.0$, the tunnels almost always squat (the vertical convergence is larger than the horizontal convergence). Therefore, it is unrealistic, in soft ground at least, to apply initial stresses with $K_0 > 1.0$ in a 2D model, since this will predict a mode of deformation completely opposed to reality. For London Clay, Wright (2013) suggests using $K_0 = 0.7$ in simple models. In a 3D numerical model, it may be possible to use $K_0 > 1.0$, because the redistribution of stresses ahead of the face can occur in the model as they do in reality. It may also be possible in a 2D numerical model to effectively use $K_0 > 1.0$ if the full stress history is modelled and if long-term drainage of the ground around the tunnel is modelled (Avgerinos et al., 2018), as squatting may be partly caused by the tunnel acting as a drain with a higher horizontal than vertical permeability in the ground around it (Wright, 2013).

In all cases, it is wise to perform sensitivity analyses. For instance, using $K_0 = 1.0$ will give higher values of axial force in the tunnel lining, but will only generate bending moments due to increase of stress with depth and non-circularity of the lining. Therefore, a lower bound value of K_0 should also be used to find a maximum bending moment. Only in very special circumstances, for example, when tectonic stresses are important and will re-establish high horizontal stresses during the design life, will it make sense to use $K_0 > 1.0$ in a 2D analysis.

7.1.3 Gap method

The gap parameter method was introduced by Rowe et al. (1983), though it is perhaps better described in Rowe & Lee (1992). The 'gap parameter' may be thought of as the distance between the crown of the tunnel and the original position of the ground at that location prior to tunnelling, and seeks to represent radial deformation and unloading of the ground ahead of the face and before lining installation in a 2D numerical model. The invert of the tunnel is assumed to rest on the ground beneath, with no gap.

To use this method in a 2D numerical model, the perimeter of the excavation must be gradually unloaded, in a similar manner to the convergence-confinement method, by applying nodal forces equal to the initial in situ stresses and gradually decreasing them until the gap is closed. At the point that the gap is closed, then the ground and the lining interact in the normal way. Applying contact rules in this way is actually not as straightforward as it sounds and most geotechnical numerical modelling programs do not have a built-in ability to do this.

For a TBM tunnel, the lower limit of the gap parameter will be the difference between the excavated diameter and the outside diameter of the lining plus grouted annulus. An allowance should also be made for deformation ahead of the face.

For a tunnel lined with shotcrete, the gap parameter will represent deformation of the ground ahead of lining installation. In a very simple model with elastic shotcrete an allowance may also be made for creep and shrinkage of the shotcrete.

Rowe et al. (1983) also looked at the effect of applying a grout pressure, though they did this after installation of the lining. A uniform pressure was applied and it was found to reduce surface settlements and to slightly narrow the settlement trough.

7.1.4 The grout pressure method

Möller & Vermeer (2008) introduced the 'grout pressure method' for the simplified analysis of TBM-driven tunnels. In a TBM tunnel in soft ground, the ground will close around the shield and tailskin, but the boundary condition at the point where the lining is installed is determined by the grout pressure. Earlier in the same paper Möller & Vermeer had demonstrated how the convergence-confinement method did not give a good estimate of surface settlements and by introducing this method they were attempting to overcome most of the limitations of 2D numerical analysis, at least for TBM-driven tunnels, without resorting to a full 3D analysis.

Figure 7.6 shows the in situ radial stress distribution applied to a tunnel when $K_0 = 0.47$ (solid circles), based on a diagram in Möller & Vermeer (2008). The radial stress is high at the crown, then it decreases around the axis level as horizontal stresses dominate, then it increases to a maximum at the invert. When the β-factor method is applied, the distribution is still strongly influenced by K_0 (open circles). When compared to surface settlements from the Second Heinenoord Tunnel, this resulted in an underprediction of maximum settlement and a settlement trough that was too wide. However, when the radial stresses around the perimeter of the excavation were relaxed to the value of the grout pressure (solid squares), the agreement with the field measurements was exceptionally good. The agreement was also reasonably good for horizontal ground movements to the sides of the tunnel, as measured by inclinometers

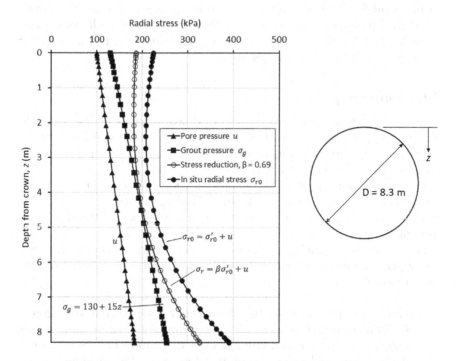

Figure 7.6 Radial stresses using the convergence-confinement (stress reduction to $\beta = 0.69$) and grout pressure method (redrawn based on Möller & Vermeer, 2008).

installed in vertical boreholes, and for lining forces and moments back-calculated from strain gauges installed in the lining.

The grout pressure method seems to be a promising simplified 2D approach to the design of TBM tunnels, perhaps to be used at an early design stage when 3D modelling is considered too expensive. Möller & Vermeer (2008) actually used a combination of the gap parameter method and grout pressure method, where the overcut and conicity of the shield were simulated by a gap. Once the gap was closed, the grout pressure was applied.

7.1.5 Surface contraction

Another modelling method is to allow a 'surface contraction' (Brinkgreve et al., 2018). The tunnel perimeter is allowed to reduce by a specified amount, usually to meet a target volume loss. In this method, the stresses applied to the lining in the final stage are not necessarily predetermined.

7.1.6 Core softening

In this modelling method the soil within the tunnel perimeter (the 'core') is softened by reducing its stiffness and/or strength, which allows stresses to redistribute in the ground around the excavation. In a further step the lining is installed and the core removed.

7.1.7 Summary

In summary, when analysing a tunnel in 2D, assumptions need to be made about how to simulate 3D effects. These might include excavation, lining installation, face pressure and grouting pressure and toolbox items such as grouted pipe arches or face dowels.

For tunnels in overconsolidated soils with $K_0 > 1.0$ it will rarely make sense to apply this stress state to the tunnel lining. In these cases sensitivity analyses should be performed with an upper limit of $K_0 = 1.0$ and a lower limit of $K_0 = 0.7$.

For open-face tunnels, the best approach seems to be to use the convergence-confinement method, but it can be misused. There are two ways to avoid this:

1. Find one or more case studies of real tunnels constructed with similar methods in similar geology to the one you are designing and try to match deformations and lining stresses in the model to the measured deformations and lining stresses in the real tunnel by varying the confinement loss factor λ. The paucity of lining stress measurements will make this difficult to achieve.

2. In the absence of empirical data from real tunnels, use a 3D numerical analysis of an advancing tunnel to determine the confinement loss factor to be used in 2D analysis.

Studies of open-face tunnels have also shown that the convergence-confinement method can either be calibrated to give the correct maximum settlement, the correct volume loss or the correct lining stresses, but rarely all at the same time (Möller & Vermeer, 2008; Jones, 2012). Therefore, the results from 2D design models should always be treated with caution and with an awareness of their limitations.

For TBM-driven tunnels, the grout pressure method proposed by Möller & Vermeer (2008) appears to provide a reasonable means of simplifying the TBM excavation process in a 2D model.

7.2 BOUNDARY CONDITIONS AT THE EDGES OF THE MODEL

In a model the top surface is always left free of any constraints. The lateral (vertical) boundaries are usually fixed in the horizontal direction but allow movement in the vertical direction. The bottom (horizontal) boundary is usually fixed in both the horizontal and vertical directions. These conditions are shown in Figure 7.7.

There is actually no particular reason why the model's bottom boundary should be fixed in the horizontal direction (after all the lateral boundaries are not fixed in the vertical direction). In reality the soil on the other side of these arbitrary boundaries will allow some movement but this will be resisted by shear stresses, so it will be somewhere between fixed and free. Möller (2006) argues that due to the tendency for soils to increase in

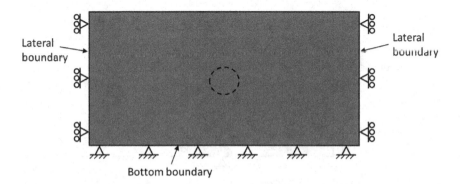

Figure 7.7 Boundary conditions at the edges of a 2D numerical model showing fixity in the horizontal direction only for the lateral boundaries and fixity in both horizontal and vertical directions for the bottom boundary.

stiffness with depth, modelling the bottom boundary as fully fixed, i.e. not allowing any lateral movement and generating shear stresses, is closer to reality.

Another possibility is to have a stress applied at the boundary rather than fixing displacements. These stresses would be set to be equal to the initial in situ stress. So a stress boundary condition keeps stress constant but allows displacement, and a displacement boundary condition keeps displacements at zero but allows stresses to change. Neither of these conditions is more realistic than the other, as the reality is somewhere in between. But if the boundary is far enough away from the tunnel, then stresses should remain approximately constant and displacements should remain at approximately zero, regardless of whether a stress boundary condition or a displacement boundary condition is selected.

Symmetry may also be exploited to reduce the size of the model and hence reduce the computation time. Often, there is a vertical plane of symmetry down the centreline of the tunnel, which allows the size of the model to be halved, as shown in Figure 7.8. This also will make handling the model and visualising results easier. Whether the left- or right-hand side is modelled is up to you. For convenience you may wish to model the side that gives you positive coordinates in your coordinate system.

For deep and/or small diameter tunnels, the change in in situ stress with depth may be small relative to the magnitude of the stress. In this case, a quarter of the tunnel could be modelled, and the model may not extend to the surface, but have a stress applied to it at a boundary, as shown in Figure 7.9. This is not a common situation for tunnels in soft ground, which tend to be relatively shallow, but we will do a variant of this in the validation section below to mimic the boundary conditions of the Curtis-Muir Wood solution.

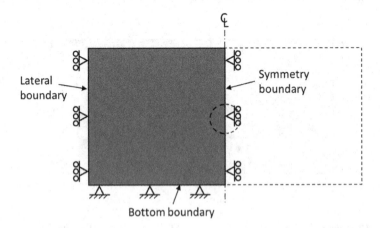

Figure 7.8 2D numerical model boundary conditions with vertical plane of symmetry on tunnel centreline.

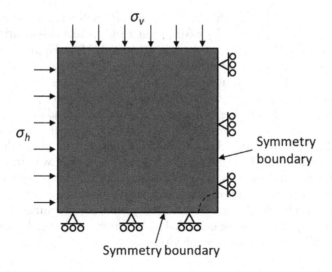

Figure 7.9 2D numerical model boundary conditions exploiting two planes of symmetry.

Where lining elements meet a symmetry boundary, they should have displacement fixity normal to the boundary, but be allowed to move parallel to the boundary. They should also have rotational fixity about the tunnel longitudinal axis direction to ensure that bending moments are mirrored. This is shown in Figure 7.10. The condition that rotation is prevented about

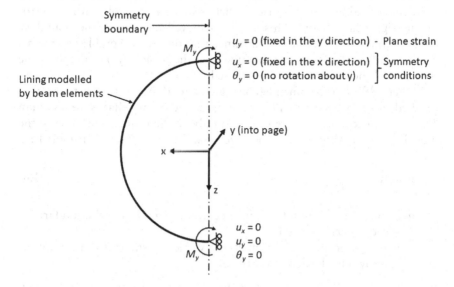

Figure 7.10 Boundary conditions for a tunnel lining on a symmetry boundary in a 2D numerical model. Displacements in the x, y, z directions are denoted by u_x, u_y, u_z and rotations about the x, y and z axes are denoted by θ_x, θ_y and θ_z.

the y-axis will ensure that the lining at this point will remain perpendicular to the symmetry axis, thereby ensuring that the correct deformation mode is produced and the correct moment is generated (M_y in Figure 7.10).

Note that when beam or shell elements are used, moments and forces in each beam or shell element will be defined by local axes that will almost always have a different orientation to the global axes. In some programs this is made clearer by labelling the local axes '1', '2' and '3' instead of 'x', 'y' and 'z'.

Also, in some software programs, 'Mx' or 'My' is not the moment *about* the local x-axis or y-axis, but the moment *along* it (i.e. it would require reinforcement parallel to the x-axis or y-axis). The only way to be sure is to read the manual and it is always good practice when starting to use new software to run some simple tests with single elements or small numbers of elements. This is all part of building your understanding of the software and the problem being modelled by beginning with simple models and incrementally adding more complexity.

7.3 BOUNDARY DISTANCES

In reality, of course there are no boundaries in the ground, and the lateral extent of the soil can be considered, to all intents and purposes, infinite. There are ways to use infinite boundary elements on the outside of a model (e.g. Reichl et al., 2003), but this is not common practice and these elements are not available in most commercially available numerical analysis packages. Therefore, in a standard continuum model, the boundaries must be finite and the aim is to make the model as small as possible without introducing significant errors. This can be achieved by varying the boundary distance and using the shortest distance that has an acceptable error. This will depend on the focus of the modelling, the geometry and depth of the tunnel, and the constitutive models used.

Möller (2006) recommended using a lateral boundary distance that resulted in surface settlements at the model boundary that were less than 1% of the maximum surface settlement above the tunnel centreline in the model. Based on this criterion he developed Equation 7.6 for his models:

$$w = 2D\left(1 + \frac{H}{D}\right) \tag{7.6}$$

w is the distance from the tunnel centreline to the lateral boundary
D is the tunnel diameter
H is the cover above the tunnel (the vertical distance from the tunnel crown to the ground surface)

It is important to remember that the optimum boundary distance will depend on the problem geometry and the constitutive model of the soil, and so it should be verified for each soil model used and each geometric configuration (i.e. for different tunnel depths or tunnel sizes). For example,

Franzius et al. (2005) found that the lateral boundary distance needed to be larger for higher values of K_0.

If there is a change to a much stiffer geological stratum below the tunnel, this is often taken as the bottom boundary, because it will be relatively rigid. For situations where soft ground extends well below the tunnel, Möller (2006) recommended a distance (from tunnel invert to the bottom boundary) that resulted in less than $2.5°$ rotation of principal stresses at the boundary after tunnel construction, which for his models was between 1.3 and 2.2 times the diameter D.

Möller's boundary distance relationships can be used as a first approximation. It is recommended that this first approximation is then tested by varying the boundary distance to higher and lower values to find the distance that gives an acceptable error in the required outputs. If the only concern is lining stresses, the boundaries may not need to be as far away as for ground movements or surface settlements (Jones, 2007).

Sometimes, it is found that increasing the boundary distance beyond a certain point will actually begin to increase the error. For example, the surface settlement may be zero at the boundary when the lateral boundary distance $w = 30$ m, but may be non-zero at the boundary when $w = 50$ m. This often happens if the number of elements is remaining constant, such that increasing the boundary distance is increasing the size of the elements, thereby reducing accuracy. Therefore, while varying the boundary distance it is important to try to keep the mesh density constant. This problem also occurs due to a sort of 'mirror effect' where displacements appear to be reflected in the bottom boundary and then have an effect on the lateral boundary. It can then be quite frustrating to find a compatible pair of lateral and bottom boundary distances as changes to the bottom boundary distance will affect the acceptability of the lateral boundary distance and vice versa. This effect usually is only a problem when linear elastic soil models are used, and when soil stiffness is not increasing with depth. If the bottom boundary is fixed in the horizontal direction, the principal stresses may be rotating at the boundary (Möller, 2006) – if this appears to be the case, consider increasing the bottom boundary distance and/or allowing displacements in the horizontal direction at the bottom boundary.

Some practitioners fix the lateral boundary in all directions, and this of course results in zero settlement at the boundary. However, if the boundary is too close to the tunnel it will be reducing settlements throughout the model and there is no way of knowing this is happening.

7.4 ELEMENT TYPES FOR THE LINING AND THE GROUND

Modern finite element software packages, particularly those that specialise in geotechnics, usually have elements suitable for modelling soil. Often there is a choice between more accurate higher order elements and less accurate lower order elements. 'Higher order' means that the elements have more nodes and more degrees of freedom, and therefore can adopt higher order deformation modes (e.g. the side of an element can follow a quadratic curve rather than a straight line). This may be important where there are high stress gradients, where a larger number of lower order elements would be needed to get the same accuracy as a smaller number of higher order elements. There is a trade-off here between accuracy and simulation time, because higher order elements mean more equations in the matrix that need to be solved, but it may be more efficient than having a larger number of lower order elements.

The way forward is to test the different elements you plan to use in a validation exercise similar to the problem you wish to solve. An excellent example of this kind of validation may be found in Pound (2006), where the number of solid elements needed to model a sprayed concrete lining is optimised by modelling a beam in flexure and comparing the stresses and displacements to simple hand-calculations.

When modelling the tunnel lining, either 1D beam elements or 2D solid elements can be used. Beam elements are usually better at modelling bending, but the constitutive models may be limited to simple elastic or elastoplastic behaviour. There is also the issue of where to place beam elements. To simulate the soil-structure interaction, they are usually placed at the perimeter of the excavation; the extrados. But this will result in an overestimate of bending moments, as effectively the radius to the centroid of the beam has been increased by half the lining thickness, as though you had increased the span of a bridge while keeping the overall load the same. For an elastic tunnel lining the overprediction can be estimated using the following equation:

$$\frac{M^*}{M} = \frac{r_0^2}{r_0\left(r_0 - t\right)} \tag{7.7}$$

M^* is the bending moment calculated by the numerical model
M is the 'true' bending moment
r_0 is the radius to the extrados
t is the thickness of the lining

7.5 MESH DENSITY AND REFINEMENT

Using a larger number of smaller size elements will generally be more accurate. However, the model will use more computer memory and will take longer to solve. As with the choice of element types and boundary distances, this can be optimised by varying the mesh density to obtain an acceptable error for the outputs we are interested in.

A higher mesh density will improve accuracy in areas with high stress gradients, which is usually in a zone around the tunnel. Therefore, it is not necessary to use the same mesh density everywhere in the model, and where stress gradients are small the mesh density can be reduced substantially. This is known as 'mesh refinement'.

For obtaining accurate lining stresses, Möller (2006) found that higher mesh density was required in 3D models than in 2D models. This was because there was significant arching in the front-to-back direction close to the face of the tunnel, resulting in high stress gradients in the longitudinal direction. This is a real phenomenon but cannot occur in a 2D model.

7.6 MODELLING GROUNDWATER

Soil is made up of solid soil particles, water and air. In tunnel modelling we normally ignore the air and model the soil as either 'fully saturated' or 'dry', though there may be situations where partial saturation is important.

In permeable soils, we can assume that below the groundwater table the soil is fully saturated. Therefore, unless there is groundwater flow, the pore pressures will increase hydrostatically from the groundwater table down. Above the water table the soil will probably be moist ('partially saturated'), and this can be taken into account in the unit weight of the soil, but essentially we will assume it is dry in terms of its pore pressure, which we will assume to be zero (e.g. Itasca, 2006).

In permeable soils, any excess pore pressures generated due to stress changes will dissipate quickly relative to the speed of construction. Therefore, with no excess pore pressures, behaviour is governed by the interaction between soil grains, i.e. the effective stress. We can model this by using an analysis option that specifies, rather than calculates, the pore pressures. If we want to include pore water in the model explicitly, we can set the fluid bulk modulus of the pore water K_f to zero. This will ensure that any changes in pore volume do not induce changes in pore pressure.

Due to capillary suction in low-permeability soils (clays), we generally assume that the clay is saturated both above and below the water table and will therefore have the same unit weight. However, as with permeable soils, unless there is a groundwater flow, the pore pressures will increase hydrostatically from the groundwater table down, and will be assumed equal to zero above the water table.

In low-permeability soils, excess pore pressures generated due to stress changes will not dissipate during construction, but may take months, years or decades. Therefore, there is a short-term 'undrained' behaviour, followed in the long term by a 'drained' behaviour. It is important that the software prevents the generation of negative pore pressures below −100 kPa, because in reality at this pressure a vacuum will form. Even though it applies to all geotechnical problems, in some software programs this cap on negative pore pressure needs to be specified by the user. This is because checking for this criterion takes computational effort.

Note that it is standard practice to express pore pressures relative to atmospheric pressure, which at sea level on Earth is approximately 100 kPa. Therefore, when pore pressure is expressed as –100 kPa, it is really 0 kPa and a vacuum has formed, an empty space with nothing in it, and the pressure cannot go any lower. This is called the 'cavitation pressure'. Incidentally, this is why no water pumps can ever suck more than a 10 m head, but they can push a lot higher if the pump is near the bottom of the pipe.

7.6.1 Undrained behaviour

During the timescale of construction, most soils with sufficient clay content can be considered to be undrained. Whether a soil behaves in a drained or undrained manner depends on the permeability of the soil, as well as the scale and speed of construction. A rule of thumb is that a soil with a permeability lower than 10^{-7} to 10^{-8} m/s will behave in an undrained manner, and soils with a higher permeability will behave in a drained manner (Anagnostou & Kovári, 1996).

If groundwater cannot flow significantly during the timescale of construction, then pore pressure changes are temporarily locked in and we call this state 'undrained'. It is often useful to model the construction phase as undrained and the long term as drained, where all pore pressures have reached either a hydrostatic equilibrium or a steady-state.

During tunnel excavation, the ground around a tunnel experiences a stress decrease in the radial direction, and a stress increase in the circumferential direction. This results in both temporary and permanent changes to the total stresses. If excess pore pressures do not have time to dissipate, the change in total stresses will mostly be experienced by the pore water, because although water can support very little shear, it is relatively incompressible compared to the soil particles and will therefore resist changes in volume. Undrained behaviour is often referred to as 'constant volume', because although an undrained soil may undergo shear deformations, the volume change depends on the pore water, which may be assumed to be incompressible and hence the overall volume of the soil will remain constant.

Elastic behaviour can be fully described by bulk modulus K and shear modulus G. Shear modulus is the stiffness in shear and is equal to the shear stress divided by the shear strain. Bulk modulus is the volumetric stiffness and is equal to the pressure divided by the volumetric strain. Alternatively, Young's modulus E and Poisson's ratio v may be used instead of K and G, but they do not separate volumetric and shear behaviour in the same convenient way. Bulk modulus and shear modulus are related to Young's modulus and Poisson's ratio by the following expressions:

$$K = \frac{E}{3(1-2v)} \qquad (7.8)$$

$$G = \frac{E}{2(1+v)} \tag{7.9}$$

K is the bulk modulus
G is the shear modulus
E is the Young' modulus
v is the Poisson's ratio

Shear modulus G is the same for both drained and undrained behaviour, but bulk modulus K can have different values for drained and undrained behaviour. This is because pore pressures are only affected by volumetric changes, which is why separating shear and volumetric behaviour using G and K is so elegant. Because drained and undrained shear modulus are equal, the following equation results:

$$G = \frac{E_u}{2(1+v_u)} = \frac{E'}{2(1+v')} \tag{7.10}$$

E' and E_u are the drained and undrained values of Young's modulus, respectively
v' and v_u are the drained and undrained values of Poisson's ratio, respectively

If we were to assume that undrained Poisson's ratio is equal to 0.5, it follows from Equation 7.10 that:

$$E_u = \frac{3E'}{2(1+v')} \tag{7.11}$$

Numerical modelling programs cannot use a Poisson's ratio of 0.5, as it results in numerical difficulties when solving the matrix equations (it tries to divide by zero), but usually a value slightly lower than 0.5, such as 0.495 or 0.499, can be used instead without introducing significant error.

Now we are going to look in more detail at how to model undrained behaviour. The notation we will use is as follows:

K' is the effective or drained bulk modulus, i.e. the bulk modulus of the soil particle assembly or 'soil skeleton'
K_f is the bulk modulus of the pore fluid, often taken as the value for pure water at room temperature, which is 2.0 GPa (note that this value can be significantly reduced by the presence of dissolved air or air bubbles in the pore water)
K_e is an equivalent bulk modulus for the soil skeleton and pore fluid combined (more explanation to follow)

There are two ways of modelling undrained behaviour:

Option 1	Pore water can be ignored and the equivalent bulk modulus K_e set to a high value relative to the drained bulk modulus K', such that the Poisson's ratio is close to 0.5. With $K_e = 100 K'$, the Poisson's ratio will be greater than 0.495, as recommended by Potts & Zdravković (1999). This is known as a 'total stress analysis' because we are not treating the effective stresses and pore pressures separately, but lumping them together.
Option 2	Pore water can be modelled by setting a fluid bulk modulus K_f to a realistic value (e.g. 2 GPa, which is the value for pure water at room temperature) and no groundwater flow is allowed. This allows the explicit use of effective stress parameters, but increases the runtime of the model.

Option 1 aims to make the Poisson's ratio as close to 0.5 as possible without introducing numerical instability. However, in reality, undrained behaviour is never strictly constant volume and undrained volume change does depend on K' for most soils.

Bishop & Hight (1977) found that the undrained Poisson's ratio approaches 0.5, *only* in the case where K' is very low (as, for example, in a normally consolidated soil under low effective stresses) and the soil is fully saturated. Bishop & Hight (1977) reported the undrained Poisson's ratio of several natural soils to be between 0.470 and 0.487. Therefore, it is simpler, and more consistent, to use Option 2 and to estimate the values of fluid bulk modulus K_f and drained bulk modulus of the soil K' and to use those in a model incorporating pore pressures, rather than trying to input values that have no physical meaning.

Option 1 also runs into difficulties because it does not calculate pore pressures. When this option is selected in a finite element program, it may be possible to plot pore pressures and effective stresses as outputs, but these pore pressures are based on the assumption that the final long-term equilibrium state after all excess pore pressures are dissipated will be the same as the initial pore pressure distribution. If the tunnel acts as a drain on the ground around it, or if the long-term groundwater situation consists of a steady-state groundwater flow, then the final pore pressure distribution is not the same as the initial and the results will be wrong. 'Excess pore pressure' should be defined as excess to the final state, not the initial state; it is what will be dissipated over time by groundwater flow, causing consolidation or swelling. A detailed explanation of this may be found in Tan et al. (2005).

Option 2 allows effective stiffness parameters to be used, for which the results of laboratory tests are often better suited, and it also allows long-term effects of groundwater flow and dissipation of excess pore pressures to be modelled. Volume changes induce stresses in both the soil skeleton and the pore water, but because the bulk modulus of the pore water is much higher than the bulk modulus of the soil skeleton, most of the stress change will be represented by a change in pore pressure. Most of the more sophisticated soil models use effective stress parameters.

The soil's failure criterion can also be expressed in terms of either drained (c' and ϕ') or undrained (c_u) parameters. This will be covered in Section 7.8.

7.6.2 Long-term effects

When a geotechnical problem is assumed to be undrained during construction, this means that in the long term we can expect the excess pore pressures to dissipate, causing the soil to experience changes in effective stresses. This will also result in strains and hence displacements. Therefore, as designers, we need to calculate these long-term effects to ensure that our design remains safe and serviceable for the full design life of the structure.

For soils that are considered undrained in the short term, we need to also calculate what happens in the long term by allowing pore water to flow in the model. This will dissipate the excess pore pressures and allow calculation of long-term stresses, strains and displacements. There are two ways this can be done: using either an 'uncoupled' or a 'coupled' fluid-mechanical analysis.

There are also two main types of problem: where the perturbation is mechanical, and where the perturbation is driven by a pore pressure change (Itasca, 2006). Tunnel construction is the former type, whereas draw-down from a well would be an example of the latter.

Most finite element modelling programs only allow 'uncoupled' fluid-mechanical analysis. 'Uncoupled' means that modelling pore water flow and modelling mechanical behaviour are two separate calculation stages performed one after the other. 'Coupled' analysis is where pore water flow and mechanical behaviour are calculated together. Therefore, as pore water flows, pore pressures change and this causes the effective stresses, strains and displacements to change, which in turn have an effect on pore pressures.

In some cases, particularly where the problem is driven by a pore pressure change, an uncoupled analysis may be considered sufficient. An uncoupled analysis will have the following stages:

1. An undrained calculation stage using effective stress parameters (Option 2 above). No fluid flow is allowed. This gives the short-term stresses and displacements for design.
2. A fluid flow calculation stage, with K_f set at a realistic value. No mechanical calculation is performed, so the effective stresses do not change. Flow continues either to simulate a period of time, or for a

long-term analysis, flow continues until either equilibrium or a steady state is reached.

3. A further mechanical calculation stage to calculate the effect of the change of pore pressures on the soil skeleton. No fluid flow is computed and K_f is set to zero. This provides the long-term stresses and displacements for design.

Uncoupled analysis is only reasonably accurate for elastic materials because plastic materials are path-dependent (Itasca, 2006). 'Path-dependent' means that the stress path the soil takes is important, not just its final stress state. Therefore, to calculate long-term behaviour in soft ground, a coupled analysis should be used, with the following stages:

1. An undrained calculation stage using effective stress parameters (Option 2 above). No fluid flow is allowed. This gives the short-term stresses and displacements for design.
2. A coupled flow-mechanical calculation stage. Depending on the software used, a lower value of K_f may be used to reduce the computation time without a significant impact on the accuracy (Itasca, 2006). Guidance should be given in the manual for the software you are using. This stage provides the long-term stresses and displacements for design.

7.7 VALIDATION AND ERROR CHECKING

Numerical models are based on many assumptions and are approximate by nature. Validation is a crucial part of the modelling process to ensure that the approximation is good enough and the results can be trusted. There are three main ways to validate a numerical model, which will be described in the following sections:

7.7.1   Comparison with an analytical solution
7.7.2   Validation by comparison with a laboratory test or experiment
7.7.3   Validation by comparison with a case history

It is always best to do as much validation as possible and of all three different types. Comparison with an analytical solution will tell us that certain aspects of the numerical model are making correct predictions, but since analytical solutions are very simple by nature, this cannot validate all aspects of the model. Comparison with a laboratory soil test is useful for validating the constitutive models, but may not represent the true ground mass behaviour. Only comparison with a case history will validate all aspects, but the case histories that are available may not be very similar to the tunnel we are designing.

Even when we are designing a tunnel in well-known geology with construction methods we have used many times before, we can never completely validate a design model, as there are always uncertainties and every design is a unique prototype. We cannot build a tunnel in exactly the same place as we have built one before. Therefore, we need monitoring during construction to verify the assumptions and simplifications made in design, and also to provide a case study for future designs to be validated against.

7.7.1 Comparison with an analytical solution

To compare a numerical model with an analytical solution, it is possible to get very good agreement if the boundary conditions and constitutive models are replicated as closely as possible. To replicate the Curtis-Muir Wood solution, for instance, we need to simulate a biaxial stress field. This is best done by building a model of a quarter of the tunnel with two planes of symmetry and applying the axis level vertical and horizontal in situ stress values (σ_v and σ_h) to the top and side of the model, as shown in Figure 7.11. To replicate the analytical solution more closely, set the unit weight of the soil and the lining to zero. This should result in close to an elliptical deformation with hoop forces and bending moments that are quite similar to the analytical solution.

Validation using an analytical solution can also be used to check the element types and mesh density. The following worked example shows how this could be done using the Curtis-Muir Wood solution.

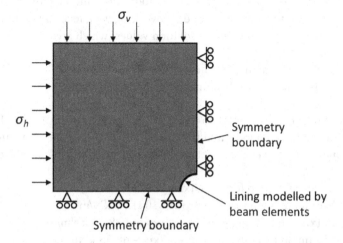

Figure 7.11 Boundary conditions for validation of a numerical modelling program using the Curtis-Muir Wood analytical solution.

WORKED EXAMPLE 7.1   VALIDATION USING THE CURTIS-MUIR WOOD ANALYTICAL SOLUTION

This Worked Example goes through the process of validation for a 2D numerical model of a 6 m diameter tunnel in soft ground by mimicking the boundary conditions of the Curtis-Muir Wood analytical solution.

A 2D plane strain model of a block of soil 50 m by 50 m is used. The water table is set to the bottom of the model so that no water pressures are calculated.

A linear elastic soil constitutive model is used with Young's modulus $E = 80$ MPa and Poisson's ratio $v = 0.2$. The unit weight of the soil is set to zero.

A tunnel lining is placed in one corner of the model, as shown in Figure 7.11, as a quarter circle arc of radius 3 m.

The lining is represented by beam elements with Young's modulus $E = 20$ GPa and Poisson's ratio $v = 0.3$. The unit weight is set to zero.

Boundary conditions are set as shown in Figure 7.11.

Apply $\sigma_v = \sigma_h = 500$ kPa to the free top and side boundaries as shown in Figure 7.11.

Set up an initial construction phase, where the 50 m × 50 m block of soil is intact. Then set up an excavation phase, where the soil elements within the tunnel boundary are deactivated and the tunnel lining beam elements are activated.

Run the numerical model. See Section 6.2 and use the Curtis-Muir Wood equations to hand-calculate the hoop force and bending moment in the lining. If the numerical model is working well, its results should be close to our Curtis-Muir Wood solution values, which are:

- 1457.99 kN/m uniform hoop force around lining
- 0 kNm/m uniform bending moment around lining

Now change the horizontal stress σ_h at the boundary to 350 kPa. Then solve the numerical model and the Curtis-Muir Wood solution again.

Compare the numerical model results to the Curtis-Muir Wood solution:

- Maximum hoop force (at axis) 1436.68 kN/m
- Minimum hoop force (at crown) 1041.90 kN/m
- Maximum bending moment (at crown) 68.33 kNm/m
- Minimum bending moment (at axis) −68.33 kNm/m

In this case, the validation is checking that the numerical modelling programme is configured correctly, and that the element types and mesh density are appropriate for modelling a tunnelling problem. It is not possible to use a model such as this to verify that boundary distances are sufficient for accurate modelling of ground movements. Also, only linear elastic behaviour of the ground and tunnel lining can be validated – as soon as we make these constitutive models more sophisticated, the numerical model will diverge from the analytical solution.

A further worked example describing the validation of a 3D numerical model of a shaft-tunnel junction using the Kirsch analytical solution is shown later in Section 7.10.3.

7.7.2 Validation by comparison with a laboratory test or experiment

Constitutive models describe the stress-strain behaviour of soil or the tunnel lining material. They are simple mathematical rules in the form of equations and algorithms.

Usually, the first step is to compare the constitutive model with different laboratory tests to ensure that the most important aspects of behaviour are modelled well. The laboratory tests should involve different stress paths, and sometimes different loading durations and stress histories, and should be relevant to the stress levels, stress paths, stress histories and magnitudes of strain expected in the actual soil during and after construction.

To validate the use of a constitutive model for the ground or tunnel lining in a numerical model, a laboratory test can be simulated in the numerical modelling program and compared to the behaviour observed in a real laboratory test. If the constitutive model is formulated correctly and adequately describes the soil behaviour, then the agreement between the two should be good.

Some numerical modelling programs allow this to be done using a built-in laboratory test module. If not, then in any case this is easy to model as laboratory tests have simple geometry. A single element could be used, but for a triaxial test Simpson et al. (1979) found that this did not work well because a triaxial specimen does not fail in either a plane strain or axisymmetric manner. Therefore, a 3D cylindrical model geometry should be used with the same dimensions as the test. The boundary conditions can be replicated in the numerical model. For example, a triaxial test can have a rigid top or bottom boundary, a cell pressure applied around the model, and either a pressure (for a stress path triaxial) or a displacement (for a standard triaxial) applied to the top or bottom, whichever is not fixed in the vertical direction.

Another method of validation for constitutive models is to compare their theoretical stress-strain behaviour, calculated by hand, with the outputs of a numerical model. This validates the implementation of the constitutive model in the software. As long as we know that the theoretical behaviour, as expressed by the constitutive model's equations, represents the behaviour of the soil in the laboratory tests, then this can be an effective validation.

WORKED EXAMPLE 7.2   VALIDATION
USING A NUMERICAL MODEL
OF A LABORATORY TEST

This is a case study, rather than a worked example, showing the procedure and results of validation of a nonlinear 'Jardine' constitutive model in FLAC3D using a cylindrical model of a triaxial test. The aims of the validation were:

- to ensure that the stress increment size was not important (i.e. that FLAC3D could follow the correct stress path to failure even when the stress increment was large)
- to reduce the calculation time without affecting accuracy by optimising the solution control method, such as varying the number of steps between stiffness updates or varying the convergence criterion (the 'mech. ratio')
- to check that the constitutive model was working correctly

Stress increment size
The results are shown in Figure 7.12. Graph (a) shows how the stiffness decreases as strain increases and Graph (b) shows the deviatoric stress versus axial strain relationship. The solid lines show the theoretical relationship from the Jardine constitutive model equations and the dashed lines show the results of a cylindrical triaxial test model in FLAC3D. The agreement is very good.

In order to check the effect of the stress increment size on the nonlinearity of the stress path, a deviatoric stress of 160 kPa was approached in various increments of stress. The difference between the value of strain at 160 kPa using 4 kPa increments and 40 kPa increments was negligible. There was a slight difference between these values and the strain calculated in one increment of 160 kPa, but this error was still very small, less than 5%. It seems that due to the way FLAC3D approaches a solution by using the velocity of gridpoints, the nonlinear stress path was followed almost perfectly.

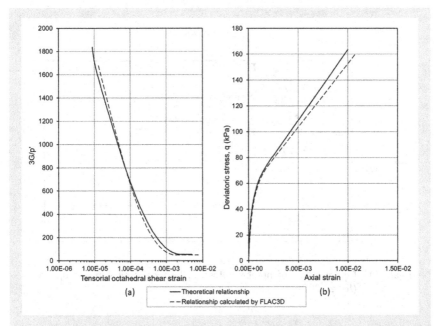

Figure 7.12 Comparison between theoretical Jardine curve and results from a FLAC3D model of a triaxial test using a Jardine constitutive model (Jones, 2007). Graph (a) shows how the stiffness decreases as strain increases and Graph (b) shows the deviatoric stress versus axial strain relationship.

Frequency of stiffness updates

The Jardine constitutive model gives a value for the bulk and shear stiffness based on the current value of the strain tensor at each point in the model. Therefore, the stiffness must be updated regularly as the strain tensor values change in the model. Each time the stiffness is updated, the constitutive model's algorithm must cycle through all the zones in the model calculating the stiffness from the strain increment vector. This takes 2–3 seconds, whereas each calculation step takes about 1 second. So, if the stiffness must be updated at every step, it will have a significant impact on the total runtime.

Figure 7.13 shows how the number of steps between updates was varied with the aim of optimising the process, for a model of a laboratory triaxial test. Updating the stiffness every ten steps results in an error of approximately 5%. When applied to the much larger 3D model of an advancing tunnel, the runtime went from 3 days when updating every 10 steps to 10 days when updating every step, so this really was a critical aspect to optimise.

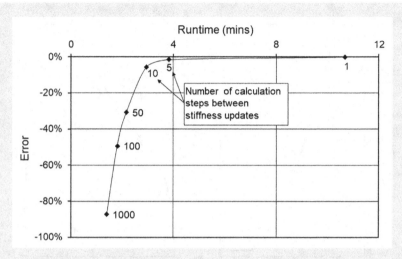

Figure 7.13 Optimisation of number of steps between soil stiffness updates (Jones, 2007).

7.7.3 Validation by comparison with a case history

Comparing a numerical model with a case history enables us to bring together all the aspects of the model and validate them together. This should only be done once boundary conditions and distances, element types and mesh density, and constitutive models, have been validated and checked individually. If the model can replicate those aspects of the case history that we are interested in, this will give us confidence that when we use the same modelling approach to design a new tunnel, the results should be reasonably accurate.

Validation to a case history also allows a model to be calibrated – in other words, for the key parameters used in the model, such as soil parameters used in the soil constitutive model or the deconfinement factor (see Section 7.1.2 for an explanation of deconfinement factor as part of the convergence-confinement method) used in a 2D plane strain model, to be varied within realistic limits to find the best value to use. This can only be done if the case history is in similar geology and involves a similar construction method to the new design situation.

Well-documented case histories are immensely valuable for validation of numerical models. This means that some case histories are used a lot and we can have some fun comparing different numerical modelling approaches to the same set of data (see for example Jones, 2013).

There are many papers that describe a numerical model, present results, and then compare these modelling results to field measurements during construction. Invariably the predictions are very good. Partly

Table 7.1 Classes of prediction (Lambe, 1973).

Prediction type	When prediction made	Results at time prediction made
A	Before event	-
B	During event	Not known
B1	During event	Known
C	After event	Not known
C1	After event	Known

this is 'publication bias', where positive results are more likely to be written up and more likely to be accepted for publication. But often it is because the model has been calibrated after construction to give as good a match to the field measurements as possible. In this case it is not a prediction at all, but a back-analysis. This is a crucial difference, easily missed.

Lambe (1973) proposed a classification system for types of predictions, which is shown in Table 7.1. Every time we are designing a new structure, we are doing a Class A prediction. However, most papers written about 'predictions' are actually Class C1 predictions. Lambe describes Class C1 predictions as 'autopsies'. He wrote, "... one must be suspicious when an author uses type C1 predictions to 'prove' that any prediction technique is correct".

The Jubilee Line Extension's running tunnels under St James's Park in London, UK, have been the subject of a lot of Class C1 back-analyses. This is because a large amount of high-quality measurements of surface and sub-surface ground movements was gathered by a team from Imperial College, both during construction and into the long term. A review of many of these attempts to obtain good agreement with the field measurements is given in Jones (2013).

7.8 CONSTITUTIVE MODELS

So far in this book we have focussed on modelling using linear elastic soils with a Mohr-Coulomb failure criterion and linear elastic tunnel linings, because this simple behaviour is widely understood by civil engineering graduates (the target audience of this book) and allows analytical solutions to be used and the first steps of numerical modelling to be understood. It should always be remembered that these constitutive models are gross approximations. True soil behaviour is very complex and nonlinear (e.g. see Chapter 4 of Potts & Zdravković, 1999). The same can be said of true shotcrete or concrete behaviour. To have any hope of accurately predicting ground movements and tunnel lining stresses, we need to use more sophisticated constitutive models.

7.9 INTERPRETATION AND PRESENTATION OF RESULTS

First of all, think about the intended audience of your report. Is it the client? An internal/external checker? Is this a scientific paper on numerical modelling? It is important to always provide sufficient information so that someone else can replicate your work independently, understand the reasons for all the steps you have taken, have confidence that you have approached the problem in a rational and efficient manner, and have confidence in your results.

Try to plot graphs when presenting results rather than hundreds of contour plots. They are pretty, but they only give the reader a qualitative view of the pattern of behaviour. This can be helpful in moderation but think carefully about what is the purpose of each table/graph/plot and ensure they are all discussed in the text and are important to whatever point you are making. Working out how best to present the results to make the points you want to make will take some time and experimentation, and is an iterative process.

7.10 3D NUMERICAL ANALYSIS

There are two main types of 3D numerical models used for soft ground tunnel design; a 3D numerical model of an advancing tunnel, and any other 3D model, for example of a tunnel-tunnel junction or a shaft-tunnel junction.

3D numerical analysis is also used for modelling complex details in the design of segmental linings. This will be covered in Chapter 9.

A big advantage of 3D analysis over 2D is that we do not need to make any assumptions about what happens ahead of the face, because it is modelled explicitly. On the other hand, a big disadvantage of 3D analysis is the larger model size and hence increased time to build, debug and run the model, and increased time to interpret and present the results.

As always, start with the simplest methods first and gradually increase complexity. Thus, simple hand calculations followed by analytical solutions, followed by 2D numerical analysis, followed by 3D numerical analysis. Thus, by the time we are ready to begin the 3D numerical analysis, we have understood the problem, have confidence in the methodology, and have a good idea what the results will tell us.

7.10.1 Modelling an advancing tunnel in 3D

In 3D, each advance of the tunnel can be modelled by an excavation stage followed by a lining stage. In fact, since the lining will not experience any increment of load until the next advance is excavated, these two stages can often be grouped together in one calculation step. Exceptions to this rule would be if you are modelling any time-dependent processes that occur between installing the lining and excavating the next step, such as consolidation or creep of the soil.

Regarding the boundary conditions, these will be similar in principle to the 2D plane strain model described in Section 7.1.7. The bottom boundary will be fixed in all directions and the vertical boundaries (comprising the lateral boundary, a lateral symmetry boundary if used, and the start and end boundaries) will be fixed in the direction normal to each plane, as shown in Figure 7.14.

Regarding the lateral and bottom boundaries, their distance should be set in a similar way to a 2D plane strain model (see previous Section 7.3). As a starting point, Möller (2006) found the lateral boundary in a 3D model needed to be the same distance as for the 2D model (c.f. Equation 7.6 in Section 7.3). The bottom boundary, however, could be closer at $1.1–1.45D$ below the invert of the tunnel. These distances should be checked and fine-tuned in each new model by varying the distances to achieve an acceptable level of accuracy.

Unlike a 2D numerical model, in a 3D model the tunnel must start somewhere. Usually it begins at a fixed boundary and is incrementally advanced through the model. This means that the first advances will be affected

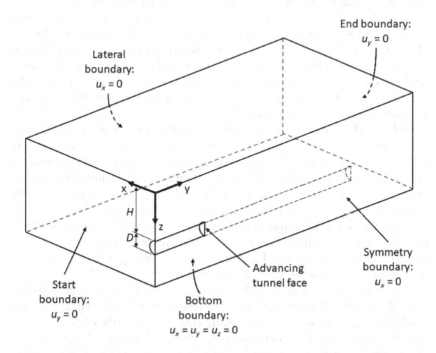

Figure 7.14 Boundary conditions for a 3D model of an advancing tunnel with a vertical symmetry boundary on the tunnel centreline. Displacements in *x*, *y* and *z* directions are denoted by u_x, u_y and u_z, respectively.

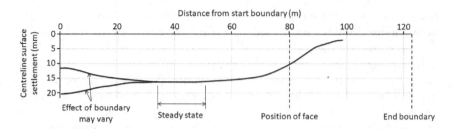

Figure 7.15 Achieving a steady state in a 3D tunnel model in terms of centreline surface settlement (redrawn based on Möller, 2006).

by the presence of the starting boundary behind, as shown in Figure 7.15. The tunnel face will also be affected as it approaches the end boundary. Therefore, the model needs to be long enough that a kind of steady-state situation occurs in a central zone of the model where the start and end boundaries are not affecting the results and the tunnel face is sufficiently far ahead that it is no longer affecting the lining forces, stresses, strains and pore pressures. This central steady-state zone needs to be long enough to allow a normal ground behaviour ahead of and behind the face, and long enough to demonstrate that a steady-state has been achieved.

Figure 7.15 shows an example of achieving a steady-state in a 3D tunnel model in terms of centreline surface settlement. As the position of the face of the tunnel advances beyond 80 m, the settlements in the steady-state zone should be unaffected, and these can be taken as the final short-term settlements. If we are also interested in lining stresses, then we need to plot a graph of bending moments or axial forces in the lining in the same way, to ensure a steady-state is achieved.

Möller (2006) also found that the first calculation stage from the start boundary has a big effect on the model before reaching steady state. In effect the start boundary has the same properties as a symmetry boundary and acts like a mirror. Therefore, starting with an excavation step and then solving to equilibrium can result in large settlements because we effectively have a two advances-long unlined cylinder with two open faces – this is the lower curve on the left-hand side of Figure 7.15. In some types of soil, this may even result in a mass failure and the calculation may not converge. If the first advance is excavated and lined before solving to equilibrium, the settlements are smaller – this is the upper curve in Figure 7.15.

Determining the model length is something that can only be achieved by trial and error. Various tricks can be employed to try to reduce the time needed to reach the central steady-state zone, particularly when departing from the start boundary. For example, you could model a 15 m length of tunnel excavated all at once and an internal pressure applied in a pseudo-convergence-confinement approach. The whole length could then be lined, and the internal pressure removed. This will get the model away from the start boundary quickly without causing ground movements that are too unrealistic and can replace 15 calculation steps with just 2.

In a tunnel model, we are often not just concerned with the final short-term settlements and lining loads well behind the face, but also what happens near the face in terms of transient settlements or possibly early age loading of shotcrete. So we need the whole response from ahead of the face to well behind the face to be in a steady-state, such that each advance resembles the one before. This can be verified by plotting a series of curves similar to the one shown in Figure 7.15, at different face positions. If they all align when plotted against distance from face position, the model has reached a steady-state.

The total model length needs to be determined for every new situation depending on the soil and tunnel lining constitutive behaviour, the tunnel geometry and depth, and the tunnel construction method. For a reasonable first estimate Möller (2006) found the following equation described the model length l needed to achieve steady-state:

$$l = D\left(13 + \frac{11}{3} \cdot \frac{H}{D}\right) \tag{7.12}$$

l is the length of the model needed to achieve steady-state
D is the tunnel diameter
H is the cover from tunnel crown to ground surface

Franzius et al. (2005) found that when coefficient of earth pressure at rest $K_0 = 1.5$, a steady-state was not achieved in their 3D model of an advancing tunnel in London Clay, even with sophisticated constitutive soil models and a total model length of 33D. Using Equation 7.12 with Franzius et al.'s model geometry of $H = 28.125$ m and $D = 4.75$ m, gives $l = 35D$, so it was slightly less than Möller would have used. However, when $K_0 = 0.5$ and an unrealistically high degree of anisotropy was used, a steady-state was achieved, though the settlements were far too large. Therefore, the constitutive model and the initial in situ stress regime have an important influence on whether a model can reach a steady-state.

Demagh et al. (2013) modelled an advancing TBM used for the Toulouse Metro with face pressures, conicity and grout pressures included. The Toulouse 'Molasse Argileuse' is a very stiff overconsolidated clay soil with $K_0 = 1.7$. In their case, settlements do stabilise behind the face using the same model length of 33D that Franzius et al. used. Perhaps the reason for the difference has something to do with the width of the model – Demagh et al. used a width of 19D whereas Franzius et al. used 16D, or perhaps it is because the Toulouse soil is stiffer and has a higher undrained shear strength than London Clay.

When modelling an advancing tunnel in 3D, quite large longitudinal tensile stresses may be generated in the lining. This is because at each advance, ground movements towards the face will pull the lining towards the face, and this tends to continue to increase with distance, as noted by Jones (2007) and Thomas (2003). It is not known whether these longitudinal tensile stresses occur in real tunnels, but it is common practice to ignore them on the basis that in a shotcrete tunnel lining creep and cracking would tend to dissipate them, and in a segmentally lined tunnel the TBM jacks will ensure the lining is in longitudinal compression. To date there is no known study of the effect of these longitudinal tensile stresses, but they may have a Poisson's ratio effect on hoop stresses in the model. This effect can be mitigated to some degree by detaching the elements representing one ring from another, in effect having two nodes at each node location on the circumferential joints and allowing them to displace independently. This, however, can cause problems with how the elements representing the soil are attached to the lining, and how they behave across any gap that is formed, as discussed for radial segment joints by Potts & Zdravković (2001: 49–50).

7.10.2 Modelling the tunnel lining

The tunnel lining may be modelled by solid 3D elements, or by 2D shell or plate elements. The same arguments discussed in Section 7.4 regarding 1D beam elements or 2D solid elements in a 2D model apply to 3D solid elements and 2D shell elements in a 3D model. The use of shell elements is far more common because fewer elements are needed to achieve the same accuracy and the outputs are in the form of forces and moments. With solid elements far more sophisticated constitutive models may be used, but outputs will be in stresses that are more difficult to interpret, and quite a large number of small elements may be needed to achieve sufficient accuracy (Pound, 2006).

The same rules regarding boundary conditions on a symmetry boundary apply to shell elements in a 3D model as for beam elements in a 2D model. In addition, at the start boundary of a 3D model of an advancing tunnel, the lining is usually fixed in the direction of tunnelling but free in the other directions.

When using shell elements to model a tunnel lining, there are usually three bending moments and five forces given in the output. The moments are usually denoted 'Mx', 'My' and 'Mxy', the shear forces 'Qx' and 'Qy', and the membrane forces 'Nx', 'Ny' and 'Nxy'. These are illustrated in Figure 7.16. In some programs, the numbers 1, 2 and 3 may be used for the local axes instead of x, y and z. Remember that in most numerical modelling programs, Mx is not the moment *about* the x-axis, but the moment *along* it. If this is the case then when we use the outputs to look at the moment-axial force interaction, then Mx is coupled with Nx and My is coupled with Ny.

Nxy is an in-plane shear force and Mxy is a twisting moment. Nxy and Mxy need to be combined with Nx, Ny, Mx and My in order to arrive at

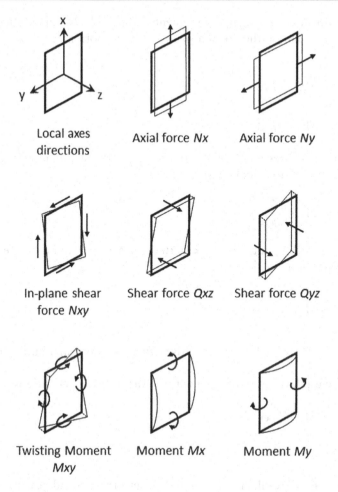

Local axes directions   Axial force *Nx*   Axial force *Ny*

In-plane shear force *Nxy*   Shear force *Qxz*   Shear force *Qyz*

Twisting Moment *Mxy*   Moment *Mx*   Moment *My*

Figure 7.16 Local axis directions and the resultant forces in a shell element.

the design forces and moments in the orientations we are interested in (for reinforced concrete this will be the primary and transverse reinforcement directions). This usually has to be done in a post-processing stage outside of the numerical modelling program. For straight tunnels modelled in 3D, Nxy and Mxy are usually small, but not insignificant, relative to the values of Nx and Ny, and Mx and My. Close to junctions Nxy and Mxy can be quite large, for instance Jones (2007) found that Mxy moments close to a shaft-tunnel junction could represent bending stresses that were equal to 40% of the hoop stress. Therefore, they always need to be considered.

Assuming the primary and transverse reinforcement directions are the same as local axes directions x and y, the design moments Mx^* and My^* according to the Wood and Armer method (Wood, 1968) are given by the following equations.

Assuming a positive bending moment in the model is defined such that it results in tension on the bottom surface, for the bottom layer of reinforcement use:

$$Mx^* = Mx + |Mxy| \qquad (7.13)$$

$$My^* = My + |Mxy| \qquad (7.14)$$

If Equation 7.13 results in $Mx^* < 0$, then set $Mx^* = 0$ and now use the following equation to calculate My^*:

$$My^* = My + \left| \frac{Mxy^2}{Mx} \right| \qquad (7.15)$$

If Equation 7.14 results in $My^* < 0$, then set $My^* = 0$ and now use the following equation to calculate Mx^*:

$$Mx^* = Mx + \left| \frac{Mxy^2}{My} \right| \qquad (7.16)$$

If both Mx^* and My^* are negative, then no bottom reinforcement is required.

For the top layer of reinforcement, use the following equations:

$$Mx^* = Mx - |Mxy| \qquad (7.17)$$

$$My^* = My - |Mxy| \qquad (7.18)$$

If Equation 7.17 results in $Mx^* > 0$, then set $Mx^* = 0$ and now use the following equation to calculate My^*:

$$My^* = My - \left| \frac{Mxy^2}{Mx} \right| \qquad (7.19)$$

If Equation 7.18 results in $My^* > 0$, then set $My^* = 0$ and now use the following equation to calculate Mx^*:

$$Mx^* = Mx - \left| \frac{Mxy^2}{My} \right| \qquad (7.20)$$

If both Mx^* and My^* are positive, then no top reinforcement is required.

For the design axial forces Nx^* and Ny^* the same procedure should be followed to include the effect of Nxy. In Equations 7.13–7.20, replace every instance of 'M' with 'N'.

7.10.3 Modelling junctions

In the past, junctions have been modelled in all sorts of approximate ways, due to the difficulty and expense of 3D numerical modelling. The most common methods are:

- using the Kirsch plane stress analytical solution to estimate axial stresses around an opening
- 3D numerical modelling of a wished-in-place junction using shell elements for the lining, applying a pressure to the outside to simulate ground loading and using model springs to provide ground reaction
- 3D numerical modelling of a wished-in-place junction using shell elements for the lining and modelling the ground explicitly using solid elements
- 3D numerical modelling of a junction constructed sequentially using shell elements for the lining and modelling the ground explicitly using solid elements

7.10.3.1 Kirsch solution

The Kirsch plane stress analytical solution is given in the following equations (Hoek & Brown, 1980; Kirsch, 1898):

$$\sigma_r = \frac{1}{2}p_z\left[(1+k)\left(1-\frac{a^2}{r^2}\right)+(1-k)\left(1-\frac{4a^2}{r^2}+\frac{3a^4}{r^4}\right)\cos 2\theta\right] \qquad (7.21)$$

$$\sigma_\theta = \frac{1}{2}p_z\left[(1+k)\left(1+\frac{a^2}{r^2}\right)-(1-k)\left(1+\frac{3a^4}{r^4}\right)\cos 2\theta\right] \qquad (7.22)$$

$$\tau_{r\theta} = \frac{1}{2}p_z\left[-(1-k)\left(1+\frac{2a^2}{r^2}-\frac{3a^4}{r^4}\right)\sin 2\theta\right] \qquad (7.23)$$

σ_r is the radial stress
σ_θ is the circumferential stress
$\tau_{r\theta}$ is the shear stress
r is the radial distance from the centre of the hole
a is the hole radius
θ is the angle from the vertical centred on the hole
p_z is the vertical applied stress in the plate
p_h is the horizontal applied stress in the plate
k is the ratio of horizontal to vertical stress, $k = p_h/p_z$

The above parameters are also defined in Figure 7.17.

Usually we will assume that p_z is in the direction of the hoop stress and that $k = 0$. The flat sheet or 'plate' that has the hole in it is assumed to extend

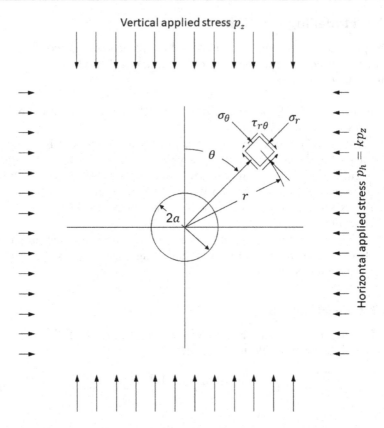

Figure 7.17 Kirsch solution diagram with notation.

an infinite distance in the two plane directions. The plate has finite thickness but this is only used to calculate the in-plane stresses. The plate is assumed to have zero stress in the out of plane direction (into or out of the page in Figure 7.17) and therefore the thickness does not feature in the equations.

There are two main types of tunnel junctions: shaft-tunnel junctions and tunnel-tunnel junctions, where we create a hole in the side of the 'parent' tunnel or shaft and begin to drive a new 'child' tunnel away from the junction. If we unwrapped and laid flat the parent tunnel lining, it would look rather like Figure 7.17, with p_z equal to the parent tunnel hoop stress prior to creating the opening, and usually we would assume that the longitudinal stress in the parent tunnel lining is zero (i.e. $p_h = 0$ in Figure 7.17).

For a shaft-tunnel junction, the hoop stress is still circumferential. So if a shaft lining were unwrapped and laid flat, it would also resemble a flat sheet or plate with a hole in it, only we would probably visualise it with the hoop stress (p_z) shown horizontally on the page, i.e. like Figure 7.17 but rotated 90°.

For a shaft-tunnel junction the Kirsch solution predicts the stress distribution shown in Figure 7.18. When the new tunnel opening is made,

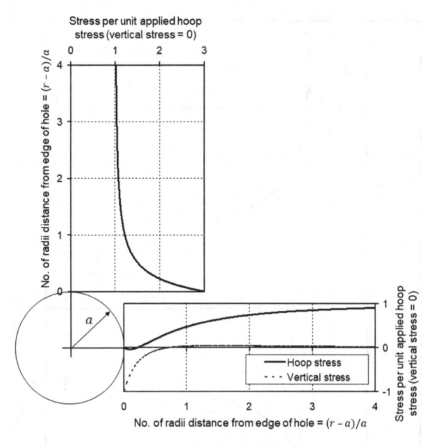

Figure 7.18 Stress concentrations predicted by the Kirsch analytical solution. The applied hoop stress is in the horizontal direction (Jones, 2007).

the hoop stress that is going circumferentially around the shaft lining has to divert around it, resulting in a higher hoop stress above and below the opening. Exactly three times higher in fact, as shown on the graph above the opening in Figure 7.18. Due to this diversion, hoop stress in the shaft lining decreases to zero at the axis level of the opening, and there is tension in the vertical direction equal to −1 times the initial hoop stress (shown on the graph to the right of the opening in Figure 7.18).

The Kirsch solution will predict axial stresses around an opening in a flat plate, but because it does not model soil-structure interaction or the curvature of the parent tunnel or shaft, these axial stresses will be very inaccurate (Jones, 2007). In addition, the Kirsch solution will not calculate any bending stresses, which have been found to be very significant (Jones, 2007).

A comparison of hoop stresses in a shaft just after an opening has been made in the side for a child tunnel is shown in Figure 7.19, along a vertical

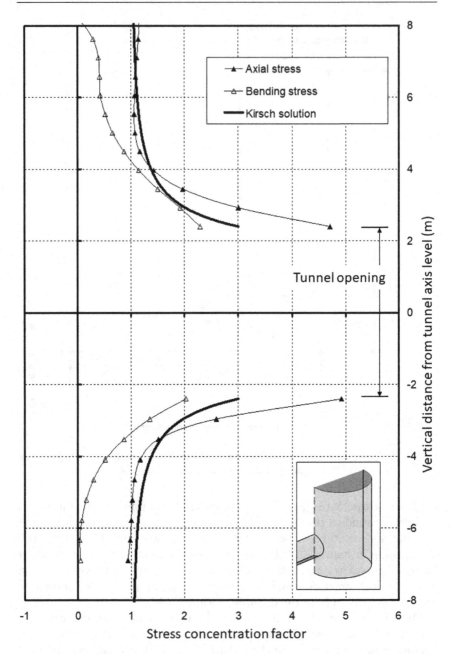

Figure 7.19 Comparison of Kirsch solution (thick line) with 3D numerical modelling (triangles), from Jones (2007).

line going through the centreline of the child tunnel (as shown in the inset figure). The Kirsch solution gives a stress concentration factor increasing from 1 far away from the opening up to a maximum value of 3 close to the opening. The 3D numerical modelling shows that hoop (axial) stress concentration can increase more rapidly and have a significantly higher maximum close to the opening compared to the Kirsch solution. This is due to curvature of the parent shaft and soil-structure interaction. The numerical modelling in FLAC3D also predicts bending stresses, which are significant in magnitude and not predicted by the Kirsch solution. More plots for vertical stresses and for hoop stresses around the shaft may be found in Jones (2007).

7.10.3.2 Wished-in-place 3D numerical model

Using a wished-in-place 3D model of the lining only, and applying a pressure to its surface, is only appropriate in situations where most of the ground or groundwater load is applied after construction. This could occur where an impermeable secondary lining is installed while the ground is dewatered, and then the pumps are switched off and the water pressure is applied. Similarly, where tunnel linings resist internal pressures during operation, such as water conveyance or sewage tunnels, then a wished-in-place model may be suitable to model stresses at junctions. However, these models will not model soil-structure interaction properly, even if springs are used, because the springs are independent and will not redistribute stresses from one part of the lining to another as the lining and the ground deform.

Hafez (1995) created a 3D finite element model of a skewed tunnel junction in the Heathrow Baggage tunnel, a shallow soft ground shotcrete-lined tunnel in London Clay within Heathrow Airport. The shotcrete junction was modelled wished-in-place without any surrounding ground, and a pressure was applied to the lining. It was found that the highest stress concentrations were at axis level at the corners of the lining at the intersection of the parent and child tunnels.

For design of primary linings in soft ground, a wished-in-place 3D model will ignore the sequential nature of construction. In reality, the most important step is when the opening is made in the parent tunnel. At this point, hoop stress is redistributed in the lining around the opening, the lining deforms leading to bending moments, and arching in the ground occurs in front of the opening as it does at an open tunnel face, leading to increased ground pressure applied to the parent tunnel around the opening. Subsequent advances of the child tunnel only cause further incremental changes to this major redistribution of stresses. Lining loads in the child tunnel close to the junction are actually lower than they are further away, as stresses in the ground arch onto the larger parent tunnel (Jones, 2007). However, a wished-in-place 3D model will share the stress concentrations between the parent and child tunnels at the junction, as Hafez (1995) found. This has led to junction designs where both the parent and the child tunnel linings have been unnecessarily thickened and reinforced close to junctions, even though only the parent tunnels need it.

7.10.3.3 3D numerical model with sequential construction

This method is the most time-consuming, because the ground must be modelled explicitly using solid elements, and many calculation steps are needed to first model the construction of the parent tunnel or shaft and then the child tunnel. Unfortunately, this is the only way to model a tunnel junction to obtain realistic values of lining stresses for design. This is because:

- the 3D nature of junctions means the problem cannot be simplified to 2D without introducing gross unconservative errors
- 3D wished-in-place models will tend to be unconservative for design of the parent tunnel or shaft and overconservative for design of the child tunnel
- in most cases, the order in which the parent and child tunnel are constructed is critical to the stress distributions, and therefore sequential excavation and lining of first the parent and then the child tunnel must be modelled
- the soil must be modelled explicitly, i.e. modelling the ground as springs or simply as an applied pressure is inaccurate and unconservative

For more detail, see Jones (2007).

Validation of a 3D numerical model using the Kirsch solution is shown in the following worked example.

WORKED EXAMPLE 7.3   VALIDATION OF A NUMERICAL MODEL USING THE KIRSCH SOLUTION

When a tunnel junction is constructed, there are large stress gradients around the opening in the parent tunnel. Therefore, it is important to check that the numerical method used to model the lining can model these stresses.

An example is shown here from Jones (2007) of the validation of a shaft-tunnel junction numerical model. The boundary conditions and elastic constitutive model used in the Kirsch solution were replicated in the numerical modelling program FLAC3D. The objective was to check that the shell element type and mesh density around the opening could model the high axial stress gradients. For the precise situation of a flat elastic plate with a hole, the Kirsch solution provides an exact solution, whereas the numerical model is approximate.

By far the biggest stress change occurs when the opening for the child tunnel is made in the shaft lining. Hoop stresses in the shaft have to divert around the opening, resulting in higher axial forces above and below the opening, and tension in the vertical direction at axis level.

Taking advantage of symmetry on two axes, we only need to model ¼ of the child tunnel opening as shown in the inset figure in Figure 7.20. The shaft lining is modelled by shell elements. The vertical stress in the shaft lining is assumed to be zero.

An additional reason for using this kind of validation is that we can compare the results to more complex models, for example introducing 3D geometry, soil-structure interaction or using more sophisticated con-stitutive behaviour, and better understand what is driving the design.

A desirable mesh density around the perimeter of the shaft lining, in terms of calculation time and detail of results, is approximately 1.5 zones/m. 'Zones' in FLAC3D are analogous to elements in a finite element calculation. Here we will look at a model in FLAC3D of a plane stress plate with a hole in it, with mesh densities of 1.5, 2, 4 and 8 zones/m. The shell elements are linear elastic. The radius of the hole is 2.4 m, the same as the external radius of the child tunnel. The boundary

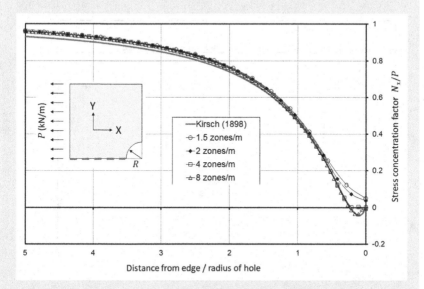

Figure 7.20 Effect of mesh density on the stress concentration factor (N_x/P) for a plane stress plate with a hole, in the direction of the applied stress along a line parallel to the direction of the applied stress (dashed line on inset figure), compared with the Kirsch analytical solution.

is set at 20 m from the edge of the hole. The agreement between the Kirsch analytical solution and the numerical models is very good, with the largest differences occurring within half a radius distance from the hole. The finer the mesh, the closer the agreement with the analytical solution.

'Stress concentration factor', which is the stress N_x divided by the applied stress P at the boundary of the model, is plotted in Figure 7.20 for the axial force in the x-direction along a line parallel to the direction of applied stress (as indicated in the location key overlaid on Figure 7.20) up to a 5 radius distance (12 m) from the edge of the hole. Also plotted on the same figure is the stress concentration factor calculated using the Kirsch analytical solution. The stress concentration factor decreases to zero as it approaches the edge of the hole. The FLAC3D models replicated this behaviour with varying degrees of accuracy, dependent on the fineness of the mesh. The stress at the edge of the hole was predicted by the model with a mesh density of 1.5 zones/m with less than 5% error.

Figure 7.21 shows the stress concentration factor for the y-component of the membrane force N_y along a line parallel to the direction of applied stress. The Kirsch solution shows that the stress concentration factor should decrease from 0 to −1 as the edge of the hole is

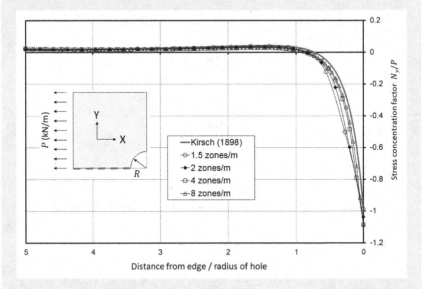

Figure 7.21 Effect of mesh density on the stress concentration factor (N_y/P) for a plane stress plate with a hole, in the direction transverse to the applied stress along a line parallel to the direction of the applied stress (dashed line on inset figure), compared with the Kirsch analytical solution.

approached. Again the FLAC3D models replicated this behaviour with varying degrees of accuracy, generally dependent on the fineness of the mesh. The mesh density of 1.5 zones/m was actually the closest to the Kirsch solution adjacent to the hole.

Figure 7.22 shows the stress concentration factor for the x-component of the axial force N_x along a line transverse to the direction of applied stress. The Kirsch solution shows that the maximum stress concentration factor of 3.0 occurs on this section, adjacent to the hole, and decreases with distance from the hole, eventually dropping towards 1.0. The FLAC3D models replicated the Kirsch solution with varying degrees of accuracy, generally dependent on the fineness of the mesh. The mesh density of 1.5 zones/m was actually the most accurate adjacent to the hole with an error of less than 0.5%.

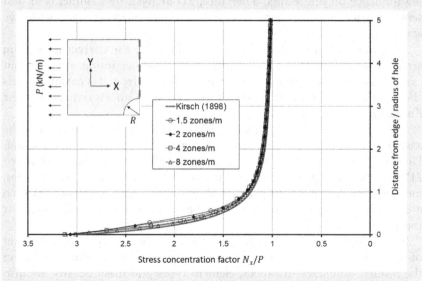

Figure 7.22 Effect of mesh density on the stress concentration factor (N_x/P) for a plane stress plate with a hole, in the direction parallel to the applied stress along a line transverse to the direction of the applied stress (dashed line on inset figure), compared with the Kirsch analytical solution.

In conclusion, the numerical model with mesh density of 1.5 zones/m predicted the maximum and minimum axial stresses to within 5% of the Kirsch analytical solution. The generally good agreement between the FLAC3D models and the Kirsch solution showed that the FLAC3D shell elements could be used to predict axial stresses in shell structures with high stress gradients with reasonable accuracy.

7.11 SUMMARY

Numerical modelling must be approached in a methodical and rational manner. Remember to always begin with simple analytical models, then proceed to simple numerical models, then add layers of complexity. Only add as much complexity as is needed to characterise the problem. In this way you will gain confidence and understanding of the problem and what is driving the design and be in a better position to assess the outputs.

We must accept that modelling a tunnel in 2D is always a gross approximation, and whether we use the convergence-confinement method or any other method of accounting for the 3D effect in a 2D model, we are fixing as an input something that is really an unknown and should be an output. In effect, although some soil-structure interaction will occur in the final stage, we are deciding beforehand what distribution of stress will be applied to the tunnel lining at the start of that final stage, which has a very big influence on the results. Therefore, 2D analysis of a tunnel is at best semi-empirical and at worst we are telling it the answer we want before we start.

2D numerical models can be calibrated to give the correct maximum settlement, the correct volume loss or the correct lining stresses, but rarely all at the same time. Therefore, the results from 2D design models should always be treated with caution and with an awareness of their limitations.

3D numerical models of advancing tunnels avoid the need to make assumptions about ground deformation prior to lining installation. However, they can still be wrong and validation is just as important as for 2D models. As long as the geometry, boundary conditions, element types, mesh density and initial stresses are all good, the main difference between a 3D numerical model and reality will be the constitutive models used for the ground and the lining.

It is important to understand the boundary conditions at the perimeter of the tunnel excavation and at the edges of the model, and how they affect the results. Boundary distances, element types and mesh density are also important. Every effort should be made to ensure that these aspects are not affecting the reliability of the results, using validation techniques and sensitivity analyses.

Numerical models must be validated in as many ways as possible. This can be achieved by comparison to analytical solutions, laboratory tests and case histories. Well-documented case histories are immensely valuable for the validation of numerical models, but care must be taken because a numerical model calibrated to give the correct surface settlements will not necessarily give the correct lining forces, and vice versa.

The validation does not end when the drawings are issued for construction. Monitoring during construction should be reviewed and design models updated to ensure that the assumptions used in design are still valid.

7.12 PROBLEMS

Q7.1. This question may be answered without needing access to numerical modelling software.

    i. Sketch the boundary conditions you would use for a 2D plane strain continuum numerical model of a circular tunnel, taking advantage of symmetry.

    ii. For a 3D numerical model of a 7 m diameter tunnel at a depth to axis $z_0 = 20$ m, estimate appropriate boundary distances to the lateral and bottom boundary using the relationships proposed by Möller (2006).

    iii. For the model in (ii), using the approach of Möller (2006) estimate the total length of the model needed to achieve steady-state surface settlements.

    iv. If the tunnel modelled in (ii) and (iii) were to be excavated by TBM in 1 m advances, estimate the number of calculation stages needed to achieve a steady-state.

    v. Using a series of sketches, show how you would model in 3D the construction of parallel 7 m diameter tunnels followed by a perpendicular 4 m diameter cross passage excavated from one tunnel to the other.

    vi. The 3D numerical model in (ii) and (iii) provides the forces in one of the shell elements modelling the lining in the steady-state zone listed in Table 7.2. 'x' is in the circumferential direction, 'y' is in the longitudinal direction, and 'z' is the direction normal to the shell. Calculate the values of Nx^* and Mx^* to use in design using the Wood and Armer method. Since the shell elements are placed at the excavation perimeter, estimate the errors introduced by their centroid not being in the correct position.

Table 7.2 Q7.1(vi) shell element internal forces.

Nx	Ny	Nxy	Qxz	Qyz	Mx	My	Mxy
	kN/m		kN/m			kNm/m	
1500	−300	120	800	200	250	70	30

Q7.2. This question may be answered without needing access to numerical modelling software.

    i. A 4.5 m diameter metro tunnel will be excavated by a TBM. There are no case histories of TBM tunnelling in this soil. Describe how you would determine the value of deconfinement factor λ to use in a 2D plane strain continuum model.

ii.   Describe the calculation stages 0, 1 and 2 needed to apply the convergence-confinement method to a 2D plane strain continuum numerical model.

iii. At the deepest point, the tunnel is 25 m deep to axis level, the water table is at the surface and the unit weight of the soil is 20 kN/m³. For a coefficient of earth pressure at rest $K_0 = 0.5$, what should the initial in situ stresses be in Stage 0 of the model (i.e. before any excavation takes place), at axis level?

iv.   Applying a deconfinement factor $\lambda = 0.4$, what in situ stresses will be applied to the tunnel lining at the start of Stage 2?

v.   Estimate the tunnel lining hoop forces and bending moments you should get at the end of Stage 2 using the Curtis-Muir Wood solution with lining stiffness parameters $E = 34$ GPa and $v = 0.2$, lining thickness 0.25 m, and soil stiffness parameters $E_c = 50$ MPa and $v = 0.25$. If you have access to a numerical modelling program, build and run a 2D model and compare the results with the Curtis-Muir Wood solution.

Q7.3. This question requires access to 2D numerical modelling software.

i.   For the tunnel in Q7.1, build a numerical model with the recommended boundary conditions and distances that you calculated based on Möller (2006).

ii.   Give the soil drained elastic material properties $E' = 30$ MPa and $v' = 0.3$, and a Mohr-Coulomb failure criterion $\phi' = 35°$ and $c' = 0$ kPa. The soil has a unit weight of 18 kN/m³. The tunnel is well above the water table. If the program requires definition of a water table, locate it at the bottom of the model. Use $K_0 = 0.43$ to establish the initial stresses. Give the lining a thickness of 0.3 m, Young's modulus $E = 30$ GPa and Poisson's ratio $v' = 0.2$. Using the convergence-confinement method with $\lambda = 0.5$, run the model.

iii. Check the in situ stresses are calculated correctly in Stage 0 and that the radial stresses are reduced to the correct values in Stage 1. Plot the final surface settlements in Stage 2 in a spreadsheet graph. Create a graph of the lining hoop force and bending moment around the tunnel, plotting them against angle from the crown.

iv.   Vary the lateral boundary distance by −20 m and +20 m and compare the surface settlements and lining forces. What is the maximum error, expressed as a percentage?

v.   Vary the bottom boundary distance by −10 m and +10 m and compare the surface settlements and lining forces. What is the maximum error, expressed as a percentage?

vi.   Increase and decrease the mesh density and compare the results for surface settlements and lining forces with those from before. If possible, also change the soil and lining element types and examine any differences in the results.

Q7.4. This is a suggested project/exploration and requires access to 2D numerical modelling software.

    i. Use a 2D plane strain model to simulate the construction of one tunnel, followed by construction of a second, identical, tunnel parallel to it. Allow 50% relaxation of overburden pressure before installation of the lining for both tunnels. Use soil and lining parameters from a previous question, or use your own.

    ii. Vary the size of the tunnels and the spacing between them and analyse the effect on surface settlements and lining deformations and stresses. What is the difference in behaviour between the first and second tunnel?

    iii. For the surface settlements, to what extent does the principle of superposition hold true (i.e. could the surface settlements due to construction of the first tunnel be superposed for the second tunnel, or does the second tunnel cause more/less settlement than the first?). How does using different soil parameters or constitutive models affect this?

    iv. Compare your results to numerical modelling or field measurements of twin tunnels in the literature.

Q7.5. This is a suggested project/exploration and requires access to 3D numerical modelling software.

    i. Use both 2D plane strain and 3D numerical models to replicate Robert Mair's stability charts (Refer to Chapter 2, Section 2.2, or to Kimura & Mair, 1981; Mair, 1979; or Mair & Taylor, 1997) based on centrifuge tests in clay. It is probably easier not to use centrifuge scale and acceleration, but to use full-scale tunnel dimensions in your model. This can be achieved by reducing undrained shear strength to failure (some programs have a function that allows this), or by increasing surcharge or decreasing an internal tunnel pressure.

    ii. Compare the results to upper bound plasticity solutions presented in Chapter 2.

    iii. Compare the results to numerical modelling of undrained stability in the literature.

    iv. Compare the results to actual reported undrained stability failures (see e.g. in Mair & Taylor, 1997).

Q7.6. This is a suggested project/exploration and requires access to 3D numerical modelling software.

    i. Build a 3D numerical model of an advancing tunnel to investigate the effect of unsupported length (TBM tunnel) or ring

   closure distance (sprayed concrete-lined tunnel). Use soil and
   lining parameters from a previous question, or use your own,
   but be aware that if the strength parameters for the soil are
   too low the heading may be unstable, and certainly if you
   are using drained parameters with no cohesion you will need
   to apply a support pressure – if in doubt, use the methods
   in Chapters 2 and 3 to check the parameters will result in a
   stable tunnel. Establish boundary conditions, boundary distances, element types and mesh density to demonstrate that a
   steady state has been achieved and that errors will be acceptable, using sensitivity analyses.
ii.   Check and validate the model in as many ways as you can
   think of.
iii.   Plot steady-state surface settlements transverse to the tunnel.
   Analyse how the settlement trough changes from ahead of the
   face all the way back to the steady-state zone.
iv.   Maintaining the advance length constant at e.g. 1 m, vary the
   distance to lining installation. Create a ground reaction curve
   similar to the one shown in Figure 7.2, plotting average radial
   convergence or volume loss on the x-axis and average hoop
   force in the lining on the y-axis. Add the results to the ground
   reaction curve, showing how as distance to lining installation
   increases, the lining forces decrease and the convergence or
   volume loss increases.
v.   Compare the effect of using different soil constitutive models on the surface settlements and the lining forces. Does the
   shape of the ground reaction curve change?

REFERENCES

Addenbrooke, T. I., Potts, D. M. & Puzrin, A. M. (1997). The influence of prefailure
   soil stiffness on the numerical analysis of tunnel construction. *Géotechnique*
   47, No. 3, 693–712.
Anagnostou, G. & Kovári, K. (1996). Face stability conditions with earth-
   pressure-balanced shields. *Tunn. Underg. Space Technol.* 11, No. 2,
   165–173.
Avgerinos, V., Potts, D. M., Standing, J. R. & Wan, M. S. P. (2018). Predicting
   tunnelling-induced ground movements and interpreting field measurements
   using numerical analysis: Crossrail case study at Hyde Park. *Géotechnique*
   68, No. 1, 31–49.
Bishop, A. W. & Hight, D. W. (1977). The value of Poisson's ratio in saturated
   soils and rocks stressed under undrained conditions. *Géotechnique* 27,
   No. 3, 369–384.
Brinkgreve, R. B. J., Kumarswamy, S. & Swolfs, W. M. (2018). *PLAXIS 3D reference manual*. The Netherlands: Plaxis bv.

Cook, R. D., Malkus, D. S., Plesha, M. E. & Witt, R. J. (2002). *Concepts and applications of finite element analysis*, 4th Edition. New York: John Wiley & Sons.

Demagh, R., Emeriault, F. & Kastner, R. (2013). Modélisation 3D du creusement de tunnel par tunnelier à front pressurisé dans les sols surconsolidés. *Revue Française de Géotechnique*, No. 142, 17–26.

Fenner, R. (1938). Untersuchungen zur Erkenntnis des Gebirgsdruckes. *Glückauf* 74, No. 32, 681–695.

Franzius, J. N., Potts, D. M. & Burland, J. B. (2005). The influence of soil anistotropy and K_0 on ground surface movements resulting from tunnel excavation. *Géotechnique* 55, No. 3, 189–199.

Gilleron, N., Bourgeois, E. & Saïtta, A. (2017). Limites de la modélisation bidimensionnelle des tunnels urbains pour la prévision des tassements. *Revue Française de Géotechnique*, No. 150, Paper 2.

Hafez, N. M. (1995). *Post-failure modelling of three-dimensional shotcrete lining for tunnelling*. PhD thesis, University of Innsbruck.

Hoek, E. & Brown, E. T. (1980). *Underground excavations in rock*. London: IMM.

Itasca. (2006). *FLAC3D manual*, version 3.1. Minneapolis, MN: Itasca Consulting Group Inc.

Janin, J. -P. (2017). Apports de la simulation numérique tridimensionnelle dans les études de tunnels. *Revue Française de Géotechnique*, No. 150, Paper 3.

Janin, J. -P., Dias, D., Emeriault, F., Kastner, R., Le Bissonnais, H. & Guilloux, A. (2015). Numerical back-analysis of the southern Toulon tunnel measurements: A comparison of 3D and 2D approaches. *Eng. Geol.* 195, 42–52.

Jones, B. D. (2007). *Stresses in sprayed concrete tunnel junctions*. EngD thesis, University of Southampton.

Jones, B. D. (2012). A loaded question. *Tunnelling Journal*, April/May issue, 26–28.

Jones, B. D. (2013). The most difficult question? *Tunnelling Journal*, June/July issue, 33–35.

Kimura, T. & Mair, R. J. (1981). Centrifugal testing of model tunnels in soft clay. *Proc. 10th Int. Conf. Soil Mech. Found. Eng.*, Stockholm, Vol. 1, pp. 319–322.

Kirsch, G. (1898). Die theorie der Elastizität und die bedürfnisse der festigkeitslehre. *Zeitschrift des Vereines deutscher Ingenieure* 42, 797–807.

Lambe, T. W. (1973). Predictions in soil engineering (Rankine Lecture). *Géotechnique* 23, No. 2, 149–202.

Mair, R. J. (1979). *Centrifugal modelling of tunnel construction in soft clay*. PhD thesis, University of Cambridge.

Mair, R. J. & Taylor, R. N. (1997). Bored tunnelling in the urban environment. Theme Lecture, Plenary Session 4. *Proc. 14th Int. Conf. Soil Mech. Found. Eng.*, Hamburg, Vol. 4.

Möller, S. (2006). *Tunnel induced settlements and structural forces in linings*. Doctoral Thesis, Universität Stuttgart. Mitteilung 54 des Instituts für Geotechnik, Herausgeber P. A. Vermeer.

Möller, S. C. & Vermeer, P. A. (2008). On numerical simulation of tunnel installation. *Tunn. Undergr. Space Technol.* 23, 461–475. Figure 8-6 reprinted with permission from Elsevier.

Pacher, F. (1964). Deformationsmessungen im Versuchsstollen als Mittel zur Erforschung des Gebirgsverhaltens und zur Bemessung des Ausbaues. *Grundfragen auf dem Gebiete der Geomechanik* (ed Müller, L.), pp. 149–161. Wien: Springer-Verlag.

Panet, M. (1995). *Le calcul des tunnels par la méthode convergence-confinement.* Paris: Presses de l'ENPC.

Panet, M. (2001). *AFTES recommendations on the convergence-confinement method.* Paris: AFTES.

Panet, M. & Guellec, P. (1974). Contribution à l'étude du soutènement derrière le front de taille. *Proc. 3rd Cong. ISRM*, Denver, Vol. 2.

Potts, D. M. & Zdravković, L. (1999). *Finite element analysis in geotechnical engineering, Vol. 1: Theory.* London: Thomas Telford.

Potts, D. M. & Zdravković, L. (2001). *Finite element analysis in geotechnical engineering, Vol. 2: Application.* London: Thomas Telford.

Pound, C. (2006). The performance of FLAC zones in bending. *Proceedings of 4th International FLAC Symposium FLAC and Numerical Modeling in Geomechnics 2006* (eds Varona, P., Hart, R.), Madrid, Spain, pp. 351–357. Minneapolis: Itasca Consulting Group, Inc.

Reichl, T., Mathis, B. & Beer, G. (2003). Advanced pre-processing methods. *Numerical Simulation in Tunnelling*, Chapter 4. Wien/New York: Springer.

Rowe, R. K. & Lee, K. M. (1992). Subsidence owing to tunnelling. II. Evaluation of a prediction technique. *Can. Geot. J.* **29**, 941–954.

Rowe, R. K., Lo, K. Y. & Kack, G. J. (1983). A method of estimating surface settlement above tunnels constructed in soft ground. *Can. Geot. J.* **20**, 11–22.

Simpson, B., O'Riordan, N. J. & Croft, D. D. (1979). A computer model for the analysis of ground movements in London Clay. *Géotechnique* **29**, No. 2, 149–175.

Tan, T. -S., Setiaji, R. R. & Hight, D. W. (2005). Numerical analyses using commercial software – a black box? *Proc. Underground Singapore 2005*, Singapore, pp. 250–258.

Thomas, A. H. (2003). *Numerical modelling of sprayed concrete lined (SCL) tunnels.* PhD thesis, University of Southampton.

Wood, R. H. (1968). The reinforcement of slabs in accordance with a predetermined field of moments. *Concrete* **2**, No. 2, February, 69–76.

Wright, P. J. (2013). Validation of soil parameters for deep tube tunnel assessment. *Proc. ICE Geot. Engrg* **166**, GE1, 18–30.

Chapter 8

Lining materials

This chapter describes the most common lining materials and their constitutive behaviours. It covers segmental linings, cast-in-place (CIP) linings and sprayed concrete linings, and both primary and secondary linings.

The United Kingdom in particular has a large stock of cast iron segmentally-lined tunnels. Many of these are over 100 years old and are mostly still in very good condition. Due to the high cost of materials and casting, cast iron is rarely used today for new construction, except occasionally where local clients require them or where other types of lining would be difficult to apply, for example, for the step-plate junction of the Northern Line extension in London (FLO/TfL, 2017).

There are many brick and masonry tunnels all over the world, which can be more than 250 years old. These materials are not used for new construction so they will not be dealt with in this book.

Concrete is strong in compression and relatively inexpensive, and so it is nowadays by far the most common material used for tunnel lining. The majority are made from conventional reinforced concrete or steel fibre-reinforced concrete (SFRC). Plain concrete is still occasionally used for smaller diameter tunnels. Due to its fluidity in its fresh state, concrete can be applied in four main ways:

- precasting segments in moulds in a factory to allow assembly of a segmental lining in the tunnel
- casting behind a shutter to form the tunnel lining in situ
- slipforming or extruding a lining, i.e. continuously casting behind a moving shutter
- spraying directly onto the tunnel wall

In all four methods of application, the concrete can be conventional steel bar-reinforced concrete, fibre-reinforced concrete (FRC; steel fibres or fibres made from other materials) or plain concrete.

DOI: 10.1201/9780429470387-8

After working through this chapter, you will understand:

- the different materials used in tunnel linings and their advantages and disadvantages
- durability, watertightness and fire considerations for lining design
- material behaviour of SFRC

After working through this chapter, you will be able to:

- determine the mean and characteristic residual strength values of SFRC from a series of beam tests

8.1 REINFORCED CONCRETE

This has been the dominant material of the last 60–70 years, though steel fibre-reinforced segmental linings are becoming more and more common. In Europe and many other countries around the world, Eurocode 2 (EN 1992-1-1: 2004) is used for design.

Since graduates of civil engineering courses should already be able to design a reinforced concrete section, the examples in this book will focus more on SFRC.

There are limits to the flexural capacity of FRC, and so bar-reinforced concrete is still used for all types of concrete tunnel linings. Sometimes FRC may be augmented by reinforcement bars at high tensile stress locations, for example, around junctions in a sprayed concrete-lined tunnel or near to joints in segmental linings.

8.2 STEEL FIBRE-REINFORCED CONCRETE

SFRC segmental linings have become increasingly common over the last 30 years, as have hybrid steel fibre plus conventional bar-reinforced concrete segmental linings. Likewise, steel fibre-reinforced shotcrete has also become more popular, as this removes the need to fix reinforcement in the tunnel. More recently, SFRC has been used for CIP secondary linings for both tunnels and shafts on the Lee Tunnel and Tideway projects (Hover et al., 2017; Psomas, 2019). This is due to the development of codes of practice and design guidance, improvements in fibre and concrete technology, as well as increasing confidence and acceptance of the use of SFRC through accumulated industry experience.

The use of SFRC has clear benefits for CIP, slipformed and sprayed concrete tunnel linings, some of which are:

- SFRC can have improved post-crack behaviour, resulting in narrower crack widths than in conventional reinforced concrete, and hence better durability (*fib* Bulletin 83, 2017)
- there is no need to fix reinforcement bars, mesh or lattice girders

Table 8.1 Embodied CO_2 of different reinforcement types per m^3 of concrete (ITAtech, 2016).

Reinforcement type	Embodied CO_2 $(kgCO_2/tonne)$	Typical dosage (kg/m^3)	Embodied CO_2 per m^3 of concrete $(kgCO_2/m^3)$
Conventional steel bars	1932	60–160	116–309
Steel fibres	2425	25–40	61–97
Polypropylene synthetic fibres	260	8–10	2–2.6

The use of SFRC also has clear benefits for segmental linings, some of which are:

- SFRC can have improved post-crack behaviour, resulting in narrower crack widths than in conventional reinforced concrete, and hence better durability (*fib* Bulletin 83, 2017).
- Since the steel fibres are mixed into the concrete prior to pouring into the segment moulds, there is no need to manufacture and place reinforcement cages. This saves on factory space, labour and materials.
- Due to the dispersion of the fibres throughout the concrete, including close to segment faces, SFRC can better resist bursting and spalling stresses caused by the application of tunnel boring machine (TBM) jacks, radial joint stresses (de Waal, 2000; Schnütgen, 2003), or accidental impacts during transportation and handling (*fib* Bulletin 83, 2017).

For all types of tunnel linings and methods of application, the total mass of steel required for SFRC is usually far less than for conventional reinforced concrete, resulting in a saving in the overall carbon footprint (see Table 8.1).

8.2.1 Codes of practice and sources of design guidance

SFRC is not covered by the current revision of Eurocode 2 (EN 1992-1-1: 2004) or the American concrete standard ACI 318-19 (2019). At time of writing, the best sources of design guidance are:

1. *fib* Model Code 2010 (2013): '*fib*' is the Fédération Internationale du Béton, or the International Federation for Structural Concrete. *fib* is an international organisation aiming to synthesise international research and experience to produce practical documents for design, so that national code commissions can take advantage of it. Model Code 1990 served as an important basis for Eurocode 2 (EN 1992-1-1: 2004) and it is likely that Model Code 2010 will feed into the Eurocode's next revision. Until Eurocode 2 is updated to include

the design of concrete reinforced by fibres, Model Code 2010 is the most authoritative and internationally recognised guidance available. Section 5.6 describes FRC material behaviour and Section 7.7 describes structural design.

2. *fib* Bulletin 83 (2017): This is a state-of-the-art report on the design of precast tunnel segments in FRC, and provides relevant load cases, guidance and example calculations. Unfortunately there is a lack of detail in some areas.

3. ITAtech Guidance for Precast Fibre-Reinforced Concrete Segments – Vol.1: Design Aspects (ITAtech, 2016). ITAtech is a committee of the International Tunnelling Association (ITA-AITES) dedicated to new technologies. This report is relatively brief but provides specific guidance on the benefits and limitations of the use of SFRC in segmental linings. Unfortunately it also contains many text formatting errors and unclear figures.

4. ACI 544.7R-16 (2016) Report on Design and Construction of Fiber-Reinforced Precast Concrete Tunnel Segments: This report presents the history of FRC precast segments in tunnelling projects throughout the world, sets out the load cases and structural design, and describes the material parameters, tests and analyses required to complete the design.

5. AFTES Recommendation GT38R1A1 (2013) Design, dimensioning and execution of precast steel fibre reinforced concrete arch segments: AFTES is the French tunnelling society.

8.2.2 Material behaviour

The presence of fibres does not significantly influence the behaviour of uncracked concrete. The point where cracking begins is known as the 'first crack', as shown in Figure 8.1. Usually SFRC exhibits deflection-softening behaviour post-crack, as the fibres bridge across the cracks and are gradually drawn out. Using an increased dosage and/or high-performance fibres can produce deflection-hardening behaviour (e.g. Hover et al., 2017), and this is sometimes referred to as 'high performance fibre-reinforced concrete' or 'HPFRC'. 'Deflection hardening' means that there is a peak stress after first crack that is higher than the first crack stress, and 'deflection softening' means that the first crack stress is the peak.

Steel fibres provide ductility in SFRC by failing in 'pull-out', rather than breaking. Therefore, the anchorage of the ends of the fibres is crucial to the post-crack behaviour. Steel fibres usually have what are called 'hooked' ends, although they are not so much hooked as stepped, as shown in Figure 8.2. Resistance to pull-out is initially provided by the bond to the concrete matrix. The fibre is gradually debonded from the crack to the tip. Once the bond has been overcome, resistance is provided by a combination of friction and the work done in continuously deforming the fibre as it is

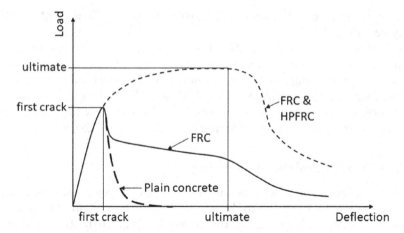

Figure 8.1 Load-deflection curves for plain concrete, strain-softening fibre-reinforced concrete ('FRC') and strain-hardening (high-performance) fibre-reinforced concrete ('FRC & HPFRC') (ITAtech, 2016).

drawn along its stepped path. This is completely different from bar reinforcement, which is expected to fail in elongation, not anchorage.

The ductile post-crack behaviour of SFRC depends not just on the type and dosage of fibres, but also on the concrete itself. A high-strength concrete will require higher strength or a higher dosage of fibres to ensure a ductile response, because otherwise the fibres will break before the bond is overcome. This means that at mature ages, when the concrete has a higher strength, the behaviour may switch from ductile to brittle. This has been called 'embrittlement' (Bernard, 2008), though this term is somewhat misleading because it seems to imply a chemical degradation (Jones, 2014). The solution is to ensure that the fibres are compatible with the concrete matrix at all ages and within the full range of expected strengths, as shown by a large number of panel tests by Bjøntegaard et al. (2014), where embrittlement was avoided by using a slightly lower dosage of a higher strength steel fibre. ITAtech (2016) recommends a fibre yield strength greater than 800 N/mm² when the concrete class is less than or equal to C40/50, but for

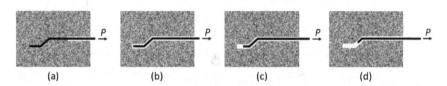

Figure 8.2 Progressive failure of a hooked-end steel fibre under pull-out load P: (a) elongation and partial debonding, (b) full debonding, (c) deformation of the fibre as it is pulled out of the bend in the concrete, (d) pull-out of straightened fibre.

higher strength concretes a fibre yield strength greater than 1500 N/mm² may be required.

In a deflection-softening SFRC, it is likely that the first crack to occur will continue to widen. SFRC tunnel linings usually rely on compressive hoop force in the ring to limit crack widths at the serviceability limit state (SLS). Deflection-hardening SFRC or hybrid SFRC with conventional bar reinforcement can develop multiple small cracks, but this can only be verified by a testing programme (Hover et al., 2017; Psomas, 2019).

8.2.3 Design assisted by testing

Testing is required to determine the flexural tensile strength parameters of SFRC for design. To design to ACI 544.FR requires four-point loading of un-notched beams. To design to the *fib* Model Code 2010 (2013) requires three-point loading of notched beams according to EN 14651: 2005.

The EN 14651: 2005 beam test set-up is shown in simplified form in Figure 8.3. The load is displacement-controlled, meaning that the load is increased or decreased by a computer to achieve a predefined displacement rate. The crack mouth opening distance ('CMOD') is measured continuously across the notch on the underside of the beam, producing a load vs CMOD curve.

SFRC may be characterised by strength classes according to the *fib* Model Code 2010 (2013). Using beam tests to EN 14651: 2005, we get a value for the 'Limit of Proportionality' (LOP), which is the stress at the tip of the notch assumed to act in an uncracked mid-span section. The 'LOP' is defined as the maximum stress at a CMOD between 0 and 0.05 mm, denoted f_L, which is assumed to be the point when the first crack occurs. We also get values for 'residual flexural tensile strengths', which are fictitious

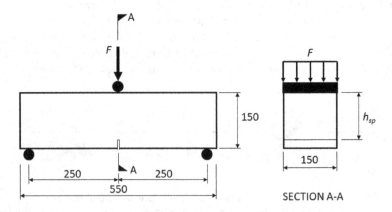

Figure 8.3 Notched beam under three-point loading test set-up according to EN 14651: 2005.

stresses at the tip of the notch assumed to act in an uncracked mid-span section with linear elastic stress distribution. These residual flexural tensile strengths are determined at CMOD values of 0.5, 1.5, 2.5 and 3.5 mm and are denoted f_{R1}, f_{R2}, f_{R3} and f_{R4}, respectively.

These stresses are all calculated based on assuming uncracked linear elastic behaviour. The bending moment M at mid-span is given by the following equation:

$$M = \frac{Fl}{4} \tag{8.1}$$

M is the mid-span bending moment in Nmm
F is the applied force in N
l is the span between the supports (nominally 500 mm) in mm

Using Euler-Bernoulli bending theory, we get the following relationship:

$$\frac{M}{I} = \frac{\sigma}{y} \tag{8.2}$$

I is the second moment of area in mm⁴
σ is the stress in N/mm²
y is the distance from the neutral axis to the extreme fibre in mm

For a linear elastic rectangular section, the second moment of area is given by:

$$I = \frac{bh_{sp}^3}{12} \tag{8.3}$$

b is the width of the beam in mm (nominally 150 mm)
h_{sp} is the section depth at mid-span from the top of the beam to the notch (nominally 125 mm)

And:

$$y = \frac{h_{sp}}{2} \tag{8.4}$$

Thus, the stress σ is given by:

$$\sigma = \frac{3}{2} \frac{Fl}{bh_{sp}^2} \tag{8.5}$$

Using Equation 8.5 gives us values for LOP and residual flexural strengths at different values of CMOD in the test, as shown in Table 8.2.

Table 8.2 Strength parameters determined by beam tests to EN 14651: 2005 for use in the *fib* Model Code 2010 (2013).

Strength parameter		CMOD (mm)
Limit of proportionality f_L		$0 \leq \text{CMOD} \leq 0.05$
Residual flexural tensile strength	f_{R1}	0.5
	f_{R2}	1.5
	f_{R3}	2.5
	f_{R4}	3.5

8.2.4 Determination of characteristic strength values

The strength parameters from multiple beam tests are statistically analysed to determine the characteristic values of these parameters, which are denoted by a subscript '*k*'. When designing to the *fib* Model Code 2010 (2013), we use f_{R1k} for SLS and f_{R3k} for ultimate limit states (ULS). These are used to define a stress block that can be either linear elastic or rigid plastic. This is similar in principle to how we design conventional reinforced concrete.

For specification of SFRC, strength classes can be used where the compressive strength, the value of f_{R1k} and the f_{R3k}/f_{R1k} ratio are defined as shown in the following example:

   FRC 40/50 – 5.0c

This shorthand notation means FRC with a characteristic compressive cylinder strength f_{ck} of 40 N/mm², a characteristic compressive cube strength of 50 N/mm², f_{R1k} = 5.0 N/mm² and the letter 'c' means that $0.9 \leq f_{R3k}/f_{R1k} \leq 1.1$. Therefore, f_{R3k} will be taken as equal to $0.9 \times f_{R1k}$ = 4.5 N/mm².

The strength interval for f_{R1k} must be two consecutive numbers from the series 1.0, 1.5, 2.0, 2.5, 3.0, 4.0, 5.0, 6.0, 7.0, 8.0. The lower value in the interval is used in the classification, so in the example above f_{R1k} is between 5.0 and 6.0 N/mm².

The letters defining the f_{R3k}/f_{R1k} ratio represent the intervals shown in Table 8.3. The letter 'a' represents the minimum level of ductility required for the SFRC to be considered 'reinforced', otherwise the design rules applicable to plain concrete must be used. The letter 'c' represents the minimum level of ductility to achieve crack width control.

More precise minimum characteristic values may be specified by the designer if desired, the strength classes do not have to be followed strictly (ITAtech, 2016).

Table 8.3 Residual strength ratio classification according to the Model Code 2010 (2013).

Residual strength ratio interval	Corresponding letter
$0.5 \leq f_{R3k}/f_{R1k} \leq 0.7$	a
$0.7 \leq f_{R3k}/f_{R1k} \leq 0.9$	b
$0.9 \leq f_{R3k}/f_{R1k} \leq 1.1$	c
$1.1 \leq f_{R3k}/f_{R1k} \leq 1.3$	d
$1.3 \leq f_{R3k}/f_{R1k}$	e

Typical values used in the design of precast SFRC segments are given in ITAtech (2016) and are as follows:

- Characteristic compressive cylinder strength f_{ck}:
  - at early age for demoulding, handling and storage: $f_{ck} \geq 12$ N/mm^2
  - at age ≥ 28 days: $f_{ck} \geq 40$ N/mm^2
  - at age ≥ 90 days: $f_{ck} \geq 50$ N/mm^2
- Residual flexural tensile strength:
  - at early age for demoulding, handling and storage: $f_{R1k} \geq 1.2$ N/mm^2
  - at age ≥ 28 days: $f_{R1k} \geq 2.2$ N/mm^2, $f_{R3k} \geq 1.8$ N/mm^2

These values are indicative and, depending on the project, specified strengths may be higher.

An example of mean and characteristic flexural tensile strengths derived from more than 30 EN 14651: 2005 beam tests is shown in Figure 8.4, from Psomas & Eddie (2016). The testing was for a 7.8 m ID water tunnel segmental lining in London, UK, designed for transient design situations only, not for long-term operational conditions. We can see that the variability of the test results must be quite high for the characteristic value to be so far below the mean. The coefficient of variation of these tests was 10% for the LOP (the peak at CMOD ≤ 0.05 mm), and varies between 21% and 27% for the residual flexural tensile strengths. This is in line with expectations for this kind of testing according to ITAtech (2016), which gives typical values of 10% for the LOP and 25% for residual strengths. Once these characteristic values have been reduced to the design value by dividing by the partial material factor for FRC in tension $\gamma_F = 1.5$, the design values will be around 2.5 times lower than the mean values.

ITAtech (2016) recommends using the approach described in EN 1990: 2002+A1: 2005 Annex D "Design assisted by testing", clause D7.1. This is for the determination of a characteristic value based on a limited number of test results. The 5th percentile characteristic value f_{Rjk}, where f_{Rjk} may represent the LOP or any of the residual flexural tensile strengths, is given

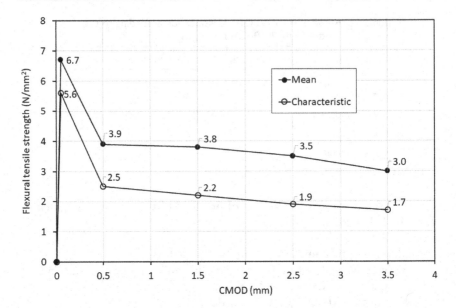

Figure 8.4 An example of mean and characteristic flexural tensile strengths derived from testing (ITAtech, 2016).

by the following equation, on the assumption that the test data follow a normal distribution:

$$f_{Rjk} = m_x \left(1 - k_n V_x\right) \tag{8.6}$$

f_{Rjk} is the 5th percentile characteristic value for the LOP (in which case replace Rj with L) or any of the residual flexural tensile strengths (in which case replace Rj with $R1$, $R2$, $R3$ or $R4$)

m_x is the mean of the flexural tensile strengths at the same CMOD measured in the tests

k_n is a statistical factor, in this case, values may be taken from Table D1 of EN 1990: 2002 (reproduced in Table 8.4)

V_x is the coefficient of variation, which is the standard deviation divided by the mean

Either the coefficient of variation V_x is 'known', i.e. a conservative upper estimate is chosen based on engineering judgement and prior experience, or

Table 8.4 Values of statistical factor k_n for use in Equation 8.6 to calculate 5th percentile characteristic values (from Table D1, EN 1990: 2002+A1: 2005).

No. of tests n	1	2	3	4	5	6	8	10	20	30	∞
'V_x known'	2.31	2.01	1.89	1.83	1.80	1.77	1.74	1.72	1.68	1.67	1.64
'V_x unknown'	-	-	3.37	2.63	2.33	2.18	2.00	1.92	1.76	1.73	1.64

it is 'unknown' and V_x is calculated directly from the test data. According to the 'Designers' Guide to Eurocode 0' (Gulvanessian et al., 2012), and ITAtech (2016), it is preferable to assume V_x is 'known' because even though the mean and standard deviation may be unknown, in most cases, particularly when the number of tests is small, a better estimate of the coefficient of variation may be estimated based on experience than calculated.

For 'V_x known', the statistical factor k_n in Equation 8.6 takes account of uncertainty in estimation of the mean related to sample size, and provides the 5th percentile value of the strength parameter based on an assumed normal distribution with coefficient of variation V_x, all with 95% confidence level (i.e. there is a 5% probability of a further test giving a lower strength than the characteristic value). This can be expressed by the following equation (Gulvanessian et al., 2012):

$$k_n = u_p \left(\frac{1}{n} + 1 \right)^{\frac{1}{2}} \tag{8.7}$$

u_p is the p fractile of the standard normal distribution. For the characteristic value, we are taking $p = 0.05$, i.e. the 5th percentile, therefore, $u_p = 1.645$

n is the sample size, i.e. the number of tests

For 'V_x unknown', k_n also takes into account uncertainty related to the coefficient of variation that has been calculated from the test results, hence the k_n values are larger, but as n approaches infinity, k_n converges on the 5th percentile value of a standard normal distribution (1.645), as shown in Table 8.4. This can be expressed by the following equation (Gulvanessian et al., 2012):

$$k_n = t_p s \left(\frac{1}{n} + 1 \right)^{\frac{1}{2}} \tag{8.8}$$

t_p is the p fractile from the Student t distribution with $(n-1)$ degrees of freedom. It is different for every value of n and can be looked up in a statistical table.

s is the standard deviation calculated from the test data

n is the sample size, i.e. the number of tests

8.2.5 Determination of the characteristic mean strength values

To determine the characteristic value of the mean $f_{R,jm}$, which is used for the SLS, we only need to take account of uncertainty in estimation of the mean related to sample size. The following equation can be used:

$$f_{R,jm} = m_x \left(1 - k_n V_x \right) \tag{8.9}$$

Table 8.5 Values of statistical factor k_n for use in Equation 8.9 to calculate characteristic mean values from test data (EN 1990: 2002+A1: 2005).

No. of tests n	1	2	3	4	5	6	8	10	20	30	∞
'V_x known'	1.64	1.16	0.95	0.82	0.74	0.67	0.58	0.52	0.37	0.30	0
'V_x unknown'	-	4.46	1.69	1.18	0.95	0.82	0.67	0.58	0.39	0.31	0

m_x is the mean of the flexural tensile strengths at the same CMOD measured in the tests

V_x is the coefficient of variation, which is the standard deviation divided by the mean

If V_x is known, then the lower bound value of the mean with 95% confidence level is given by Equation 8.9 using the following value for k_n:

$$k_n = \frac{u_p}{\sqrt{n}} \tag{8.10}$$

u_p is the p fractile of the standard normal distribution. We are taking $p = 0.95$, i.e. the 95th percentile, therefore, $u_p = 1.645$

If V_x is unknown, then the lower bound value of the mean with 95% confidence level is given by Equation 8.9 using the following value for k_n:

$$k_n = \frac{t_p}{\sqrt{n}} \tag{8.11}$$

Values for k_n are given in Table 8.5 based on Equations 8.10 and 8.11.

For a more detailed explanation of how the statistics are calculated, see the 'Designers' Guide to Eurocode 0', Appendix C (Gulvanessian et al., 2012). For 5th percentile characteristic values, see the section on the 'Prediction method' on p. 175, and for mean characteristic values, see 'Estimation of the mean' on p. 172 of Gulvanessian et al. (2012).

8.2.6 Durability

There are several potential durability concerns regarding SFRC:

- chloride-induced corrosion of the steel fibres
- carbonation of the concrete
- alkali-aggregate reaction in the concrete
- freeze-thaw attack
- stray current-induced corrosion

In a literature review of chloride corrosion of SFRC, *fib* Bulletin 83 (2017) found that SFRC could resist chloride penetration much better than conventional reinforced concrete. For uncracked high-quality SFRC, any corrosion of steel fibres is limited to the outer 2–5 mm of uncracked concrete, and therefore has no effect on the long-term flexural capacity. In addition, steel fibres do not cause spalling of the concrete when they corrode, unlike steel bar reinforcement.

For cracked SFRC, *fib* Bulletin 83 (2017) found much more variation in the results of tests by different researchers. Recommended crack width limits proposed in the literature varied between 0 and 0.5 mm. Furthermore, it is difficult to extrapolate results of accelerated laboratory tests on such a wide variety of different concretes and steel fibres to real structures that are usually in far less onerous exposure conditions.

Tests by Nordström (2005) found that galvanised steel fibres had a delayed onset of corrosion and less loss of fibre diameter compared to low-carbon steel fibres, and that stainless steel fibres were completely resistant to corrosion.

In order to design an SFRC tunnel lining for durability, an assumption needs to be made about the maximum allowable crack width for the SLS. There does not seem to be a scientific consensus on what this value should be (*fib* Bulletin 83, 2017) and examples of values used on real projects are scarce. The SLS crack width used on the Lee Tunnel project for the 300 mm thick CIP secondary lining was 0.15 mm (Hover et al., 2017). For the Tideway project SFRC shotcrete shaft primary linings, the SLS crack width was 0.3 mm, and for the CIP SFRC secondary lining, it was 0.2 mm (Psomas, 2019).

8.2.7 Watertightness

Virtually all tunnels and shafts are required to limit leakage from the ground into the structure. Tunnels and shafts that are designed to contain liquid need to be designed to limit leakage out into the ground.

EN 1992-3: 2006 classifies all concrete structures according to the amount of leakage that is allowed:

- 'Tightness class 0': Some leakage acceptable, or leakage irrelevant. Durability considerations will govern the design. The exposure class according to Section 4 of EN 206-1: 2000 and the loading type will determine the maximum crack width according to Section 7.3 of EN 1992-1-1: 2004. In most tunnel linings this will be 0.2 mm.
- 'Tightness class 1': Leakage to be limited to a small amount, some surface staining or damp patches acceptable. Any cracks expected to pass through the full thickness of the lining should be limited to a value w_{k1}. If cracks are not expected to penetrate the full thickness, then durability will govern the design as above.
- 'Tightness class 2': Leakage to be minimal, no staining. If cracks are expected to pass through the whole thickness of the lining, then

additional measures such as waterproof membranes or waterbars will need to be specified.
- 'Tightness class 3': No leakage permitted. Waterproof membranes or other special measures will be needed.

The limiting crack width w_{k1} for a watertight concrete structure without additional measures such as a waterproof membrane is determined by the hydraulic gradient across the lining thickness, i.e. the hydraulic head difference h_D divided by the lining thickness h. Values of w_{k1} are given by National Annexes. The recommended values are 0.2 mm for $h_D/h \leq 5$ and 0.05 mm for $h_D/h \geq 35$. Interpolation is used for intermediate values of h_D/h. For ease of reference:

w_{k1} is the limiting crack width in mm
h_D is the hydraulic head difference from one side of the lining to the other in m
h is the lining thickness in m
h_D/h is the hydraulic gradient across the lining thickness

For most tunnel linings, we do not expect cracks to pass through the full thickness because some part of the section will usually be in compression. EN 1992-3: 2006 specifies that a minimum depth of the section needs to be in compression otherwise cracks are assumed to pass all the way through. The recommended values are the lesser of 50 mm or $0.2h$, where h is the lining thickness in mm. Therefore, for SFRC tunnel linings, it may be important to ensure that the compression zone at the SLS meets this criterion, for characteristic design situations at least.

In some situations, tunnel linings will have cracks passing all the way through. One example is where a secondary lining is resisting an internal liquid pressure, which may result in the full thickness being in tension. Another example is the construction joints between bays of a CIP tunnel lining, which are usually circumferential. These will need additional measures such as a waterbar installed if a waterproof membrane is not provided.

Remember that serviceability of the structure is not the only consideration when specifying watertightness. For example, a sewer passing through a zone where groundwater is abstracted as drinking water may have strict requirements to prevent sewage leaking out of the tunnel. A potable water tunnel may have strict requirements on leakage to avoid wasting precious drinking water, or to prevent untreated groundwater entering the system. Also, a tunnel acting as a drain on the ground around it may cause settlement due to consolidation, may incur expensive pumping costs over many years or may cause environmental damage by lowering the groundwater level and draining peat bogs or ponds.

8.2.8 Fire resistance

When sufficient heat is applied to the surface of a tunnel lining, water in the concrete will be turned into steam. The pressure that is built up by the steam in the voids in the concrete near the surface breaks chunks off and this is known as 'explosive spalling'. This then exposes a new surface and it happens again and again until the lining collapses, the full thickness of the lining is gone, the fire is put out or the fire has consumed all the available fuel.

In most tunnel linings it is sufficient to include a low dosage of microfilament polypropylene fibres in the concrete mix. These very fine fibres melt at approximately 160°C, leaving space for steam to expand and in this way preventing explosive spalling. Full-scale fire tests in a testing laboratory are normally needed to prove the design is adequate.

The material properties of concrete are known to deteriorate with temperature, and it is possible that this could cause structural failure. The thermal properties of SFRC, i.e. the thermal conductivity and specific heat capacity, are primarily determined by the type of coarse aggregates and are not affected by the presence of steel fibres (Li et al., 2016) and so the research and experience related to concrete and reinforced concrete behaviour in a fire can be applied.

EN 1992-1-2: 2004 gives the following equation for the design value of concrete compressive strength during a fire:

$$f_{cd,fi} = \frac{k_\theta f_{ck}}{\gamma_{c,fi}} \tag{8.12}$$

$f_{cd,fi}$ is the design value of compressive strength during a fire in N/mm^2

k_θ is a reduction factor that is dependent on the temperature θ and the type of aggregates. Values are given in Table 3.1 of EN 1992-1-2: 2004.

f_{ck} is the characteristic value of compressive strength in N/mm^2

$\gamma_{c,fi}$ is the material partial factor for concrete in compression during a fire. Values for $\gamma_{c,fi}$ are given in the National Annexes. The recommended value is 1.0.

The relationship between the residual flexural tensile strength f_{R3k} and temperature also needs to be known so that the flexural capacity of the section can be calculated at different temperatures. This is not in the *fib* Model Code 2010 (2013), but *fib* Bulletin 83 (2017) recommends the use of a relationship in the Italian guideline (CNR-DT 204, 2006), which gives values of a reduction factor $k_{\theta t}$ that are 1.0 at 20°C, 0.5 at 200°C, 0.25 at 400°C and 0 at 600°C, with linear interpolation between these values.

8.3 CONCRETE REINFORCED WITH OTHER FIBRES

Although the use of structural synthetic fibres, colloquially referred to as 'plastic fibres', is common in shotcrete, particularly in deep mines where large deformations are expected, it is not common in segmental linings, where serviceability requirements usually require small crack widths that are difficult to achieve with structural synthetic fibres.

The *fib* Model Code 2010 (2013) may also be used for concrete reinforced with structural synthetic fibres, as long as their Young's modulus is not "significantly affected by time and/or thermo-hygrometrical phenomena". This means that the fibres should not experience significant creep, particularly if subjected to humidity or temperature. The reason for this caveat is that the Model Code is mainly based on experience and testing of SFRC.

Structural synthetic fibres tend to fail in elongation, not in pull-out. Since they can generally elongate to high strains before breaking, this does not in itself cause brittle behaviour, and embrittlement has not been known to occur (Bjøntegaard et al., 2014).

Since structural synthetic fibres have a much lower density than steel, about seven to eight times less mass is required to achieve the same volume of fibres; 6 kg/m^3 of synthetic fibres is equivalent in volume to about 45 kg/m^3 of steel fibres.

There are other types of fibres under development, from nanofibres of graphene (Papanikolaou et al., 2019) to alkali-resistant glass or basalt (Sandbakk et al., 2018). These are not yet widely used and are not covered by existing guidelines or standards.

8.4 PLAIN CONCRETE

Historically, segmental linings have been made from plain concrete, and there are many still in service today. These are generally small diameter tunnels that were excavated by open-face TBMs, ensuring that jacking loads were relatively light. The segments were small and so handling loads and bending moments were reduced. As long as the design showed that the section was mainly in compression under service loads, problems only arose at the joints, where bursting or shear stresses may have caused cracking or spalling. Radial and circumferential joint capacity was often verified by full-scale testing, though sadly much of this was not published.

In conventional reinforced concrete design, it is usual to ignore the tensile strength of the concrete itself. Concrete does have a tensile strength, but without reinforcement by bars or fibres, failure will be brittle and sudden (c.f. Figure 8.1), so a more generous safety margin is required.

Design values of tensile and compressive strength of plain concrete may be determined using Section 12 of EN 1992-1-1: 2004. The plain concrete design strength in tension f_{ctd} is given by:

$$f_{ctd} = \frac{\alpha_{ct,pl} f_{ctk}}{\gamma_c} \tag{8.13}$$

f_{ctd} is the design value of concrete tensile strength in N/mm².

$\alpha_{ct,pl}$ is a coefficient to take account of "long-term effects on the tensile strength and of unfavourable effects resulting from the way the load is applied" (EN 1992-1-1: 2004). For plain concrete, EN 1992-1-1: 2004 recommends $\alpha_{ct,pl} = 0.8$ and the UK National Annex requires $\alpha_{ct,pl} = 0.6$ (NA to BS EN 1992-1-1: 2004).

f_{ctk} is the characteristic value of concrete tensile strength in N/mm². For plain concrete, the characteristic value f_{ctk} can be obtained by splitting tests to EN 12390-6: 2009. For preliminary design, the value from Table 3.1 of EN 1992-1-1: 2004 can be used.

γ_c is the partial material factor for concrete and the recommended value is 1.5.

The plain concrete design strength in compression f_{cd} is given by:

$$f_{cd} = \frac{\alpha_{cc,pl} f_{ck}}{\gamma_c} \tag{8.14}$$

f_{cd} is the design value of concrete compressive (cylinder) strength in N/mm².

$\alpha_{cc,pl}$ is a coefficient to take account of "long-term effects on the compressive strength and of unfavourable effects resulting from the way the load is applied" (EN 1992-1-1: 2004). For plain concrete, EN 1992-1-1: 2004 recommends $\alpha_{cc,pl} = 0.8$ and the UK National Annex requires $\alpha_{cc,pl} = 0.6$ (NA to BS EN 1992-1-1: 2004).

f_{ck} is the characteristic value of concrete compressive strength in N/mm².

8.5 CAST IRON

Cast iron has good strength in compression and good corrosion resistance and is therefore a good choice of material for tunnel linings. However, it is more expensive than the various types of concrete linings, can be time-consuming to install, and temporary works are required to support the exposed ground at the face (unless a TBM is used). Cast iron tunnel linings consist of segments that are always bolted together at both the radial and circumferential joints. Radial joints are usually staggered. Any void outside the rings is usually grouted with cementitious grout through grout ports built into the segments. Special tapered segments may be used to go round curves, sometimes with very tight radii.

Older cast iron is known as 'grey cast iron'. Properties can vary for historic grey cast iron before 1928, but for grey cast iron made after 1928, several Grades are defined in BS EN 1561: 2011, from which compressive strength, tensile strength, Young's modulus and Poisson's ratio may be derived (LUL, 2007). Strengths are usually expressed as permissible design stresses, with no need for partial factors.

Figure 8.5 shows an example of relatively modern cast iron rings built in 2007. Note the staggered radial joints and the solid key segments in the crown. These were a modern form of cast iron called 'spheroidal graphite iron' (SGI), for which there is a European Standard BS EN 1563: 2018.

Since cast iron segments are not solid rectangular sections, the second moment of area must be calculated using the parallel axis theorem. Also, the position of the neutral axis must be found by taking moments of area.

A good overview of the properties of cast iron tunnel linings, common defects and common durability issues may be found in the 'London Underground Manual of Good Practice for Deep Tube Tunnels and Shafts' (LUL, 2007).

Figure 8.5 Cast iron rings in the Piccadilly Line Access passageway at King's Cross under construction.

Laboratory testing of both grey cast iron and SGI was reported in detail by Thomas (1977). More recent testing and finite element modelling were described in Tsiampousi et al. (2017). 3D numerical modelling of cast iron tunnel linings may be found in Li et al. (2014, 2015).

8.6 SUMMARY

The most common materials used as tunnel linings were described, with particular focus on SFRC.

The procedure for deriving characteristic residual flexural strength values for SFRC was described in detail, as well as durability, watertightness and fire resistance.

REFERENCES

ACI 318-19 (2019). *Building code requirements for structural concrete*. Farmington Hills: American Concrete Institute.

ACI 544.7R-16 (2016). *Report on design and construction of fiber-reinforced precast concrete tunnel segments*, Reported by ACI Committee 544. Farmington Hills: American Concrete Institute.

AFTES Recommendation GT38R1A1 (2013). Design, dimensioning and execution of precast steel fibre reinforced concrete arch segments. *Tunnels et Espace Souterrain*, No. 238, 312–324.

Bernard, E. S. (2008). Embrittlement of fibre-reinforced shotcrete. *Shotcrete*, Summer issue, 16–20.

Bjøntegaard, Ø., Myren, S., Klemetsrud, K. & Kompen, R. (2014). Fibre reinforced sprayed concrete (FRSC): Energy absorption capacity from 2 days age to one year. *Proceedings of 7th International Symposium on Sprayed Concrete – Modern Use of Wet Mix Sprayed Concrete for Underground Support* (eds Beck T., Woldmo, O. & Engen, S.), Sandefjørd, Norway, 16th–19th June, pp. 88–97. Oslo, Norway: Norwegian Concrete Society/Norsk Betongforening.

BS EN 1561: 2011. *Founding. Grey cast irons*. London, UK: British Standards Institution.

BS EN 1563: 2018. *Founding. Spheroidal graphite cast irons*. London, UK: British Standards Institution.

CNR-DT 204 (2006). *Guidelines for design, construction and production control of fiber reinforced concrete structures*. Rome: National Research Council of Italy.

de Waal, R. G. A. (2000). *Steel fibre reinforced tunnel segments for the application in shield driven tunnel linings*. PhD Thesis. Technische Universiteit Delft.

EN 14651: 2005. *Test method for metallic fibered concrete – Measuring the flexural tensile strength (limit of proportionality (LOP), residual)*. Brussels: European Committee for Standardization.

EN 12390-6: 2009. *Testing hardened concrete – Tensile splitting strength of test specimens*. Brussels: European Committee for Standardization.

EN 1990: 2002+A1: 2005. *Eurocode – Basis of structural design*, incorporating corrigenda December 2008 and April 2010. Brussels: European Committee for Standardization.

EN 1992-1-1: 2004. *Eurocode 2: Design of concrete structures – Part 1-1: General rules and rules for buildings*. Brussels: European Committee for Standardization.

EN 1992-1-2: 2004. *Eurocode 2: Design of concrete structures – Part 1-2: General rules – Structural fire design*. Brussels: European Committee for Standardization.

EN 1992-3: 2006. *Design of concrete structures – Part 3: Liquid retaining and containment structures*. Brussels: European Committee for Standardization.

EN 206-1: 2000. *Concrete – Part 1: Specification, performance, production and conformity*, incorporating Corrigendum No. 1 and Amendment No. 1. Brussels: European Committee for Standardization.

fib Bulletin 83 (2017). *Precast tunnel segments in fibre-reinforced concrete*, State-of-the-art Report, fib WP 1.4.1, October. Lausanne: Fédération internationale du béton (*fib*).

fib Model Code 2010 (2013). *fib Model Code for concrete structures 2010*. Lausanne: Fédération internationale du béton (*fib*).

FLO/TfL (2017). *Northern Line Extension Kennington Step-Plate Junction Update*. Available at: http://content.tfl.gov.uk/step-plate-junction-surface-works-possession.pdf [last accessed 17th September 2019].

Gulvanessian, H., Calgaro, J. -A. & Holický, M. (2012). *Designers' guide to Eurocode: Basis of structural design*, 2nd Edition. London: ICE Publishing.

Hover, E., Psomas, S. & Eddie, C. (2017). Estimating crack widths in steel fibre-reinforced concrete. *Proc. Inst. Civ. Engrs Construction Materials* **170**, No. 3, 141–152.

ITAtech (2016). *Design guidance for precast fibre reinforced concrete segments – Vol. 1: Design aspects*, ITAtech Report No. 7. Lausanne: ITA-AITES. Reproduction of figures with permission of the Committee on new technologies of the International Tunnelling and Underground Space Association – ITAtech.

Jones, B. D. (2014). Roundup of the 7th International Symposium on Sprayed Concrete. *Tunnelling Journal*, August/September issue, 25–29.

Li, S., Jones, B., Thorpe, R. & Davis, M. (2016). An investigation into the thermal conductivity of hydrating sprayed concrete. *Constr. Build. Mater.* **124**, 363–372.

Li, Z., Soga, K., Wang, F., Wright, P. & Tsuno, K. (2014). Behaviour of cast-iron tunnel segmental joint from the 3D FE analyses and development of a new bolt-spring model. *Tunn. Undergr. Space Technol.* **41**, 176–192.

Li, Z., Soga, K. & Wright, P. (2015). Behaviour of cast-iron bolted tunnels and their modelling. *Tunn. Undergr. Space Technol.* **50**, 250–269.

LUL (2007). *Manual of good practice – Civil Engineering – Deep tube tunnels and shafts*, G-055 A1. London, UK: London Underground Limited.

NA to BS EN 1992-1-1: 2004. *UK National Annex to Eurocode 2: Design of concrete structures – Part 1-1: General rules and rules for buildings*. London: British Standards Institution.

Nordström, E. (2005). *Durability of sprayed concrete – Steel fibre corrosion in cracks*. Doctoral thesis, Luleå University of Technology, Sweden.

Papanikolaou, I., Al-Tabbaa, A. & Goisis, M. (2019). An industry survey on the use of graphene-reinforced concrete for self-sensing applications. *Proceedings of International Conference on Smart Infrastructure and Construction* (eds De Jong, M. J., Schooling, J. M. & Viggiani, G. M. B.), pp. 613–622. London: ICE Publishing.

Psomas, S. (2019). Service limit state design for pressurised steel fibre reinforced concrete tunnel linings. *Proceedings of WTC2019 – Tunnels and Underground Cities: Engineering and Innovation meet Archaeology, Architecture and Art* (eds Peila, D., Viggiani, G. & Celestino, T.), Naples, Italy, pp. 2898–2908. London, UK: Taylor & Francis Group.

Psomas, S. & Eddie, C. M. (2016). SFRC segmental lining design for a pressurised tunnel. *Proceedings of WTC2016*, San Francisco, USA, Paper 0118. Englewood, CO: SME.

Sandbakk, S., Miller, L. W. & Standal, P. C. (2018). MiniBarsTM – A new durable composite mineral macro fiber for shotcrete, meeting the energy absorption criteria for the industry. *Proceedings of 8th International Symposium on Sprayed Concrete – Modern Use of Wet Mix Sprayed Concrete for Underground Support*, Trondheim, Norway, 11th–14th June (eds Beck, T., Myren, S. & Engen, S.), pp. 272–282. Oslo, Norway: Norwegian Concrete Society/Norsk Betongforening.

Schnütgen, B. (2003). Design of precast steel fibre reinforced tunnel elements. *International RILEM Workshop on Test and Design Methods for Steel Fibre Reinforced Concrete* (eds Schnütgen, B. & Vandewalle, L.), Bochum, Germany, pp. 145–152. Paris, France: RILEM Publications SARL.

Thomas, H. S. H. (1977). Measuring the structural performance of cast iron tunnel linings in the laboratory. *Ground Engineering*, July, 29–36.

Tsiampousi, A., Yu, J., Standing, J., Vollum, R. & Potts, D. (2017). Behaviour of bolted cast iron joints. *Tunn. Undergr. Space Technol.* **68**, 113–129.

Chapter 9

Segmental lining design

In Chapters 6 and 7 we learnt how to undertake simple 2D plane strain analysis of a tunnel lining and the basic principles of numerical modelling in 2D and 3D. We can use these methods to design segmental linings for ground and water loads if we adjust the lining stiffness to take account of the joints. The presence of joints also means there will be contact eccentricities, which themselves cause bending moments and shear forces, as well as bursting and spalling stresses.

This chapter will first deal with joints in a global analysis of a segmental lining. Then it will go through the design of a segmental lining for service loads from the ground and groundwater. It is assumed that most readers of this book will already be familiar with reinforced concrete design, and so the focus in the worked examples will be on fibre-reinforced concrete design.

After working through this chapter, you will understand:

- how flat and curved joints behave when rotated and the effect of packers
- contact stress concentrations caused by geometrical inaccuracies
- how global analysis of a tunnel lining can incorporate joint rotation behaviour

After working through this chapter, you will be able to:

- include joint stiffness in an analytical or numerical model of a segmental lining
- design for bursting stresses near to joints
- estimate the effect of joint rotation on axial force eccentricity and hence on bending moments
- design segmental linings for ground and groundwater loads in service, with a particular focus on steel fibre-reinforced concrete (SFRC)

This chapter is not intended to be a comprehensive manual for design of segmental tunnel linings, but a teaching aid to demonstrate the principles

DOI: 10.1201/9780429470387-9

and most important aspects of design. Every project is unique and has different load cases and performance requirements, some of which may not be covered here. But the understanding gained from working through this chapter, and careful reading of the papers, guidelines and standards cited, will give you a solid foundation.

In this chapter we will look at how to design a segmental lining taking account of the effect of the joints. Having joints in a ring will reduce its overall flexural stiffness compared to a monolithic (solid, jointless) ring. This is beneficial, as it will reduce the bending moments in the ring. You can demonstrate this yourself using the Curtis-Muir Wood method (see Chapter 6) in a spreadsheet – reduce the moment of inertia of the lining and see what happens. There are different methods available to take account of this effect, depending on the type of analysis used. These will be presented and critiqued in the following sections.

As usual in design there is a trade-off; increasing the number of segments may slow down the rate of tunnelling, as it takes longer to build a ring with more segments. It also increases the number of joints, and thus the potential for defects to allow water ingress, which could be an issue for operation and maintenance. On the other hand, reducing the number of segments will increase the size and weight of each segment, increasing the handling loads the segment needs to resist during demoulding, storage, transportation, handling and erection. This will be covered in Chapter 10, where we will cover transient loads during the whole life of the segments before they are in their final position.

9.1 TAKING ACCOUNT OF THE EFFECT OF JOINTS IN 2D PLANE STRAIN ANALYSES

In preliminary analysis of ground and groundwater loads on a segmental tunnel lining, Muir Wood (1975) proposed that the effective moment of inertia I_e of the tunnel lining should be reduced to take account of the presence of radial joints using the following equation:

$$I_e = I_j + I\left(\frac{4}{n}\right)^2 \text{ with } I_e \not> I, \, n > 4 \tag{9.1}$$

I_e is the effective moment of inertia of the tunnel lining
I_j is the moment of inertia of the joints
I is the moment of inertia of a segment
n is the number of equal segments in a ring. Where a key segment is used, it is usually counted as one joint, not two, since the two top segments plus key usually have the same arc as two ordinary segments. So a ring with six ordinary segments, two top segments and a key would have $n = 8$.

Note that this equation was incorrectly printed in Muir Wood's paper, though it is clear from his example in the text that Equation 9.1 above is what was intended.

The equation is only applicable if there are more than four joints. In reality a ring can deform with four joints if the joints are aligned with the principal stress directions in the ground (i.e. joints at 3, 6, 9 and 12 o'clock positions). But if they are rotated 45°, then the joints have no effect. When a ring has four joints, their orientation makes a huge difference, but this effect decreases as the number of joints increases. Hefny & Chua (2006) compared Muir Wood's equation to analyses using continuum finite element models and found that as the number of joints increases beyond six, the influence of joint position relative to the principal stress directions becomes negligible.

Judgement is required to determine I_j. We know that I_j cannot be less than zero, and that I_e must be less than I of a segment section. If $I_j = 0$, this is the same as assuming that the joints are completely free to rotate. If $I_e = I$, this is the same as assuming the lining is monolithic. Free rotation is virtually impossible in practice, as rotation will always result in an eccentricity of thrust across a radial joint, resulting in a bending moment that resists further rotation. It also assumes that an individual ring's deformation is unimpeded by the adjacent rings, which would only be the case if the circumferential joint had no shear keys, dowels or friction, or if the radial joints were all aligned down the length of the tunnel (Klappers et al., 2006). It is usual practice to stagger joints and to encourage shear resistance, so this assumption is unrealistic. The difference between aligned and staggered radial joints, and monolithic rings, is illustrated in Figure 9.1.

Our objective is to estimate a realistic but conservative value of I_e so that we can design a more efficient tunnel lining. The two limiting situations are either that the radial joints are pin joints, providing no rotational

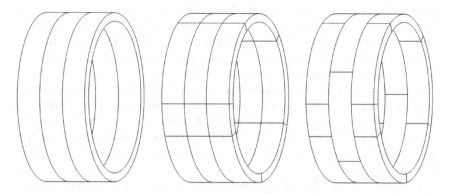

Figure 9.1 From left to right: monolithic rings, aligned radial joints, staggered radial joints.

stiffness, or that the lining is monolithic (i.e. as though there were no joints at all). de Waal (2000) showed, and perhaps it is intuitively obvious, that the truth always lies somewhere between these two limiting situations. Fei et al. (2014) also proved this using scale models. Kavvadas et al. (2017) found that when the radial joints are staggered and have a high rotational stiffness, and the shear stiffness at the circumferential joint is also high, the behaviour may be similar to a continuous monolithic lining with no joints at all. Their work suggests that for a conservative first approximation it is probably best to ignore the effect of joints altogether.

Moreover, Equation 9.1 is effectively smearing the effect of the joints around the whole ring. This allows us to calculate 'average' bending moments, but in reality, bending moments near to the joints will have a different value (usually higher) than bending moments mid-segment and will depend on the joint geometry, bolt details, packer and gasket. Also, in staggered rings with friction and/or shear keys across the circumferential joints, forces are diverted around joints in a complex 3D manner.

One way to conservatively estimate the local bending moment (described in greater detail in Section 9.2) is to assume a joint rotation based on the maximum specified ovalisation and then to use the joint geometry and packer properties to calculate the eccentricity of the hoop force and hence a bending moment. The designer can then use the greater of the 'average' and this local value for design, and this approach should be conservative. From experience, however, I know that this approach is not commonplace and the usually less conservative 'average' values are used for design.

9.2 ROTATIONAL RIGIDITY – FLAT JOINTS

In reality, bending moments at the joints depend on the geometry of the joint and the properties of the packer, if present. As a joint rotates, the line of thrust across it moves and this eccentricity of the thrust transfers moment to the adjacent segment (Figure 9.2). The moment is equal to the hoop force multiplied by the eccentricity. In addition, as eccentricity increases and the hoop force gets closer to the edge, the shearing resistance of the corner reduces as the potential shear plane reduces in length. This can result in damage to the edges of the joints and if it occurs on the intrados, requires repair. If a shear failure occurs on the extrados, it cannot be seen or repaired, and it may also compromise the watertightness of the gasket. Therefore, designing to minimise eccentricity is important.

The properties of the packer have a big influence on the eccentricity of the hoop force. A packer that is too thin or too compressible will result

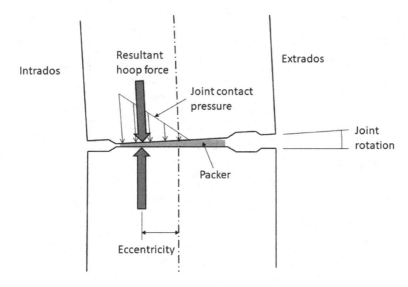

Figure 9.2 Eccentricity of hoop thrust caused by joint rotation.

in a concrete-to-concrete point contact and a high stress concentration. A packer that is too stiff will also generate a large eccentricity as stress is mobilised at the edge of the packer first. Understanding the stress-strain behaviour of a packer and how it interacts with joint rotation is absolutely essential.

Sometimes radial joints are designed with curved surfaces, and in theory, curved joints have a lower rotational stiffness than flat joints. This is because when flat joint surfaces rotate relative to each other, they will quickly generate a larger eccentricity by tending to hinge about the edge of the contact area. As curved joints rotate, the line of thrust stays closer to the centreline of the segment. This will be discussed later in this chapter. Notice also that as flat joints in a ring rotate, the perimeter of the ring increases. As far as I know, no one has researched this effect, but it should be expected to slightly increase the hoop force. If curved joints have a radius approximately equal to one half the arc length of a segment, then the perimeter does not change.

Eccentricity in radial joints is also caused by 'lipping', where segments are misaligned and have a step between them. In this case, the eccentricity is equal to half the misalignment, as shown in Figure 9.3.

What we want to estimate is a value for rotational stiffness. This is the bending moment per unit length (in the tunnel longitudinal direction) required to cause a unit rotation angle along a radial joint. According to full-scale tests by Teachavorasinskun & Chub-uppakarn (2010), typical values are 1000–3000 kNm/rad. In the following sections we will

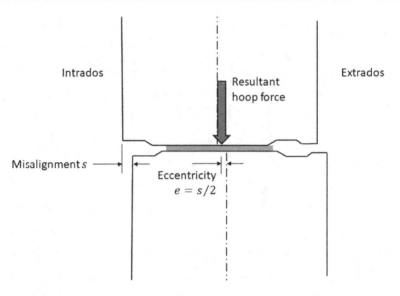

Figure 9.3 Effect of misalignment on eccentricity.

look at the following different packer types and their effect on rotational stiffness:

- linear elastic packers, which do not exist in the real world, but the assumption of linear elastic behaviour allows for a simple calculation of the contact stress distribution, which helps us to understand the behaviour of a rotating joint and the calculation steps
- nonlinear bituminous and plastic packers, which exhibit an increase in stiffness with strain
- nonlinear plywood and medium density fibreboard (MDF) packers, which exhibit a decrease in stiffness with strain

This will be followed by a comparison of the different types and a summary.

9.2.1 Linear elastic packers

Figure 9.4 shows a contact stress distribution that one might expect from a linear elastic packer, and this is sometimes assumed in calculations because it is easy to solve, even though linear elastic packers do not exist in the real world. Nowadays, packers are most often made from bituminous or plastic (polyethylene or polyurethane based) materials. MDF or plywood is becoming less popular due to durability concerns (Cavalaro & Aguado, 2012). None of these materials behave in a linear elastic manner so to get a reasonable estimate of rotational stiffness we will then need to do something a bit more sophisticated, but let's start with a simple linear elastic packer

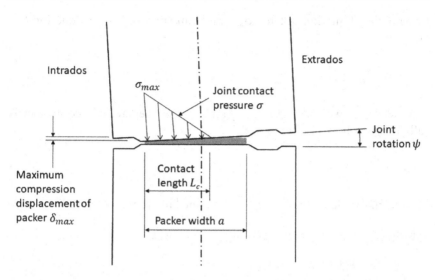

Figure 9.4 Calculation of contact stress distribution – linear elastic packer with triangular stress distribution.

model to explain the steps. Figure 9.4 shows the geometry and symbols we are going to use.

For simplicity it is assumed that the packer is symmetrically positioned on the centreline of the joint. The packer is linear elastic, with Young's modulus E_p, so it follows from Hooke's Law that the maximum compression displacement of the packer δ_{max} is given by:

$$\delta_{max} = \frac{\sigma_{max} t_p}{E_p} \tag{9.2}$$

δ_{max} is the maximum compression displacement of the packer in mm
σ_{max} is the maximum stress as shown in Figure 9.4 in N/mm²
t_p is the thickness of the packer in mm
E_p is the Young's modulus of the packer in N/mm²

Also, the area of the triangular contact stress distribution in Figure 9.4 must equal the hoop force N across the joint for there to be equilibrium:

$$NL_s = \frac{\sigma_{max} L_c L_p}{2} \tag{9.3}$$

N is the hoop force per metre length of tunnel in N/m
L_s is the segment length in the tunnel longitudinal direction in mm
L_c is the contact length as shown in Figure 9.4 in mm
L_p is the length of the packer in the tunnel longitudinal direction (as it won't be covering the full length of the segment) in mm

Rearranging Equation 9.2 for σ_{max} and substituting it into Equation 9.3 gives:

$$N = \frac{\delta_{max} E_p L_c L_p}{2 t_p L_s} \tag{9.4}$$

Now we have two unknowns, δ_{max} and L_c, but they are related geometrically by the following Equation 9.5:

$$L_c = \frac{\delta_{max}}{\psi} \tag{9.5}$$

ψ is the angle of joint rotation, as defined in Figure 9.4, in radians (rad)

Substituting Equation 9.5 into Equation 9.4 gives:

$$N = \frac{\delta_{max}^2 E_p L_p}{2 t_p \psi L_s} \tag{9.6}$$

Rearranging for δ_{max} gives us:

$$\delta_{max} = \sqrt{\frac{2 t_p \psi L_s N}{E_p L_p}} \tag{9.7}$$

Using Equation 9.7, we can find the value of δ_{max} that gives us equilibrium for a given value of hoop force N, packer properties E_p, L_p and t_p, and joint rotation ψ. We can then use this value of δ_{max} in Equation 9.5 to find L_c. Once L_c is known, the position of the resultant hoop force can be found, in this case, a triangular contact stress distribution, it will be $L_c/3$ from the position of σ_{max}.

If the calculated contact length L_c is greater than the packer width a, then it means that the contact pressure distribution is trapezoidal rather than triangular, as shown in Figure 9.5.

Equation 9.2 still stands but we now reason as follows:

The hoop force per segment is now in equilibrium with a trapezoidal pressure distribution varying from σ_{max} to σ_{min} over the full width of the packer a:

$$NL_s = \left(\frac{\sigma_{max} + \sigma_{min}}{2} \right) L_p a \tag{9.8}$$

Now δ_{min} has a similar relationship to σ_{min} as δ_{max} has to σ_{max} in Equation 9.2, so we end up with:

$$NL_s = \frac{E_p L_p a}{2 t_p} (2\delta_{max} - a\psi) \tag{9.9}$$

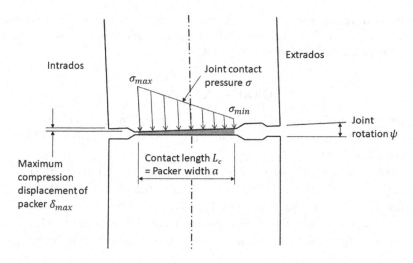

Figure 9.5 Calculation of contact stress distribution – linear elastic packer with trapezoidal stress distribution.

Solving for δ_{max} gives us:

$$\delta_{max} = \frac{NL_s t_p}{E_p L_p a} + \frac{a\psi}{2}$$ (9.10)

Because we know δ_{max} and the rotation angle ψ, δ_{min} can be calculated:

$$\delta_{min} = \delta_{max} - a\psi$$ (9.11)

By taking moments about the centreline, we can calculate the eccentricity:

$$NL_s e = \frac{(\sigma_{max} - \sigma_{min})aL_p}{2} \times \left(\frac{a}{2} - \frac{a}{3}\right)$$ (9.12)

The moment due to eccentricity of hoop force is given by:

$$M = Ne$$ (9.13)

  M is the moment in Nmm
  e is the eccentricity of the hoop force N from the centreline of the segments in mm

The rotational stiffness is defined as the bending moment per unit length (in the tunnel longitudinal direction) required to cause a unit rotation angle along a radial joint, and is given by:

$$k = \frac{M}{\psi}$$ (9.14)

  k is the rotational stiffness in kNm/rad

Thus we have calculated the eccentricity of hoop force due to joint rotation, and from that we have calculated the moment induced by that eccentricity and the rotational stiffness of the joint.

WORKED EXAMPLE 9.1

Use a spreadsheet to calculate the eccentricity of hoop force for joint rotations from $\psi = 0°$ to $1°$ and for hoop forces $N = 500, 1000, 1500, 2000$ and 2500 kN/m.

The linear elastic packer has the following properties: width $a = 150$ mm, thickness $t_p = 3$ mm, packer length in tunnel longitudinal direction $L_p = 900$ mm and Young's modulus $E_p = 40$ MPa.

The segment has the following properties: internal diameter $ID = 6000$ mm, segment thickness $t_s = 250$ mm, length in tunnel longitudinal direction $L_s = 1000$ mm.

From the values of eccentricity, calculate the bending moments. Plot a graph of bending moment (on the y-axis) vs joint rotation (on the x-axis).

The relationship between bending moment and joint rotation is not linear, even though the packer is linear elastic. Investigate this by plotting a graph of contact stress (on the y-axis) vs distance across the packer (on the x-axis) for hoop force $N = 1000$ kN/m and explain what is going on. An example of the spreadsheet calculation for $\psi = 0.5°$ ($= 0.008727$ rad) and $N = 1500$ kN/m ($= 1500$ N/mm) is presented below.

Assuming a triangular contact stress distribution, we can use Equation 9.7:

$$\delta_{max} = \sqrt{\frac{2t_p\psi L_s N}{E_p L_p}} = \sqrt{\frac{2 \times 3 \times 0.008727 \times 1000 \times 1500}{40 \times 900}} = 1.477 \text{ mm}$$

Then using Equation 9.5:

$$L_c = \frac{\delta_{max}}{\psi} = \frac{1.477}{0.008727} = 169.257 \text{ mm}$$

Now $L_c > a$, so the contact stress distribution must be trapezoidal. Start again using Equations 9.10 and 9.11 instead:

$$\delta_{max} = \frac{NL_s t_p}{E_p L_p a} + \frac{\psi a}{2} = \frac{1500 \times 1000 \times 3}{40 \times 900 \times 150} + \frac{0.008727 \times 150}{2} = 1.488 \text{ mm}$$

$$\delta_{min} = \delta_{max} - a\psi = 1.488 - 150 \times 0.008727 = 0.179 \text{ mm}$$

Using Equation 9.2 rearranged for σ_{max} and σ_{min}:

$$\sigma_{max} = \frac{\delta_{max}E_p}{t_p} = \frac{1.488 \times 40}{3} = 19.838 \text{ N/mm}^2$$

$$\sigma_{min} = \frac{\delta_{min}E_p}{t_p} = \frac{0.179 \times 40}{3} = 2.384 \text{ N/mm}^2$$

To check the answer is correct, we can insert σ_{max} and σ_{min} into the equilibrium Equation 9.8:

$$N = \left(\frac{\sigma_{max} + \sigma_{min}}{2}\right)\frac{L_p a}{L_s} = \left(\frac{19.838 + 2.384}{2}\right) \times \frac{900 \times 150}{1000} = 1500 \text{ kN/m}$$

To find the eccentricity e of the resultant hoop force N from the joint centreline, we resolve moments about the centreline as follows:

$$NL_s e = \frac{(\sigma_{max} - \sigma_{min})aL_p}{2} \times \left(\frac{a}{2} - \frac{a}{3}\right)$$

Therefore:

$$e = \frac{(19.838 - 2.384) \times 150 \times 900}{2 \times 1500 \times 1000} \times \left(\frac{150}{2} - \frac{150}{3}\right) = 19.635 \text{ mm}$$

So in this case the eccentricity $e = 19.64$ mm, and using Equation 9.13 the bending moment M resisting rotation is $M = Ne = 29.45$ kNm/m.

We can then calculate that the (secant) rotational stiffness, using Equation 9.14, is $M/\alpha = 3375$ (kNm/m)/rad.

Figure 9.6 shows the relationship between joint rotation and bending moment for a linear elastic packer as calculated in a spreadsheet. At higher values of joint rotation, the line changes from a straight line to a curve, and this corresponds exactly to the point at which the contact stress switches from a trapezoidal distribution to a triangular distribution. As one might expect, this transition from trapezoidal to triangular stress distribution occurs later at higher values of hoop force.

The rotational stiffness, which is the gradient of the curves in Figure 9.6, therefore is only a constant while there is a trapezoidal stress distribution, but beyond the transition to a triangular distribution, it decreases (or becomes less stiff) as rotation increases.

In order to illustrate the transition, the contact stress distributions for the hoop force of 1000 kN/m, as rotation is increased, are shown in Figure 9.7.

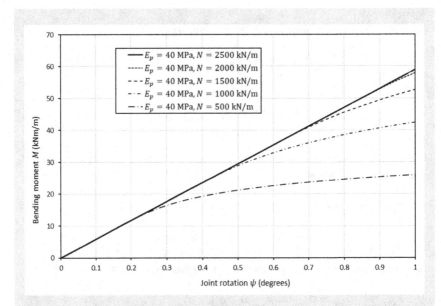

Figure 9.6 Bending moment – joint rotation relationship for a linear elastic packer with Young's modulus 40 MPa at different levels of hoop force.

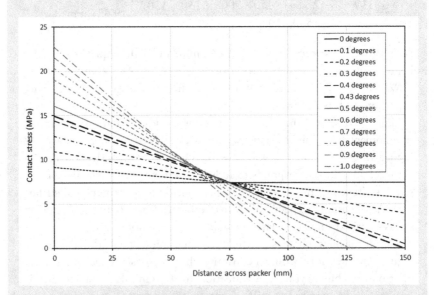

Figure 9.7 Contact stress distribution across a 150 mm wide linear elastic packer as joint rotates under 1000 kN/m hoop force (note transition from trapezoidal to triangular stress distribution at approximately 0.43°).

At approximately 0.43 degrees, shown by the thick black dashed line, the stress distribution switches from a trapezoidal shape to a triangular shape. Perhaps counterintuitively, the rate of change of the eccentricity (i.e. how far the resultant hoop force is from the centreline) decreases after the transition to a triangular distribution, and this results in the shape of the curves in Figure 9.6.

9.2.2 Nonlinear packers

Linear elastic packers do not exist in the real world. All real packers are nonlinear.

Cavalaro & Aguado (2012) performed tests on several types of packers used in projects in Spain. They presented stress-strain data for a 1.92 mm thick bituminous packer and a 2.15 mm thick plastic packer, both used for Barcelona Metro Line 9 tunnels. I have fitted a stress-strain relationship to their data, which has the form:

$$\sigma = A\left[\exp\left(\varepsilon^n\right) - 1\right] \tag{9.15}$$

σ is the stress in the packer in N/mm^2
ε is the strain in the packer
A and n are curve-fitting constants

Cavalaro & Aguado (2012) did propose an equation $\left[\sigma = A\left(1 - e^{-\varepsilon^n}\right)\right]$ for the stress-strain relationship but it did not fit their data.

An important effect, particularly for bituminous packers, is that the first load cycle has a much softer response and on subsequent load cycles the material has a much stiffer behaviour. This is known as the 'Mullins effect' and is due to densification of the material.

The stress-strain curves based on Equation 9.15 and using the curve-fitting parameters in Table 9.1 are shown in Figure 9.8.

Table 9.1 Parameters from curve-fitting of tests on plastic and bituminous packers (based on test data from Cavalaro & Aguado, 2012).

	Code	Thickness t_p	A	n
Plastic 1st loading cycle	P1	2.15 mm	0.028	0.398
Plastic 2nd loading cycle	P2		0.047	0.394
Plastic 3rd loading cycle	P3		0.048	0.394
Bituminous 1st loading cycle	B1	1.92 mm	0.013	0.366
Bituminous 2nd loading cycle	B2		0.079	0.425
Bituminous 3rd loading cycle	B3		0.063	0.434

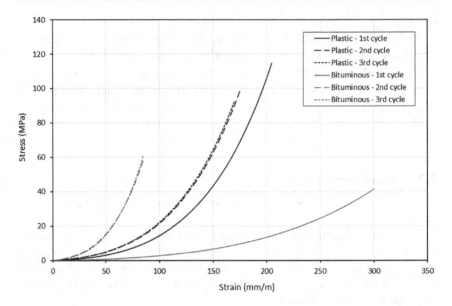

Figure 9.8 Stress-strain curves of plastic and bituminous packers plotted using Equation 9.15.

As an example, we will use the parameters for only the first loading cycle of the plastic packer 'P1' as given in Table 9.1.

This calculation is now a bit more complicated than the one in Section 9.2.1. Nevertheless, it can still be done in a spreadsheet by dividing the width of the packer into slices and calculating the stress in each slice. The rotation angle is fixed and the maximum compression displacement δ_{max} is varied using a 'goal seek' function until the integral of the stresses is in equilibrium with the hoop force.

The contact stress distributions as rotation is increased for the plastic packer with parameters P1 is shown in Figure 9.9. Compared to the linear elastic packer in Figure 9.7 the stress moves much more quickly to the edge of the packer and attains much higher levels, but once there it changes less and less. This is because as the strain increases, the plastic packer becomes stiffer and stiffer, as was shown in Figure 9.8.

Other types of packer become less stiff as strain increases. This is the opposite of the plastic or bituminous materials we have just seen, but is typical behaviour of most engineering materials such as steel and concrete, and indeed other materials used as packers, such as plywood and MDF. Based on a compression test on a 3 mm plywood packer, the following bilinear relationship could be assumed:

$E = 170$ MPa for $0 \leq \varepsilon \leq 0.05$

$E = 23$ MPa for $0.05 \leq \varepsilon \leq 0.5$

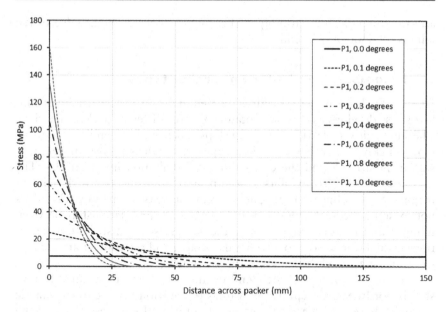

Figure 9.9 Contact stress distribution for nonlinear plastic packer with parameters P1.

The contact stress distribution for a 3 mm thick plywood packer is shown in Figure 9.10 for different values of joint rotation. The stress distribution here is completely different from Figure 9.9, being much more spread out and avoiding the very high values found at the edge of the plastic packer. Maximum compressive stresses are lower than for the linear elastic packer

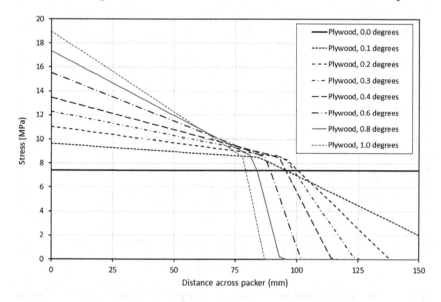

Figure 9.10 Contact stress distribution for nonlinear 3 mm plywood packer.

and the potential for crushing or shearing of concrete at the edge of the joint is therefore much reduced.

9.2.3 Comparison of different packer types

Figure 9.11 shows how the bending moment induced by eccentricity of contact stress develops as the joint rotates for three different packer materials. The joint with a plastic packer develops much higher bending moments more quickly, as perhaps one might expect given the stress-strain relationship in Figure 9.8 and the contact stress distribution in Figure 9.9. The plywood packer also exhibits a nonlinear relationship but with a lower magnitude. The linear elastic packer (remember, this type of packer does not exist in the real world!) begins with a linear relationship between moment and rotation angle, but as the contact stress distribution transitions from a trapezoidal to a triangular shape, and the contact area begins to shrink, it also exhibits a nonlinear relationship.

Since in all the cases the bending moment-joint rotation relationship is not linear (except at low values of rotation for the linear elastic packer), there is not a single value of rotational stiffness that can be used for a particular packer type. In Figure 9.12 the rotational stiffness is plotted against rotation angle for the segment geometry in Worked Example 9.1, with the three different packers, and with a hoop

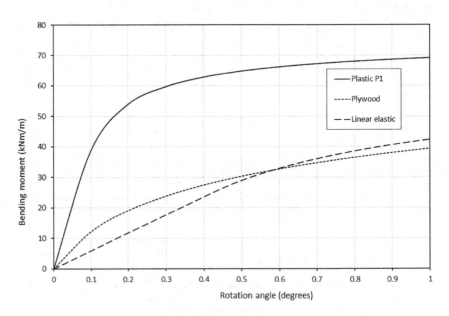

Figure 9.11 Bending moment vs joint rotation for plastic, plywood and linear elastic packers, at a hoop force of 1000 kN/m.

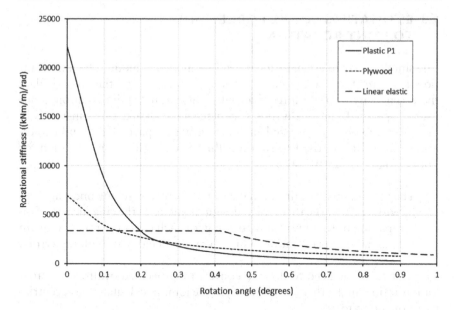

Figure 9.12 Variation of rotational stiffness of joints with plastic, plywood and linear elastic packers with rotation angle, at a hoop force of 1000 kN/m.

force of 1000 kN/m. All packer types settle into a rotational stiffness of around 1000–3000 kNm/rad as found by Teachavorasinskun & Chubuppakarn (2010), but at lower levels of joint rotation, rotational stiffness can be much higher.

It is difficult to briefly summarise the overall effect of these very different packer behaviours on how a segmental lining ring will behave. For instance, a higher radial joint rotational stiffness will result in a higher overall ring stiffness, which will result in less deformation but higher bending moments. A lower rotational stiffness will result in a more flexible ring, with larger deformations but lower bending moments.

On the other hand, the high stress concentrations at the edge of the packer that result from using plastic or bituminous packers that become stiffer at higher strain levels, may have a knock-on effect on the design of the concrete segment itself in the vicinity of the joint, as bursting, crushing or shear failure may occur at the intrados or extrados due to the larger eccentricity of the hoop force.

For detailed design it may be appropriate to feed these relationships between rotational stiffness and rotation angle into a model of a tunnel (either bedded beam or finite element) to investigate the effect. It is clear, however, that a material that has a stiffness that decreases with strain level, such as concrete (i.e. no packer) or plywood, is preferable to a material whose stiffness increases with strain level, such as bituminous or plastic packers.

9.3 ESTIMATING LOCAL FORCES DUE TO JOINT ROTATION

The simplest way to estimate local bending moments induced by joint rotation is to specify the maximum rotation as an input and then to calculate the eccentricity of axial force that will be generated. Joint rotation is not usually specified, but ovalisation (the percentage change of diameter) is in most cases explicitly specified in the contract as part of the build tolerances. For instance, the British Tunnelling Society (BTS) Specification for Tunnelling (BTS, 2010) says:

> The maximum and minimum measured diameters in any one ring shall be within 1% of the theoretical design diameter of the ring measured on completion of ring build and grouting, or such other tolerance stated in the Particular Specification. This tolerance includes all building errors.

The first thing we need to do is to convert this specified ovalisation into a joint rotation angle, then we can analyse the joint to calculate the eccentricity of the hoop force.

As an aside, the specification of a maximum ovalisation as 1% of the diameter results in rather large tolerances for tunnels larger than approximately 6.5 m diameter. Therefore, for larger tunnels, a value rather than a percentage should be given.

9.3.1 Calculating joint rotation from a specified ovalisation

The ovalisation is assumed to be elliptical (as we assumed in Chapter 6 for the Curtis-Muir Wood analytical solution), as shown in Figure 9.13.

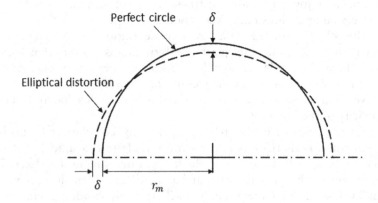

Figure 9.13 Elliptical distortion or 'ovalisation', shown for half a tunnel ring in squatting mode.

The change in radius in Figure 9.13 is given by:

$$\delta = \rho r_m \qquad (9.16)$$

δ is the change in radius in mm
ρ is the percentage ovalisation/100, so for BTS (2010) this would be 0.01
r_m is the radius to the centroid of the lining in mm

The worst case, in terms of joint rotation, would be for a joint to be located at either the axis or the crown. Due to the symmetry of an ellipse, the rotation in both these locations will be equal and opposite, so we only need to calculate for one.

A segment to one side of a joint at the axis is shown in Figure 9.14, and a further simplification of the geometry is shown in Figure 9.15. In Figure 9.15, AB represents the original segment chord and DE represents the new segment chord position. The segment is assumed to be rigid and hence the chord length does not change.

The segment arc angle θ shown in Figures 9.14 and 9.15 is given by:

$$\theta = \frac{360°}{n} \qquad (9.17)$$

θ is the angle of the segment arc in °
n is the number of segments in the ring, assuming they are of equal size

From Figure 9.15, the joint rotation ψ will be given by the following equation, which is the difference between angles α and β, multiplied by 2 because of symmetry (the adjacent segment below axis level will rotate as well):

$$\psi = 2(\alpha - \beta) \qquad (9.18)$$

ψ is the joint rotation
α and β are the angles defined in Figure 9.15

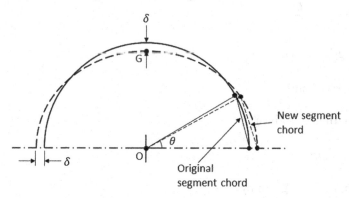

Figure 9.14 Geometrical representation of a segment when the ring undergoes elliptical distortion.

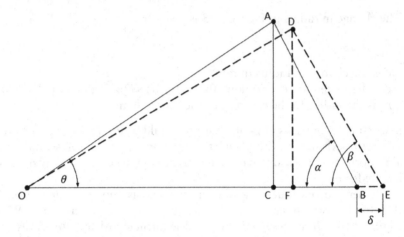

Figure 9.15 Geometrical simplification of segment rotation.

It can be seen from Figure 9.14 that the distance OG is given by:

$$OG = r_m - \delta \tag{9.19}$$

For the undeformed perfectly circular lining we get the following relationship:

$$OA = OB = r_m \tag{9.20}$$

The triangle OAB is an isosceles triangle, so the angle α in Figure 9.15 is given by:

$$\alpha = \frac{180° - \theta}{2} \tag{9.21}$$

The length AC is given by:

$$AC = OA\sin\theta \tag{9.22}$$

The length OC is given by:

$$OC = OA\cos\theta \tag{9.23}$$

The length CB is given by:

$$CB = OB - OC \tag{9.24}$$

The chord lengths AB and DE are assumed to be equal and are given by:

$$AB = DE = \sqrt{AC^2 + CB^2} \tag{9.25}$$

The ellipse major axis OE is:

$$OE = OB + \delta \tag{9.26}$$

The equation for an ellipse gives us:

$$\frac{DF^2}{OG^2} + \frac{OF^2}{OE^2} = 1 \tag{9.27}$$

Rearranged this gives:

$$DF^2 = \left(1 - \frac{OF^2}{OE^2}\right)OG^2 \tag{9.28}$$

Now Pythagoras's equation for right-angled triangle DEF gives us:

$$(OE - OF)^2 + DF^2 = DE^2 \tag{9.29}$$

Combining Equations 9.28 and 9.29 gives:

$$\left(1 - \frac{OG^2}{OE^2}\right)OF^2 - 2 \times OE \times OF + (OE^2 + OG^2 - DE^2) = 0 \tag{9.30}$$

This is a quadratic equation of the form:

$$a \times OF^2 + b \times OF + c \tag{9.31}$$

a, b, c are coefficients

To find OF, we first find the coefficients a, b and c:

$$a = \left(1 - \frac{OG^2}{OE^2}\right) \tag{9.32}$$

$$b = -2 \times OE \tag{9.33}$$

$$c = (OE^2 + OG^2 - DE^2) \tag{9.34}$$

Now to solve the quadratic equation:

$$OF = -b \pm \frac{\sqrt{b^2 - 4ac}}{2a} \tag{9.35}$$

There are two solutions to the quadratic equation, one being a spurious result that will be much too large and can be discarded.

Now EF is given by:

$$EF = OE - OF \qquad (9.36)$$

And using a trigonometric identity for triangle DEF:

$$\beta = \arccos\left(\frac{EF}{DE}\right) \qquad (9.37)$$

arccos is the inverse cosine

Once the angle β has been calculated, the joint rotation may be calculated using Equation 9.18. Then we can calculate the eccentricity of the hoop thrust across the joint, taking account of the joint geometry and packer properties by using the procedure described in the previous Section 9.2. Once the eccentricity is calculated, several aspects then need to be checked including local compression (crushing of the concrete), bending stresses, shear and bursting stresses.

WORKED EXAMPLE 9.2

Calculate the joint rotation for the precast concrete segmental lining described in Worked Example 9.1, then calculate the eccentricity of the hoop force.

The segmental lining has eight segments plus a key. It is to be built with an ovalisation tolerance of 1%. The ring geometry is:

Internal diameter, $ID = 6.0$ m
Outside diameter, $OD = 6.5$ m
Lining thickness, $t = 0.25$ m

Usually the hoop force would be estimated using an analytical solution, such as the Curtis-Muir Wood solution described in Section 6.2, or a bedded beam model. For this example we will assume the design value of the hoop force $N_{Ed} = 1000$ kN/m.

A plywood packer is used in the joint, of thickness 3 mm and a bilinear stress-strain relationship as described in Section 9.2, i.e. $E_p = 170$ MPa for $0 \leq \varepsilon \leq 0.05$ and $E_p = 23$ MPa for $0.05 \leq \varepsilon \leq 0.5$.

First we calculate the radius to the lining centroid:

$$r_m = \frac{ID+t}{2} = \frac{6+0.25}{2} = 3.125 \text{ m}$$

Each segment arc angle $\theta = 360°/8 = 45°$

Maximum change in radius, $\delta = 0.01 r_m = 0.01 \times 3.125 = 0.03125 \text{ m}$

A segment to one side of a joint at the axis (with arc of 45°) is shown in Figure 9.16, and a further simplification of the geometry is shown in Figure 9.17. In Figure 9.17, AB represents the original segment chord and DE represents the new segment chord position. The segment is assumed to be rigid and hence the chord length does not change.

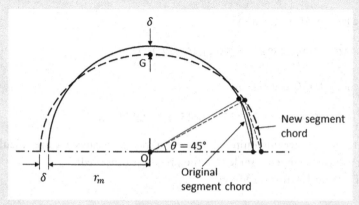

Figure 9.16 Worked Example 9.2 – Geometrical representation of a segment when the ring undergoes elliptical distortion.

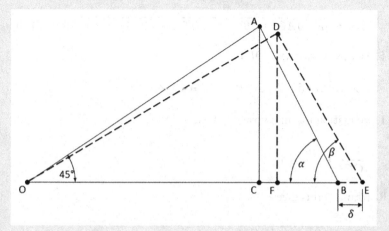

Figure 9.17 Worked Example 9.2 – Geometrical simplification of segment rotation.

It can be seen from Figure 9.16 that:

$$OG = r_m - \delta = 3.125 - 0.03125 = 3.09375 \text{ m}$$

For the undeformed lining:

$$OA = OB = r_m = 3.125 \text{ m}$$

The triangle OAB is an isosceles triangle, so the angle α is given by:

$$\alpha = \frac{180° - 45°}{2} = 67.5°$$

The length AC is given by:

$$AC = OA\sin 45° = 3.125 \times \sin 45° = 2.209709 \text{ m}$$

The length OC is given by:

$$OC = OA\cos 45° = 3.125 \times \cos 45° = 2.209709 \text{ m}$$

Note we are carrying quite a few decimal places because we are calculating a small change in angle and the precision is needed.
  The length CB is given by:

$$CB = OB - OC = 3.125 - 2.209709 = 0.915291 \text{ m}$$

The chord lengths AB and DE are equal and are given by:

$$AB = DE = \sqrt{AC^2 + CB^2} = \sqrt{2.209709^2 + 0.915291^2} = 2.391771 \text{ m}$$

The ellipse major axis OE is:

$$OE = OB + \delta = 3.125 + 0.03125 = 3.15625 \text{ m}$$

The equation for an ellipse gives us:

$$\frac{DF^2}{OG^2} + \frac{OF^2}{OE^2} = 1$$

Rearranged this gives:

$$DF^2 = \left(1 - \frac{OF^2}{OE^2}\right)OG^2$$

Now Pythagoras's equation for right-angled triangle DEF gives us:

$$(OE - OF)^2 + DF^2 = DE^2$$

Combining these two equations gives:

$$\left(1 - \frac{OG^2}{OE^2}\right)OF^2 - 2 \times OE \times OF + \left(OE^2 + OG^2 - DE^2\right) = 0$$

This is a quadratic equation of the form:

$$a \times OF^2 + b \times OF + c$$

To find OF, we first find the coefficients a, b and c:

$$a = \left(1 - \frac{OG^2}{OE^2}\right) = \left(1 - \frac{3.09375^2}{3.15625^2}\right) = 0.039212$$

$$b = -2 \times OE = -2 \times 3.15625 = -6.3125$$

$$c = \left(OE^2 + OG^2 - DE^2\right) = \left(3.15625^2 + 3.09375^2 - 2.391771^2\right) = 13.812632$$

Now to solve the quadratic equation for OF:

$$-b \pm \frac{\sqrt{b^2 - 4ac}}{2a} = 6.3125 \pm \frac{\sqrt{(-6.3125)^2 - 4 \times 0.039212 \times 13.812632}}{2 \times 0.039212}$$

$$= 2.218719 \text{ m}$$

The second solution to the quadratic equation (158.7658 m) is spurious and can be discarded.

Now EF is given by:

$$EF = OE - OF = 3.15625 - 2.218719 = 0.937531 \text{ m}$$

And using a trigonometric identity for triangle DEF:

$$\cos\beta = \frac{EF}{DE} = \frac{0.937531}{2.391771} = 0.391982$$

And therefore the angle β is 66.922117°, and the joint rotation $\psi = 2(67.5° - \beta) = 1.156°$.

Now we need to use this joint rotation to calculate the eccentricity of the hoop thrust.

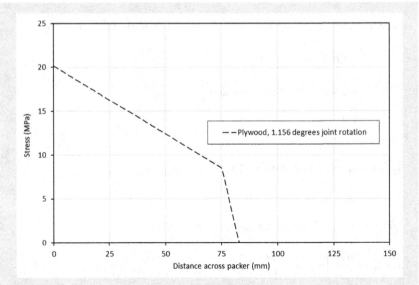

Figure 9.18 Worked Example 9.2 – Plywood packer contact stress distribution at joint rotation of 1.156° corresponding to an ovalisation of 1%.

The contact stress distribution may be calculated using the method described in Section 9.2, and the result of the spreadsheet calculation is shown in Figure 9.18.

By finding the centroid of the stress distribution, the eccentricity of the hoop thrust across the joint is 43.7 mm from the segment centreline, and therefore the bending moment induced by this eccentricity is 43.7 kNm/m (the hoop thrust of 1000 kN/m multiplied by the eccentricity). This 'local' bending moment can be compared to the 'average' bending moment obtained by a global analysis. The larger of the two should be used for design of the segment body.

Note that this calculation assumes that the packer is centred on the joint centreline. This may not always be the case, and adjustments need to be made to the methodology to allow for this.

9.3.2 Using eccentricity of hoop force to check joint capacity in crushing and shear

The next step is to use the contact stress distribution and eccentricity of the hoop force at the maximum ovalisation to check the joint for concrete crushing or shear.

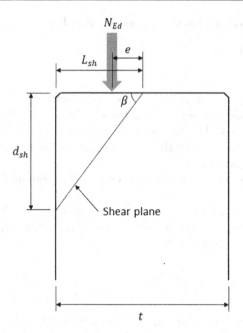

Figure 9.19 Shear plane across a segment joint.

From the contact stress distribution we know the maximum compressive stress applied to the concrete joint surface through the packer. Eurocode 2 for concrete design, EN 1992-1-1: 2004 Cl. 5.10.2.2(5), requires this to be limited to less than $0.6f_{ck}$ to avoid local crushing, where f_{ck} is the characteristic cylinder compressive strength.

Knowing the eccentricity of the hoop force, we can also calculate whether the corner of the segment could be sheared off. This kind of damage is quite common in segmentally lined tunnels.

Figure 9.19 shows the shear plane. We need to calculate the shear stress on the shear plane, then check that it is within the shear resistance capacity of the segment. The shear stress will depend on the angle β assumed for shear failure and the position of shear plane initiation on the joint surface L_{sh}.

We could use the contact stress distribution rather than the resultant hoop force. Then if the shear plane begins closer to the edge of the segment, it will be shorter but the force acting on it will be lower. The position of the shear plane L_{sh} could be varied to find the worst case. However, it is simpler and more conservative to assume the shear plane begins under the resultant hoop force, such that:

$$L_{sh} = \frac{t}{2} - e \qquad\qquad (9.38)$$

The shear stress along the shear plane is given by:

$$\tau = \frac{N_{Ed}\sin\beta\cos\beta}{L_{sh} \times L_s} \tag{9.39}$$

τ is the design value of shear stress in N/mm²
N_{Ed} is the design value of hoop force in N
β is the angle of the shear plane (see Figure 9.19)
L_{sh} is the distance from the beginning of the shear plane to the edge of
the segment in mm
L_s is the length of the segment in the tunnel longitudinal direction in mm

The maximum shear stress always occurs when $\beta = 45°$, because $\sin\beta\cos\beta$ is at a maximum when $\beta = 45°$. Therefore, since $\sin45°\cos45° = 0.5$, Equation 9.39 can be rewritten as:

$$\tau = \frac{N_{Ed}}{2L_{sh}L_s} \tag{9.40}$$

For a conventional steel bar-reinforced concrete, EN 1992-1-1: 2004 can be used to check the shear capacity of the section using the following equation to calculate shear resistance:

$$v_{Rd,c} = C_{Rd,c}k\left(100\rho_l f_{ck}\right)^{1/3} + k_1\sigma_{cp} \tag{9.41}$$

$v_{Rd,c}$ is the shear resistance as a stress in N/mm²
$C_{Rd,c}$ has a recommended value of $0.18/\gamma_c$ or a value defined in the country's National Annex
γ_c is the partial material factor for concrete in compression and the
recommended value is 1.5

$k = 1 + \sqrt{200/d} \leq 2.0$

d is the depth of the section in shear
ρ_l is the proportion of the shear area that is longitudinal steel bar reinforcement, where $\rho_l = A_s/b_w d$
b_w is the width of the member, in this case, it is the segment length in
the longitudinal tunnel direction, which we have denoted L_s
f_{ck} is the characteristic cylinder compressive strength of the concrete in
N/mm²
$k_1 = 0.15$ or a value defined in the country's National Annex
$\sigma_{cp} = N_{Ed}/b_w d$, which is the axial stress in the member in N/mm²

For plain concrete, lightly reinforced concrete or SFRC, Section 12.6.3 of EN 1992-1-1: 2004 can be used.

The axial stress σ_{cp} is defined as for reinforced concrete. The shear stress calculated in Equation 9.40 must be less than the 'concrete design strength in shear and compression' f_{cvd}.

The concrete design strength in tension f_{ctd} is needed for the calculation of f_{cvd}. It is given by:

$$f_{ctd} = \frac{\alpha_{ct} f_{ctk}}{\gamma_c} \tag{9.42}$$

f_{ctd} is the design value of concrete tensile strength in N/mm²

α_{ct} is a coefficient to take account of "long-term effects on the tensile strength and of unfavourable effects resulting from the way the load is applied" (EN 1992-1-1: 2004). For reinforced concrete, EN 1992-1-1: 2004 recommends $\alpha_{ct} = 1.0$ and this is also the value in the UK National Annex (UK NA to BS EN 1992-1-1: 2004). For plain concrete, EN 1992-1-1: 2004 recommends $\alpha_{ct,pl} = 0.8$ and the UK National Annex requires $\alpha_{ct,pl} = 0.6$. The reinforced concrete value could be used for SFRC if ductility requirements of the *fib* Model Code 2010 (2013) are met.

f_{ctk} is the characteristic value of concrete tensile strength in N/mm². For plain or SFRC concrete, the characteristic value f_{ctk} can be obtained by splitting tests to EN 12390-6: 2009. For preliminary design, the value from Table 3.1 of EN 1992-1-1: 2004 can be used, and we will use that in the worked examples here.

γ_c is the partial material factor for concrete in tension and the recommended value is 1.5

The concrete design strength in compression f_{cd} is needed for the calculation of f_{cvd}. It is given by:

$$f_{cd} = \frac{\alpha_{cc} f_{ck}}{\gamma_c} \tag{9.43}$$

f_{cd} is the design value of concrete compressive (cylinder) strength in N/mm².

α_{cc} is a coefficient to take account of "long-term effects on the compressive strength and of unfavourable effects resulting from the way the load is applied" (EN 1992-1-1: 2004). For reinforced concrete, EN 1992-1-1: 2004 recommends $\alpha_{cc} = 1.0$, but the value in the UK National Annex (UK NA to BS EN 1992-1-1: 2004) is $\alpha_{cc} = 0.85$. For plain concrete, EN 1992-1-1: 2004 recommends $\alpha_{cc,pl} = 0.8$ and the UK National Annex requires $\alpha_{cc,pl} = 0.6$. The reinforced concrete value could be used for SFRC if ductility requirements of the *fib* Model Code 2010 (2013) are met.

f_{ck} is the characteristic value of concrete compressive strength in N/mm².

One of the following two expressions is used to obtain f_{cvd}:

If $\sigma_{cp} \le \sigma_{c,lim}$; $f_{cvd} = \sqrt{f_{ctd}^2 + \sigma_{cp} f_{ctd}}$ $\qquad$ (9.44)

If $\sigma_{cp} > \sigma_{c,lim}$; $f_{cvd} = \sqrt{f_{ctd}^2 + \sigma_{cp} f_{ctd} - \left(\dfrac{\sigma_{cp} - \sigma_{c,lim}}{2}\right)^2}$ $\qquad$ (9.45)

f_{cvd} is the concrete design strength in shear and compression in N/mm²
$\sigma_{cp} = N_{Ed}/b_w d$, which is the axial stress in the member in N/mm²
$\sigma_{c,lim}$ is a limiting compressive stress, where $\sigma_{c,lim} = f_{cd} - 2\sqrt{f_{ctd}(f_{ctd} + f_{cd})}$, in N/mm²
f_{cd} is the concrete design strength in compression in N/mm²

The following Worked Example will continue from where Worked Example 9.2 left off, using the previously calculated hoop force and eccentricity to check for local crushing and shear of the concrete local to the joint.

WORKED EXAMPLE 9.3

Check for local crushing and shear local to the joint face for an SFRC segment. The same geometry as in Worked Examples 9.1 and 9.2 is used. The SFRC segment is FRC 50/60 5.0c according to the fib Model Code 2010 (2013).

Checking for local crushing:

The maximum contact stress in Figure 9.18 is 20.15 MPa. The European Standard EN 1992-1-1: 2004 Cl. 5.10.2.2(5) requires this to be limited to less than $0.6f_{ck}$ to avoid local crushing, where f_{ck} is the characteristic cylinder compressive strength, in this case 50 MPa. Therefore the contact stress is OK and there will be no local crushing failure.

Checking for local shear:

This kind of situation is not covered by the fib Model Code 2010 (2013), so the provisions of EN 1992-1-1: 2004 for plain concrete should be used. For a concrete with a characteristic compressive strength $f_{ck} = 50$ N/mm², Table 3.1 of EN 1992-1-1: 2004 gives a characteristic tensile strength $f_{ctk} = 2.9$ N/mm².

For the values of design tensile and compressive strength of concrete, we will use the values of α_{cc} and α_{ct} for reinforced concrete, not plain concrete, because the SFRC should prevent a brittle failure.

Taking $\alpha_{ct} = 1.0$, the EN 1992-1-1: 2004 'recommended value' for reinforced concrete in tension, and $\gamma_{ct} = 1.5$, the design value of tensile strength is given by:

$$f_{ctd} = \frac{\alpha_{ct}f_{ctk}}{\gamma_{ct}} = \frac{1.0 \times 2.9}{1.5} = 1.93 \text{ N/mm}^2$$

Taking $\alpha_{cc} = 0.85$, the UK National Annex (UK NA to BS EN 1992-1-1: 2004) value for reinforced concrete in compression, and $\gamma_{cc} = 1.5$, the design value of compressive strength is given by:

$$f_{cd} = \frac{\alpha_{cc}f_{ck}}{\gamma_{cc}} = \frac{0.85 \times 50}{1.5} = 28.33 \text{ N/mm}^2$$

Now, according to Section 12.6.3 of EN 1992-1-1: 2004:

$$\sigma_{c,lim} = f_{cd} - 2\sqrt{f_{ctd}\left(f_{ctd} + f_{cd}\right)} = 28.33 - 2\sqrt{1.93(1.93 + 28.33)} = 13.05 \text{ N/mm}^2$$

Similarly to a reinforced concrete calculation, the axial stress in the member is given by:

$$\sigma_{cp} = N_{Ed}/b_w d = 1000 \times 10^3 /1000 \times 81.3 = 12.3 \text{ N/mm}^2$$

Therefore, $\sigma_{cp} \leq \sigma_{c,lim}$ and we use Equation 9.44:

$$f_{cvd} = \sqrt{f_{ctd}^2 + \sigma_{cp}f_{ctd}} = \sqrt{1.93^2 + 12.3 \times 1.93} = 5.24 \text{ N/mm}^2$$

Using units of N and mm, the shear stress is given by Equation 9.40:

$$\tau = \frac{N_{Ed}}{2L_{sh}L_s} = \frac{1000 \times 10^3}{2 \times 81.3 \times 1000} = 6.15 \text{ N/mm}^2$$

This is more than the value of f_{cvd}, so the segment corner will fail in shear. There may be ways to improve the geometry of the joint or a different geometry or type of packer could be used. A thicker lining is also a possibility, or alternatively, bar reinforcement could be provided close to the radial joints.

9.4 BURSTING STRESSES

When ram loads are applied to the circumferential joint by the tunnel boring machine (TBM) jacks, and when hoop force is transferred across a radial joint, the force is not applied evenly across the whole segment

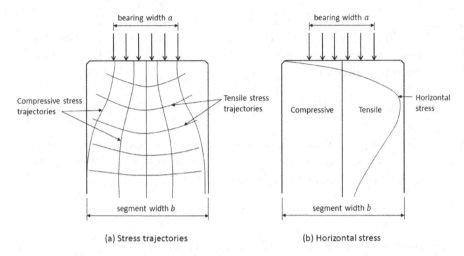

(a) Stress trajectories   (b) Horizontal stress

Figure 9.20 Bursting stress visualisation. (a) Stress trajectories, and (b) Horizontal stress.

thickness, but concentrated on a loading area defined by the contact width of the packer or the size of the ram shoe.

For TBM jacks, usually pairs of hydraulic rams have a 'shoe' on the end, which pushes on the circumferential joint. There are typically two of these shoes on each segment with a gap between them, and so the loading area also does not cover the full segment arc and bursting stresses can develop in both directions, radially and circumferentially.

Whenever a compressive stress is applied to an area smaller than the surface area of a segment face, then as this compressive stress spreads out with depth into the section, orthogonal tensile bursting stresses are induced. This is illustrated in Figure 9.20.

We can calculate bursting force F_{bst} using Leonhardt's equations (reproduced in BS EN 1992-1-1: 2004 as Equation 9.14):

$$F_{bst} = 0.25\left(1 - \frac{a}{b}\right)N_{Ed} \qquad (9.46)$$

F_{bst} is the bursting force in kN, which can be assumed to act between a depth of $0.1b$ and b.

a is the bearing width in mm, as shown in Figure 9.20.

b is the width of the segment face in mm, as shown in Figure 9.20 for a centrally applied load. For an eccentric load, it is 2× the distance to the nearest segment edge.

N_{Ed} is the design value of axial force in kN, which could be the TBM jacking force for a circumferential joint or the hoop force for a radial joint.

This solution assumes the segment material is elastic.

For reinforced concrete, bars need to be detailed to resist this bursting force. This is usually done by providing U-shaped returning bars at some distance from the joint face.

For SFRC, it is more difficult because the SFRC is not elastic in tension, except at very low stresses. Therefore, if the design stress exceeds the characteristic value of the residual tensile strength at the limit of proportionality divided by its partial material factor, then nonlinear numerical modelling and/or full-scale laboratory testing will be needed. For example, Nogales & de la Fuente (2020) describe laboratory testing and 3D numerical modelling of bursting due to TBM ram load application on an SFRC segment, with the load applied both centrally and with an eccentricity.

For plain concrete, the bursting stress cannot exceed the design tensile strength f_{ctd}.

9.5 CURVED JOINTS

Curved joints can be either convex-convex or convex-concave. Convex-convex joints usually have the same radius on both joint faces. Convex-concave, or 'knuckle' joints, will always have a larger radius on the concave face compared to the convex face.

Usually it is only the radial joints that are curved, and the circumferential joints are flat. An example of a curved radial joint is shown in Figure 9.21, which uses joint dimensions from the Storebælt Eastern Railway Tunnel (Elliott et al., 1996). The 7.7 m ID ring was made up of six segments plus a key. The reinforced concrete segments were 400 mm thick and 1650 mm long and were bolted and gasketed. There was no packer in the convex-convex radial joints but there was a bituminous packer in the flat circumferential joints.

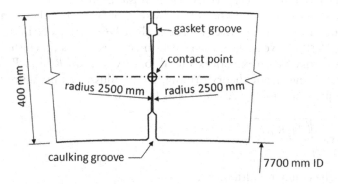

Figure 9.21 Convex-convex radial joint for the Storebælt Eastern Railway Tunnel (joint dimensions from Elliott et al., 1996).

It has been known for segments to have curved circumferential joints, or even radial joints that are curved in the longitudinal direction as well as across the thickness, but these are thankfully rare. These doubly curved joints were used on the Jubilee Line in London in the 1970s and were very difficult to build within tolerance. They also gradually deteriorated over time and began moving and cracking, until they eventually needed to be replaced with steel segments (TfL, 2015).

It is sometimes said (e.g. Luttikholt, 2007) that curved joints do not transfer any bending moment. This is not true. When curved joints rotate, there is an eccentricity produced, but it is smaller than for a flat joint. The eccentricity due to curved joint rotation may be calculated using the following equation:

$$e_j = \left(\frac{R_1 R_2}{R_1 + R_2} \right) \theta \qquad (9.47)$$

e_j is the eccentricity of thrust due to joint rotation in mm
θ is the joint rotation in radians
R_1 and R_2 are the joint radii, assuming they are convex-convex, in mm

The eccentricity of a curved joint e_m due to a misalignment s may be calculated using the following equation:

$$e_m = \left(\frac{R_1}{R_1 + R_2} \right) s \qquad (9.48)$$

e_m is the eccentricity of hoop force across a curved radial joint in mm
s is a misalignment (or 'lip') between one segment and another in mm, at a radial joint

As for flat joints, the total eccentricity is the sum of the eccentricity due to joint rotation and the eccentricity due to misalignment.

Curved joints usually do not have packers, or bolts, but they can have a gasket and a caulking groove (Elliott et al., 1996; Psomas & Eddie, 2016; Nirmal, 2019). If we assume the two segments are elastic, then the contact width b across a convex-convex radial joint with radii R_1 and R_2 may be calculated using elastic theory (Young, 1989), illustrated in Figure 9.22.

The equation for contact width is as follows:

$$b = 1.6 \sqrt{\left(\frac{N_d K_d C_E}{L_c} \right)} \qquad (9.49)$$

b is the contact width in mm
N_d is the design hoop force in N
L_c is the contact length in mm

K_d is given by:

$$K_d = \frac{2R_1 R_2}{R_1 + R_2} \tag{9.50}$$

R_1 and R_2 are the radii of the curved radial joint surfaces of segment 1 and segment 2, respectively, in mm

C_E is given by:

$$C_E = \frac{1 - v_1^2}{E_1} + \frac{1 - v_2^2}{E_2} \tag{9.51}$$

E_1 and v_1 are the Young's modulus in N/mm² and Poisson's ratio of segment 1
E_2 and v_2 are the Young's modulus in N/mm² and Poisson's ratio of segment 2

The contact length L_c will not necessarily be the length of the segment, as there are usually chamfers at the ends to take account of.

The contact width allows us to calculate a contact stress that we can check against concrete crushing, and also determines the bursting stress that will be generated deeper in the segments below the contact. The contact width and eccentricity allow us to check for local shear and bending in the segment.

Segment 1 elastic
parameters E_1, v_1

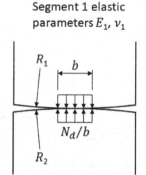

Segment 2 elastic
parameters E_2, v_2

Figure 9.22 Contact width for a convex-convex radial joint.

9.6 MODELLING JOINTS IN BEDDED BEAM ANALYSES

When a bedded beam model is used, the radial joints may be modelled by rotational hinges, which is a slight improvement on the 2D plane strain analytical solutions, where the effect of joints is smeared or averaged around the whole ring. If a finite element or frame analysis program is used, then these hinges can be assigned a rotational stiffness. However, knowing what the value of this rotational stiffness should be is not straightforward. de Waal (2000) showed that the results will always be between the two limiting situations of a monolithic ring with no joints and a ring with free (or pin) joints. Lee & Ge (2001) estimated that it is between 1/10 and 1/4 of the segment stiffness. An analysis like that described in Section 9.2 could be used to estimate rotational stiffness, or the rotational stiffness could be calibrated to measured deformations of previously constructed tunnels (similar to what Lee & Ge (2001) did for effective bending rigidity).

A 3D bedded beam model can be used to take account of the shear stiffness of circumferential joints by introducing 'shear springs' into the model between one ring and another, and the staggering of radial joints in adjacent rings. An example is shown in Figure 9.23 for a tunnel lining with six segments and staggered radial joints. Two rings have been modelled to allow the coupling effect between adjacent rings to be modelled, but only half of the segment width is considered in each ring. The boundary conditions do not allow displacement in the tunnel longitudinal direction, so this

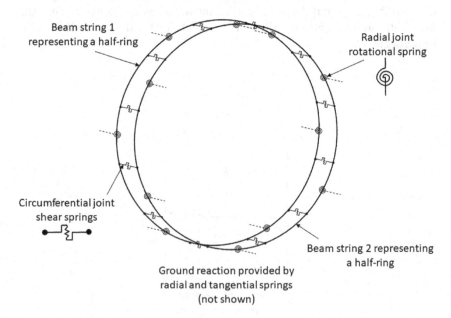

Figure 9.23 3D bedded beam (or 'beam-spring') model of two six-segment half-rings with staggered radial joints.

means the model behaves as though the structure were infinitely mirrored in the longitudinal direction, forwards and backwards, and this is why it makes sense to only consider half the segment width. The segments are modelled by curved 1D beams. This kind of model can be easily built in most 3D frame analysis programs.

In this kind of model, ground and water loads are applied to the beams as distributed loads. The ground reaction is modelled by radial and tangential springs, which are not shown in Figure 9.23. This is an approximation of the soil-structure interaction, but the ground reaction springs are independent and will not allow redistribution of stresses caused by arching within the ground. For instance, a more flexible zone of lining would result in more deformation in the ground locally and a local reduction in ground pressure, which would result in arching and hence an increase in the ground pressure either side of this zone. This becomes increasingly important as stress gradients in the ground become larger, for example, in situations where the tunnel is passing near to an existing tunnel, or perhaps when local grouting pressures are applied.

Guidance on spring stiffness values to use can be found in Bakhshi & Nasri (2013a, 2013b and 2013c). Kavvadas et al. (2017) found by parametric study that for the radial joints only the rotation about the longitudinal tunnel axis direction is important and for the circumferential joints only shear displacement is important. For the other degrees of freedom, high stiffness values were used. For the radial joint rotational stiffness, values could be based on full-scale testing or the methods described in Section 9.2. For the circumferential joints, Gijsbers & Hordijk (1997) found from full-scale tests that the shear stiffness value in the elastic zone could be assumed to be 10^6 kN/m. The peak coefficient of friction varied between 0.4 and 0.7, with higher values tending to be for lower normal forces and vice versa.

9.7 MODELLING JOINTS IN 2D OR 3D NUMERICAL ANALYSIS

Using the finite element or finite difference method in 2D or 3D allows the ground to be modelled by continuum elements, which is an improvement over bedded beam models as it allows soil-structure interaction to be modelled properly, although there will be a cost in terms of computation time. It also allows complete flexibility of geometry in terms of modelling nearby structures, varying surface or geological strata levels, fault zones or otherwise nonuniform soil properties, as well as nonlinear soil or structure behaviour and staged construction.

The simplest method of modelling a segmental tunnel lining is to use beam elements in 2D, or shell elements in 3D. This means that the thickness is not modelled explicitly but is taken account of implicitly in the axial and flexural stiffness parameters. Therefore, joints cannot be modelled explicitly, but can be represented by introducing rotational and shear springs between beam or shell elements, in the same way as for a bedded beam model. When these kinds of models are used, it is usually assumed that the lining shell elements are placed at the extrados. This will model the ground pressures more accurately, but the lining will behave as though its centroid is at the extrados. This means it has a radius that is half the segment thickness larger than it should be. This will slightly reduce bending stiffness, and this effect can be estimated using the following equation:

$$\frac{M^*}{M} = \frac{r_0^2}{r_0(r_0 - t)} \tag{9.52}$$

M^* is the bending moment calculated by the model
M is the bending moment in the real situation
r_0 is the radius to the extrados
t is the thickness of the lining

Kavvadas et al. (2017) used 3D numerical modelling of an advancing earth pressure balance (EPB) TBM to compare the effect of aligned radial joints, staggered radial joints and a continuous monolithic lining without radial or circumferential joints. The results, in terms of bending moments and axial forces in the lining, are shown in Figure 9.24. They showed that aligned radial joints resulted in significantly lower bending moments but similar axial forces to the continuous monolithic lining in their models. For staggered radial joints, the internal forces in the lining varied from ring to ring and so the results were presented as an envelope of forces that encompassed the aligned radial joints and continuous lining forces.

Kavvadas et al. (2017) did not explicitly model the joint contacts but used joint rules in ABAQUS to apply a rotational stiffness. The rotational stiffness they used was based on concrete-to-concrete contact with no packer

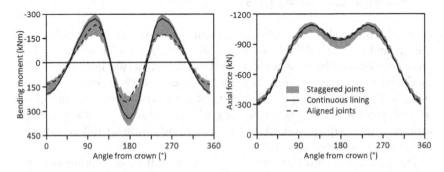

Figure 9.24 Lining forces from a 3D finite element analysis with staggered joints, continuous lining and aligned joints (redrawn from Kavvadas et al., 2017). For the axial force, compression is negative.

(described in detail in Litsas et al., 2015), using the method of Blom (2002) and was therefore more than an order of magnitude higher than values presented earlier in this chapter where the properties of the packer were used. This may be why the envelope for staggered joints has an upper bound of bending moment close to the value for a continuous lining.

It is theoretically possible to model joints explicitly in a 3D numerical model, by using continuum elements for the tunnel lining, and paying close attention to the contact rules and geometry of the joints. It would very likely be necessary to have a much finer mesh for the segments, particularly in the vicinity of the joints, than for the ground. Very few commercially available software packages will allow sudden changes of mesh size, or even contact rules (it is possible in ABAQUS and to some extent in FLAC3D), while also being able to model the ground explicitly and with a realistic constitutive model. A model built in this way will be very large and complex. It is unlikely to be feasible at present for anything other than an advanced research project. It is possible to model the lining with solid elements in 3D in many structural finite element analysis programs with nonlinear contacts modelled, but there will be limitations in the way the ground is modelled.

Attempts have been made to model joints using finite elements and contact rules, but without modelling the ground, in order to obtain rotational stiffness values. These kinds of models are usually used to back-analyse full-scale segment tests, to check on stress concentrations around details, such as rebates, shear keys, gasket grooves, grout holes or bolt holes, or to find values of joint rotational and shear stiffness for use in simpler models.

9.8 IN SERVICE LOADS – ULTIMATE LIMIT STATE DESIGN

The ground and water loads applied to the segmental lining should be the most onerous loads in terms of design of the lining thickness and material strength. In preliminary design, the lining bending moment and hoop force may be obtained from an analytical solution, such as the Curtis–Muir Wood solution (see Chapter 6), whereas in detailed design it is common practice to use numerical modelling in 2D or 3D to obtain the lining forces (see Chapter 7).

Usually, the design models will have various load cases and combinations, which may include:

- a representative sample of cross-sections along the route, representing the highest and lowest overburden or other geometric or geological variations
- short-term and long-term groundwater levels
- undrained and drained soil behaviour

- for twin tunnels, the distance between them or their level relative to each other may vary along the route
- sensitivity analyses, e.g. of a realistic range of coefficient of earth pressure at rest, soil strength or stiffness parameters, or confinement loss factor (see Section 7.1.2)

These load cases and combinations will each result in lining forces, usually expressed as a hoop thrust N, a bending moment M and shear force V, which will vary around the lining. Therefore, we will have a large number of points at which we have calculated a set of N, M and V and we need an efficient way of dealing with all of these sets. We cannot just take the maximum N and pair it with the maximum M and maximum V, because they may not have occurred at the same point in the lining or in the same design model.

An important aspect to bear in mind is that a high value of N is not necessarily a bad thing, because it can increase the moment capacity at low to moderate load levels. This is why we should check the lowest overburden and the lowest groundwater table level as well as the highest.

9.8.1 Application of partial factors to the lining forces

Each set of lining forces N, M and V must have a partial factor of 1.35 applied to them according to EN 1997-1: 2004 +A1: 2013, if they are unfavourable loads. Since it is difficult to ascertain whether applying the partial factor will be beneficial or not, the lining should be checked for both factored and unfactored loads.

To illustrate this, the moment-hoop force interaction diagram for a 200 mm thick, 1000 mm long SFRC segment with a characteristic compressive cylinder strength $f_{ck} = 40$ N/mm^2 and a characteristic value of flexural tensile residual strength at crack mouth opening distance (CMOD) = 2.5 mm, $f_{R3k} = 4.5$ N/mm^2 is shown in Figure 9.25. The eye-shaped line is the capacity envelope and points anywhere inside this line are within the capacity of the section. The unfactored loads are shown by the solid marker. In this case, if the hoop force is factored it is favourable because the moment capacity is increased. The worst case would be to factor M but not N.

Most practitioners argue that N and M are interdependent because each set is from the same point in the lining in the same design model and therefore constitutes a 'single source'. Therefore, both N and M should be factored or unfactored together. However, the partial factor on actions or the effects of actions is to take account of "the possibility of unfavourable deviations of the action values from the representative values" (EN 1990: 2002 +A1: 2005, Section 6.3.1), and it is possible to imagine that M could increase without also affecting N. Situations that could cause this include: unforeseen heterogeneity of ground or drainage conditions, a value

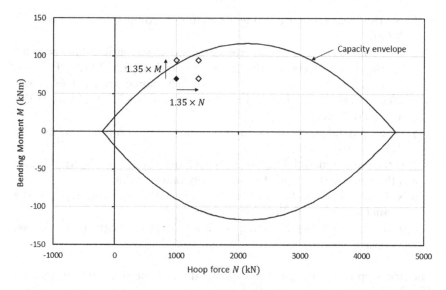

Figure 9.25 An example moment-hoop force interaction diagram showing the effect of application of partial factors.

of K_0 outside the range assumed, soil strength or stiffness anisotropy not included in the model, 3D effects not included in the design model, geometrical imperfections, adjacent construction, variations in grouting procedures. Therefore, it is my personal view that as well as applying the partial factor to both M and N, the combination where a partial factor is applied to M but not to N should also be checked. Alternatively, the traditional method of applying a partial factor of 1.0 or 1.35 to each pair of M and N could be used, as long as consideration is given to design situations that may cause additional bending moments to arise.

9.8.2 Constructing a moment-axial force interaction diagram

The key part of segmental lining design for geotechnical actions is to construct a moment-axial force interaction diagram. This is commonplace for the design of reinforced concrete columns, but for SFRC segments it is not described in the *fib* Model Code 2010 (2013) nor in detail in any of the supporting guidance for segment design such as *fib* Bulletin 83 (2017) or ITAtech (2016). The procedure is described here, for SFRC segments without bar reinforcement.

The inputs are:

f_{R3k} is the characteristic value of the SFRC flexural tensile residual strength at CMOD = 2.5 mm, obtained from a suite of EN 14651: 2005 beam tests, in N/mm²

f_{ck} is the characteristic value of cylinder compressive strength in N/mm²

α_{cc} is a coefficient that reduces the compressive strength to take account of long-term effects and the application of load (EN 1992-1-1: 2004, Section 3.1.6). The value depends on the National Annex. The recommended value is 1.0 but in the UK the value is 0.85 (UK NA to BS EN 1992-1-1: 2004).

γ_c is the partial material factor for concrete in compression, usually 1.5

γ_F is the partial material factor for SFRC, usually 1.5 (see Section 8.2)

λ is the neutral axis depth factor, usually taken as 0.8

b is the member width, which in this case is the segment length in the tunnel longitudinal direction, in mm

h is the member depth, which in this case is the segment thickness, in mm

The first step is to calculate the design values of the strength parameters. The design compressive strength is given by:

$$f_{cd} = \frac{\alpha_{cc} f_{ck}}{\gamma_c} \tag{9.53}$$

f_{cd} is the design compressive strength in N/mm²

The design value of flexural tensile residual strength at CMOD = 2.5 mm is given by:

$$f_{R3d} = \frac{f_{R3k}}{\gamma_F} \tag{9.54}$$

f_{R3d} is the design value of flexural tensile residual strength at CMOD = 2.5 mm, in N/mm²

The design value of the ultimate post-cracking tensile strength of the SFRC is given by:

$$f_{Ftud} = \frac{f_{R3d}}{3} \tag{9.55}$$

f_{Ftud} is the design value of the ultimate post-cracking tensile strength of the SFRC, in N/mm²

The behaviour in bending can be simplified to two rigid plastic stress blocks, as shown in Figure 9.26. We will not include bar reinforcement in this procedure, though it is fairly straightforward to add it in.

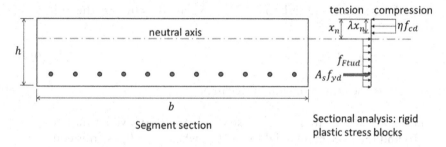

Figure 9.26 Rigid-plastic stress block approach for ultimate limit state (ULS) design of a fibre-reinforced concrete segment.

The symbols used in Figure 9.26 that we have not already defined are:

   x_n is the distance from the compressive extreme fibre to the neutral axis in mm.
   η is a factor that is used to reduce the design compressive strength at high strength values. For $f_{ck} \leq 50$ MPa, $\eta = 1$.
   A_s is the area of steel bar reinforcement (if present) in mm².
   f_{yd} is the design yield strength of the steel bar reinforcement (if present) in N/mm².

If we have no axial force, then we can directly calculate the depth to the neutral axis x_n using horizontal force equilibrium:

$$\lambda x_n \eta f_{cd} = (h - x_n) f_{Ftud} \tag{9.56}$$

Rearranging gives:

$$x_n = \frac{h f_{Ftud}}{\lambda \eta f_{cd} + f_{Ftud}} \tag{9.57}$$

Now we know x_n we can take moments about the neutral axis to calculate the resistance moment of the section at the ULS M_u:

$$M_u = f_{Ftud} \frac{b(h - x_n)^2}{2} + \eta f_{cd} b \lambda x_n \left(x_n - \frac{\lambda x_n}{2} \right) \tag{9.58}$$

   M_u is the ULS moment in Nmm

In order to construct a moment-axial force interaction diagram, set up a spreadsheet with the first column containing values of axial force. We will assume compression is positive and tension is negative. To find the minimum axial force N_{min} when $M_u = 0$, use the following equation:

$$N_{min} = -f_{Ftud}bh \tag{9.59}$$

N_{min} is the minimum axial force in N

The answer will be in units of N, so divide by 1000 if you want kN.
    To find the maximum axial force N_{max} when $M_u = 0$, use this equation:

$$N_{max} = f_{cd}bh \tag{9.60}$$

N_{max} is the maximum axial force in N

Set the values of axial force to be separated by equal increments between the minimum and maximum values. Approximately 20 increments will give sufficient resolution in the diagram.
    In the second column, divide the axial force in the first column by the cross-sectional area bh to give the axial stress σ_a as follows:

$$\sigma_a = \frac{N}{bh} \tag{9.61}$$

In a third column, calculate the stress capacity available for tension in bending $\sigma_{t,avail}$:

$$\sigma_{t,avail} = \sigma_a + f_{Ftud} \tag{9.62}$$

And in a fourth column, calculate the stress capacity available for compression in bending $\sigma_{c,avail}$:

$$\sigma_{c,avail} = \eta f_{cd} - \sigma_a \tag{9.63}$$

Now, in a fifth column, calculate the depth to the neutral axis x_n using Equation 9.57, with f_{Ftud} replaced with $\sigma_{t,avail}$ and ηf_{cd} replaced with $\sigma_{c,avail}$, as follows:

$$x_n = \frac{h\sigma_{t,avail}}{\lambda\sigma_{c,avail} + \sigma_{t,avail}} \tag{9.64}$$

Now, in a sixth column, M_u may be calculated using Equation 9.58, again with f_{Ftud} replaced with $\sigma_{t,avail}$ and ηf_{cd} replaced with $\sigma_{c,avail}$, as follows:

$$M_u = \sigma_{t,avail}\frac{b(h-x_n)^2}{2} + \sigma_{c,avail}b\lambda x_n\left(x_n - \frac{\lambda x_n}{2}\right) \tag{9.65}$$

Now plot a graph with values of axial force N from the first column on the x-axis and values of bending moment M_u from the sixth column on the y-axis. It should have a similar shape to Figure 9.25.

WORKED EXAMPLE 9.4

A precast steel fibre-reinforced concrete segmental lining is to be designed for a 4.5 m internal diameter tunnel. The concrete is of grade FRC 40/50 – 5.0c. Take $\alpha_{cc} = 0.85$, $\lambda = 0.8$, $\gamma_c = 1.5$ and $\gamma_F = 1.5$.

Construct a moment-axial force interaction diagram for the ultimate limit state, taking the lining thickness $t = 250$ mm and the segment length $b = 1000$ mm.

The concrete is grade FRC 40/50 – 5.0c, which means that:

$$f_{ck} = 40 \text{ N/mm}^2$$

$$f_{R1k} = 5.0 \text{ N/mm}^2$$

The letter 'c' means that $0.9 \le f_{R3k}/f_{R1k} \le 1.1$, so we will take the lower value:

$$f_{R3k} = 0.9 f_{R1k} = 0.9 \times 5.0 = 4.5 \text{ N/mm}^2$$

Now $\gamma_F = 1.5$, so:

$$f_{R3d} = \frac{f_{R3k}}{\gamma_F} = \frac{4.5}{1.5} = 3.0 \text{ N/mm}^2$$

And:

$$f_{Ftud} = \frac{f_{R3d}}{3} = \frac{3.0}{3} = 1.0 \text{ N/mm}^2$$

Also, $\alpha_{cc} = 0.85$ and $\gamma_c = 1.5$, so:

$$f_{cd} = \frac{\alpha_{cc} f_{ck}}{\gamma_c} = \frac{0.85 \times 40}{1.5} = 22.67 \text{ N/mm}^2$$

The range of axial force we need to calculate is bounded by:

$$N_{min} = -f_{Ftud} bh = -1.0 \times 1000 \times 250 = -250 \times 10^3 \text{N} = -250 \text{ kN}$$

$$N_{max} = f_{cd} bh = 22.67 \times 1000 \times 250 = 5667 \times 10^3 \text{N} = 5667 \text{ kN}$$

This calculation shows ten equal intervals of axial force N, but you can do more if you want to. The calculation results are shown in Table 9.2.

Table 9.2 Calculation of values to construct a moment-axial force interaction diagram.

Axial force	Axial stress	Available for tension	Available for compression	Depth to neutral axis	Ultimate moment
N	σ_a	$\sigma_{t,avail}$	$\sigma_{c,avail}$	x_n	M_u
(kN)	(N/mm²)	(N/mm²)	(N/mm²)	(mm)	(kNm)
−250.00	−1.000	0.000	23.667	0.00	0.00
341.67	1.367	2.367	21.300	30.49	66.52
933.33	3.733	4.733	18.933	59.52	118.07
1525.00	6.100	7.100	16.567	87.21	154.56
2116.67	8.467	9.467	14.200	113.64	176.03
2708.33	10.833	11.833	11.833	138.89	182.61
3300.00	13.200	14.200	9.467	163.04	174.48
3891.67	15.567	16.567	7.100	186.17	151.87
4483.33	17.933	18.933	4.733	208.33	115.05
5075.00	20.300	21.300	2.367	229.59	64.32
5666.67	22.667	23.667	0.000	250.00	0.00

The values can be plotted on a graph, as shown in Figure 9.27. Note that because we have a symmetrical section, the negative moments (hogging) are the same but with the sign reversed.

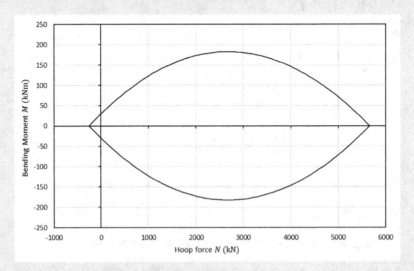

Figure 9.27 Worked Example 9.4 moment-axial force interaction diagram.

9.8.3 Design for shear

Coccia et al. (2015) proposed a method of including the effect of shear in the ULS check for moment and axial force, by reducing the design value of the ultimate post-cracking tensile strength f_{Ftud} as follows:

$$f_{Ftud}^{red} = f_{Ftud}\left[1-\left(\frac{\tau}{f_{Ftud}}\right)^2\right]$$

(9.66)

f_{Ftud}^{red} is the reduced value of f_{Ftud} in N/mm^2

f_{Ftud} is the design value of the ultimate post-cracking tensile strength in N/mm^2

τ is the shear stress, given by $\tau = V/bh$, in N/mm^2

The moment-axial force interaction diagram should then be constructed with the reduced value f_{Ftud}^{red} in place of f_{Ftud}.

9.9 IN SERVICE LOADS – SERVICEABILITY LIMIT STATE DESIGN

So far we have only looked at the ULS design under geotechnical loads. We also need to design the tunnel for the serviceability limit state (SLS), to demonstrate that it will meet whatever criteria are needed for it to remain serviceable for the whole design life.

9.9.1 Estimating crack widths

Specifying the maximum allowable crack width for the SLS was discussed in Sections 8.2.6 and 8.2.7. Here we are concerned with how to estimate crack widths during design of a tunnel lining to ensure they are smaller than this specified maximum allowable value.

Estimating crack widths in reinforced concrete is a subject that is well covered in design guides and codes, such as EN 1992-1-1: 2004. For SFRC it is still an area that requires further research, and really we can only be sure of any estimate by undertaking full-scale laboratory tests. This is because for a given strain in an SFRC segment, the size of cracks will depend on the spacing between the cracks. In other words, a given tensile strain could result in either multiple small cracks or one big one.

Methods of crack width estimation are therefore based on simplifying assumptions that require verification by testing (e.g. Johnson et al., 2017). Without testing, calculations can only provide an average crack width based on a mean tensile strain.

Since most tunnel linings have significant hoop forces and the section is therefore mainly in compression, some tunnel designers advocate a simple method where the lining is assumed to be elastic in the analysis and any tensile stresses are then checked to ensure they are below the first crack value. This avoids the need to model the tunnel lining as a nonlinear material.

9.9.2 Constructing a moment-axial force interaction diagram for the SLS

It can be useful to define a capacity envelope for moment-axial force interaction at the SLS. This is much more complicated than for the ULS and guidance is scarce.

fib Bulletin 83 (2017) proposes one method, but the explanation is very brief and does not explain all the steps. The strain distribution is assumed to be linear, with a maximum compressive stress equal to $0.6f_{ck}$ at the compressive extreme fibre and the maximum tensile strain defined by the specified maximum crack width for serviceability.

To construct a moment-axial force interaction diagram, we must calculate the strain distribution, convert this into a stress distribution and then calculate the moment and axial force.

To calculate the maximum compressive strain, we need to estimate the Young's modulus of the concrete to convert $0.6f_{ck}$ into a compressive strain. This should probably include an allowance for plasticity and creep.

$$\varepsilon_{c,max} = \frac{\sigma_{c,max}}{E_c} = \frac{0.6f_{ck}}{E_c} \tag{9.67}$$

$\varepsilon_{c,max}$ is the maximum compressive strain
$\sigma_{c,max}$ is the maximum compressive stress in N/mm²
E_c is the Young's modulus of the concrete in compression in N/mm²
f_{ck} is the characteristic compressive strength of the concrete N/mm²

To calculate the maximum tensile strain, we need to use the following equation to convert crack width to strain:

$$\varepsilon_{t,max} = \frac{w_d}{s} \tag{9.68}$$

$\varepsilon_{t,max}$ is the maximum tensile strain
w_d is the maximum allowable crack width in mm
s is the crack spacing in mm

The key assumption here is the crack spacing s. There is currently no clear guidance available on what this should be, since the *fib* Model Code 2010 (2013) does not consider the influence of axial force. The characteristic crack spacing could be determined by testing (for an example, see Hover et al., 2017). A potential design method is proposed by Johnson et al. (2017), where crack spacing is assumed to be equal to the depth of the crack, but this needs validation by testing. The *fib* Bulletin 83 (2017) states that crack spacing may be assumed to be equal to the segment thickness.

9.10 SUMMARY

Joints in segmental linings have an effect on the overall stiffness of the lining. Muir Wood (1975) proposed a simple expression to reduce the moment of inertia of the lining to account for the number of radial joints in a ring. However, studies have shown that tunnel linings with staggered radial joints have a behaviour that is closer to a monolithic ring.

The behaviour of joints undergoing rotation may be analysed using relatively simple methods to determine the rotational stiffness. This could then be used in a bedded beam analysis or finite element analysis program to model the interaction of joints in adjacent rings.

On most projects, a maximum ovalisation is specified in the contract. This can be used to calculate the maximum joint rotation, which can in turn be used to calculate the eccentricity of axial force across the joint. We can then use this eccentricity of axial force to calculate a bending moment, and to check for local crushing, local shear and bursting stress at or near the joint.

Curved radial joints are also used, and generally speaking they tend to reduce eccentricity of hoop force across the joint. The bearing area and eccentricity may be calculated, and from these values the same design checks on bending moment, local crushing, local shear and bursting stress may be performed as for flat joints with packers.

In service, the tunnel lining is loaded by the ground and groundwater pressures. This imposes hoop forces, bending moments and shear forces on the lining. For concrete segments, the capacity at the ULS may be verified by plotting a moment-axial force interaction diagram. For plain or reinforced concrete, EN 1992-1-1: 2004 may be used and for SFRC, *fib* Model Code 2010 (2013) and *fib* Bulletin 83 may be used to construct the interaction diagram. To take account of shear force for SFRC segments, a reduction of the design value of the ultimate post-cracking tensile strength f_{Ftud} was proposed by Coccia et al. (2015).

For the in service loading of SFRC segments at the SLS, the limiting parameter is crack width. There is much research still to be done before design rules can be codified, and the current methods are exceedingly complex. It was decided that this was beyond the scope of this textbook, which only seeks to cover the basic principles of design using well-established and proven methods.

9.11 PROBLEMS

Q9.1. An 8 m ID, 400 mm thick segmental lining is to be designed with ten precast concrete segments per ring. The ring length is 2 m in the tunnel longitudinal direction. The design value of hoop force is 2200 kN/m. The radial joints are flat joints with dimensions as shown in Figure 9.28.

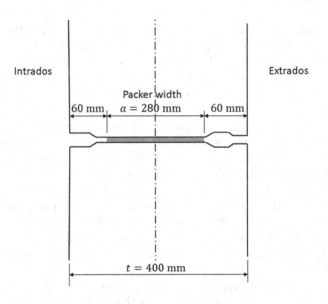

Figure 9.28 Q9.1 radial joint dimensions.

  i. The specified ovalisation is 1%. Calculate the maximum radial joint rotation.
  ii. Calculate the eccentricity of hoop force caused by a misalignment of the radial joint of 10 mm.
  iii. Assuming the packer is linear elastic, with Young's modulus $E_p = 40$ MPa, length $L_p = 1900$ mm and thickness $t_p = 4$ mm, calculate the eccentricity of hoop force caused by the maximum radial joint rotation from part (i).
  iv. Calculate the total eccentricity of hoop force.
  v. Calculate the bending moment induced by the total eccentricity.
  vi. Calculate the bursting force at the total eccentricity.
  vii. [Advanced] Calculate the eccentricity if the packer is bituminous, using the parameter set 'B2' from Table 9.1 and with the same length $L_p = 1900$ mm and thickness $t_p = 4$ mm as for the linear elastic packer. Then calculate the bending moment induced by the total eccentricity.

Q9.2. A 6.6 m ID, 300 mm thick segmental lining has eight precast concrete segments per ring. The ring length is 1.5 m long in the tunnel longitudinal direction. The radial joints are flat with 3 mm thick plywood packers that are 1400 mm long and 200 mm wide, positioned centrally. The plywood packers are 4 mm thick and have a bilinear stress-strain relationship with $E = 170$ MPa for $0 \leq \varepsilon \leq 0.05$ and $E = 23$ MPa for $0.05 \leq \varepsilon \leq 0.5$. The design value of hoop force is 1500 kN/m.

   i. The specified ovalisation is 0.7%. Calculate the maximum joint rotation.

   ii. Using a spreadsheet, plot a graph of the contact stress distribution at the maximum joint rotation. Check it by integrating the stresses and comparing the result with the hoop force.

   iii. Calculate the eccentricity of hoop force caused by the maximum joint rotation by taking moments of the stress distribution about the centre of the segment thickness.

   iv. Calculate the bending moment induced by the eccentricity of hoop force.

   v. The precast concrete is of class FRC 50/60 – 5.0d according the *fib* Model Code 2010 (2013). Check the concrete bearing stress under the packer using the maximum stress calculated in the contact stress distribution found in part (ii). If you skipped the earlier parts, then use a maximum stress of 18 N/mm^2.

   vi. Check also that the eccentricity at the maximum joint rotation will not cause local shear failure of the corner of the segment. If it fails, what would you do to improve the design?

Q9.3. A segmental lining is designed with FRC 50/60 – 5.0d of thickness 300 mm.

   i. Construct a moment-axial force interaction diagram at the ULS. Use $\alpha_{cc} = 0.85$, $\lambda = 0.8$, $\gamma_c = 1.5$ and $\gamma_F = 1.5$. Do the calculations per metre length of tunnel, i.e. take the segment length in the tunnel longitudinal direction as 1000 mm.

   ii. The results of numerical modelling of the tunnel are in Table 9.3. These pairs of actions have not been factored. Plot them on the moment-axial force interaction diagram created in (i) as both unfactored and factored values. Do they exceed the ULS?

Table 9.3 Q9.3 Axial force and bending moment values from numerical modelling.

N (kN/m)	M (kNm/m)
1997	185
1912	130
1657	163
2339	205
1321	144
2027	233

iii. The last pair of values in Table 9.3 should have exceeded the capacity of the segment. Change the segment design so that all the pairs of values are within the capacity curve.

REFERENCES

Bakhshi, M. & Nasri, V. (2013a). Structural design of segmental tunnel linings. *3rd International Conference on Computational Methods in Tunnelling and Subsurface Engineering: EURO:TUN 2013*. Ruhr University Bochum, 17–19 April.

Bakhshi, M. & Nasri, V. (2013b). Latest developments in design of segmental tunnel linings. *Canadian Society for Civil Engineering General Conference*. Montréal, Québec, 29 May–1 June.

Bakhshi, M. & Nasri, V. (2013c). Practical aspects of segmental tunnel lining design. *Underground – the way to the future*, Proc. World Tunnel Congress 2013, Geneva, Switzerland (eds Anagnostou, G. & Ehrbar, H.). London: Taylor & Francis Group.

Blom, C. B. M. (2002). *Design philosophy of concrete linings for tunnels in soft soils*. PhD thesis, Technical University of Delft.

BTS (2010). *Specification for tunnelling*, 3rd Edition, including 2012 errata. London: Thomas Telford.

Cavalaro, S. H. P. & Aguado, A. (2012). Packer behavior under simple and coupled stresses. *Tunn. Undergr. Space Technol.* 28, 159–173.

Coccia, S., Meda, A. & Rinaldi, Z. (2015). On shear verification according to *fib* Model Code 2010 in FRC elements without traditional reinforcement. *Struct. Concr.* 16, No. 4, 518–523.

de Waal, R. G. A. (2000). *Steel fibre reinforced tunnel segments for the application in shield driven tunnel linings*. PhD Thesis. Technische Universiteit Delft.

Elliott, I. H., Odgård, A. S. & Curtis, D. J. (1996). Storebælt eastern railway tunnel: design. *Proc. ICE Civil Eng. Supp.* 114, Special Issue 1, 9–19.

EN 12390-6: 2009. *Testing hardened concrete – Tensile splitting strength of test specimens*. Brussels: European Committee for Standardization.

EN 14651: 2005 (2005). *Test method for metallic fibered concrete – Measuring the flexural tensile strength (limit of proportionality (LOP), residual)*. Brussels: European Committee for Standardization.

EN 1990: 2002 +A1: 2005 (2010). *Eurocode – Basis of structural design*, incorporating corrigenda December 2008 and April 2010. Brussels: European Committee for Standardization.

EN 1992-1-1: 2004 (2004). *Eurocode 2: Design of concrete structures – Part 1-1: General rules and rules for buildings*. Brussels: European Committee for Standardization.

EN 1997-1: 2004 +A1: 2013. *Eurocode 7: Geotechnical design – Part 1: General rules*, incorporating corrigendum February 2009. Brussels: European Committee for Standardization.

EN 206-1: 2000 (2004). *Concrete – Part 1: Specification, performance, production and conformity*, incorporating Corrigendum No. 1 and Amendment No. 1. Brussels: European Committee for Standardization.

Fei, Y., Chang-fei, G., Hai-dong, S., Yan-peng, L., Yong-xu, X. & Zhuo, Z. (2014). Model test study on effective ratio of segment transverse bending rigidity of shield tunnel. *Tunn. Undergr. Space Technol.* **41**, 193–205.

fib Bulletin 83 (2017). *Precast tunnel segments in fibre-reinforced concrete*, State-of-the-art Report, fib WP 1.4.1, October. Lausanne: Fédération internationale du béton (*fib*).

fib Model Code 2010 (2013). *fib Model Code for concrete structures 2010.* Lausanne: Fédération internationale du béton (*fib*).

Gijsbers, F. B. J. & Hordijk, D. A. (1997). *Experimenteel onderzoek naar het afschuifgedrag van ringvoegen*, 1ˢᵗ November, TNO Bouw No.97-CON-R1337, COB Reference K111-W-001, Delft, The Netherlands.

Hefny, A. M. & Chua, H. -C. (2006). An investigation into the behaviour of jointed tunnel lining. *Proceedings of the ITA-AITES 2006 World Tunnel Congress – Safety in the Underground Space*, Seoul, Korea, 22–27 April.

Hover, E., Psomas, S. & Eddie, C. (2017). Estimating crack widths in steel fibre-reinforced concrete. *Proc. Inst. Civil Eng. Constr. Mater.* **170**, No. 3, 141–152.

ITAtech (2016). *Design guidance for precast fibre reinforced concrete segments – Vol. 1: Design aspects*, ITAtech Report No. 7. Lausanne: ITA-AITES. Reproduction of figures with permission of the Committee on new technologies of the International Tunnelling and Underground Space Association – ITAtech.

Johnson, R. P., Psomas, S. & Eddie, C. M. (2017). Design of steel fibre reinforced concrete tunnel linings. *Proc. Inst. Civil Eng. Struct. Build.* **170**, SB2, February, 115–130.

Kavvadas, M., Litsas, D., Vazaios, I. & Fortsakis, P. (2017). Development of a 3D finite element model for shield EPB tunnelling. *Tunn. Undergr. Space Technol.* **65**, 22–34. Figure 10-24 reprinted with permission from Elsevier.

Klappers, C., Grübl, F. & Ostermeier, B. (2006). Structural analyses of segmental lining – coupled beam and spring analyses versus 3D-FEM calculations with shell elements. *Proceedings of the ITA-AITES 2006 World Tunnel Congress – Safety in the Underground Space*, Seoul, Korea, 22–27 April.

Lee, K. M. & Ge, X. W. (2001). The equivalence of a jointed shield-driven tunnel lining to a continuous ring structure. *Can. Geotech. J.* **38**, 461–483.

Litsas, D., Paterianaki, G. & Kavvadas, M. (2015). Investigation of the influence of cracking on the stiffness and capacity of segmental tunnel lining. *Proceedings of the World Tunnel Congress* (ed. Kolić, D.), Dubrovnik, Croatia, 22–28 May. Zagreb, Croatia: HUBITG.

Luttikholt, A. (2007). *Ultimate limit state analysis of a segmented tunnel lining – Results of full-scale tests compared to finite element analysis.* Master's thesis, Delft University of Technology.

Muir Wood, A. M. (1975). The circular tunnel in elastic ground. *Géotechnique* **25**, No. 1, 115–117.

Nirmal, S. S. (2019). Design of steel fibre reinforced concrete segment with curved radial joints. *Proceedings of the WTC 2019 – Tunnels and Underground Cities: Engineering and Innovation meet Archaeology, Architecture and Art* (eds Peila, D., Viggiani, G. M. B. & Celestino, T.), pp. 2804–2809. London: Taylor & Francis Group.

Nogales, A. & de la Fuente, A. (2020). Crack width design approach for fibre reinforced concrete tunnel segments for TBM thrust loads. *Tunn. Undergr. Space Technol.* **98**, Paper 103342.

Psomas, S. & Eddie, C. M. (2016). SFRC segmental lining design for a pressurised tunnel. *Proceedings of the WTC2016*, San Francisco, USA, Paper 0118. Englewood, CO: SME.

Teachavorasinskun, S. & Chub-uppakarn, T. (2010). Influence of segmental joints on tunnel lining. *Tunn. Undergr. Space Technol.* **25**, 490–494.

TfL (2015). *Jubilee Line Tunnel Works*, Rail and Underground Panel paper, 16th July. Available at: http://content.tfl.gov.uk/rup-20150716-part-1-item06-jubilee-line-tunnel-works.pdf (last accessed 11th August 2019).

UK NA to EN 1992-1-1: 2004. *UK National Annex to Eurocode 2: Design of concrete structures – Part 1-1: General rules and rules for buildings.* London: British Standards Institution.

Young, W. C. (1989). *Roark's formula for stress and strain*, 6th Edition. New York: McGraw-Hill.

Chapter 10

Segment design
for transient loads

As well as designing segments for their long-term condition as part of a tunnel lining, we also need to consider the transient load cases that occur before then.

This chapter will describe how segments may be designed for demoulding, storage, transportation, handling and erection. These are different to the in-service loading, because a single segment with external loads applied to it in these ways is a structurally determinate load case. This means that the internal forces in the segment may be determined by equilibrium of the external forces. Another difference is that there is no hoop force and the segments must be able to resist bending moments and shear forces without the (usually) beneficial effect of axial force.

The chapter will then discuss installation loads, such as gasket compression, bolt loads, application of the tunnel boring machine (TBM) jacks and grouting pressures.

After working through this chapter you will understand:

- the main load cases that apply to segments from casting in a factory to installation in a tunnel
- the complex interaction of TBM jacking loads and grouting pressures and their effect on longitudinal bending of the tunnel lining

After working through this chapter you will be able to:

- design a concrete segment for transient loads, made from reinforced concrete, plain concrete or steel fibre reinforced concrete

This chapter is not intended to be a comprehensive manual for design of tunnel segments, but a teaching aid to demonstrate the principles and most important aspects of design. Every project is unique and has different load cases and performance requirements, some of which may not be included here. But with the understanding gained from working through this chapter, and with careful reading of the papers, guidelines and standards that are cited, you will have a solid foundation.

DOI: 10.1201/9780429470387-10

10.1 DEMOULDING

Assuming that the moulds and curing process of the precast concrete segments are designed to minimise adhesion or trapping of the segment in the mould, the load case here is the lifting of a segment against its self-weight. The concrete is young, perhaps with an age between 3 and 12 hours, depending on the concrete mix and the curing process (*fib* Bulletin 83, 2017). Cycle times in the factory and hence the size of the factory and the number of segment moulds required depend on the time from casting to demoulding, so this could have an impact on cost and programme. There are a number of factors at play and the rate of strength gain also depends on the concrete mix ingredients and proportions. This will have an impact on durability and carbon footprint, so it is important that the designer communicates with the segment manufacturer to achieve the optimum solution.

Usually the designer will specify a concrete strength required and it will be up to the segment factory to ensure the concrete has attained this strength before demoulding, usually by using standard cube or cylinder compressive strength tests as an index test. For SFRC, a relationship between the flexural tensile strength required to avoid cracking and the compressive strength needs to be established at early ages to determine a value of compressive strength at which it is safe to demould the segment (*fib* Bulletin 83, 2017).

Designing for demoulding requires knowledge of the lifting equipment. For smaller segments, lifting points may be cast into the segment. In most cases nowadays, and particularly for larger segments, demoulding will be done by vacuum lifting. Vacuum lifters usually have one to three areas where a vacuum is applied. Segments are usually cast intrados-down, so the vacuum lifter is applied to the extrados as shown in Figure 10.1.

The calculation of bending moments and shear forces due to lifting are a simple case of equilibrium of vertical loads. Vacuum lifters will spread the load more and bending moments will be reduced because the length of cantilevering segment outside the vacuum pads will be smaller compared to using one or more lifting points.

This is a statically determinate load case, i.e. there is no need to consider deflections or compatibility of strains. The self-weight is the same regardless of how the segment deforms under load. When we looked at a segmental lining being loaded by the ground and groundwater this was a statically indeterminate load case and was treated slightly differently.

The self-weight of a segment as a function of its arc length, is given by:

$$w_{arc} = tL_s\gamma_{seg} \tag{10.1}$$

w_{arc} is the self-weight per unit arc length in kN/m
t is the segment thickness in m

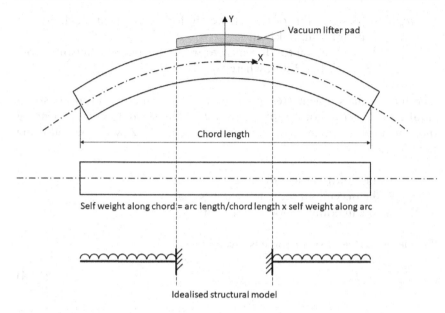

Figure 10.1 Segment demoulding idealisation for design.

L_s is the segment length in the tunnel longitudinal direction in m (the dimension into the page in Figure 10.1)

γ_{seg} is the unit weight of the segment concrete in kN/m³ (usually assumed to be 25 kN/m³ but a more exact value may be known from testing)

If we simplify the problem to a straight beam, as shown in the middle diagram of Figure 10.1, the self-weight may be expressed as:

$$w_{chord} = \frac{L_{arc}}{L_{chord}} w_{arc} \tag{10.2}$$

w_{chord} is the self-weight per unit length of the segment simplified as a straight beam in kN/m

L_{arc} is the arc length of the segment in m

L_{chord} is the chord length, i.e. the length of the segment idealised as a straight beam, in m

The design value of self-weight is the self weight multiplied by the partial factor for unfavourable permanent actions from EN 1992-1-1: 2004:

$$w_d = w_{chord} \gamma_G \tag{10.3}$$

w_d is the design value of the self-weight for the segment idealised as a straight beam, in kN/m

γ_G is the partial factor for unfavourable permanent actions from EN 1992-1-1: 2004, where $\gamma_G = 1.35$

The ends of the segment are cantilevering as shown in the idealised structural model at the bottom of Figure 10.1. The shear force and bending moment may be calculated as shown in Figure 10.2 with the following notation:

L_{cant} is the cantilever length in m
V_{max} is the maximum shear force in kN
M_{max} is the maximum moment in kNm

The design value of shear force is given by:

$$V_{max} = w_d L_{cant} \tag{10.4}$$

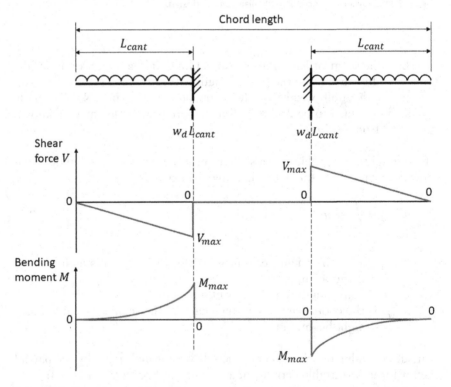

Figure 10.2 Shear force and bending moment diagrams for segment demoulding load case.

The design value of bending moment is given by:

$$M_{max} = \frac{w_d L_{cant}^2}{2} \tag{10.5}$$

We can check the design moment and shear force against the capacity of the segment if we know the compressive strength and flexural tensile strength development at early age. It is not acceptable to assume that the relationship between compressive strength and flexural tensile strength will be the same at early age as it will be at 28 days (ACI 544.7R-16, 2016; *fib* Bulletin 83, 2017). Therefore, specific early age unconfined compressive strength and EN 14651: 2005 flexural beam tests will be required.

WORKED EXAMPLE 10.1

During preliminary design, the designer assumes a segment is to be demoulded using a vacuum lifter that covers the central third of the segment extrados. The geometry of the problem is similar to that shown in Figure 10.1. There are six segments in the ring, so the segment arc is 60°. The length of the ring L_s (in the tunnel longitudinal direction) is 1 m, the thickness of the segment t is 200 mm and the internal diameter of the ring is 3.0 m.

Calculate the maximum shear force and bending moment due to demoulding.

The arc length along the centroid of the segment (mid-thickness) is given by:

$$L_{arc} = \frac{\pi(3+0.2)}{6} = 1.676 \text{ m}$$

Because the segment arc is 60°, the chord length in this case is one side of an equilateral triangle with the other two sides being the radius to the centroid. Therefore the chord length is:

$$L_{chord} = 1.6 \text{ m}$$

The segment has a uniform thickness along the arc length and therefore assuming a unit weight of 25 kN/m³ the segment has a self weight $w_{arc} = 5.0$ kN per metre of arc length, hence 5.0 kN/m.

If we simplify the problem to a straight beam, the self-weight per metre of chord length w_{chord} is given by:

$$w_{chord} = \frac{L_{arc}}{L_{chord}} w_{arc} = \frac{1.676}{1.6} \times 5.0 = 5.236 \text{ kN/m}$$

The design value of self-weight w_d is the self-weight multiplied by the partial factor for permanent actions γ_G:

$$w_d = w_{chord}\gamma_G = 5.236 \times 1.35 = 7.069 \text{ kN/m}$$

The vacuum lifter in this example covers the central third of the segment. The ends of the segment are therefore cantilevering as shown in the idealised structural model at the bottom of Figure 10.1. The shear force and bending moment may be calculated as shown in Figure 10.2. The cantilever length $L_{cant} = L_{chord}/3 = 1.6/3 = 0.533$ m.

The design value of shear force is given by:

$$V_{Ed} = w_d L_{cant} = 7.069 \times 0.533 = 3.768 \text{ kN}$$

The design value of bending moment is given by:

$$M_{Ed} = \frac{w_d L_{cant}^2}{2} = 1.005 \text{ kNm}$$

For an SFRC segment, we can check the design moment and shear force against the capacity of the segment if we know the compressive strength and flexural tensile strength development at early age. The procedure will be similar to that shown in the next section for segment stacking, which is also a statically determinate load case with no axial force.

For a reinforced concrete or plain concrete segment, we can check the design moment and shear force against the capacity of the segment using EN 1992-1-1: 2004.

10.2 STORAGE

Segments need to be stacked to minimise space requirements at the factory and site. Usually timber battens are used between segments to avoid damage. A photograph of segments stacked on site is shown in Figure 10.3. In this case there were six trapezoidal segments per ring with a universal taper, so each stack is one ring. If the rings have a key, this will be at the top of the stack. Sometimes with large diameter tunnels, rings are divided into two or more stacks to reduce stacking loads and lifting weight.

Figure 10.3 Segments stacked on site.

10.2.1 Actions

The positioning of the timber battens should be specified by the designer on the segment drawings. If they were positioned perfectly then there would be no bending moments induced in the segments except for those induced by self-weight of an individual segment, because loads from segments above travel straight down through the battens, as shown in Figure 10.4. However, when the battens are not aligned vertically, additional bending moments and shear forces will be induced. Segment stacking is a statically determinate load case.

In the case of Figure 10.4, bending moments and shear forces in the segments will depend on self-weight with two supports, as shown in Figure 10.5. Similar to the demoulding load case, we have simplified each segment to a straight beam.

The notation in Figure 10.5 is as follows:

L_{chord} is the chord length, i.e. the length of the segment idealised as a straight beam, in m

L_b is the distance between the timber battens in m

L_s is the segment length in the tunnel longitudinal direction in m

W is the weight of one segment in kN

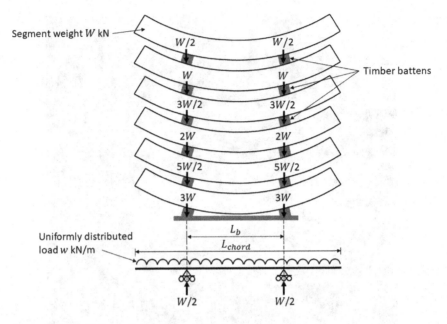

Figure 10.4 Perfect segment stacking using timber battens.

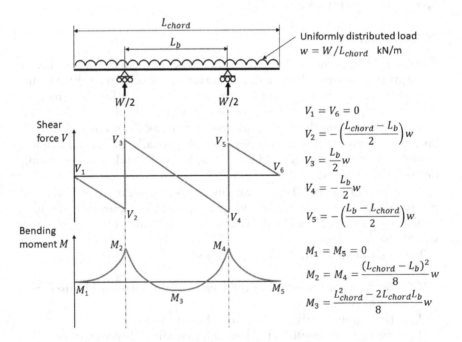

$$V_1 = V_6 = 0$$

$$V_2 = -\left(\frac{L_{chord} - L_b}{2}\right)w$$

$$V_3 = \frac{L_b}{2}w$$

$$V_4 = -\frac{L_b}{2}w$$

$$V_5 = -\left(\frac{L_b - L_{chord}}{2}\right)w$$

$$M_1 = M_5 = 0$$

$$M_2 = M_4 = \frac{(L_{chord} - L_b)^2}{8}w$$

$$M_3 = \frac{L_{chord}^2 - 2L_{chord}L_b}{8}w$$

Figure 10.5 Shear force and bending moment diagrams for perfect segment stacking.

w is the self-weight of one segment as a uniformly distributed load per unit chord length, in kN/m

V_1 to V_6 are the shear forces per unit segment length in the tunnel longitudinal direction, in kN, at the locations shown in Figure 10.5

M_1 to M_5 are the bending moments per unit segment length in the tunnel longitudinal direction, in kNm, at the locations shown in Figure 10.5

The design shear force and moment will be the maximum values calculated in Figure 10.5, multiplied by the partial safety factor for permanent unfavourable actions, $\gamma_G = 1.35$.

The designer also needs to take account of tolerances in the placement of battens. The worst case is if a pair of battens is placed as far apart as the tolerances allow, with the battens of the segment above being as close together as the tolerances allow (*fib* Bulletin 83, 2017). The tolerance is usually taken to be 100 mm (*fib* Bulletin 83, 2017; ACI 544.7R-16, 2016), so the adjacent battens could be up to 200 mm apart.

The moment due to eccentric batten placement in the bottom two segments can be added to the self-weight moment calculated above. Since poor batten placement is an accidental action, a partial factor of 1.00 can be used for this action (*fib* Bulletin 83, 2017). Figure 10.6 shows

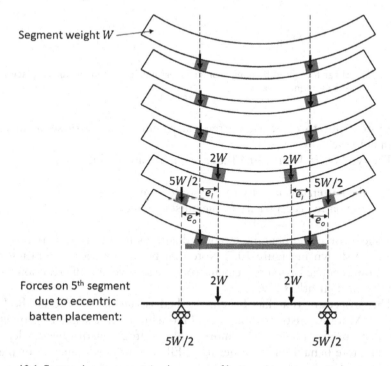

Figure 10.6 Forces due to eccentric placement of battens in segment stack.

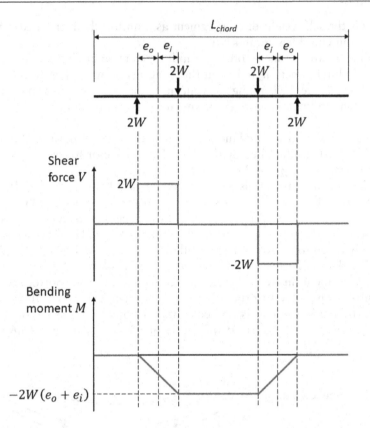

Figure 10.7 Shear force and bending moment diagrams for eccentric batten placement in a segment stack.

the forces due to eccentric batten placement on the fifth segment in a segment stack.

The symbols in Figure 10.6 that are not already defined are:

e_i is the batten eccentricity to the inside in m
e_o is the batten eccentricity to the outside in m

The shear force and bending moment diagrams for eccentric batten placement are shown in Figure 10.7. Note that because we are separating the effect of eccentric batten placement from the self-weight, the reaction from the battens beneath is $2W$.

The procedure that has been described is the most basic method of design. More sophisticated numerical modelling of the segment geometry could be used, along with a more sophisticated constitutive model for the concrete behaviour, whether it is SFRC, reinforced concrete or plain concrete.

10.2.2 SFRC segment ultimate limit state design for bending moment

In the previous section, we calculated the loads, so now we need to calculate the capacity of the segment at the ultimate limit state. If the loads are within the capacity, then the design is safe.

Ultimate limit state (ULS) and serviceability limit state (SLS) design to the *fib* Model Code 2010 (2013) is based on the load-displacement curve shown in Figure 10.8.

Figure 10.8 uses the following notation:

P_{cr} is the load at first crack
P_{ser} is the load based on service actions calculated from a linear elastic analysis with the assumption of uncracked concrete and initial elastic Young's modulus
P_u is the ultimate load, which is often driven by the maximum deformation requirement of the structure
P_{max} is the maximum load
δ_{ser} is the displacement at P_{ser}
δ_{peak} is the displacement at P_{max}
δ_u is the displacement at P_u

Note that P_u and δ_u are not the ULS load and delection, but are the 'ultimate load' and the 'ultimate deflection', respectively. The ULS load is the ultimate load divided by a partial factor, and represents the resistance of the structure.

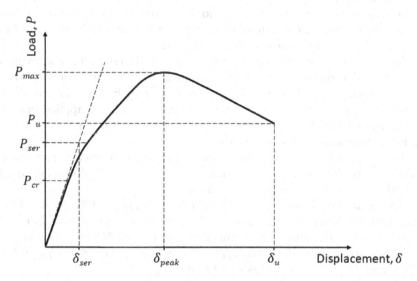

Figure 10.8 Idealised load-displacement curve with definitions according to the *fib* Model Code 2010 (2013) (based on a diagram in ITAtech, 2016).

Note also that P_{ser} and δ_{ser} are written as P_{SLS} and δ_{SLS} in the *fib* Model Code 2010 (2013). However, this is misleading as they are not the SLS load and deflection, but are the load and deflection due to service actions (as defined in Section 7.7.2 of the *fib* Model Code 2010). The subscript 'SLS' is not in the list of notation of the *fib* Model Code 2010, and appears to have been used in error, so here we will use P_{ser} and δ_{ser}, which are in the list of notation.

If fibres are used as the only reinforcement or the minimum conventional reinforcement to achieve ductility is not present, the *fib* Model Code 2010 (2013) Section 7.7.2 requires that one of the following conditions is satisfied:

1. $\delta_u \geq 20\delta_{ser}$
2. $\delta_{peak} \geq 5\delta_{ser}$

fib Bulletin 83 (2017) says that δ_u need only be greater than δ_{ser}, rather than at least 20 times greater as the Model Code requires. In this case, the designer needs to ensure that the segment has sufficient ductility and rotation capacity.

These conditions impose a requirement for minimum deformability in statically indeterminate structures to allow bending moment redistribution to occur. They do not apply to segment demoulding or stacking, which are statically determinate situations. However, they will apply to ground and water loads on the tunnel lining later in this chapter.

A further condition is that the ultimate load P_u shall be higher than both the cracking load P_{cr} and the service load P_{ser}. This is to avoid a brittle failure after cracking. The load cases we have seen so far for demoulding and stacking are 'statically determinate' situations. In other words, static equilibrium equations are sufficient to determine the internal forces in the segment. Therefore, the requirement that $P_u \geq P_{cr}$ can also be expressed in terms of bending moments, such that $M_u \geq M_{cr}$.

We also want to ensure that $P_{cr} \geq P_{ser}$ (or $M_{cr} \geq M_{ser}$) so that the segment is not cracked at the service load, as recommended by *fib* Bulletin 83 (2017). This is different to how we design the segmental lining once installed, where we allow some cracking when the service loads are applied but try to keep crack widths below 0.1–0.3 mm (depending on the specification) to allow autogenous healing to occur.

First we need to calculate what bending moment the section can resist. This may be done using a sectional analysis approach. To do this we need to calculate the design values of resistance.

The segment concrete is specified by a grade, for example 'FRC 40/50 – 5.0c' (see Chapter 8 Section 8.2 for more details). This means fibre reinforced concrete with a compressive cylinder strength of 40 N/mm², a compressive cube strength of 50 N/mm², $f_{R1k} = 5.0$ N/mm² and $0.9 \leq f_{R3k}/f_{R1k} \leq 1.1$. Therefore f_{R3k} will be taken as equal to $0.9 \times f_{R1k} = 4.5$ N/mm².

The segment concrete will also be required to meet the requirement $f_{R1k}/f_{Lk} \geq 0.4$ (Equation 5.6-2, *fib* Model Code 2010, 2013).

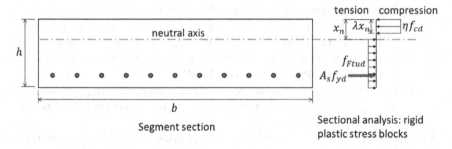

Figure 10.9 Rigid-plastic stress block approach for ultimate limit state design of a fibre-reinforced concrete segment.

To check the ultimate limit state, a rigid-plastic stress block approach may be used, as shown in Figure 10.9.

Figure 10.9 uses the following notation:

f_{cd} is the design compressive strength in N/mm^2

f_{Ftud} is the design value of the ultimate post-cracking tensile strength in N/mm^2

x_n is the distance from the compressive extreme fibre to the neutral axis in mm

λ defines the depth of the compressive stress block as a factor of x_n

η is a factor that is used to reduce the design compressive strength at high strength values

A_s is the area of steel bar reinforcement (if present) in mm^2

f_{yd} is the design yield strength of the steel bar reinforcement (if present) in N/mm^2

We are only going to consider SFRC segments here, the bar reinforcement in Figure 10.9 is only shown for completeness as it is possible to use both fibre and bar reinforcement together and this is included in the *fib* Model Code 2010 (2013) design methodology.

The moment that can be resisted at the ultimate limit state will depend on the values of the design compressive strength f_{cd} and the design value of the ultimate post-cracking tensile strength f_{Ftud}, as well as the distance to the neutral axis x_n.

If the characteristic cylinder compressive strength $f_{ck} \le 50$ MPa, then $\lambda = 0.8$ and $\eta = 1.0$. If $f_{ck} > 50$ MPa then η and λ need to be determined by equations 7.2–15 to 7.2–18 in the *fib* Model Code 2010 (2013). As an example, for FRC 40/50 – 5.0c, $f_{ck} = 40$ MPa, so $\eta = 1.0$ and $\lambda = 0.8$.

The design compressive strength f_{cd} is given by Section 3.1.6 of EN 1992-1-1: 2004, and the value for FRC 40/50 – 5.0c concrete is calculated like this for our numerical example:

$$f_{cd} = \frac{\alpha_{cc} f_{ck}}{\gamma_c} = \frac{0.85 \times 40}{1.5} = 22.67 \text{ N/mm}^2 \qquad (10.6)$$

f_{cd} is the design compressive strength in N/mm^2.

α_{cc} is a coefficient to take account of long-term loading and other factors and is between 0.8 and 1.0, with its value given in the National Annex to EN 1992-1-1. In the UK the value is 0.85 (UK NA to BS EN 1992-1-1: 2004).

f_{ck} is the characteristic value of the cylinder compressive strength in N/mm^2.

γ_c is the partial material factor for concrete in compression, and is 1.5.

The design value of ultimate post-cracking tensile strength f_{Ftud} is given by the following equation:

$$f_{Ftud} = \frac{f_{R3d}}{3} \tag{10.7}$$

f_{Ftud} is the design value of the ultimate post-cracking tensile strength in N/mm^2.

f_{R3d} is the design value of the residual tensile strength at CMOD = 2.5 mm. The design value is the characteristic value f_{R3k} divided by the material partial factor, therefore $f_{R3d} = f_{R3k}/\gamma_F$, where γ_F is the material partial factor for fibre reinforced concrete in tension, usually taken as 1.5 (*fib* Bulletin 83, 2017) but can be as low as 1.3 if this can be justified by 'improved control procedures' (Section 5.5.6, *fib* Model Code 2010, 2013). For example, Psomas & Eddie (2016) used $\gamma_F = 1.35$.

The following equation demonstrates a numerical calculation for the example of FRC 40/50 – 5.0c concrete:

$$f_{Ftud} = \frac{f_{R3d}}{3} = \frac{f_{R3k}/\gamma_F}{3} = \frac{4.5/1.5}{3} = 1.0 \text{ N/mm}^2 \tag{10.8}$$

From the values of $f_{Ftud} = 1.0$ N/mm^2 and $f_{cd} = 22.67$ N/mm^2 we can immediately see from Figure 10.9 that, for our numerical example, without conventional bar reinforcement the height of the compression stress block will be 22.67 times smaller than the height of the tension stress block in order for there to be horizontal force equilibrium. Therefore the neutral axis at the ultimate limit state will be very close to the upper surface of the segment.

This can also be expressed by the equation for horizontal force equilibrium:

$$\lambda x_n f_{cd} = (h - x_n) f_{Ftud} \tag{10.9}$$

Which can be rearranged such that:

$$x_n = \frac{h f_{Ftud}}{(\lambda f_{cd} + f_{Ftud})} \tag{10.10}$$

Using Equation 10.10 we can calculate the size of the stress blocks, and from that we can calculate the ULS resistance moment M_{Rd}, which is the design value of the ultimate moment. The ULS resistance moment may be obtained by taking moments about the neutral axis of the design values of compressive and tensile strength (c.f. Figure 10.9):

$$M_{Rd} = f_{Ftud}b(h - x_n)\left(\frac{h - x_n}{2}\right) + f_{cd}b\lambda x_n\left(x_n - \frac{\lambda x_n}{2}\right) \tag{10.11}$$

M_{Rd} is the ULS resistance moment in Nmm

The ULS resistance moment can be either positive or negative. Unless bar reinforcement is present and not the same in the top and bottom layers, moment capacity is always symmetrical in SFRC members if the section geometry is symmetrical.

If the ULS resistance moment M_{Rd} is greater than the design moment M_{Ed}, the segment passes the ultimate limit state check for this load case. The design moment M_{Ed} is calculated from the characteristic actions multiplied by a partial factor.

To ensure there is no brittle failure after cracking, we need to check that $M_u \geq M_{cr}$. This is because this is a statically determinate load case, and after cracking the self-weight load cannot be redistributed. The *fib* Bulletin 83 (2017) states that unfactored mean strengths should be used for this check, though this is not mentioned in the *fib* Model Code 2010 (2013). If design values are used then we can check that $M_{Rd} \geq M_{crd}$ instead.

M_{cr} is the cracking bending moment in Nmm
M_{crd} is the design value of cracking bending moment in Nmm

The design value of cracking bending moment M_{crd} may be calculated as follows, assuming an elastic stress-strain relationship (*fib* Bulletin 83, 2017):

$$M_{crd} = \frac{bh^2}{6}f_{ctd,fl} \tag{10.12}$$

M_{crd} is the design value of cracking bending moment in Nmm
b is the width in mm
h is the segment thickness in mm
$f_{ctd,fl}$ is the design flexural tensile strength at first crack in N/mm²

In the absence of experimental data, the design flexural tensile strength at first crack $f_{ctd,fl}$ can be calculated using the equations in EN 1992-1-1: 2004 for concrete, as follows.

First we get the characteristic mean value of axial tensile strength f_{ctm} from Table 3.1 in EN 1992-1-1: 2004. For our example FRC 40/50 – 5.0c, which has a characteristic compressive cylinder strength $f_{ck} = 40$ MPa, $f_{ctm} = 3.5$ MPa.

Next we can calculate the mean flexural tensile strength using Equation (3.23) from EN1992-1-1: 2004, as follows:

$$f_{ctm,fl} = \max\left\{\left(1.6 - \frac{h}{1000}\right)f_{ctm}; f_{ctm}\right\}$$

(10.13)

$f_{ctm,fl}$ is the mean flexural tensile strength at first crack in N/mm²
h is the segment thickness in mm
f_{ctm} is the mean axial tensile strength in N/mm²

A very similar value may be obtained using equations 5.1–3a, 5.1–8a and 5.1–8b in the *fib* Model Code 2010 (2013).

Equation 10.13 may also be used to obtain the characteristic flexural tensile strength $f_{ctk,fl}$ from a value of characteristic axial tensile strength f_{ctk}.

Alternatively, if EN 14651: 2005 flexural beam test data exist (and this is preferred for SFRC segment design), then the mean flexural tensile strength $f_{ctm,fl}$ may be taken as the characteristic mean value of flexural tensile strength at the limit of proportionality f_{Lm}:

$$f_{ctm,fl} = f_{Lm}$$

(10.14)

f_{Lm} is the characteristic mean value of flexural tensile strength at the limit of proportionality in N/mm²

Then the design value may be calculated by dividing the mean value by the partial material factor for fibre-reinforced concrete in tension γ_F, usually taken as 1.5.

$$f_{ctd,fl} = \frac{f_{ctm,fl}}{\gamma_F}$$

(10.15)

γ_F is the partial material factor for fibre-reinforced concrete

Thus we can now calculate M_{crd} using Equation 10.12, and if $M_{Rd} \geq M_{crd}$ then there will not be a brittle failure after cracking. If $M_{Rd} \leq M_{crd}$ then the *fib* Bulletin 83 (2017) recommends applying an additional partial factor of 1.2 to reduce M_{Rd}, to provide additional safety against a brittle failure.

Note that if the segments are stacked before they have achieved their specified 28 day strength, then a design check will need to be made for an earlier age strength.

10.2.3 SFRC segment ULS design for shear force

The *fib* Model Code 2010 (2013) allows a simplified method of checking for tensile-hardening FRC, that the design value of the ultimate residual tensile strength f_{Ftud} $(= f_{Ftuk}/\gamma_F)$ must be greater than the principal tensile stress in the segment. However, we have a complex situation where we have both bending stresses and shear stresses. If a continuum numerical model of the segment were used, then the outputs could be in the form of principal stresses, and then this check would be straightforward.

Coccia et al. (2015) proposed a method for including the effect of shear in the ULS check for moment and axial force, by reducing the design value of the ultimate post-cracking tensile strength f_{Ftud} as follows:

$$f_{Ftud}^{red} = f_{Ftud} \left[1 - \left(\frac{\tau}{f_{Ftud}} \right)^2 \right] \tag{10.16}$$

f_{Ftud}^{red} is the reduced value of f_{Ftud} in N/mm^2

f_{Ftud} is the design value of the ultimate post-cracking tensile strength in N/mm^2 and is given by $f_{Ftud} = f_{R3d}/3$ (see Equation 10.8)

τ is the shear stress, given by $\tau = V/bh$, in N/mm^2

The moment capacity check described in the previous section should then be repeated with the reduced value f_{Ftud}^{red} in place of f_{Ftud}.

Note that if the segments are stacked before they have achieved their specified 28 day strength, then a design check will need to be made for an earlier age strength.

10.2.4 SFRC segment SLS design

We must also check the SLS, because even though we have the opportunity to discard segments that are damaged by storage, we do not want to throw away lots of segments either. For transient situations such as demoulding, storage and transportation it is common to impose a condition that no cracking should occur (*fib* Bulletin 83, 2017). This means that the minimum moment at which cracking occurs must be greater than the moment due to service actions, i.e.:

$$M_{cr,min} \geq M_{ser} \tag{10.17}$$

$M_{cr,min}$ is the minimum moment at which cracking may occur in kNm
M_{ser} is the moment due to service actions in kNm

The minimum cracking moment $M_{cr,min}$ is calculated assuming an elastic stress-strain distribution, as shown in Figure 10.10:

$$M_{cr,min} = \frac{bh^2}{6} f_{ctk,fl} \tag{10.18}$$

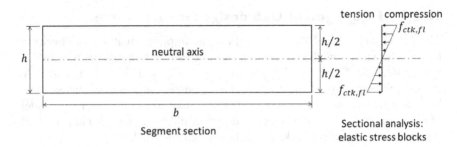

Figure 10.10 Elastic stress blocks for SLS design of a fibre-reinforced concrete segment.

b is the segment length in the tunnel longitudinal direction in mm
h is the segment thickness in mm
$f_{ctk,fl}$ is the characteristic flexural tensile strength in N/mm²

Note that for the SLS the partial material factor is 1.0.
The characteristic flexural tensile strength is given by Equation 10.13:

$$f_{ctk,fl} = \max\left\{ \left(1.6 - \frac{h}{1000}\right) f_{ctk}; f_{ctk} \right\} \tag{10.19}$$

f_{ctk} is the characteristic (5th percentile) axial tensile strength in N/mm²

The characteristic axial tensile strength, in the absence of experimental data, may be obtained from Table 3.1 of EN 1992-1-1: 2004, and is given by:

$$f_{ctk} = 0.7 f_{ctm} \tag{10.20}$$

f_{ctm} is the characteristic mean value of axial tensile strength in N/mm², which may be found in Table 3.1 of EN 1992-1-1: 2004

The moment due to service actions M_{ser} is calculated in the same way as the design moment M_{Ed}, but with a partial factor on actions of 1.0.

WORKED EXAMPLE 10.2

We will use the same segment dimensions as in Worked Example 10.1, i.e. the length of the ring (in the tunnel longitudinal direction) $L_s = 1$ m, the thickness of the segment $t = 200$ mm and the internal diameter of the ring is 3.0 m. There are six segments in a ring and all six will form a single stack. The unit weight of the segment concrete $\gamma_{seg} = 25$ kN/m³ and consists of SFRC classified as FRC 40/50 – 5.0c.

The design drawings specify a batten placement at quarter points measured horizontally, with a maximum batten eccentricity $e_i = e_o = 100$ mm. The lowest level of support battens between the 6th segment and the ground are joined together and so cannot be out of tolerance. The geometry is shown in Figure 10.11.

a. Calculate the design values of the maximum shear force and bending moment.
b. Check the capacity of the segment at the ULS.
c. Check for the effect of shear on the resistance moment at the ULS.
d. Check the capacity of the segment at the SLS.

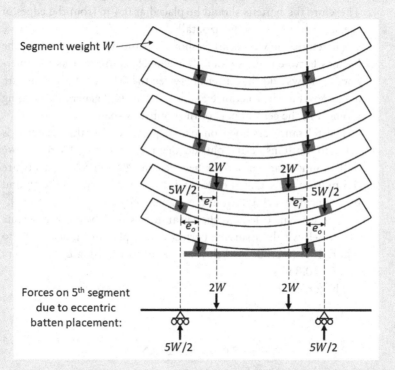

Figure 10.11 Forces due to eccentric placement of battens in segment stack.

a. Calculate the design values of the maximum shear force and bending moment.

The arc length L_{arc} of a segment is given by:

$$L_{arc} = r_m \theta$$

r_m is the radius to the lining centroid in m, which is 1.6 m

θ is the segment angle in radians, which is $2\pi/6$ rad

Therefore:

$$L_{arc} = r_m\theta = 1.6\frac{2\pi}{6} = 1.676 \text{ m}$$

Now we will calculate the self weight of one segment W:

$$W = \gamma_{seg}L_sL_{arc}t = 25 \times 1.0 \times 1.676 \times 0.2 = 8.38 \text{ kN}$$

Since the segment arc is 60°, then the segment chord and two radii make an equilateral triangle and the segment chord length is 1.6 m. Therefore the battens should be placed at 0.4 m from the edges of the segment measured horizontally.

Therefore, the worst case is shown for the 5th segment where the battens between the 4th and 5th are as close together as the tolerance e_i allows and the battens between the 5th and 6th segments are as far apart as the tolerance e_o allows. This will generate a sagging moment in the central part of the segment's span.

The tolerance assumed on batten placement in this example is 0.1 m for both inside and outside tolerances e_i and e_o. Therefore, we will assume that the battens between the 4th and 5th segments are 0.5 m from the segment edges and the battens between the 5th and 6th segments are 0.3 m from the segment edges.

First we will calculate the shear forces and bending moments due to self-weight assuming perfect batten placement, according to Figure 10.5 (reproduced below as Figure 10.12) using $L_{chord} = 1.6$ m and $L_b = 0.8$ m.

The forces are:

$$V_1 = V_6 = 0$$

$$V_2 = -\left(\frac{1.6 - 0.8}{2}\right) \times 7.069 = -2.828 \text{ kN}$$

$$V_3 = \frac{0.8}{2} \times 7.069 = 2.828 \text{ kN}$$

$$V_4 = -\frac{0.8}{2} \times 7.069 = -2.828 \text{ kN}$$

$$V_5 = -\left(\frac{0.8 - 1.6}{2}\right) \times 7.069 = 2.828 \text{ kN}$$

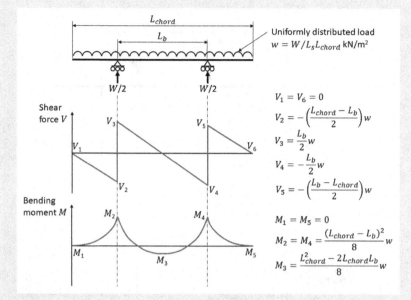

The uniformly distributed load $w = W/L_s L_{chord}$ kN/m^2

$$V_1 = V_6 = 0$$

$$V_2 = -\left(\frac{L_{chord} - L_b}{2}\right)w$$

$$V_3 = \frac{L_b}{2}w$$

$$V_4 = -\frac{L_b}{2}w$$

$$V_5 = -\left(\frac{L_b - L_{chord}}{2}\right)w$$

$$M_1 = M_5 = 0$$

$$M_2 = M_4 = \frac{(L_{chord} - L_b)^2}{8}w$$

$$M_3 = \frac{L_{chord}^2 - 2L_{chord}L_b}{8}w$$

Figure 10.12 Shear force and bending moment diagrams for perfect segment stacking.

$$M_1 = M_5 = 0$$

$$M_2 = M_4 = \frac{(1.6 - 0.8)^2}{8} \times 7.069 = 0.566 \text{ kNm}$$

$$M_3 = \frac{1.6^2 - 2 \times 1.6 \times 0.8}{8} \times 7.069 = 0 \text{ kNm}$$

Next we can look at the eccentric batten placement. The shear force and bending moment diagrams for this load case were shown in Figure 10.7, reproduced here as Figure 10.13.

In Figure 10.13, the maximum shear force in the 5th segment due to eccentric batten placement is equal to half the weight of the segments above. The segment weight will have a partial factor of 1.0, because eccentric batten placement is considered an accidental action and so the design value of the action is given by:

$$W_d = W \times 1.0 = 8.38 \times 1.0 = 8.38 \text{ kN}$$

The maximum design value of shear force due to batten eccentricity is therefore:

$$V_{Ed,ecc} = \pm 2 \times W_d = \pm 16.76 \text{ kN}$$

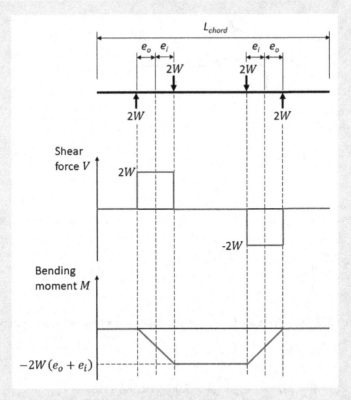

Figure 10.13 Shear force and bending moment diagrams for eccentric batten placement in a segment stack.

The maximum design value of bending moment due to batten eccentricity is given by:

$$M_{Ed,ecc} = -2 \times W_d \left(e_o + e_i\right) = -2 \times 8.38 \times 0.2 = -3.352 \text{ kNm}$$

The shear force and bending moment induced by eccentric batten placement are quite a lot higher than those induced in a perfect stack, and this illustrates the importance of considering tolerances.

Adding together the self-weight value multiplied by a load factor of 1.35 and the eccentric batten placement design value, the design value of shear force is:

$$V_{Ed} = 16.76 + 2.828 = 19.59 \text{ kN}$$

Superposing the self-weight and eccentric batten placement, the worst combination in sagging will occur at mid-span, where the self-weight with perfect batten placement moment is zero.

$$M_{Ed,sag} = M_3 + M_{Ed,ecc} = -3.352 \text{ kNm}$$

The worst combination in hogging will be over the supports due to self-weight. We will ignore the sagging moment provided by the battens at these locations as it is favourable.

$$M_{Ed,hog} = M_2 = 0.566 \text{ kNm}$$

b. Check the capacity of the segment at the ultimate limit state.
   This is an SFRC design calculation according to the *fib* Model Code 2010 (2013).

   Structural and serviceability design is based on the load-displacement curve shown in Figure 10.8. The ultimate moment M_u shall be higher than the cracking moment M_{cr} to avoid a brittle failure after cracking. We also want to ensure that $M_{cr} \geq M_{ser}$ so that the segment is not cracked at the service load, as recommended by *fib* Bulletin 83 (2017).

   We need to calculate the ultimate resistance moment of the section, M_u. To do this we first need to calculate the design values of resistance.

   The segment concrete is FRC 40/50 – 5.0c. This means fibre reinforced concrete with a compressive cylinder strength of 40 N/mm², a compressive cube strength of 50 N/mm², $f_{R1k} = 5.0$ N/mm² and $0.9 \leq f_{R3k}/f_{R1k} \leq 1.1$. Therefore f_{R3k} will be taken as equal to $0.9 \times f_{R1k} = 4.5$ N/mm².

   The segment concrete will also be required to meet the requirement $f_{R1k}/f_{Lk} \geq 0.4$ (Equation 5.6-2, *fib* Model Code 2010, 2013).

   To check the ultimate limit state, a rigid-plastic stress block approach may be used, as shown in Figure 10.9.

   The characteristic cylinder compressive strength $f_{ck} \leq 50$ MPa, so we will use $\lambda = 0.8$ and $\eta = 1.0$. The design compressive strength f_{cd} is given by Equation 10.6:

$$f_{cd} = \frac{\alpha_{cc} f_{ck}}{\gamma_c} = \frac{0.85 \times 40}{1.5} = 22.67 \text{ N/mm}^2$$

Ultimate post-cracking tensile strength f_{Ftud} is given by Equation 10.8:

$$f_{Ftud} = \frac{f_{R3d}}{3} = \frac{f_{R3k}/\gamma_c}{3} = \frac{4.5/1.5}{3} = 1.0 \text{ N/mm}^2$$

Considering horizontal force equilibrium of the stress blocks (Equation 10.10), for $h = 200$ mm we get:

$$x_n = \frac{h f_{Ftud}}{(\lambda f_{cd} + f_{Ftud})} = \frac{200 \times 1.0}{(0.8 \times 22.67 + 1.0)} = 10.45 \text{ mm}$$

The resistance moment at the ultimate limit state M_{Rd} is calculated by taking moments about the neutral axis (Equation 10.11):

$$M_{Rd} = f_{Ftud} b (h - x_n)\left(\frac{h - x_n}{2}\right) + f_{cd} b \lambda x_n \left(x_n - \frac{\lambda x_n}{2}\right)$$

$$= 1.0 \times 1000 \times (200 - 10.45) \times \left(\frac{200 - 10.45}{2}\right) + 22.67 \times 1000$$

$$\times 0.8 \times 10.45 \times \left(10.45 - \frac{0.8 \times 10.45}{2}\right)$$

$$= 19.15 \times 10^6 \text{ Nmm} = 19.15 \text{ kNm}$$

Since $M_{Rd} \geq |M_{Ed}|$ in both hogging and sagging, the segment passes the ultimate limit state check for storage.

To ensure there is no brittle failure after cracking, we need to check that $M_{Rd} \geq M_{crd}$.

First we get the value of mean axial tensile strength f_{ctm} from Table 3.1 in EN 1992-1-1: 2004. For a characteristic compressive cylinder strength $f_{ck} = 40$ MPa, $f_{ctm} = 3.5$ MPa.

Then we calculate the mean flexural tensile strength using Equation 10.13:

$$f_{ctm,fl} = \max\left\{\left(1.6 - \frac{h}{1000}\right) f_{ctm}; f_{ctm}\right\} = \max\left\{\left(1.6 - \frac{200}{1000}\right) \times 3.5; 3.5\right\}$$

$$= 4.9 \text{ N/mm}^2$$

Then the design value may be calculated by dividing the mean value by the partial material factor for fibre reinforced concrete in tension γ_F, usually taken as 1.5.

$$f_{ctd,fl} = \frac{f_{ctm,fl}}{\gamma_F} = \frac{4.9}{1.5} = 3.27 \text{ MPa}$$

Thus the design value of the cracking bending moment is given by:

$$M_{crd} = \frac{bh^2}{6}f_{ctd,fl} = \frac{1000 \times 200^2}{6} \times 3.27 = 21.78 \text{ kNm}$$

Thus $M_{Rd} \leq M_{crd}$ and so there could be a brittle failure after cracking. *fib* Bulletin 83 (2017) argues that in this situation an additional partial factor of 1.2 should be applied to reduce the risk of cracking and brittle failure, and strongly recommends full-scale bending tests on segments.

Applying the additional partial factor:

$$M_{Rd} = \frac{19.15}{1.2} = 15.96 \text{ kNm}$$

Since $M_{Rd} \geq |M_{Ed}|$ and $M_{crd} \geq |M_{Ed}|$ in both hogging and sagging, the segment still passes the ultimate limit state check for storage.

c. Check for the effect of shear on the resistance moment at the ULS.

Using Coccia et al. (2015)'s method of including the effect of shear in the ULS check for moment and axial force, by reducing the design value of the ultimate post-cracking tensile strength f_{Ftud}. The shear stress is given by:

$$\tau = \frac{V_{Ed}}{bh} = \frac{19.59 \times 10^3}{1000 \times 300} = 0.098 \text{ N/mm}^2$$

And the reduced value of f_{Ftud} is given by Equation 10.16:

$$f_{Ftud}^{red} = f_{Ftud}\left[1 - \left(\frac{\tau}{f_{Ftud}}\right)^2\right] = 1.0\left[1 - \left(\frac{0.098}{1.0}\right)^2\right] = 0.990 \text{ N/mm}^2$$

The ULS check in (c) and (d) should be repeated with the reduced value f_{Ftud}^{red} in place of f_{Ftud}. We can see that in this case the effect will be negligible.

d. Check the capacity of the segment at the serviceability limit state.

For transient situations such as demoulding, storage and transportation it is usual practice to impose a condition that no cracking should occur.

The characteristic axial tensile strength, in the absence of experimental data, may be obtained from Table 3.1 of EN 1992-1-1: 2004, and is given by:

$$f_{ctk} = 0.7f_{ctm} = 0.7 \times 3.5 = 2.45 \text{ N/mm}^2$$

And the characteristic flexural tensile strength is given by Equation 10.13:

$$f_{ctk,fl} = \max\left\{\left(1.6 - \frac{h}{1000}\right)f_{ctk}; f_{ctk}\right\} = \max\left\{\left(1.6 - \frac{200}{1000}\right) \times 2.45; 2.45\right\}$$

$$= 3.43 \text{ N/mm}^2$$

The minimum cracking moment $M_{cr,min}$ is given by:

$$M_{cr,min} = \frac{bh^2}{6}f_{ctk,fl} = \frac{1000 \times 200^2}{6} \times 3.43 = 22.87 \text{ kNm}$$

The moment due to service actions M_{ser} is calculated in the same way as the design moment M_{Ed}, but with a partial factor on actions of 1.0. Therefore the maximum values are:

$$M_{ser,sag} = M_3 \times \frac{1.0}{1.35} + M_{Ed,ecc} \times \frac{1.0}{1.0} = -3.352 \text{ kNm}$$

$$M_{ser,hog} = M_2 \times \frac{1.0}{1.35} = 0.566 \times \frac{1.0}{1.35} = 0.419 \text{ kNm}$$

We need to check that:

$$M_{ser} \leq M_{cr,min}$$

So for both the sagging and hogging service moments, the SLS check is therefore passed.

10.3 TRANSPORTATION AND HANDLING

The segments must be moved from the factory to storage yard, then to site on a railway wagon, barge or lorry, and then to the TBM on a multi-service vehicle or rail wagon. There are also all the crane and telehandler lifts between these modes of transport to consider. For each of these, the loading on the segments or a stack of segments must be considered.

There are also situations where the segments need to be turned over. Most segments are cast extrados-upwards, whereas it is usual to stack segments extrados-down. Sometimes segments are cured or stored standing on their circumferential joint. All these manipulations and the loads they may induce must be considered.

In most cases, the bending moments are important, but the axial and shear forces are relatively small (*fib* Bulletin 83, 2017).

fib Bulletin 83 (2017) recommends that where dynamic loads are expected, for example during road or rail transport, an additional partial factor of $\gamma_d = 2.0$ should be applied to the self-weight (as well as the usual partial factor of 1.35).

In all these cases, the design model will be of lifting points or areas and cantilevering sections, with a design procedure similar to that for demoulding and storage.

If the segments are lifted using cast-in lifting pins, or by using a device screwed into the grout plug, then these lifting devices need to be checked. Also, pull-out of a cone of concrete or edge failure around the lifting device also needs to be considered.

10.4 ERECTION

Within the TBM, segments may be lifted off the transport by a hoist or rollers and moved through to the front. Then they will be lifted by the segment erector, which could have lifting points or use vacuum lifting.

In all these cases, the design model will be of lifting points or areas and cantilevering sections, with a design procedure similar to those for demoulding, storage, transportation and handling.

10.5 INSTALLATION LOADS

TBM jacks shove the machine forward by pushing against the circumferential joint of the last ring built. These jacks are usually equally spaced around the circumference and are often in pairs with a shared ram shoe that applies the load to the circumferential joint. Although the aim is to spread the load as evenly as possible, bending moments and other kinds of stress concentrations may result from imperfections in ring build and segment geometry.

10.5.1 Temporary TBM jacking loads

Burgers et al. (2007) used 3D numerical modelling to investigate the effect of jacking forces on SFRC segments for the Barcelona Line 9 tunnels. Each ring consisted of seven segments plus a key segment, the tunnel was about 12 m diameter, and each segment had an arc length of 4.70 m. Each segment was loaded by two pairs of thrust rams. The trailing edge of each segment had four bearing pads to help transfer jacking loads to the previous ring. As jacking loads were increased in the model, the first cracks were formed between the loading surfaces (the ram shoes) due to spalling. At higher loads, the model showed splitting cracks developing under the loading surfaces. The steel fibres ensured that cracks propagated in a controlled manner and increased load resulted in more cracks rather than widening of a single crack.

This numerical model ignored the effect of imperfections in the segments or geometrical tolerances in ring assembly. In real life the ring joint may

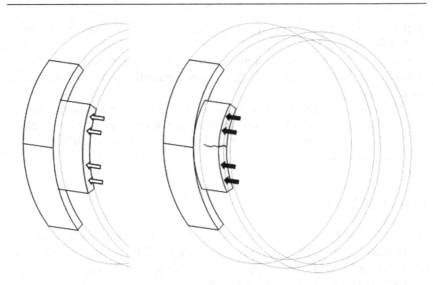

Figure 10.14 Cracking induced by eccentricity of jacking load towards the extrados turning segment outwards.

not be plane and the jacking forces may not be applied perfectly. Therefore, Burgers et al. (2007) went on to analyse the effects of eccentric placement of a thrust jack in the radial direction by applying a jacking force in the model with a triangular pressure distribution across the segment thickness. This caused a significant reduction in the failure load from 30.7 MN to 23.6 MN and cracking began much earlier. The (exaggerated) deformation of a segment due to eccentricity of the jacking load towards the extrados of the circumferential joint is shown in Figure 10.14. The eccentricity causes the segment to bend outwards about a tangential axis, which concentrates the bearing stress on the previous ring into the centre and induces bending about a radial axis causing a longitudinal tensile crack on the leading edge of the segment between the two pairs of jacks, due to the uneven support on the trailing edge. Cracks exactly like this one were observed in real segments during the construction of Line 9.

Another possible cause of longitudinal cracks between the ram shoes is uneven support due to steps in the circumferential joint plane (de Waal, 2000). Cavalaro et al. (2011) say that this is the most common cause of this type of cracking. An illustration of this is shown in Figure 10.15.

The effect of this uneven support depends on its magnitude, on the thickness and material properties of the segments, and on the thickness and material properties of the packing material. For a small step between segments in the previous ring, the new segment will deform until contact is made across the gap, with only a small reduction in capacity. For a larger step critical damage may occur before contact is made. Cavalaro et al. (2011) found that thinner and longer rings could cope with larger steps.

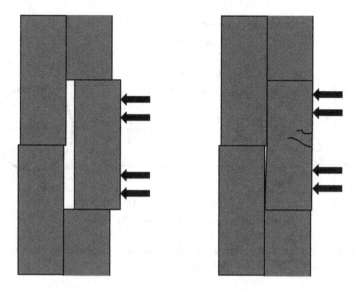

Figure 10.15 Longitudinal cracks caused by uneven support.

Thinner rings have a lower stiffness, allowing more deformation to occur to close the gap before critical tensile stresses are induced. Longer rings, although stiffer, allow more load to be transferred to the first contact and therefore result in lower tensile stresses between the ram shoes. A lower packer stiffness was found to help too. Heijmans & Jansen (1999) looked at this load case for the design of the Pannerdensch Kanaal Tunnel and found that the gap needed to be less than a few tenths of a millimetre to avoid cracking.

Bilotta & Russo (2012) installed strain gauges in segments used for the Line 1 tunnels in Naples. Each ring consisted of six segments plus a key. Four independent rings were instrumented with a large number of strain gauges. In each ring, five vibrating wire strain gauges were cast into each segment, all on the circumferential centreline, i.e. equidistant from both circumferential joints. A pair of strain gauges, one near the intrados and one near the extrados, was installed near to each radial joint of every segment (except the key) to measure hoop strain, and one was installed in the centre of each segment to measure longitudinal strain. They found that the jacking forces caused a high longitudinal compression, as one would expect, but also an extension in the circumferential direction, which they attributed to a Poisson's ratio effect and the lack of significant restraint to the ring within the tailskin. This circumferential extension was later reversed by grouting and ground load.

By using a back-calculation method to convert the strains into stresses, Bilotta & Russo (2012) also calculated internal forces around the lining. A further calculation, which they did not explain in the paper, enabled

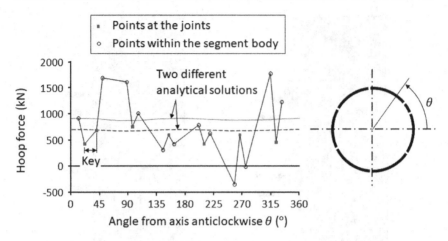

Figure 10.16 Hoop forces around a segmental lining back-calculated from strain gauges (from Bilotta & Russo, 2012).

the forces across the radial joints to be estimated. The assembly of the lining, jacking forces and grout pressures resulted in a highly variable hoop axial force and bending moment distribution, particularly if the longitudinal strains were taken into account in the calculation, as shown in Figure 10.16. Hoop axial forces were lower at joints than in the body of the segment. Bilotta & Russo explained that the segments were placed quite smoothly, but didn't explain where the hoop force was going. It is possible that the hoop force could have been transferred via dowels and friction in the circumferential joints to the adjacent ring's segments rather than across the joints.

If the results of Bilotta & Russo's back-calculation are taken at face value, the most interesting aspect is how much real hoop thrusts vary within segments and around the ring compared to conventional calculations using analytical solutions. The effects of small geometric misalignments, asperities, variable and/or eccentric jacking forces and grout pressures in the real situation clearly have a large impact on the forces in the segments in the medium- to long-term.

We need to be able to design a segmental lining for jacking forces, when small geometrical misalignments can cause such large changes in stress. These effects can only really be modelled using 3D numerical analysis, with careful attention paid to modelling the expected geometrical tolerances and the contact between segments. It is also wise to consider full-scale testing of segments in a laboratory at the tolerance limits.

Usual practice in design is to make assumptions, usually in consultation with the TBM manufacturer and the contractor, about the worst case distribution of jacking forces around the ring and their potential eccentricities.

Groeneweg (2007) recommends a partial factor of 2.0 on jacking loads for comparison with the ultimate limit state. Alternatively, *fib* Bulletin 83

(2017) suggests using the maximum capacity of the rams, based on the maximum pressure of the hydraulic system.

Since we do not want to cause excessive cracking of the segments, *fib* Bulletin 83 (2017) recommends applying a partial factor of 1.5 on the nominal working load of the thrust rams for comparison with the SLS.

The following load cases should be considered:

- local crushing of the concrete under the ram shoes
- bursting stress within the segment beneath the ram shoes, in both the radial and circumferential directions (see Section 9.4), also considering potential eccentricity of the ram shoe
- shear failure of the circumferential joint due to ram load
- spalling near the surface of the circumferential joint between the ram shoes
- geometric imperfections, such as steps in the circumferential joint plane

10.5.2 Longitudinal effects of TBM jacking loads

Hoefsloot (2009) used an analytical model and also strain gauge measurements from the Groene Hart Tunnel to show that jacking forces remain in the tunnel lining permanently. At a distance of about 40 m behind the TBM, average longitudinal axial force in the lining had decreased to approximately 70% of the jacking load, and this stayed in the lining permanently. Talmon et al. (2009), in an accompanying paper, calculated that the residual axial force acted at 1.5m above the axis level of the Groene Hart Tunnel, meaning a significant residual longitudinal bending moment is also left in the tunnel permanently.

Figure 10.17, taken from Hoefsloot (2009), shows the longitudinal bending moments (about the horizontal axis of the tunnel) generated by jacking loads. These bending moments are caused by the fact that jacking forces tend to be higher at the invert to counteract the tendency for TBMs to dive, perhaps due to the weight of the cutterhead and higher friction around the bottom part of the shield. The rings within the TBM are also to some extent cantilevering off the rings further back where the grout has hardened (Hoefsloot & Verweij, 2006). With distance behind the TBM, these bending moments are gradually reversed by grout pressures and the weight of the backup equipment.

Predicting these longitudinal bending moments is complex, but can be done analytically or numerically, as shown by Hoefsloot (2009), Hoefsloot & Verweij (2006) and Talmon et al. (2009). They depend mainly on the distribution of jacking forces, restraint provided by the shield/tailseal brushes, the properties of the grout and the position and weight of the backup.

Hoefsloot (2009) also found that the bending moment distribution in Figure 10.17 means that the last ring built nearly always has overhang,

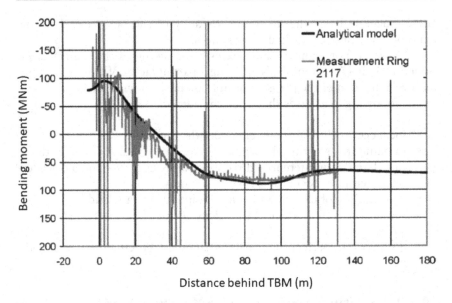

Figure 10.17 Strain gauge data and analytical model of the Groene Hart Tunnel (from Hoefsloot, 2009).

i.e. the crown is slightly further inbye than the invert. The new ring has to be placed to the design alignment, hence there will always be a change in plane of the rings and perhaps stepping. This can cause eccentric loads across the circumferential joint and increases the risk of cracking. Another problem is that the TBM itself often has 'lookup' (i.e. the axis of the TBM is pointing slightly upwards) while the rings in the tailskin have overhang, which results in wear to the tailseal brushes or excessive forces applied to the segments at the crown and invert, similar to what happens on a tight curve but rotated 90°.

The effect of ringbuilding tolerances, stiffness of tailseal brushes and TBM attitude were investigated in detail by Mo & Chen (2008) using a 3D numerical model. They found that the key segment was the most vulnerable to dislocation or overstressing and this meant that the TBM diving downwards was the worst case (assuming the key is positioned at or near the crown, which is not always the case).

As anyone who has seen a bent pipe will know, tubes flatten in bending, and so longitudinal bending of a tunnel causes it to squat, as shown by Huang et al. (2012).

Koyama (2003) used pressure cells on the extrados of segments to measure the pressure applied by the tailseal brushes and found that it could be twice as high as the pressure later exerted by the ground and that tight curves could result in permanent loads and deformations in the lining.

10.5.3 Grout pressures and tailseal brushes

Initially, when grout is pumped into the tail void using grout ports through the tailskin, the principal direction of flow is circumferential (Talmon et al., 2006). Further back from the injection point, the velocity is slower and the grout flow is longitudinal. By modelling the flow using a Bingham model for the grout based on laboratory tests, Talmon et al. found that the grout pressures exerted on the tunnel lining depended on the grout ports used. When all six grout ports were used uniformly (Figure 10.18), the pressure distribution in the model increased linearly with depth, whereas when only the top three grout ports were used, the downwards flow made the pressure distribution approximately uniform (Figure 10.19). Talmon et al. (2009) later showed that the 'centre of gravity' of the grout injection relative to the axis of the tunnel could be used to calculate the uplift force for situations between these two limiting cases. The type of grout used in these cases, as is common practice in the Netherlands, was an inert grout. Where semi-active or active grouts are used (i.e. grouts that set within seconds or minutes of injection), the patterns of behaviour may be different.

The linear increase of grout pressure with depth in Figure 10.18 is not hydrostatic, but increases at a lower rate than the unit weight of the grout. This is due to buoyancy of the tunnel lining, which applies an uplift force on the grout. By taking this into account, Talmon et al. (2006) found good agreement with field measurements of grout pressure.

As fresh grout exerts an uplift force on the tunnel lining, this is resisted by the rings within the TBM, and by the rings further back where the grout has already hardened (Hoefsloot & Verweij, 2006). It is also resisted by the self-weight of the rings and the weight of the TBM backup and depends on the properties of the grout. The effect is obviously worse if inert or

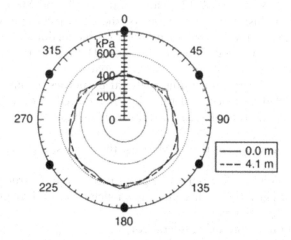

Figure 10.18 Calculated grout pressures at 0 m and 4.1 m from the rear of the TBM using all six injection ports uniformly (from Talmon et al., 2006).

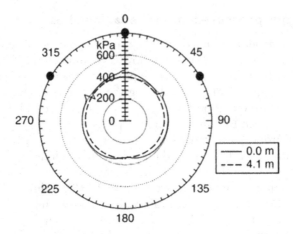

Figure 10.19 Calculated grout pressures at 0 m and 4.1 m from the rear of the TBM using the top three injection ports only (from Talmon et al., 2006).

slow-setting grouts are used. The tunnel is therefore like a pipe supported at both ends with an upwards pressure in the middle, which induces longitudinal bending moments in the tunnel, exacerbating the deformations due to jacking loads. A basic conclusion from this Dutch research is that by injecting more grout through the upper injection ports, uplift forces on the lining, and hence longitudinal bending moments, can be reduced.

10.5.4 Design for permanent effects of installation loads

It is clear from Figure 10.16, as well as the work of Bakker & Bezuijen (2009), Bilotta & Russo (2012), Blom et al. (1999) and Molins & Arnau (2011), that the installation of segments, the TBM jacking forces and the injection of grout have a permanent effect on the medium- and long-term loads in the tunnel lining. The use of simple 2D analytical solutions or 2D bedded beam models will not predict these loads.

As well as the segmental lining material properties, geometry and tolerances, detailed knowledge of the TBM is required in order to properly design a segmental tunnel lining and this may include:

- precise geometry of the TBM jacks, their shoes, the tailskin and the tailseal brushes
- the location of grout ports, the grout rheology and hardening parameters, and grouting strategy (i.e. how much to inject through each port)
- likely range of advance rates
- segment manufacturing tolerances
- ringbuilding tolerances
- jacking forces and their potential eccentricity

This information is not always available at the design stage, so conservative assumptions may need to be made.

Current practice usually considers TBM jacking forces as a transient load only, i.e. in isolation and only for a short duration. The maximum jacking force, the ram shoe geometry and the maximum eccentricity of jacking force are assumed, and segments are checked for crushing, bursting stresses, shear and bending, in a similar way to how the radial joints are designed. Designing for tensile stresses between ram shoes (c.f. Figure 10.14 and Figure 10.15) can only be done using 3D numerical analysis.

10.6 GASKET COMPRESSION AND BOLT LOADS

Gaskets consist of strips of extruded rubber that are placed all around each segment on the radial and circumferential joints. Therefore, each joint will have two gaskets in contact. Their purpose is to prevent water ingress to the tunnel. There are two ways that watertightness may be achieved. One is by compression; when a segment is pushed against another segment, the gaskets are compressed until the concrete faces of the segments or their packers make contact and are compressed. Compression of the gaskets prevents water from pushing its way in between them. The other way is by using a compound that swells in contact with water, known as a 'hydrophilic' gasket. Hydrophilic gaskets for precast concrete segmental linings usually consist of a compression gasket with a superficial layer or bead of hydrophilic rubber.

Gaskets are either glued into a preformed gasket groove (Figure 10.20), or cast-in to the segment by attaching them inside the moulds (Figure 10.21).

Gaskets may be needed to keep groundwater and grout from leaking into the tunnel, or to keep a fluid that is in the tunnel from leaking out.

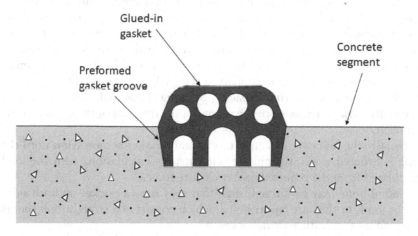

Figure 10.20 Glued-in segment gasket (based on VIP028 type gasket manufactured by VIP-Polymers Ltd, 2020).

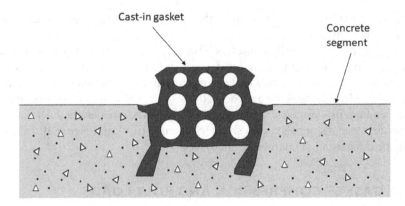

Figure 10.21 Cast-in segment gasket (based on VIP028CI type gasket manufactured by VIP-Polymers Ltd, 2020).

Sometimes, they need to achieve both, for instance in a potable water tunnel below the groundwater table with fluctuating internal water pressure. We do not want potable water to leak into the ground, because this is wasteful, and we do not want untreated groundwater entering the tunnel. In this case we also have the requirement that the gasket materials are safe to use in contact with drinking water. In the case of a sewage tunnel, we do not want sewage leaking into the ground, but we may be less concerned with groundwater infiltration into the tunnel, except that pumping and treatment costs may be increased because of the additional flow. The gaskets will need to be chemically resistant to the groundwater and to any fluid within the tunnel. It is not possible to replace segment gaskets, so they are required to last for the full design life of the tunnel.

The applied load for segment design is the force required to compress the gasket. Laboratory tests can determine what this closing force should be. An example is shown in Figure 10.22. When the gap reaches zero, the concrete faces of the segments are in contact and very little further deformation is possible. Therefore, this can be taken as the design load. The maximum gasket compression force is applied to the gasket groove and the concrete can be checked for bearing capacity, bursting stress and, since the gasket is usually close to the edge of the segment, shearing off of the corner of the segment.

Usually the bolts are designed to maintain gasket compression until the ring is grouted using a simple tensile capacity calculation, with the load calculated from the required gasket closure force.

Radial bolts may be curved, such that they have a pocket on both sides of the joint, or they are 'spear bolts', which consist of a straight bolt that is screwed into a plastic cast-in socket.

For some tunnels, it may be necessary to design the bolts and connectors to hold the weight of the segment, if the TBM jacks are to be removed for

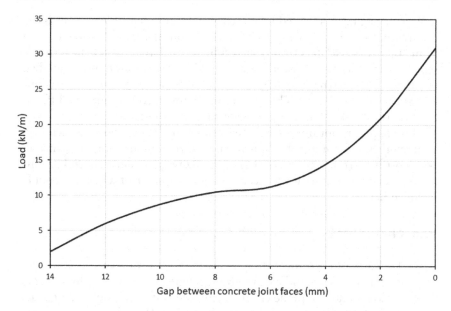

Figure 10.22 An example of gasket compression force versus joint closure (VIP-Polymers Ltd, 2020).

some reason. In this case, the bolts and connectors will have to resist tension and shear forces, and sometimes bending.

It is common nowadays to use plastic dowels across the circumferential joints and steel bolts across the radial joints. The plastic dowels are also helpful in aligning the new segments with the previous ring.

In recent years, designers have begun to question the need for radial bolts, because the TBM jacks and the confinement provided by the tailskin brushes should be sufficient to keep the segments in position and the joint faces in contact until the ring is grouted. On some projects bolts or dowels across the radial joints have been omitted and guide rods have been used to aid positioning.

Concrete around bolt pockets or connector sockets may need to be checked for local bearing stresses, because there is a possibility of local crushing or shear failure. For SFRC segments, more detailed guidance is available in *fib* Bulletin 83 (2017).

10.7 SUMMARY

Segments undergo various load cases during their life, including demoulding, storage, transportation, erection and then installation loads from the TBM jacks and grouting pressures. The most common load cases were described, with worked examples used to demonstrate the ultimate limit state and SLS calculations.

When a segment is loaded by its self-weight or external forces, such that the internal forces may be calculated from consideration of force equilibrium, then this is 'structurally determinate' behaviour. When a ring of segments is in its final position and loaded by the ground and groundwater, then this is 'structurally indeterminate' because as the ring deforms, stresses will redistribute. The requirements for SFRC design are different for structurally determinate or indeterminate load cases.

This chapter is not intended to be a comprehensive manual for design of segmental tunnel linings, but a teaching aid to demonstrate the principles and most important aspects of design. Every project is unique and has different load cases and performance requirements, some of which may not be included here.

10.8 PROBLEMS

Q10.1. A 6.5 m ID tunnel lining consists of six ordinary segments, two top segments and a key. The segments are 325 mm thick, 1.2 m long, and are made from steel fibre reinforced concrete of grade FRC 40/50 – 5.0d and of unit weight 25 kN/m³.
  i.   The segments are to be stacked in the yard in a single stack for each ring, using timber battens between each segment. The six ordinary segments are at the bottom of each stack and the key is at the top. The battens are to be placed at ¼ of the segment chord length from each end, with a maximum eccentricity of 100 mm, except for the ground level battens, which are joined together. Calculate the bending moments and shear forces in the second segment up from the bottom, assuming the worst possible eccentricities of batten placement. If the stack does not work, try a smaller stack.
  ii. Check the capacity of the segment at the ULS for these loads according to the *fib* Model Code 2010 (2013), taking account of shear using the method of Coccia et al. (2015).
  iii. Check the capacity of the segment at the SLS.
  iv. Each ring is transported to site on a lorry in three stacks of three segments. For a stack of three ordinary segments, calculate the design values of bending moments and shear forces using the same method as in part (i), and applying the additional partial factor for dynamic loads recommended by *fib* Bulletin 83 (2017).

Q10.2. An 8.5 m ID tunnel lining consists of ten trapezoidal segments. The segments are 400 mm thick, 2 m long, and are made from steel fibre reinforced concrete of grade FRC 50/60 – 5.0d and of unit weight 25 kN/m³.
  i.   The segments are to be erected in the TBM using a vacuum lifter that covers the central 2/3 of the segment intrados.

Calculate the bending moments and shear forces when the segment is being lifted vertically.

ii. Check the capacity of the segment at the ULS for these loads according to the fib Model Code 2010 (2013), taking account of shear using the method of Coccia et al. (2015).

iii. Check the capacity of the segment at the SLS.

iv. [Advanced] What is the worst case orientation for the segment erector to be holding the segment?

REFERENCES

ACI 544.7R-16 (2016). *Report on Design and Construction of Fiber-Reinforced Precast Concrete Tunnel Segments*, Reported by ACI Committee 544. Farmington Hills: American Concrete Institute.

Bakker, K. J. & Bezuijen, A. (2009). Ten years of bored tunnels in The Netherlands: Part II, structural issues. *Geotechnical aspects of underground construction in soft ground* (eds Ng, C. W. W., Huang, H. W. & Liu, G. B.), pp. 249–254. London: Taylor & Francis Group.

Bilotta, E. & Russo, G. (2012). Backcalculation of internal forces in the segmental lining of a tunnel: The experience of Line 1 in Naples. *Geotechnical aspects of underground construction in soft ground* (ed. Viggiani, G.), pp. 213–221. London: Taylor & Francis Group.

Blom, C. B. M., van der Horst, E. J. & Jovanovic, P. S. (1999). Three-dimensional structural analyses of the shield-driven green heart tunnel of the high-speed line south. *Tunn. Undergr. Space Technol.* 14, No. 2, 217–224.

Burgers, R., Walraven, J., Plizzari, G. A. & Tiberti, G. (2007). Structural behavior of SFRC tunnel segments during TBM operations. *Underground space – the 4th dimension of metropolises* (eds Barták, J., Hrdina, I., Romancov, G. & Zlámal, J.), pp. 1461–1467. London: Taylor & Francis Group.

Cavalaro, S. H. P., Blom, C. B. M., Walraven, J. C. & Aguado, A. (2011). Structural analysis of contact deficiencies in segmented lining. *Tunn. Undergr. Space Technol.* 26, 734–749.

Coccia, S., Meda, A. & Rinaldi, Z. (2015). On shear verification according to *fib* Model Code 2010 in FRC elements without traditional reinforcement. *Struct. Concret.* 16, No. 4, 518–523.

de Waal, R. G. A. (2000). *Steel fibre reinforced tunnel segments for the application in shield driven tunnel linings*. PhD Thesis. Technische Universiteit Delft.

EN 14651: 2005 (2005). *Test method for metallic fibered concrete - Measuring the flexural tensile strength (limit of proportionality (LOP), residual)*. Brussels: European Committee for Standardization.

EN 1992-1-1: 2004 (2004). *Eurocode 2: Design of concrete structures – Part 1-1: General rules and rules for buildings*. Brussels: European Committee for Standardization.

fib Bulletin 83 (2017). *Precast Tunnel Segments in Fibre-Reinforced Concrete*, State-of-the-art Report, fib WP 1.4.1, October 2017. Lausanne: Fédération internationale du béton (*fib*).

fib Model Code 2010 (2013). *fib Model Code for Concrete Structures 2010*. Lausanne: Fédération internationale du béton (*fib*).

Groeneweg, T. W. (2007). *Shield driven tunnels in ultra high strength concrete – Reduction of the tunnel lining thickness*. Ph.D. Thesis, Technical University of Delft.

Heijmans, R. W. M. G. & Jansen, J. A. G. (1999). Design features of the Pannerdensch Kanaal Tunnel in the Betuweroute. *Tunn. Undergr. Space Technol.* **14**, No. 2, 151–160.

Hoefsloot, F. J. M. (2009). Analytical solution of longitudinal behaviour of tunnel lining. *Geotechnical aspects of underground construction in soft ground* (eds Ng, C. W. W., Huang, H. W. & Liu, G. B.), pp. 775–780. London: Taylor & Francis Group.

Hoefsloot, F. J. M. & Verweij, A. (2006). 4D grouting pressure model PLAXIS. *Geotechnical aspects of underground construction in soft ground* (eds Bakker, K. J., Bezuijen, A., Broere, W. & Kwast, E. A.), pp. 529–541. London: Taylor & Francis Group.

Huang, X., Huang, H. & Zhang, J. (2012). Flattening of jointed shield-driven tunnel induced by longitudinal differential settlements. *Tunn. Undergr. Space Technol.* **31**, 20–32.

ITAtech (2016). *Design Guidance for Precast Fibre Reinforced Concrete Segments – Vol.1: Design Aspects*, ITAtech Report No.7. Lausanne: ITA-AITES. Reproduction of figures with permission of the Committee on new technologies of the International Tunnelling and Underground Space Association – ITAtech.

Koyama, Y. (2003). Present status and technology of shield tunneling method in Japan. *Tunn. Undergr. Space Technol.* **18**, 145–159.

Mo, H. H. & Chen, J. S. (2008). Study on inner force and dislocation of segments caused by shield machine attitude. *Tunn. Undergr. Space Technol.* **23**, 281–291.

Molins, C. & Arnau, O. (2011). Experimental and analytical study of the structural response of segmental tunnel linings based on an in situ loading test. Part 1: Test configuration and execution. *Tunn. Undergr. Space Technol.* **26**, 764–777.

Psomas, S. & Eddie, C. M. (2016). SFRC Segmental Lining Design for a Pressurised Tunnel. *Proc. WTC2016*, San Francisco, USA, Paper 0118, p. 13. Englewood, Colorado, USA: SME.

Talmon, A. M., Aanen, L., Bezuijen, A. & van der Zon, W. H. (2006). Grout pressures around a tunnel lining. *Geotechnical aspects of underground construction in soft ground* (eds Bakker, K. J., Bezuijen, A., Broere, W. & Kwast, E. A.), pp. 77–82. London: Taylor & Francis Group.

Talmon, A. M., Bezuijen, A. & Hoefsloot, F. J. M. (2009). Longitudinal tube bending due to grout pressures. *Geotechnical aspects of underground construction in soft ground* (eds Ng, C. W. W., Huang, H. W. & Liu, G. B.), pp. 357–362. London: Taylor & Francis Group.

UK NA to EN 1992-1-1: 2004. *UK National Annex to Eurocode 2: Design of concrete structures – part 1-1: general rules and rules for buildings*. London: British Standards Institution.

VIP-Polymers Ltd (2020). *Tunnel Segment Gaskets*. Brochure available at: https://www.vip-polymers.com/wp-content/uploads/2020/01/VIP-TSG-Brochure-2017-Lo.pdf (last accessed 3rd November 2020).

Chapter 11

Sprayed concrete lining design

Sprayed concrete linings differ from the segmental linings covered in the previous two chapters in the following ways:

- Precast or prefabricated linings already have their long-term design strength and stiffness at the time of installation, while sprayed concrete gains strength and stiffness while also being increasingly loaded by the ground as the tunnel advances.
- Sprayed concrete quality cannot be verified prior to installation as it depends on workmanship, variability of materials and environmental conditions.
- Unlike segmental linings, there are no intentional articulating joints in a sprayed concrete lining. There may be construction joints, but usually every effort is made to ensure these behave as monolithically as possible.
- Sprayed concrete linings do not have to be circular.

Sprayed concrete allows much more flexibility in terms of geometry than other lining types. However, there are still constraints on the shapes we can use. In soft ground we still need the lining to act as an arch in order to have an efficient structural form, but the profile does not have to be circular. Depending on the space envelope required inside the tunnel, the profile can be modified to reduce the volume of excavation.

There is a trade-off between volume of excavation, lining thickness and reinforcement. The less circular the profile of a sprayed concrete lining is, the lower the volume of excavation will be, but the bending moments will be larger, and the lining will need to either be thicker or contain more reinforcement. If we make the lining thicker, it will be stiffer, and will therefore attract higher bending moments, possibly requiring an even thicker lining. Therefore, it is really important to find the right balance between minimising volume of excavation and minimising lining materials to get the most sustainable solution.

This chapter is not intended to be a comprehensive manual for design of sprayed concrete linings, but a teaching aid to demonstrate the most

DOI: 10.1201/9780429470387-11

important principles. For sprayed concrete, even more than for other construction methods, the design and construction are inextricably interlinked. The design is not finished when construction starts, but must be verified by monitoring and back-analysis. For brevity, this book focuses on design and includes very little discussion of construction, but make no mistake: you should find out as much as you can about construction before embarking on the design of a real sprayed concrete tunnel. Designers who do not take the trouble to understand construction are inefficient and dangerous, and this could equally be said of site engineers who do not understand design. Diverse teams of people with construction and design backgrounds, who are willing to learn from each other, produce the best and safest solutions.

After working through this chapter you will understand:

- the principles of design for sprayed concrete tunnel linings
- the importance of collaboration between design and construction teams
- the complex factors that will affect the real stresses and strains in a sprayed concrete lining
- how basic toolbox measures work

After working through this chapter you will be able to:

- optimise the profile and face division of a sprayed concrete lining to fit a required space envelope
- design toolbox measures to improve stability and reduce ground movements
- design a sprayed concrete lining using 2D or 3D numerical analysis and a simple ageing elastic or elastoplastic constitutive model

II.I PRIMARY AND SECONDARY LININGS

The traditional approach to design of a sprayed concrete lining was to assume that the primary lining is 'temporary works' only and does not contribute any load-carrying capacity in the long term. A waterproof membrane is then installed followed by a cast-in-place (CIP) reinforced concrete secondary lining. This was because of concerns about the long-term durability of sprayed concrete, but was also due to the traditional contractual separation of temporary and permanent works. The secondary lining was designed for the full long-term ground and groundwater pressures, which were assumed to act on the outside of the waterproof membrane.

As quality control of sprayed concrete improved and confidence about its long-term durability increased, it became common practice to assume the primary lining had at least some long-term load-bearing capacity (ITA-AITES, 2020). This can make the overall design of the lining system more efficient. It is also possible to spray the secondary lining, which in some situations may be more cost-effective than a CIP lining.

There are many variants, which depend on the design philosophy regarding water ingress, durability and the final finish at the intrados of the lining system. These will be described in the following sub-sections. The main types are summarised in Table 11.1.

11.1.1 Single-pass lining

If the primary lining is durable, has sufficient capacity for the long-term loads, is considered sufficiently watertight for the ground conditions and intended usage of the tunnel, and an adequate internal finish can be provided, then no secondary lining is required.

11.1.2 Single-shell lining

In soft ground, a primary lining that is designed for the temporary loads during construction is often sufficient, or almost sufficient, for the

Table 11.1 Sprayed concrete lining system types.

Lining type	Waterproofing system	Primary lining	Secondary lining
Single-pass lining	No waterproof membrane	Permanent, takes all long-term ground and groundwater loads	None
Single-shell lining	Sprayed waterproof membrane	Permanent, takes all long-term ground and groundwater loads	For protection of the membrane and to take fixings only
Traditional approach	Sheet waterproof membrane	Temporary loads only	CIP, designed for full long-term ground and groundwater loads
Traditional approach, no membrane	No waterproof membrane	Temporary loads only	CIP or sprayed, designed for full long-term ground and groundwater loads
Double-shell lining – permanent primary	Sheet waterproof membrane	Permanent	CIP, designed for a share of long-term ground pressures and 100% of groundwater loads
Composite shell lining – no membrane	No waterproof membrane	Permanent	CIP or sprayed, designed for a share of long-term ground and groundwater loads utilising partial or full composite action
Composite shell lining – sprayed membrane	Sprayed waterproof membrane	Permanent	CIP or sprayed, designed for a share of long-term ground and groundwater loads utilising partial or full composite action

long-term loads as well. If a sprayed waterproof membrane is used to provide enhanced watertightness, and is bonded to the primary lining, then in theory as long as the bond strength is higher than the expected long-term groundwater pressure, no secondary lining is needed, except to protect the membrane from damage and to provide a substrate for internal fixings.

Therefore, there is little or no additional cost to designing the primary lining for the long-term loads, and providing a 100–150 mm thick secondary lining that is not designed for any ground or groundwater loads. The secondary lining is there only to protect the waterproof membrane (whether sprayed or sheet) and to allow penetration of fixings for things like cladding, services or equipment without puncturing the membrane. In this case, the design loads on the secondary lining are only its self-weight and equipment or cladding loads on the fixings. The secondary lining may also be designed as a fire protection layer.

11.1.3 Traditional approach

The traditional method of design of a primary and secondary lining system is to assume that the primary lining is only temporary, and the secondary lining takes all the loads in the long term. In a numerical model, the primary lining is there for the short-term construction period, then for the long-term analysis it is switched off and the secondary lining is switched on.

For this case, it does not matter whether there is a waterproof membrane between the primary and secondary linings, or not. The full long-term ground and groundwater loads will be conservatively applied to the secondary lining in the design.

11.1.4 Double-shell lining system

When the primary lining is considered permanent and there is also a secondary lining inside it, the simplest assumption to make is that there is no shear interaction between the primary and secondary linings. This is referred to as a 'double-shell lining system'. This is a valid assumption when a sheet waterproof membrane is used between the two linings, but may not be realistic when a bonded sprayed membrane is used.

If the primary lining is considered permanent and we wish to take advantage of its ability to share the load in the long-term, then we need to know about the ultimate limit state. In other words, what is the ultimate resistance of the double-shell lining system? Will the primary lining fail before load can be shared with the secondary lining? This seems unlikely, but it all comes down to compatibility of strains. All the strains – elastic, plastic, creep, shrinkage and thermal, and how they evolve over time – must be considered to find the answer to this question.

As we saw in Chapter 1, the primary lining will experience elastic and plastic strains due to ground and groundwater loads, as well as creep, shrinkage and thermal strains, early in its life. When the secondary lining is installed, it also experiences shrinkage and thermal strains. The secondary lining contracts and moves away from the primary lining, and this means that a significant increment of load is needed before the secondary lining will even begin to share the load applied to the primary lining. Jones (2018) estimated for a case study that this load increment would need to be at least 10 MPa, and therefore it is unrealistic to assume any load-sharing for most tunnels in soft ground. Exceptions to this may include when another tunnel is excavated nearby or where a junction is broken out through the primary lining after the secondary lining has been installed.

The groundwater pressure may be assumed to be acting on the second-ary lining, particularly if a sheet waterproof membrane is present between the primary and secondary linings. Some bending moment will be induced by the hydrostatic increase in groundwater pressure with depth around the membrane, and if the secondary lining is non-circular.

If there is the potential for chemical attack from the ground, it is some-times assumed that in the long term a certain thickness of the primary lining is degraded, and is not considered in the long-term analysis. This is called a 'sacrificial layer'. In these ways, even with a permanent primary lining, the secondary lining can be designed for some load.

11.1.5 Composite shell lining system

In some cases, the designer may consider the primary and secondary lining to be sufficiently watertight so that no waterproof membrane is required. Then the secondary lining may be either CIP or sprayed, and should have good bond and interlock with the primary lining, thus providing interface shear and some composite action.

If a sprayed waterproof membrane is used between the primary and sec-ondary lining and it is bonded to both of them, then to some degree the two linings may act together as a composite.

The importance of composite action to flexural capacity may be illustrated by considering an elastic beam, as shown in Figure 11.1. If the interface is perfectly smooth and frictionless, then there will be no composite action,

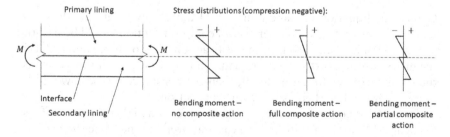

Figure 11.1 Varying degrees of composite action of a primary and secondary lining assuming elastic behaviour.

and the primary and secondary lining act independently. If the interface has a perfect bond, then there will be full composite action, where the primary and secondary lining act as though they were one monolithic lining. If there is some bond and/or friction between the primary and secondary linings, providing an interface shear stiffness, then some degree of partial composite action will occur, which could be anywhere between the two limiting cases of no composite action or full composite action (Jäger, 2016).

Figure 11.1 shows, and it can be demonstrated by the calculation that follows, that the same bending moment will induce higher bending stresses in the system with no composite action. Therefore, partial or full composite action will increase the moment capacity.

Assuming the primary and secondary linings have the same thickness t, the moment with no composite action may be given by:

$$M = 2 \times \frac{I\sigma_{max}}{y} = \frac{4t^3\sigma_{max}}{12t} = \frac{t^2\sigma_{max}}{3} \tag{11.1}$$

M is the bending moment in Nmm
σ_{max} is the maximum stress at the extreme fibre in N/mm^2
t is the thickness of each lining in mm
I is the second moment of area in mm^4, which per metre length is $t^3/12$
y is the distance from the neutral axis to the extreme fibre, which is equal to $t/2$

For full composite action, the primary and secondary linings act like a monolithic beam of thickness $2t$, and thus the moment is given by:

$$M = \frac{I\sigma_{max}}{y} = \frac{8t^3\sigma_{max}}{12t} = \frac{2t^2\sigma_{max}}{3} \tag{11.2}$$

I is the second moment of area in mm^4, which per metre length is $\left(2t\right)^3/12$
y is the distance from the neutral axis to the extreme fibre, which is equal to t

Comparing Equation 11.1 and Equation 11.2, if the moment were the same, then the value of σ_{max} when there is full composite action would be half the magnitude of the value for no composite action, as shown in Figure 11.1. If the maximum stress σ_{max} were the same in both cases, i.e. if it represented the strength of the lining material, then the resistance moment would be twice as large when there is full composite action.

Jäger (2016) noted that composite action means a higher overall stiffness (compare the values of I in the two cases above) and therefore the bending moments obtained from a soil-structure interaction analysis will be higher as a result.

Su et al. (2019) showed that if fully composite behaviour is assumed, where there is a hogging moment (tension at the intrados), the secondary lining will have a lower bending moment compared to a double-shell lining system, but it will also have a lower hoop force. When the moment – axial force interaction was plotted, they found that the secondary lining was outside the capacity curve when full composite action was applied, but inside it when there was no composite action. Therefore, fully composite behaviour does not necessarily result in a thinner overall lining system.

The simple model in Figure 11.1 assumes the two linings are elastic and have no pre-existing strains. However, in reality they are not elastic and there are plastic, creep, shrinkage and thermal strains. These non-elastic strains will affect the primary lining first, before the secondary lining is installed. Shrinkage and thermal strains in the secondary lining will be restrained by bond with the primary lining, and this may lead to significant tensile hoop strains in the secondary lining. Alternatively, the secondary lining may debond from the primary lining if the bond strength is not high enough. Su et al. (2019) included only shrinkage and self-weight of a secondary lining in a numerical model and found that the fully composite lining developed significant tensile stresses and induced compression in the primary lining, whereas a double-shell lining system did not. Jäger (2016) considered pre-loading of the primary lining and shrinkage of the secondary lining and showed that the primary lining failed before the axial load capacity of the secondary lining could be reached.

The ability of sprayed membranes to bond to the primary lining means that the water pressure can be assumed to be applied to the primary lining, and as long as the water pressure is not so high that this bond strength is exceeded, this can be beneficial because in most cases the primary lining will have sufficient capacity for the long-term loads. It is therefore possible, and this approach has been used on recent projects, to design the secondary lining only to support its self-weight and to confine and protect the sprayed membrane. Nasekhian & Feiersinger (2017) proposed a minimum secondary lining thickness of 150 mm to allow bolt penetrations for services and equipment.

11.1.6 Initial layer

It is common practice in soft ground to spray an initial or 'sealing' layer of shotcrete onto the exposed ground immediately after excavation. This is typically 50–75 mm thick and covers the exposed ground of the tunnel profile and the face. Because it is thin, it can be applied relatively quickly. The purpose of the initial layer is to prevent low cohesion soils from drying out or ravelling, or to prevent local block failures in clay. It cannot do much to prevent global instability, though the face is typically domed to provide some structural arching effect in the ground and the initial layer.

One advantage of applying an initial layer to the face is that if there is significant movement indicating the beginning of a failure, cracks in the initial layer will be easily seen. Without the initial layer, the early signs of face collapse may go unnoticed.

The initial layer at the excavation profile is usually included as part of the primary lining for structural analysis purposes, but the initial layer on the face is usually ignored in 3D numerical models (e.g. Jones et al., 2008).

11.1.7 Regulating layer

When a sprayed membrane is used, it is usual practice to spray a 'regulating layer' over the primary lining first. The regulating layer consists of a sprayed concrete without fibre reinforcement and without larger aggregates greater than 4 mm, and is typically 30–50 mm thick. This creates a smoother surface for the sprayed membrane and means that less volume of the sprayed membrane product is needed to achieve the coverage required. It also covers any fibres that are protruding from the primary lining, which may be difficult to spray the membrane around without shadowing.

11.1.8 Finishing layer

If the sprayed concrete forms the internal surface of the tunnel (i.e. there is no secondary lining, or the secondary lining is sprayed concrete), then a finishing layer may be applied. This is a sprayed concrete without large aggregates or fibres, and with a reduced accelerator dosage.

Immediately after spraying, the surface is trowel-finished by hand to give a smooth internal surface. This requires a lot of manual labour, but may be cost-effective for non-public areas such as equipment rooms, emergency intervention/escape passages or ventilation tunnels that do not require cladding, but need a smooth internal surface that can be washed down. A typical sprayed concrete lining without a finishing layer will be quite bumpy and will collect dust, and if it has steel fibre reinforcement, protruding fibres can cause injuries to people who accidentally lean against or brush past the lining.

Concrete tunnel linings often have microfilament polypropylene fibres added to prevent explosive spalling during a fire. In some cases, these fibres are added only to the finishing layer if it is of sufficient thickness to also act as a fire protection layer.

11.2 DESIGNING THE PROFILE

Most sprayed concrete linings have a vertical line of symmetry through the centreline. It is possible to design non-symmetrical linings, but for most purposes, symmetry provides the right fit.

Usually we assume that the upper part of the lining (the 'crown') above axis level is a single circular arc. This is because in the temporary case during construction it tends to take more load and therefore needs the strongest shape. This has practical benefits in highway or railway tunnels, where above the kinematic envelope of the vehicles some space is often needed for ventilation, lighting or overhead line electrification. For pedestrian passageway tunnels, sometimes the required space envelope is chamfered in the top corners, making a semi-circular crown more efficient.

In many tunnels for vehicle or pedestrian use, the finished floor is horizontal, and although there is some underfloor space needed for drainage and cable ducts, much of the invert would be wasted space if the tunnel were circular. Therefore, the invert may be flattened to reduce the volume of excavation. A flatter invert is also easier to excavate using mobile excavators.

11.2.1 Three arc profile

The most commonly-used non-circular shape is formed by four circular arcs with three different radii for the crown, bench (two arcs, one each side) and invert, and we will call this a 'three arc profile' because there are three types of arc, i.e. three different radii are used. An example is shown in Figure 11.2 for a pedestrian passageway tunnel. To avoid attracting large bending moments at the points where the circular arcs meet, we need to ensure they have common tangents, and therefore no vertices. A common tangent is achieved by ensuring the centres of adjacent arcs are on the same radial line. For example, where the crown arc meets the bench arc, the centre C_B of the bench arc with radius R_B is on a radius R_C of the crown arc,

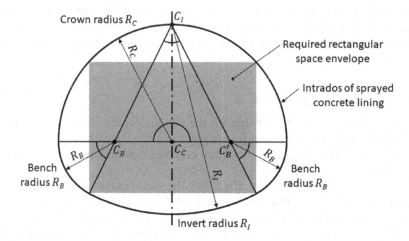

Figure 11.2 Three arc sprayed concrete lining profile.

which has its centre at C_C. Similarly, C_B is also on an invert radius R_I that has its centre at C_I.

It is important to understand the origin of the three arc profile. Originally, lattice girders or steel ribs were used between two layers of welded wire mesh within a sprayed concrete lining, and if set out correctly they enabled the sprayer to spray to the correct profile. Each lattice girder or steel rib could be easily fabricated to a single radius, with bolted joints at the interface between construction stages, i.e. between the top heading, bench and invert. Therefore, it was convenient if the transitions between arcs were at the levels of the construction stage interfaces. Nowadays, in most soft ground tunnels, steel fibre reinforced concrete has replaced lattice girders or steel ribs and welded wire mesh to provide the necessary flexural strength and ductility, and reflectorless laser total stations are used during spraying to get the correct profile. Therefore, there is no real need to define the profile using three arc radii, or for the arc transitions to be at the construction stage interfaces. However, using a three arc profile does keep things simple and provides a reasonable shape in most situations.

In Figure 11.2, the crown arc is semi-circular, the centre of the invert arc is at the apex of the crown, and the transitions from the bench arc to the invert arc occur at the vertices of the required rectangular space envelope. These constraints enable a reasonable shape to be found algebraically to fit the rectangular envelope, but none of them have to be fixed. For instance, some designers prefer to make the crown arc less than 180°, particularly for wide or tall caverns. Similarly, the positions of the bench and invert arc centres can vary, as long as the arcs have common tangents at the points where they meet to avoid vertices.

The profile shown in Figure 11.2 would be the intrados of the sprayed concrete lining. In order to draw the extrados, simply offset the curves by the thickness of the lining. If drawing by hand, use the same centres, but increase the radii by the thickness of the lining. The common tangents of the extrados arcs will be on the same radial lines as the common tangents of the intrados arcs.

11.2.2 Two arc profiles

Another simple non-circular shape is the two arc profile. This can be useful when the space envelope required is particularly wide or tall, as shown in Figure 11.3. This profile is also quite common for noncircular shotcrete-lined shafts.

The crown arc with centre at C_C has the same radius R_{CI} as the invert arc with centre at C_I. The side arcs have the same radius R_S and centres at C_{S1} for the left side and C_{S2} for the right side. The profile has two lines of symmetry, as shown by the dash-dot lines in Figure 11.3. The profile can be

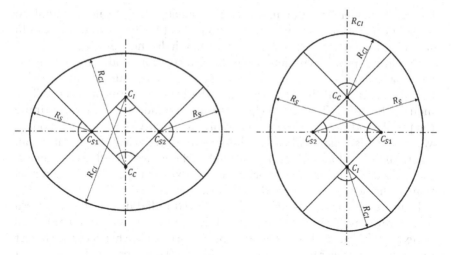

Figure 11.3 Two arc sprayed concrete lining profile.

made more circular by moving the centres closer to the profile centre (where the two lines of symmetry cross).

11.2.3 Other profile shapes

There are many possible variants using a number of circular arcs joined at common tangents. For instance, the invert could be flattened in the wider profile in Figure 11.3 by continuing the side arcs further down and having a larger invert radius, making it a three-arc profile (but unlike Figure 11.2 it would not have a semi-circular crown arc). Four arc profiles are also possible, and could be used to flatten the invert of the taller profile in Figure 11.3 by introducing short radii arcs between the side arcs and the invert arc.

Ellipse profiles have been used in the past, but they are actually quite mathematically complex to use for setting out, since the radius is continuously changing around the perimeter.

11.2.4 Optimising the profile

With the exception of the need to avoid vertices by having common tangents and avoid large differences in arc radii to minimise bending moments, it is difficult to come up with generic rules or constraints for optimisation. The optimal shape depends not just on the required space envelope, but also on the lining forces. Usually a 2D bedded beam model or a 2D or 3D numerical model is used and the profile is refined iteratively. A more efficient profile in terms of excavation volume often means larger differences between the arc radii and hence larger bending moments, which may require a thicker lining or more reinforcement. A good rule of thumb to use as a first estimate before optimisation is that the bench and invert radii should not be more than double, or less than half, the crown radius.

The lining profile will determine the deformation mode, and hence the bending moments. Assuming the ground pressure is uniform, it is possible to sketch the radial displacements and the bending moments that will be induced by them. This can help us to understand how best to optimise the profile and can be a useful check on the validity of numerical models. Where there is a short radius, the lining will behave in a stiffer manner than where there is a long radius. Accordingly, where there is a flattened invert, the midspan of the invert will deflect more towards the centre of the tunnel than at the bench arcs where there is a shorter radius.

Figure 11.4 shows a sketch of the expected radial displacement pattern for the three arc profile from Figure 11.2. Around the semi-circular crown we expect the radial displacement to be uniform, at the benches the shorter radius has a stiffer response and so radial displacement is reduced, and at the invert the longer radius behaves more flexibly and so radial displacement is increased.

This ignores plenty of complex effects, such as the construction sequence, shear between the lining and the ground, and anisotropy of in situ stress and ground stiffness, but it approximates well to the pattern we see in field data and numerical models.

Figure 11.5 shows a sketch of the bending moments that might be induced by the pattern of radial displacements in Figure 11.4. If the bending moments calculated in a numerical model are too high in the invert, then we could reduce the invert radius. If the bending moments are too high at the benches, then we could increase the bench radius. The closer we make the profile to a perfect circle, the lower the bending moments should be.

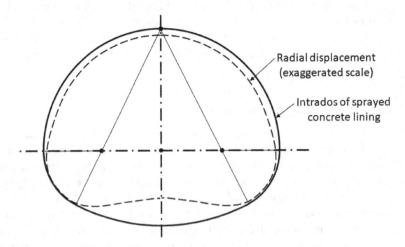

Figure 11.4 Sketch of radial displacement pattern for a three arc profile.

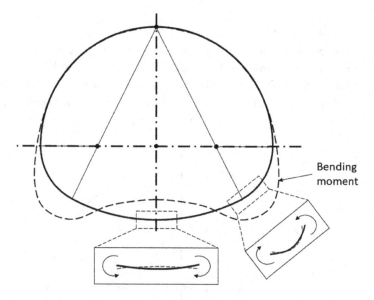

Figure 11.5 Sketch of bending moment pattern for a three arc profile.

WORKED EXAMPLE 11.1

A rectangular required space envelope 3.65 m high and 3.2 m wide is required for a metro station passageway.

a. Using a spreadsheet or iterative hand-calculation find a best fit three arc profile assuming the crown arc is semi-circular, the invert arc centre is at the crown and the transition from bench to invert occurs at the vertex of the rectangle.

b. Using a CAD program or drawing by hand using a compass, create a four arc profile by adding a long radius arc between the crown and bench arcs to see if the excavation volume can be reduced. Some trial and error may be required.

c. Sketch the radial displacement and bending moment distributions you would expect to see for this four arc profile, assuming it is wished-in-place in a homogeneous isotropic soil with $K_0 = 1.0$. Comment on how the construction sequence might influence the radial deformations and bending moments and what changes could be made if this were a problem.

a. Three arc profile

The geometry of a three arc profile is shown in Figure 11.6. The required space envelope has height $H = 3.65$ m and width $B = 3.2$ m. We need to find the position of the crown centre C_C below the centreline of the required space envelope, denoted A in Figure 11.6, then we can find the rest of the geometry.

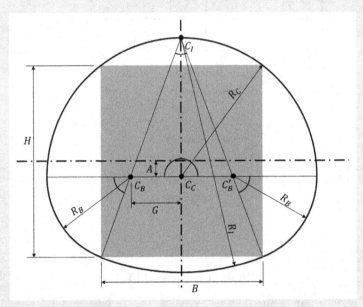

Figure 11.6 Worked Example 11.1 three arc geometry.

Using Pythagoras's theorem for R_C:

$$R_C = \sqrt{\left(\frac{H}{2} + A\right)^2 + \left(\frac{B}{2}\right)^2} \qquad (11.3)$$

And again for R_I:

$$R_I = \sqrt{\left[R_C + \left(\frac{H}{2} - A\right)\right]^2 + \left(\frac{B}{2}\right)^2} \qquad (11.4)$$

Let the distance between the centres C_B and C_C be denoted G. It is given by:

$$G = R_C - R_B = \frac{R_C}{R_C + \left(\frac{H}{2} - A\right)} \times \frac{B}{2} \qquad (11.5)$$

And

$$R_B = R_C - G \tag{11.6}$$

The bench radius R_B may also be expressed in terms of the invert radius R_I:

$$R_B = \frac{\dfrac{H}{2} - A}{R_C + \left(\dfrac{H}{2} - A\right)} \times R_I \tag{11.7}$$

By putting these equations into a spreadsheet, we can vary A until Equation 11.6 and Equation 11.7 give the same value of R_B. Alternatively, keep trying different values of A in a hand calculation. This is the optimal geometry and occurs when $A = 0.293$ m and $R_B = 1.640$ m.

From this value of A we can use Equation 11.3 and Equation 11.4 to calculate the other two radii.

$$R_C = \sqrt{\left(\frac{H}{2}+A\right)^2 + \left(\frac{B}{2}\right)^2} = \sqrt{\left(\frac{3.65}{2}+0.293\right)^2 + \left(\frac{3.2}{2}\right)^2} = 2.655 \text{ m}$$

$$R_I = \sqrt{\left[R_C + \left(\frac{H}{2} - A\right)\right]^2 + \left(\frac{B}{2}\right)^2}$$

$$= \sqrt{\left[2.655 + \left(\frac{3.65}{2} - 0.293\right)\right]^2 + \left(\frac{3.2}{2}\right)^2} = 4.482 \text{ m}$$

b. Four arc profile.

Using a CAD program and some trial and error, the four arc profile shown in Figure 11.7 was obtained. It is not perfect, and does not quite touch the required space envelope, but it is close. The three arc profile is also shown for comparison. The four arc profile has a smaller cross-sectional area and hence a reduced excavation volume.

c. Sketch of radial displacement and bending moment.

A sketch of the radial displacement (left side) and bending moment (right side) is shown in Figure 11.8.

Note that there is very little bending moment in the crown, because the ground load is assumed to be radial and uniform. If a top heading–invert or top heading–bench–invert construction sequence were used (these are described in the next section), then we would expect bending moments to be induced in the crown arc

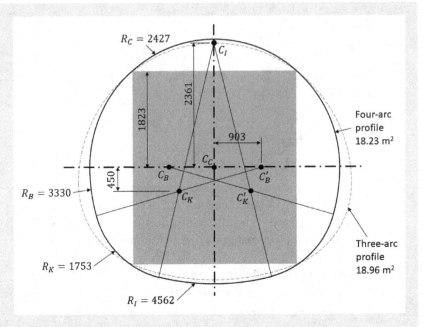

Figure 11.7 Worked Example 11.1 four arc profile geometry.

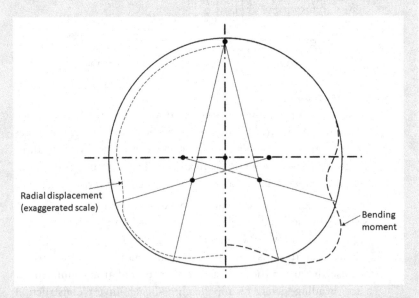

Figure 11.8 Worked Example 11.1 four arc profile sketch of radial displacement and bending moment.

because the bottom parts of the arc near axis level will deform inwards. This effect could be reduced by a temporary invert acting as a strut across the base of the top heading.

Note also that if the coefficient of earth pressure at rest $K_0 \neq 1$ then the radial displacements will also be anisotropic even for a perfect circle and bending moments will be induced, as was discussed in Chapter 6 for the Curtis-Muir Wood analytical solution.

11.3 DIVIDING THE FACE

For smaller tunnels, it may be possible to advance with a full-face excavation. For tunnels larger than approximately 5–6 m, it is normal to divide the face. This may be done to improve face stability, but is also dictated by the size and type of plant used to excavate the tunnel. Even in very good ground where face stability is not an issue, the reach of the excavator will limit the height of the top heading. At the detailed design stage it is very important to discuss face division with the contractor. For preliminary design purposes, at the very least it is necessary to consult someone who has construction expertise.

Where there are cohesionless soils exposed during excavation, particularly if they are very dry or saturated, toolbox measures will almost certainly be needed to avoid uncontrolled loss of ground. Very dry sand or saturated sand will both be unstable even when only small areas are exposed, whereas moist sand may stand up in small exposures for a short time. The sizes of partial headings need to have enough space to allow the planned toolbox measures to be undertaken. For this it is necessary to consult with the contractor or a construction expert.

As well as considering the limitations of construction plant, the maximum allowable size of a heading may be determined using an undrained or drained stability calculation, according to the methods described in Chapters 2 and 3. Minimising ground movements may also be important, and if so, then the methods described in Chapter 12 should be used.

The size of a sprayed concrete tunnel, since the diameter is not constant, may be expressed as the maximum width of the excavation profile. Typical minimum and maximum sizes of tunnel for each face division method are shown in Table 11.2.

It would be very difficult to construct a sprayed concrete tunnel smaller than the minimum size of 3 m, because of the typical sizes of small excavation and spraying plant. Once a 3 m tunnel is lined, it may be more like 2.5 m. Even if excavation and spraying plant could access the space, it would be impossible for them to pass each other and one would need to back all the way out before the other could enter, and drive lengths would be limited.

Table 11.2 Sprayed concrete tunnel face division methods and typical sizes.

Face division method	Shorthand	Typical minimum and maximum size	Notes
Full-face		3–6 m	
Top heading–invert	TH-I	4–7 m	
Top heading–bench–invert	TH-B-I	5–10 m	
Sidewall drift and enlargement*		9–15 m	Min. limited by width of plant
Twin sidewall drift*		14–18 m	Min. limited by width of plant
Pilot tunnel then enlargement by top heading–bench–invert		8–11 m	

* Note that sidewall drifts and enlargements are usually advanced with a TH-B-I method, so that a twin sidewall drift may be divided into nine partial faces.

Smaller tunnels are possible with hand excavation and/or hand spraying, but below 3 m timber support may be preferable to sprayed concrete. In these situations the design and construction team should consider whether the tunnel should be made larger to allow mechanisation, and hence safer and faster tunnelling.

The following sections discuss each of these face division methods in more detail.

11.3.1 Full-face excavation

Full-face excavation means that an advance length of the whole face is excavated then covered by a full ring of sprayed concrete lining, as shown in Figure 11.9. In reality, the upper part of the face is usually excavated followed by an initial layer of 50–75 mm sprayed concrete on the perimeter and the face, then the invert is excavated followed by the full primary lining sprayed from the invert up to the crown. Note also that the face is usually domed, as shown in Figure 11.9.

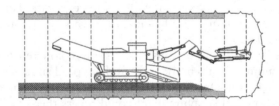

Figure 11.9 Full-face excavation. An advance has just been excavated and is ready to be sprayed.

Advance lengths are typically 1 m, but can be up to 1.5 m if plant, logistics, stability and ground movements allow.

Full-face excavation can achieve advance rates up to 8 m/day. It is limited in size by the maximum face area that will remain stable and by the reach of construction plant. A big advantage is that there are no radial joints in the lining.

Full-face excavation is, in some respects, the simplest to analyse in design. For 3D models, there is only one advancing face and no longitudinal staggering of the construction stages. For 2D models, we only have to make one assumption about the proportion of confinement loss that occurs before installation of the lining, whereas with more complex partial face excavation methods we have to make several assumptions about the proportion of confinement loss for each stage. Modelling will be discussed in more detail later in this chapter.

Lining stresses measured by pressure cells installed in the Storm Water Outfall Tunnel (SWOT) frontshunt tunnel at Heathrow Terminal 5, UK (Jones et al., 2008) showed that a full-face excavation can result in high stresses at early age. This is because the ring was closed immediately, and so when the peak temperature of shotcrete hydration occurred 10–15 hours later, the thermal expansion induced a compressive stress higher than the long-term stress in the lining. For this tunnel, the peak stress was fortunately within the capacity of the lining, but for larger full-face tunnels, or at higher advance rates, this effect could cause overstressing.

The face can be inclined to improve safety by providing a canopy of hardened shotcrete above someone standing near the face. This is usually an inclination of about 20° from the vertical, as shown in Figure 11.10.

11.3.2 Top heading–invert (TH-I) excavation

An example of a TH-I excavation is shown in Figure 11.11. In this example, the excavator has just finished a second top heading advance and the primary lining is about to be sprayed. Then a double advance length of invert will be excavated and sprayed. Then the invert needs to be re-covered with

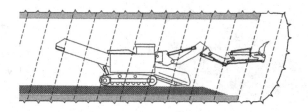

Figure 11.10 Full-face excavation with an inclined face.

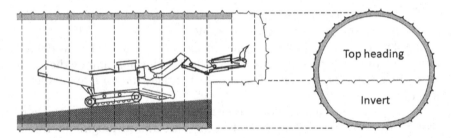

Figure 11.11 Top heading–invert (TH-I) excavation.

a ramp of muck so that the excavator can reach the next top heading. This covering and uncovering of the invert is inefficient, and that is why a double advance length of the invert is usually excavated and sprayed after every two top heading advances.

There are many possible variations on this sequence. For instance, the invert could be staggered, so that it always lags behind the top heading by one or two advance lengths. Although this increases the ring closure distance and hence should increase ground movements, it means that the joint between the top heading and the invert shotcrete can be cleaned and prepared more safely because the operatives are under a canopy of hardened shotcrete, rather than recently sprayed 'green' shotcrete.

The most recently sprayed top headings are overhanging the last completed invert. As illustrated by the case studies described in Chapter 1, the overhanging top heading experiences some loading from the ground, which will tend to make it move downwards. This downwards displacement is arrested by the completion of the invert, from which point the lining begins to behave like a ring in compression and further ground movements should be relatively small. Sometimes the top heading is given 'elephant's feet', a thickening of the lining at the footings of the top heading, which has the aim of spreading the lining load and hence increases the bearing capacity. Another option is to give the top heading a 'temporary invert', which is a shallow arch of shotcrete (Figure 11.12). This is removed when the invert excavation advances.

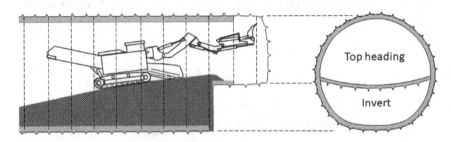

Figure 11.12 TH-I excavation, with 'temporary invert' in the base of the top heading.

Similar to the full-face excavation method shown in Figure 11.10, the top heading could be inclined. However, it is not common practice to incline the invert excavation.

11.3.3 Top heading–bench–invert (TH-B-I) excavation

An example of a TH-B-I excavation is shown in Figure 11.13. Stage 1 has been excavated and sprayed, then 2, 3, 4 and 5 have each in turn been excavated and sprayed, completing a cycle. The first stage of the next cycle (Stage 6 in Figure 11.13) has just been excavated and is about to be sprayed. This will be followed by a bench excavation, another top heading, another bench, and then a double invert. These are not shown on the diagram but follow the same pattern as Stages 2-5.

There are many variations of the TH-B-I sequence. Sometimes the advance lengths themselves are staggered by a half advance length so that the circumferential joints of the top heading, bench and invert do not occur at the same chainage. Sometimes two top headings are followed by two benches, rather than alternating. Sometimes the ring is closed earlier by completing the invert closer to the face. The optimum method will be the one that balances the competing demands of controlling stability and ground movements with quality, productivity and safety.

Unlike in rock tunnels where the bench and invert excavations may be hundreds of metres behind the top heading, in soft ground the invert is usually closed within approximately 5 m of the face. This is because the lining is not very effective at supporting the ground until the ring is closed, at which point continuing deformations should be small. Each time the invert is excavated and sprayed, the backfill then has to be replaced on top of the invert shotcrete to allow the excavator to advance and attain sufficient height to excavate the next top heading advance. This adds time to each cycle, so invert advances are usually twice as long as top heading and bench advances so they only need to be done half as often.

Similar to the full-face excavation method shown in Figure 11.10, the top heading and/or the bench advances could be inclined.

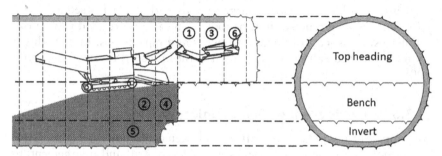

Figure 11.13 Top heading–bench–invert (TH-B-I) excavation.

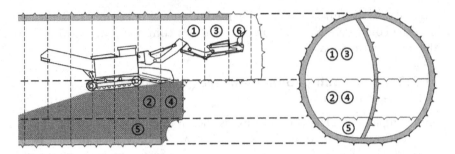

Figure 11.14 Single sidewall drift and enlargement excavation.

11.3.4 Single sidewall drift and enlargement

An example of a single sidewall drift and enlargement is shown in Figure 11.14. First the sidewall drift (the lenticular shape on the left of the cross-section in Figure 11.14) is excavated, in a TH-B-I sequence. This means there is a completed ring in the temporary case, limiting ground movements, and meaning that the enlargement can follow some distance behind the sidewall drift. The enlargement is also excavated in a top heading – bench – invert sequence. The central wall is not removed until the full ring is completed and has gained sufficient strength. Access is not usually allowed to the sidewall drift while enlargement works are ongoing.

The single sidewall drift method requires joints at the crown and invert to be formed, which need to be designed carefully to ensure they have sufficient rotation capacity. This is usually achieved using reinforcement bars, and so miners need access to the joint to place these, which involves working at height, where the risk of falls of ground or green shotcrete must be managed. Alternatively, an inner shotcrete lining may be sprayed after the sidewall is removed to ensure structural continuity without requiring access.

The main advantage of sidewall drifts is that the width and area of the open excavation are reduced, thereby improving stability in very wide caverns. It is possible to construct large caverns with a TH-B-I sequence if a pilot tunnel is constructed first, or toolbox measures are used to improve stability, such as jet grouting, face dowels and pipe arches. In fact, this is common practice in Italy and is known as the 'Adeco-RS system' (Lunardi, 2008).

11.3.5 Twin sidewall drift and enlargement

An example of a twin sidewall drift and enlargement is shown in Figure 11.15. The principles are the same as for a single sidewall drift and enlargement. In

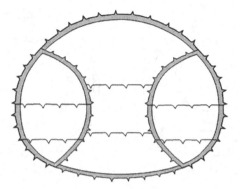

Figure 11.15 Twin sidewall drift and enlargement excavation.

this case the two sidewall drifts are constructed and then the enlargement, or 'core excavation' is done last.

11.3.6 Pilot tunnels

Pilot tunnels are another form of face division, and improve stability by reducing the area of exposed ground during excavation. During excavation of the enlargement, the remaining pilot tunnel will act as a large dowel in the ground ahead of the face, potentially reducing ground movements and improving stability. However, this does mean that the ground is disturbed twice by an advancing tunnel, and incremental demolition of the pilot tunnel can cause a lot of vibration, which may have a negative impact on stability.

A pilot tunnel is like a very large horizontal borehole, and can be used to learn more about the ground conditions to be encountered along the route of the tunnel. It can also be used for grouting or dewatering of the ground ahead of the enlargement.

There may also be logistical advantages to constructing pilot tunnels. Being full-face and of a relatively small area, they can be advanced much more quickly than a larger tunnel. This may provide early access to construct another part of the works that is critical to complete early in the programme.

For metro systems where the running tunnels are to be constructed by TBM and the stations by sprayed concrete, the TBMs may run right through creating the pilots for the platform tunnels. This depends on the programme, but has been shown to be safer and quicker than shotcrete pilot tunnels and, if a closed-face TBM is used, the overall ground movements should be smaller (St John et al., 2015). It is also easier to separate concrete segments from the spoil than to separate fragments of sprayed concrete. A potential disadvantage is that the enlargement cannot begin

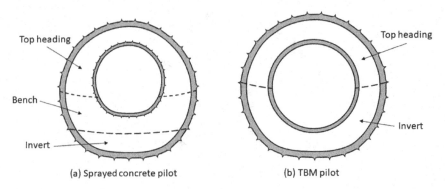

Figure 11.16 Sprayed concrete tunnel enlarged from (a) a sprayed concrete pilot, and (b) a TBM segmental lining pilot.

until the TBM has reached its final destination and all the backup systems have been removed from the tunnel.

An example from the design of Crossrail Farringdon Station platform tunnels is illustrated in Figure 11.16, which shows a comparison of two pilot tunnels, one a sprayed concrete pilot and the other a TBM-driven pilot tunnel. Note that because the TBM pilot was larger the enlargement also had to be larger to ensure sufficient space for toolbox measures, such as spiling, grouting and depressurisation of sand channels, and for excavating around and breaking out the segmental lining. Breaking out is much easier if the length of the segmental lining rings is the same as the advance length of the sprayed concrete enlargement, and for Farringdon special shorter rings were used through the station section of the TBM drives for this purpose (St John et al., 2015).

11.3.7 Binocular caverns

A question is often asked by those planning underground infrastructure projects about how close parallel tunnels can be constructed to each other. The answer is that tunnels can be constructed as close as you like, as long as the lining is designed for the higher forces and bending moments that will result. At the extreme, a pair of tunnels may have no ground at all between them, in which case they become essentially a wide cavern with a wall down the middle.

There are two main ways in which such binocular caverns may be built. One method is to drive a pilot tunnel and then to construct the strong central wall within it, usually structural steel or reinforced concrete. The main tunnels are then driven to each side, with the lining connected to the central wall. This central wall may be continuous, or have openings, or be a series of columns with a longitudinal beam along the top and bottom. The second method is to construct one of the tunnels, then to cast an internal lining with the strong central wall on one side of it. Then the second tunnel may be driven, utilising the central wall for both the temporary and permanent support.

11.3.8 Trinocular caverns

We do not have to stop at binocular; this method can be used to construct three parallel caverns, which we could call 'trinocular'. These are quite common in metro stations in Prague, Kiev, Moscow and elsewhere, where there is a central concourse tunnel with platform tunnels either side (Figure 11.17).

There is no reason why we should stop at three tunnels joined together, any number of parallel caverns could be constructed using similar methods. In Budapest there is a metro station with five (Mecsi, 2002).

11.4 TOOLBOX MEASURES

Toolbox measures are additional measures that may be employed in order to improve face stability, either to prevent collapse or excessive ground movements. They are usually designed in advance to mitigate geological risks that may or may not be realised, and the equipment and materials needed to implement them must be ready to be used on site. For example, if the ground is stiff clay but probing identifies a saturated sand lens ahead of the face, then grouting, dewatering or depressurisation may be employed to reduce the risk of running sand flowing uncontrollably into the tunnel.

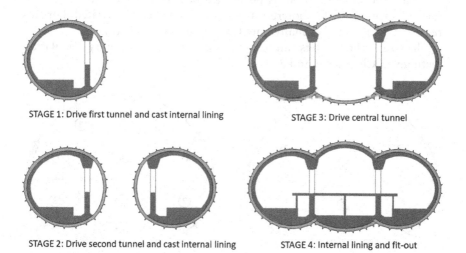

STAGE 1: Drive first tunnel and cast internal lining   STAGE 3: Drive central tunnel

STAGE 2: Drive second tunnel and cast internal lining   STAGE 4: Internal lining and fit-out

Figure 11.17 Trinocular cavern construction sequence.

Note that when using a sprayed concrete lining in soft ground, we do not reduce the basic support design if the ground is better than foreseen, but we will employ toolbox measures if a geological risk is realised.

The principles of undrained and drained stability in Chapters 2 and 3 can be used to predict the need for, and effectiveness of, different toolbox measures.

11.4.1 Pocket excavation

In either very dry or saturated drained soils with very low cohesion, where the ground has limited stand-up time and will fall down from the crown or the face soon after exposure, pocket excavation may be used (e.g. Aagaard et al., 2017; Gall et al., 2017). Two examples of pocket excavation are shown in Figure 11.18. Each pocket is excavated and then sprayed with a shotcrete sealing layer on all the exposed ground in the pocket until the top heading arch has been excavated (1–3 on the left of Figure 11.18, 1–5 on the right), then a primary lining layer is sprayed over the whole top heading arch. Then the core may be excavated and the core face sprayed with a sealing layer.

If the stand-up time of the ground at the crown is very short and a chimney failure is possible, then pocket excavation should be used in conjunction with a means of preventing loss of ground at the perimeter of the excavation, such as spiles, permeation grouting, jet grouting, ground freezing or a pipe arch.

Note that pocket excavation still requires some stand-up time. Ground that will fail immediately, particularly saturated cohesionless soils, cannot be controlled using this method.

Ideally the spraying robot is positioned next to the excavator and each pocket is sprayed with shotcrete as soon as it has been excavated. For this reason, this method is usually only feasible in quite wide tunnels.

The size of the pockets can be determined using the stability calculation methods in Chapters 2 and 3.

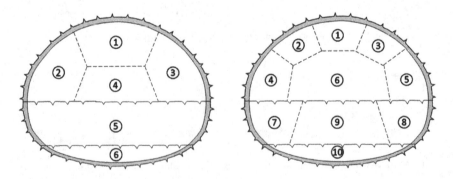

Figure 11.18 Two examples of pocket excavation.

11.4.2 Reduction of advance length or ring closure distance

The advance length is usually set at 1 m in soft ground, but has been known to be reduced to 0.8 m or increased to 1.2 or 1.5 m. If the advance length is changed, then consideration will need to be given to logistics and the capacity of spoil handling plant such as loaders, dumpers and skips, as well as the capacity of the batching plant and shotcrete delivery trucks to provide the correct amount of shotcrete.

As we saw in Chapter 2, the unsupported length P of a heading has an effect on undrained stability. The smaller P is, the more stable the heading will be.

For a full-face excavation, the value of P is equal to the advance length, plus an allowance for excavation of the face beyond the advance length, which is usually assumed to be 0.5 m. For undrained soil, a change from 1 m advance length to 0.5 m will only make a small difference to the P/D ratio (where D is the equivalent excavation diameter), even at the smallest feasible sprayed concrete tunnel diameter of 3 m, so the improvement to stability will be negligible.

For TH-I or TH-B-I excavation, the effective value of P is more difficult to determine. A conservative assumption is that it is equal to the ring closure distance, which is the maximum distance from the face to a completed invert. For the estimation of volume loss, Macklin (1999) recommended using only the excavation ahead of the leading edge of top heading shotcrete (this is covered in Chapter 12), because the unexcavated ground below the top heading provides some weight to counter instability, like a berm.

For the sequence shown in Figure 11.13 (also shown on the left side of Figure 11.19), and assuming that the advance length is 1 m, the maximum distance occurs just after excavation of the invert (Stage 5) but before it is sprayed, and is therefore 5 m. With an allowance for excavation of the face ahead of the top heading lining of 0.5 m, the ring closure distance is 5.5 m.

Reducing the ring closure distance may be achieved by reducing the advance length. However, this is likely to be inefficient in terms of productivity. Another solution is to move the bench and invert excavation closer to the face, as shown in Figure 11.19, where on the left is shown the sequence

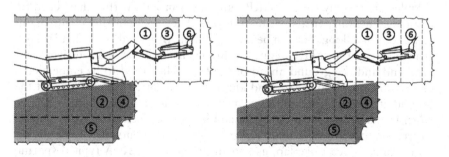

Figure 11.19 Reducing the ring closure distance in a TH-B-I excavation.

from Figure 11.13, and on the right is shown a modified sequence with a shorter ring closure distance. In this example the ring closure distance has been reduced from 5.5 to 4.5 m.

11.4.3 Ground improvement

Ground improvement means improving the properties of the ground, either by grouting or ground freezing. The aim is to improve face stability by increasing the soil's cohesion. It will also reduce the soil's permeability and this will improve working conditions by reducing groundwater inflows.

Grouting can be targeted at zones of ground with low cohesion, for instance sand lenses within an otherwise competent clay, or it can be a systematic procedure if all the ground has low cohesion. It can be done from the surface, from a shaft, or from the tunnel itself. The wedge-prism method described in Chapter 3 can be used to determine the cohesion required to make the face stable.

This book will not go into detail on grouting methods. These are covered by EN 12715: 2000 'Execution of special geotechnical work – Grouting', and there is plenty of guidance in the CIRIA guide *Grouting for Ground Engineering* (Rawlings et al., 2000) or in the book *Chemical Grouting and Soil Stabilization* (Karol, 2007).

11.4.4 Support ahead of the face

In both drained and undrained soils, tunnel headings collapse in a wedge-shaped mechanism that is loaded by the weight of the ground above. Therefore, if the ground above the face can be supported before the face is excavated, this gravity load will not be applied to the face wedge, improving stability. Even if the face collapses, the pre-support may prevent upwards migration of the collapse, where it might have an impact on buildings or utilities. This is illustrated in Figure 11.20.

Pre-support may be in the form of spiles, pipes, or could be jet-grouted or frozen soil columns. Typically this pre-support is only around the crown over an arc of 120°, as shown in Figure 11.21 for spiles.

Spiles are usually steel or GRP reinforcement bars with a sharpened tip to aid driving, but can also be hollow rods. They are typically 32 mm diameter and 3–4 m long, but can be up to 6 m long (Aagaard et al., 2017). They span between the sprayed concrete lining that will be sprayed over their back end and the ground ahead of the face. They need to be embedded a certain distance into the ground ahead to make sure they can support the ground at the perimeter of the excavation, so they will stop being effective when the face approaches and the end bearing capacity begins to reduce. Therefore, they are driven into the ground at a slight outwards angle so that arrays of spiles can overlap, as shown in Figure 11.21. A typical spacing around the perimeter is 200–300 mm (Holzleitner et al., 2005).

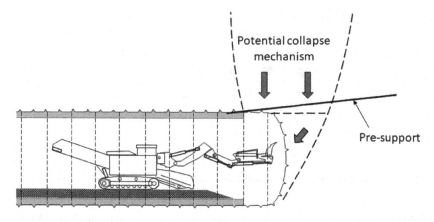

Figure 11.20 How pre-support improves stability.

In soils with very low cohesion, support ahead of the face will help to prevent the soil in the crown from falling down, since it only needs to arch over the spacing between the spiles or pipes, rather than arching front-to-back over the advance length, as shown in Figure 11.22. Usually the soil will break back to the line of the spiles and may in some cases continue to ravel back until an arch is formed.

Pipe arches typically consist of self-drilling steel pipes of 114 mm diameter (they can be any size, though normally they are between 60 and 200 mm) and 15–20 m long. If the soil around them is to be grouted, they can be fitted with tube à manchette valves at intervals along their length. They can be self-drilling, or installed in pre-drilled boreholes (Lopez et al., 2019). Because of their length, pipe arches need to be installed at a shallower angle than spiles, and so an enlargement of the tunnel of approximately 20% is required to create a headwall to allow installation (Holzleitner et al., 2005). The tunnel

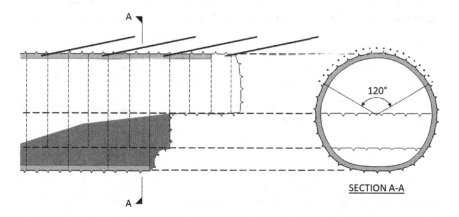

Figure 11.21 Spiles as pre-support.

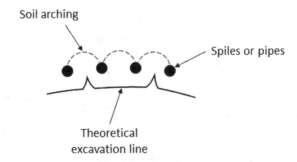

Figure 11.22 Soil arching between spiles or pipes.

therefore has a sawtooth profile in long-section, as shown in Figure 11.23, which means more secondary lining concrete is used. As with spiles, pipe arches need to overlap and this is normally by at least 0.4 times the width of the tunnel (Peila & Pelizza, 2003), but may be larger depending on the ground conditions and potential failure mechanisms. Pipes are installed with an inclination away from the tunnel of between 3–14° (Janin et al., 2015; Lopez et al., 2019). The pipes are typically spaced at 250–500 mm centres.

Spiles do not help much with face stability and are mostly used to limit overbreak at the perimeter of the excavation. Pipes, on the other hand, will have a small effect and this can be quantified. To calculate the drained stability using the wedge-prism method, we can calculate the silo pressure acting on the wedge using the Janssen silo equation (see Appendix A, Equation 15.27), with the width of the silo being the spacing of the pipes. Remember to take account of the fact that because they are angled outwards, the spacing and the offset from the theoretical excavation line increase with length.

To check the spiles or pipes have sufficient structural capacity, a number of design methods are available. The spiles or pipes could be put into a 3D numerical model of the advancing tunnel as beam elements (e.g. Yamamoto et al., 2005; Janin et al., 2015; Janin, 2017; Syomik et al., 2019).

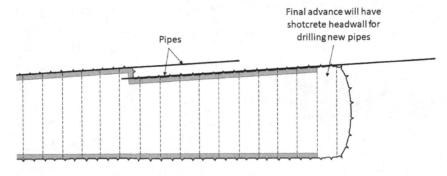

Figure 11.23 Sprayed concrete tunnel long section showing pipe arch pre-support.

Alternatively, a simple model can be used with a pin support (or a series of springs) at one end representing the shotcrete lining and spring supports over the embedded length representing the ground, with an assumed ground load applied over the span (Ischebeck, 2005; Volkmann & Schubert, 2010).

11.4.5 Face dowels

While pre-support consisting of spiles or a pipe arch do improve face stability, they only have a limited effect, and will not prevent face collapse in soils with low cohesion and/or where there is groundwater present, or where the face area is large. To stabilise the face wedge, face dowels can be used to anchor it to the undisturbed ground ahead. Face dowels usually consist of steel or GRP bars, and should be grouted into place.

Anagnostou & Perazzelli (2015) provide an excellent literature review of design methods for face dowels, and present design nomograms based on the wedge-prism method adapted with a method of slices to calculate the stabilising effect. Face dowels will act as a reinforcement to strengthen the ground, and also will apply a support pressure to restrain the face deformations, in a similar manner to a soil nailed wall restraining a slope. They primarily act in tension, and so their capacity is dependent not only on their tensile strength, but also the bond strength of the bar/grout interface and the resistance provided by the grout/soil interface.

Design and installation guidance for face dowels may be found in CIRIA Report C637 'Soil Nailing – best practice guidance' and partial factors and design rules may be found in BS 8006-2: 2011 +A1: 2017 'Code of practice for strengthened/reinforced soils – Part 2: soil nail design'.

11.4.6 Dewatering/depressurisation

Dewatering or depressurisation may be used as a toolbox measure to reduce the risk of seepage-induced instability. The further this is done from the face of the tunnel, the more effective it will be and the less it will disrupt excavation and spraying, so if it is possible to do this from the surface, a shaft, another tunnel or from a pilot tunnel, then this may be more efficient.

The effect of dewatering or depressurisation on drained soils may be calculated by using a 3D numerical seepage model to calculate the pore pressures, and then to use these in the wedge-prism method presented in Chapter 3, or to use a 3D numerical model for both the seepage and stability analysis.

11.5 CONSTRUCTION DETAILS AND TOLERANCES

Construction joints between parts of the lining sprayed at different times are important for structural stability and watertightness of the tunnel. There are two main types of joints; circumferential joints and radial joints.

Tolerances for setting out and spraying of the lining, including an allowance for deformation, need to be included in the design drawings so that the lining does not encroach on the required space envelope.

11.5.1 Circumferential construction joints

Circumferential construction joints are not structurally important, but to make the tunnel as durable and watertight as possible, the joints should be carefully cleaned and prepared prior to spraying. The leading edge of the previous advance is tapered to make it more difficult to trap rebound in a corner. In some cases, the initial layer has been staggered so that it does not coincide with the primary layer to make any potential water path more tortuous (Jones et al., 2008).

11.5.2 Radial construction joints

Radial construction joints are structurally important, and can be a weak point in the lining where a failure may occur. Where lattice girders or steel ribs and welded wire mesh are used this is not a problem, because there will always be connections in the lattice girders or steel ribs and lapping of the mesh between construction stages.

In fibre-reinforced shotcrete linings there will be very little continuity of reinforcement across the joint, and the bond strength of shotcrete across a construction joint is not usually as good as within a part of the lining sprayed in one stage. Typically, bar reinforcement is placed across the joint to ensure shear may be transferred and to provide rotation capacity in bending. This may be achieved by drilling and resin-grouting steel reinforcement bars into the exposed joint before spraying the adjoining part of the lining, or by using 'Kwikastrip', which are bent reinforcement bars housed within a casing that can be removed, allowing the bars to be straightened (Figure 11.24).

In some cases, designers and constructors working together have found ways of removing the need for bar reinforcement across radial construction joints, improving safety by avoiding the need for operatives to enter the face of the tunnel. As long as joints are cleaned and spraying is done carefully, the bond across the joints can be very good. This usually requires careful attention to mix design and pre-construction testing of joints in cylinder and flexural beam tests, followed by careful supervision and quality control during construction (Eddie & Neumann, 2004). An alternative used on the Bank Station Capacity Upgrade project was to use lapped joints, where the lowest 1 m of the primary lining of the top heading was only sprayed to half thickness. This was then sprayed to full thickness at the same time as the bench primary lining (Anthony et al., 2020). Compressive testing of cores through the joint showed no reduction of compressive strength.

For sidewall drifts, the joint at the crown and invert between the sidewall drift and the enlargement is critical. Due to the size of these tunnels and the

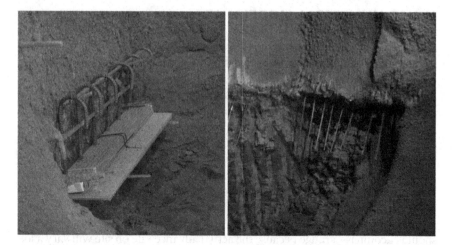

Figure 11.24 'Kwikastrip' installed at the axis level prior to spraying the top heading (left). The bars hidden in the box are later exposed and bent downwards ready for invert spraying (right). This tunnel was a 5.3 m diameter TH-1 excavation.

high shear forces and bending moments that can be induced at these joints, they typically require reinforcement. Installing this reinforcement in the crown during excavation exposes operatives to the risk of falling ground or shotcrete. It is possible to design this out by making the sidewall drift slightly oversize and then, during enlargement excavation, spraying a second layer of primary lining that is continuous across the sidewall drift and enlargement. This will require careful 3D numerical modelling of the construction stages to ensure that the shotcrete lining has sufficient capacity at each stage of construction for the various temporary and permanent loads.

11.5.3 Tolerances

To ensure that the sprayed concrete lining does not impinge on the required space envelope, a tolerance needs to be provided between the theoretical intrados and the required space envelope. This will account for setting out and spraying tolerances and for deformation of the lining, and will be based on what the contractor believes is achievable. A lower tolerance means a smaller overall volume of excavation, but it needs to be achievable most of the time to avoid the need for milling back the lining.

On recent projects in the United Kingdom, a tolerance of 100 mm has been used with approximately 75 mm being required as a spraying tolerance and 25 mm required as a deformation tolerance. For particularly large caverns >10 m wide the deformation tolerance may be larger, and for areas that are tricky to spray accurately, a larger tolerance may be necessary.

The tolerance is normally included in the drawings as a dashed line offset inside the theoretical intrados line.

11.6 3D NUMERICAL MODELLING

For the basics of numerical modelling of tunnels, refer to Chapter 7. In this section and the next we will discuss the principles of numerical modelling of sprayed concrete tunnels, starting with 3D numerical modelling because it involves fewer assumptions and is closer to the real situation, and is therefore easier to understand than 2D numerical modelling.

Since the excavation and lining sequence for a shotcrete-lined tunnel is usually quite complex, 3D numerical modelling is often used. A 3D model allows the explicit modelling of age-dependent shotcrete properties, such as its compressive and flexural tensile strength, plasticity, creep, shrinkage and temperature effects.

The simplest constitutive model is to assume the shotcrete has a constant elastic stiffness. This stiffness can be made to increase with age as the tunnel advances. Assumptions about the speed of construction need to be made, which should encompass a range because the actual advance rate on site will vary a lot.

In a numerical model, unless time is being explicitly modelled (which is rare), the stress state only changes when the soil is excavated. Therefore, each time an advance has been excavated (i.e. the elements representing the soil are removed/deactivated in the model), there are out of balance forces, so we need to perform a calculation to find the new equilibrium. Then the lining is sprayed (i.e. lining elements are added/activated at the excavation perimeter), but if a further calculation were performed after installing the lining elements, nothing would happen because the model is still in equilibrium. Therefore, the model is solved after each excavation step but not after each lining installation step.

11.6.1 Linear elastic sprayed concrete constitutive model

The simplest constitutive model that can be used for a sprayed concrete lining is a linear elastic model, and the simplest elements are shell elements located at the perimeter of the excavation with no interface (i.e. the lining shell element nodes and soil solid element nodes cannot move independently of each other). For a linear elastic model, only the lining thickness, Young's modulus and Poisson's ratio need to be defined. As mentioned in Chapter 7, positioning shell elements at the extrados results in an overestimate of bending moments, which is conservative.

The outputs from any model, in terms of axial forces, shear forces and bending moments, need to be checked against the capacity of the section. If the constitutive model for the lining is linear elastic, then the model itself will not fail if it is overstressed, so this check is essential. For fibre-reinforced concrete, it may be simpler to output the maximum tensile and compressive stresses and strains, and to compare these to the limiting values.

If the capacity of the section is exceeded at any point, then either the lining thickness is increased, the degree of reinforcement (whether mesh or fibres) is increased, or the profile may be changed to a more circular one. Alternatively, a more sophisticated shotcrete constitutive model could be used, because it is likely this will reduce peak stresses.

11.6.2 Ageing sprayed concrete properties

Each lining and excavation stage in a numerical model happens instantaneously, but represents a period of time in the real tunnel. If a full-face excavation were proceeding at four advances per day, then each lining and excavation stage would be six hours. During the time period of each advance, the stiffness and strength of the sprayed concrete (and other properties, if modelled) will be changing rapidly, therefore it is not obvious what single value to use for each property to represent its behaviour during each time period. In reality, excavation and spraying are not immediate and take some time. There are also pauses between excavation and spraying activities as plant is moved around, invert backfill is excavated or placed, hoses are connected, shotcrete is delivered, joints in the lining are cleaned and prepared, etc. Common practice is to use the value of the property at the midpoint of the time period of each advance. An example of this is shown in Figure 11.25 for an advance rate of 4 m/day.

The designer should use the maximum advance rate the contractor believes they can achieve at peak production, as this will be the worst case. If this results in overstressing of the lining in the model, then either the lining needs to be strengthened or a limit can be placed on the maximum advance rate.

The variation of properties with age is usually linked to the compressive strength development. This should be obtained from testing of the shotcrete used on site at different ages, but if this is not yet available, the designer may use values from the literature for preliminary design, which can be verified at a later stage when test data is available.

An example of compressive strength development used for numerical analysis by Jones et al. (2008), based on shotcrete testing for the Heathrow Terminal 5 works, is shown in Table 11.3.

On some projects, the design is based on one of the strength development curves in EN 14887-1: 2005, known colloquially as the 'J' curves, because they originate from the Austrian sprayed concrete guideline where the curves were called 'J1', 'J2' and 'J3'. However, the strengths given in the standard are minimum requirements and do not follow the typical strength development of sprayed concrete.

For a linear elastic constitutive model, we need to know how the Young's modulus develops with age. Chang & Stille (1993) proposed the following equation that relates Young's modulus with compressive strength:

$$E_c = 3.86 f_c^{0.6} \tag{11.8}$$

E_c is the Young's modulus of the sprayed concrete in GPa
f_c is the compressive strength of the sprayed concrete in MPa

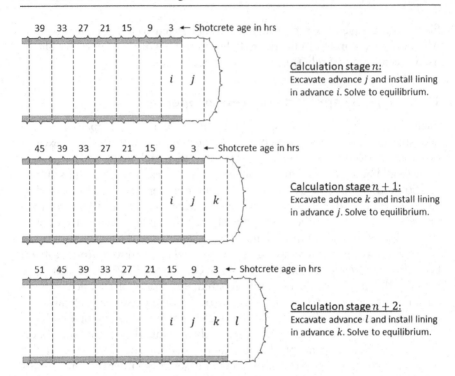

Figure 11.25 Ageing shotcrete in a 3D numerical model of an advancing full-face excavation, at an advance rate of 4 m/day.

Table 11.3 Example of compressive strength development with age.

Age	Sprayed concrete characteristic compressive strength (MPa)
0 hours	0
0.1 hours	0.2
1 hours	0.5
3 hours	1.0
6 hours	3.0
12 hours	8.0
1 day	15.0
3 days	25.0
5 days	28.0
7 days	30.0
10 days	31.5
15 days	33.0
20 days	34.0
28 days	35.0

Since a higher stiffness will attract more ground load and will result in higher bending moments, it is important to consider both the lower and upper bound characteristic values. The lower bound value will give a more conservative estimate of ground movements, whereas an upper bound value will give a more conservative estimate of lining forces.

Poisson's ratio is usually taken as 0.2 for shotcrete or concrete.

11.6.3 Linear elastic – perfectly plastic sprayed concrete constitutive model

This kind of constitutive model is particularly useful for fibre-reinforced concrete, where it is very likely that the tensile strength will be exceeded and a linear elastic model would therefore result in an unrealistic stress distribution in the lining at these locations.

Not all numerical modelling programs are able to model plasticity in shell elements, and it may be necessary to use solid elements for the lining instead. Nasekhian & Feiersinger (2017) describe the use of a 'concrete damaged plasticity model' for shell elements in the finite element program ABAQUS to model elastic – perfectly plastic behaviour in tension. The yield stress was set at the design value of the ultimate post-cracking tensile strength f_{Ftud} (*fib* Model Code 2010, 2013), as shown in Figure 11.26.

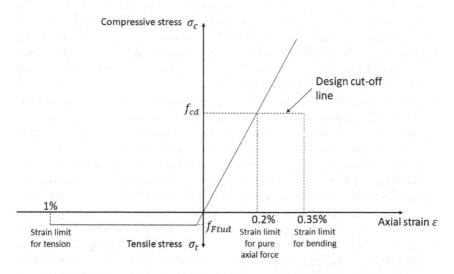

Figure 11.26 Elastic model with perfect plasticity at the design value of ultimate post-cracking tensile strength.

Figure 11.26 uses the following notation:

f_{cd} is the design value of compressive strength for the fibre-reinforced sprayed concrete according to EN 1992-1-1: 2004 or the *fib* Model Code 2010 (2013)

f_{Ftud} is the design value of the ultimate post-cracking tensile strength for the fibre reinforced sprayed concrete according to the *fib* Model Code 2010 (2013)

After running a numerical model using the constitutive relationship shown in Figure 11.26, the tensile and compressive strains were checked against the strain limits to verify that all parts of the lining were within the ultimate limit state. As long as tensile strains in the model were less than 1%, and compressive strains less than 0.2%, the lining was assumed to be within its capacity. These strain limit values were determined based on the properties of the sprayed concrete used on this particular project and included partial factors on material strengths and the partial factor for actions.

Geotechnical numerical modelling programs often have limited shell element constitutive models, in which case using solid elements for the lining may be the only way of including shotcrete plasticity. An example is given in Jones et al. (2008), where solid elements were used with a Mohr-Coulomb model, with the parameters fit to pre-construction shotcrete testing data.

11.6.4 More complex constitutive models

Thomas (2003) used a series of 3D numerical models to investigate the effect of different ageing shotcrete constitutive models, including linear elastic, nonlinear elastic, hypothetical modulus of elasticity, elastoplastic, creep and shrinkage. Replacing linear elastic behaviour with 'softer' models, such as the model including creep, resulted in reduced stresses in the lining, with bending stresses reduced more than axial stresses because creep is stress-dependent and there was more stress relaxation where stresses were larger.

Schütz (2010) and Schütz et al. (2011) proposed an elastoplastic model for shotcrete that takes account of creep, shrinkage and thermal deformation. It can also allow for cracking using a smeared crack approach.

All of the models we have discussed so far use the age of the shotcrete as the basis for the development of all parameters with time. However, the properties of all concretes including shotcrete actually depend on maturity rather than age. Maturity can be thought of as the proportion of total hydration (the chemical reactions between cement and water) that have occurred. The rate of hydration is highly dependent on temperature, and approximately doubles every 10°C. Therefore, at higher temperatures concrete strength will increase much faster with time than at lower temperatures. This has led to the development of shotcrete models based on maturity rather than age by Hellmich et al. (2001), heavily based on the experimental work of Byfors (1980) and the theoretical framework of Ulm & Coussy (1995).

The problem for the designer in using these more complex models is that at the design stage there is considerable uncertainty about the shotcrete behaviour, because every shotcrete is different, and some aspects of these models are based on experiments performed more than 20 years ago when shotcrete technology was very different. Even if pre-construction testing allows the use of parameters specific to the shotcrete being used, the environmental conditions on site, and in particular the temperature, will have a huge impact on how the properties of the shotcrete vary with time. It is also very difficult to predict the rate that a tunnel will be advanced.

For all these reasons, it is important to undertake sensitivity analysis, where all the parameters used in the constitutive model are varied within realistic limits, so that the sensitivity of the design to each of them can be understood. It is also important to monitor the deformations and stresses within the tunnel during construction, so that the design assumptions can be verified.

11.7 2D NUMERICAL MODELLING

In a 2D numerical model, we need some way to take account of the 3D nature of tunnel excavation, with stress redistribution occurring as arching develops in the ground ahead of and around the face and unsupported perimeter. In Chapter 7, various methods for simulating the relaxation of the ground ahead of lining installation in a 2D numerical model were discussed. The most common method used for sprayed concrete tunnels is the convergence-confinement method.

Essentially, the convergence-confinement method involves removing the elements representing the ground within the tunnel perimeter and replacing them with nodal forces at the perimeter equal to some proportion of the in situ stress in the ground. The lining is then installed and the nodal forces are removed. For sprayed concrete tunnels, however, (except when full-face excavation is used) this can become quite complex, as for each excavation stage an assumption needs to be made about the amount of relaxation that will occur.

An example for a 2D numerical model of a TH-B-I excavation is shown in Figure 11.27. A total of six values of either confinement loss factor or β factor (see Chapter 7) need to be specified to obtain the nodal forces p_{TH1}, p_{TH2}, p_{TH3}, p_{B1}, p_{B2} and p_I. For a sidewall drift and enlargement excavation, 12 or more values may be needed.

Designers will often assume that the nodal forces are removed when the lining is installed, i.e. that $p_{TH2} = p_{TH3} = p_{B2} = 0$ in Figure 11.27. However, it is important to remember that assuming they are equal to zero is still an assumption.

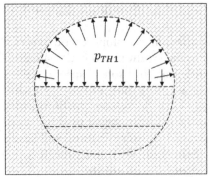

STAGE 1: Remove soil elements from top heading and replace with nodal forces p_{TH1}.

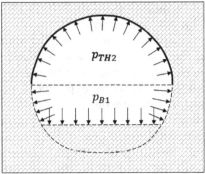

STAGE 2: Install top heading lining and reduce nodal forces from p_{TH1} to p_{TH2}. Remove soil elements from bench and replace with nodal forces p_{B1}.

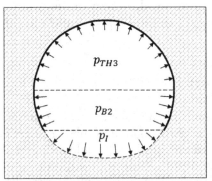

STAGE 3: Reduce p_{TH2} to p_{TH3}. Install bench lining and reduce p_{B1} to p_{B2}. Remove soil elements from invert and replace with nodal forces p_I.

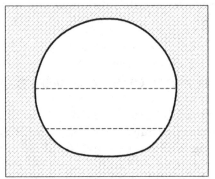

STAGE 4: Install invert lining and remove all nodal forces.

Figure 11.27 2D numerical model of a TH-B-I excavation.

The values of these nodal forces may be estimated using engineering judgement, or preferably calibrated to a case study of a similar construction method in similar ground conditions. Another option is to calibrate the nodal forces using a 3D numerical model. This is not as silly as it sounds, because the 2D model can be used more easily for sensitivity analyses, or for slightly different sizes or shapes of tunnels, and it takes much less time to run a series of 2D numerical models than to do the same in 3D.

Whichever calibration method is used, there will always be a large amount of judgement required. In addition, calibration to a case study is likely to only be a comparison with ground movements and not lining stresses, because there are very few case studies that include measurements of lining stresses. As discussed in Chapter 7, a model calibrated to ground movements will not necessarily predict accurate lining forces. For this reason, a sensitivity analysis varying the amount of relaxation prior to lining installation should be performed, and the effect on the lining forces understood. This is also one of the reasons why it is important to verify the design during construction using monitoring, and why designers should always be given access to the site and the monitoring data.

11.8 SUMMARY

Sprayed concrete linings allow much more flexibility in terms of geometry than other lining types. However, there are constraints, and we cannot use just any shape. In soft ground we still need the lining to act as an arch in order to have an efficient structural form, but the shape does not have to be a circle. Depending on the space envelope required inside the tunnel, the shape can be modified to reduce the volume of excavation, though this must be balanced against the thickness and reinforcement requirements of the lining. The less circular the profile of a sprayed concrete lining is, the lower the volume of excavation will be, but the bending moments will be larger, and the lining will need to either be thicker or contain more reinforcement.

As the tunnel advances, the sprayed concrete lining is gaining strength and stiffness while at the same time being increasingly loaded by the ground, unlike precast or prefabricated linings, which have their long-term design strength and stiffness at the time of installation.

Sprayed concrete quality cannot be fully verified prior to installation and is dependent on workmanship, variability of materials and environmental conditions. In particular, the temperature of the shotcrete will affect its rate of strength development.

3D numerical modelling is essential for accurately modelling the shape and construction sequence of a sprayed concrete lining. 2D numerical models can be useful and efficient if carefully calibrated, but the results should be used with caution.

Decisions about face division, toolbox measures, construction details and tolerances require the input of the construction team. Ideally, sprayed concrete tunnels are designed and planned by an integrated design and construction team.

The traditional contractual separation of temporary and permanent works is inappropriate for any tunnelling project, but especially so for sprayed concrete tunnels where the temporary and permanent works are essentially the same structure, and the way that they are built affects the loads they have to withstand. Even if the (inefficient) traditional view is taken that the primary lining is temporary and the secondary lining is permanent, the construction methods used for the primary lining will affect the loads in the secondary lining and it cannot be designed in isolation.

Constitutive models for sprayed concrete are available that simulate the most important aspects of behaviour: the variation of properties with age or maturity, elastoplasticity, creep, shrinkage and thermal effects. Every effort should be made to base the model parameters on the actual shotcrete being used, but even so, there will remain considerable uncertainties due to the inherent variability of shotcrete as a material, the variability of environmental conditions, the actual rate of advance of the tunnel, and potential for variability in both profile and thickness. For these reasons, as with other tunnelling methods but even more so; for sprayed concrete the design and construction are inextricably interlinked. The design is not finished when construction starts, but must be verified by monitoring and back-analysis.

11.9 PROBLEMS

Q11.1. The required space envelope for a large pedestrian travellator passageway is shown in Figure 11.28. The bottom of the required space envelope is 18 m below ground level. The ground is soft clay with undrained shear strength $c_u = 40$ kPa and bulk unit weight $\gamma = 18$ kN/m³.
  i. Fit a sprayed concrete lining profile around this required space envelope.
  ii. Allowing 100 mm construction tolerance and allowing 300 mm for a secondary lining, draw a primary lining 300 mm thick.
  iii. Calculate the excavation area.

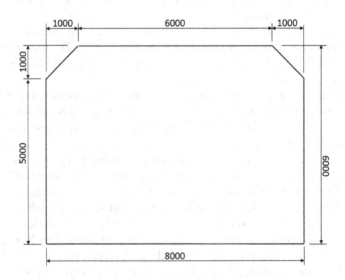

Figure 11.28 Problem Q11.1 required space envelope for a large pedestrian travellator passageway.

iv. From the excavation area in (iii), calculate an equivalent diameter. Use this to estimate the undrained stability for a full-face excavation, remembering to use a factor of safety of 1.4.

v. Assuming the maximum reach of the excavator is 5 m, and taking account of undrained stability, divide the face in an appropriate manner. Draw a cross-section and long-section showing the construction sequence.

Q11.2. The required space envelope for an upper lift lobby in a metro station is rectangular, 5.3 m high and 3.6 m wide. The permanent structure will be placed inside this envelope. The ground is slightly clayey medium dense fine and medium sand with drained cohesion $c' = 5$ kPa, drained angle of friction $\phi' = 35°$ and unit weight $\gamma = 17$ kN/m^3. The bottom of the required space envelope is 18 m below ground level. The tunnel is above the water table.

i. Fit a sprayed concrete lining profile around this required space envelope.

ii. Allowing 100 mm construction tolerance, draw a primary lining 250 mm thick.

iii. Sketch the likely radial deformation assuming a uniform radial ground pressure applied to the lining.

iv. From the radial deformation in (iii), sketch the likely bending moment distribution in the lining.

v. Taking account of drained stability using the wedge-prism method, and remembering to apply a partial factor of 1.25 on $\tan\phi'$ and c', design a viable face division and construction sequence.

vi. Design toolbox measures in case the soil cohesion is lower than expected.

Q11.3. The required space envelope for a road tunnel is given in Figure 11.29. It is to be constructed using a sprayed concrete lining. The bottom of the required space envelope is at 20 m below ground level. Assume the tunnel is in stiff clay with a unit weight of 20 kN/m^3, a characteristic value of undrained shear strength of 140 kPa and an undrained Young's modulus of 84 MPa. The pore pressure is hydrostatic with a piezometric level at the surface. Coefficient of earth pressure at rest $K_0 = 0.8$.

i. Fit a sprayed concrete lining profile around the required space envelope.

ii. Allowing 100 mm construction tolerance and 300 mm for a secondary lining, draw the primary lining assuming it is 300 mm thick.

iii. Sketch the likely radial deformation assuming a uniform radial ground pressure applied to the lining.

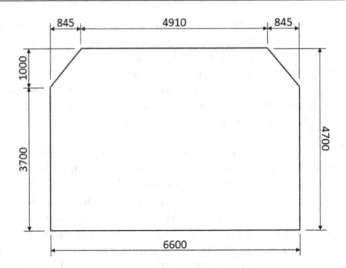

Figure 11.29 Problem Q11.3 required space envelope for a road tunnel.

iv. From the radial deformation in (iii), sketch the likely bending moment distribution in the lining.

v. Assuming the maximum vertical reach of the excavator is 5 m, and taking account of undrained stability, divide the face in an appropriate manner. Draw a cross-section and long-section showing the construction sequence.

vi. Using the compressive strength development in Table 11.3 and Chang & Stille's relationship between compressive strength and Young's modulus, calculate the stiffness values that need to be used in a 3D numerical analysis, if the advance rate is 4 advances (whether they are top headings, benches or inverts) per day.

vii. (Advanced) Build and run a 2D numerical model of the tunnel primary lining. Experiment with different assumptions about nodal forces for each calculation stage and lining thickness. Based on your answer to (vi), you will need to estimate the Young's modulus of the sprayed concrete at each calculation stage.

viii. (Advanced) Build and run a 3D numerical model of the tunnel primary lining. Compare the results in terms of ground movements and lining forces with the 2D numerical modelling for (vii).

ix. (Advanced) Assuming the lining is steel fibre reinforced shotcrete grade FRC 40/50 – 4.0c, plot the results from your numerical analysis in (vii) and/or (viii) on a moment-force interaction diagram, remembering to apply partial factors. Are the lining forces within the ultimate limit state capacity curve? If not, adjust the profile or the thickness of the lining and re-run the numerical model.

REFERENCES

Aagaard, B., Gylland, A. S., Schubert, P. & Løne, B. (2017). The Joberg Tunnel – Successful tunnelling in moraine. *Proc. World Tunnel Congress 2017, Surface Challenges – Underground Solutions* (eds Nilsen, B., Holter, K. G. & Jakobsen, P. D.), Bergen, Norway, pp. 1038–1047. Norwegian Tunnelling Society NFF.

Anagnostou, G. & Perazzelli, P. (2015). Analysis method and design charts for bolt reinforcement of the tunnel face in cohesive-frictional soils. *Tunn. Undergr. Space Technol.* **47**, 162–181.

Anthony, C., Kumpfmüller, S., Feiersinger, A. & Ares, J. (2020). Improving safety through design at London Underground's Bank station capacity upgrade. *Proc. Instn of Civ. Engrs – Civil Engrg* **173**, No. 5, 41–47.

BS 8006-2: 2011 +A1: 2017. *Code of practice for strengthened/reinforced soils – Part 2: Soil nail design.* London, UK: British Standards Institution.

Byfors, J. (1980). *Plain concrete at early ages,* CBI Forskning Fo 3:80, p. 464. Stockholm: Swedish Cement and Concrete Research Institute.

Chang, Y. & Stille, H. (1993). Influence of early age properties of shotcrete on tunnel construction sequences. *Proc. shotcrete for underground support VI* (eds. Wood, D. F. & Morgan, D. R.), Niagara-on-the-lake, Canada, pp. 110–117. USA: ASCE.

CIRIA (2005). *Soil nailing – best practice guidance,* C637. London, UK: CIRIA.

Eddie, C. and Neumann, C. (2004). Development of LaserShell™ Method of Tunneling. *Proc. of the North American Tunneling Conference 2004,* Atlanta. Rotterdam: Balkema.

EN 12715: 2000. *Execution of special geotechnical work – Grouting.* Brussels: European Committee for Standardization.

EN 14887-1: 2005. *Sprayed concrete – Part 1: definitions, specifications and conformity.* Brussels: European Committee for Standardization.

EN 1992-1-1: 2004. *Eurocode 2: Design of concrete structures – Part 1-1: general rules and rules for buildings.* Brussels: European Committee for Standardization.

fib Model Code 2010 (2013). *fib Model Code for Concrete Structures 2010.* Lausanne: Fédération internationale du béton (fib).

Gall, V., Pyakurel, S. & Munfa, N. (2017). Conventional tunnelling in difficult grounds. *Proc. World Tunnel Congress 2017, Surface Challenges – Underground Solutions* (eds Nilsen, B., Holter, K. G. & Jakobsen, P. D.), Bergen, Norway, pp. 2519–2528. Norwegian Tunnelling Society NFF.

Hellmich, C., Macht, J., Lackner, R., Mang, H. A. & Ulm, F.-J. (2001). Phase transitions in shotcrete: from material modelling to structural safety assessment. *Shotcrete: Engineering Developments, Proc. Int. Conf. Engineering Developments in Shotcrete,* Hobart, Tasmania, Australia (ed. Bernard, E. S.), pp. 173–184. Rotterdam, The Netherlands: Balkema.

Holzleitner, W., Deisl, F. & Holzer, W. (2005). Tunnel support ahead of the face with pipe roofs or forepoling. *Rational Tunnelling 2nd Summerschool* (eds Kolymbas, D. & Laudahn, A.), Innsbruck, Austria, pp. 147–164. Berlin, Germany: Logos Verlag.

Ischebeck, E. F. (2005). New approaches in tunnelling with composite canopies – installation-design-monitoring. *Rational Tunnelling 2nd Summerschool* (eds Kolymbas, D. & Laudahn, A.), Innsbruck, Austria, pp. 165–171. Berlin, Germany: Logos Verlag.

ITA-AITES (2020). *Permanent sprayed concrete linings.* ITA Report No. 24, ITA Working Group no. 12 and ITAtech. Lausanne, Switzerland: ITA-AITES.

Jäger, J. (2016). Structural design of composite shell linings. *Proc. World Tunnel Congress 2016,* 22nd –28th April, San Francisco, USA. Englewood, CO, USA: Society for Mining, Metallurgy and Exploration.

Janin, J. P. (2017). Apports de la simulation numérique tridimensionnelle dans les études de tunnels. *Revue Française de Géotechnique* 150, No. 3, 1–13.

Janin, J. P., Dias, D., Emeriault, F., Kastner, R., Le Bissonnais, H. & Guilloux, A. (2015). Numerical back-analysis of the southern Toulon tunnel measurements: A comparison of 3D and 2D approaches. *Eng. Geol.* **195**, 42–52.

Jones, B. D. (2018). A 20 year history of stress and strain in a shotcrete primary lining. Proc. 8th Int. Symp. Sprayed Concrete - Modern use of wet mix sprayed concrete for underground support, Trondheim, Norway, 11th - 14th June (eds Beck, T., Myren, S. A. & Engen, S.), pp.206-222. Oslo, Norway: Norwegian Concrete Society/Norsk Betongforening.

Jones, B. D., Thomas, A. H., Hsu, Y. S. & Hilar, M. (2008). Evaluation of innovative sprayed-concrete-lined tunnelling. *Proc. Instn of Civ. Engrs – Geotech. Engrg* **161**, GE3, 137–149.

Karol, R. H. (2007). *Chemical grouting and soil stabilization,* 3rd revised edition. New York: Dekker.

Lopez, F., von Havranek, F. & Severi, G. (2019). Alternative umbrella arches: The use of composite pile roofs. *Proc. World Tunnel Congress 2019, Tunnels and Underground Cities: Engineering and Innovation meet Archaeology, Architecture and Art* (eds Peila, D., Viggiani, G. & Celestino, T.) Naples, Italy, pp. 2527–2535. London, UK: Taylor & Francis Group.

Lunardi, P. (2008). *Design and construction of tunnels. Analysis of controlled deformation in rocks and soils (ADECO-RS).* Springer.

Macklin, S. R. (1999). The prediction of volume loss due to tunnelling in overconsolidated clay based on heading geometry and stability number. *Ground Eng.* No. 4, 30–33.

Mecsi, J. (2002). Ground surface movements resulting from metro station and tunnel construction in Budapest (Hungary). *Proc. Int. Symp. Geotechnical Aspects of Underground Construction in Soft Ground,* Toulouse, France (eds Emeriault, F., Kastner, R., Dias, D. & Guilloux, A.), pp. 349–354. Lyon, France: Spécifique JLP.

Nasekhian, A. & Feiersinger, A. (2017). SCL Design Optimisation at Bank - A combined lining approach. *Proc. World Tunnel Congress 2017, Surface Challenges – Underground Solutions* (eds Nilsen, B., Holter, K. G. & Jakobsen, P. D.), Bergen, Norway, pp. 658–665. Norwegian Tunnelling Society NFF.

Peila, D. & Pelizza, S. (2003). Ground reinforcing and steel pipe umbrella system in tunnelling. *Rational Tunnelling Summerschool* (ed. Kolymbas, D.), Innsbruck, Austria, pp. 93–132. Berlin, Germany: Logos Verlag.

Rawlings, C. G., Hellawell, E. E. & Kilkenny, W. M. (2000). *Grouting for ground engineering,* CIRIA C514. London: CIRIA.

Schütz, R. (2010). *Numerical modelling of shotcrete for tunnelling.* PhD Thesis, Imperial College London.

Schütz, R., Potts, D. M. & Zdravkovic, L. (2011). Advanced constitutive modelling of shotcrete: Model formulation and calibration. *Comput. Geotech.* **38**, 834–845.

St John, A., Potts, V., Perkins, O. & Balogh, Z. (2015). Use of TBM pilots for large diameter SCL caverns: Crossrail C300/C410. *Proc. World Tunnel Congress 2015, SEE Tunnel: Promoting tunnelling in SEE Region* (ed. Kolić, D.), 22nd–28th May, Dubrovnik, Croatia, p. 15. Zagreb, Croatia: HUBITG.

Su, J., Bedi, A. & Bloodworth, A. G. (2019). On composite action and tunnel lining design; is composite action beneficial in reducing lining thickness? *Tunnelling Journal*, April/May, 38–46.

Syomik, A., Rex, R., Zimmermann, A. & Anagnostou, G. (2019). Evaluation and numerical interpretation of measured pipe umbrella deformations. *Proc. World Tunnel Congress 2019, Tunnels and Underground Cities: Engineering and Innovation meet Archaeology, Architecture and Art* (eds Peila, D., Viggiani, G. & Celestino, T.) Naples, Italy, pp. 6249–6257. London, UK: Taylor & Francis Group.

Thomas, A. H. (2003). *Numerical modelling of sprayed concrete lined (SCL) tunnels.* PhD thesis, University of Southampton.

Ulm, F. -J. & Coussy, O. (1995). Modeling of thermochemomechanical couplings of concrete at early ages. *J. Engrg Mech. ASCE* **121**, No.7, pp. 785–794.

Volkmann, G. M. & Schubert, W. (2010). A load and load transfer model for pipe umbrella support. *Proc. Eurock 2010, Rock Mechanics in Civil and Environmental Engineering* (eds. Zhao, J., Labiouse, V. Dudt, J. -P. & Mathier, J. -F.), pp. 379–382. London, UK: CRC Press.

Yamamoto, T., Murakami, K., Shirasagi, S., Negishi, K., Nishioka, K. & Kobayashi, T. (2005). Application of the tunnel face reinforcement and forepiling. *Proc. World Tunnel Congress 2005*, Istanbul, Turkey, pp. 391–396. London: Taylor & Francis Group.

Chapter 12

Estimating ground movements

In Chapter 1, we looked at how the ground moves when a tunnel is constructed, and how this depends on the construction method, soil type and groundwater. In this chapter empirical equations and numbers will be introduced to characterise these patterns of ground movements, and to allow predictions to be made for future tunnelling projects. The word 'prediction' is used here advisedly, as it is not possible to predict ground movements with any degree of accuracy, even in greenfield conditions. The word 'estimate' should perhaps be used, as suggested by Burland (2001) and Pound (2003).

After working through this chapter you will be able to estimate:

- transient and short-term transverse and longitudinal surface settlements in different soil types
- volume loss
- long-term surface settlements in clay soils
- subsurface settlements
- surface and subsurface horizontal ground movements
- ground slopes, curvatures and strains
- ground movements due to shaft construction

'Transient' means the settlement is changing as the tunnel face is passing. It does not mean the ground movements are temporary, their cumulative effect is permanent.

'Short-term' settlement means after the tunnel face has advanced sufficiently far beyond the point in question that construction is no longer having an effect and settlement has stabilised to a relatively constant value. This short-term value may change in the long-term due to creep or consolidation effects.

12.1 TRANSVERSE SURFACE SETTLEMENTS

Peck (1969) and Schmidt (1969) found that ground surface settlements transverse to the tunnel may be represented by a Gaussian curve. Surface settlements above real tunnels have repeatedly been found to take approximately

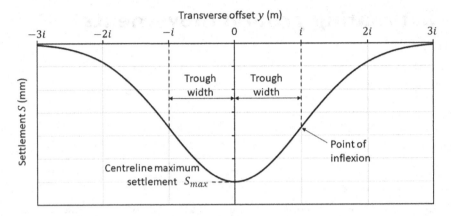

Figure 12.1 Gaussian surface settlement trough.

this shape, though as we will see the approximation tends to be better for clays than for coarse-grained soils such as sands and gravels. This Gaussian settlement trough is mathematically an 'error function' or Gaussian 'normal distribution function' upside-down. It is shown in Figure 12.1.

The distance from the centreline to the point of inflexion of the curve, which in a normal distribution would be the standard deviation, is known as the 'trough width', i. The maximum settlement over the centreline of the tunnel S_{max} is analogous to the frequency of the mean of a normal distribution. Settlement S at offset y from the tunnel centreline is therefore given by:

$$S = S_{max}\exp\left(\frac{-y^2}{2i^2}\right)$$

(12.1)

S is the settlement at offset y, in m
S_{max} is the maximum settlement at $y = 0$, in m
y is the offset, or transverse distance, from the tunnel centreline, in m
i is the 'trough width', or the offset to the point of inflexion, in m

The area under the curve is defined as the 'volume loss', V_s. This may be found by integrating Equation 12.1, as shown in the following equation:

$$V_s = \int_{-\infty}^{\infty} S\,dy = \int_{-\infty}^{\infty} S_{max}\exp\left(\frac{-y^2}{2i^2}\right)dy = \sqrt{2\pi}\,.iS_{max}$$

(12.2)

V_s is the volume loss per unit face advance, in m³/m or m²

Using Equation 12.1, Equation 12.2 or combinations thereof, the curve may be defined by any two of the parameters V_s, S_{max} or i.

O'Reilly & New (1982) did a big meta-analysis of settlements induced by tunnelling in the United Kingdom, and found that the trough width may be estimated by multiplying the height above the tunnel by a 'trough width parameter' K, which is estimated by consideration of case histories. Trough width parameter K is related to trough width i by the following equation:

$$i = K(z_0 - z) \tag{12.3}$$

K is the trough width parameter
z_0 is the depth to axis of the tunnel under construction in m
z is the depth to the point under consideration, where $z \leq z_0$, in m. At the surface, $z = 0$ and $i = Kz_0$.

Note that Equation 12.3 can also be used to estimate the subsurface settlement trough width, and we will look into this in the following Section 12.5.

The Gaussian curve is a useful tool, as it allows comparison of settlements at different locations and on different projects by a small number of practically relevant parameters. This then facilitates the empirical prediction of the magnitude of settlement (as described by volume loss V_s or maximum settlement S_{max}) and its extent (described by trough width i). Empirical prediction is usually based on a meta-analysis of case histories. This meta-analysis may be refined by restricting the list of case histories considered to consist of, for example, only those tunnels using the same excavation and support method at similar depth and with similar geology, or by using correlations that take some or all of these factors into account. O'Reilly & New (1982) provided meta-analyses for most soil types in the UK, and Lake et al. (1996) and then Mair & Taylor (1997) extended this to tunnels in a wide range of soil types around the world. All these studies validated the use of the Gaussian curve.

In order to better make comparisons between tunnels of different sizes, the volume loss is often expressed as a percentage of the excavation volume. In other words, V_s is divided by the cross-sectional area of the excavation. This is also referred to as the 'volume loss' but is given the symbol V_l.

$$V_l = \frac{V_s}{A} \% \tag{12.4}$$

V_l is the volume loss as a percentage of face area, in %
V_s is the volume loss per unit face advance, in m^2
A is the face area, in m^2

Based on their meta-analysis of tunnels in the United Kingdom, O'Reilly & New (1982) gave typical ranges of values of trough width parameter K and volume loss V_l in different soil types as shown in Table 12.1.

Table 12.1 does not include coarser, granular deposits such as sand or gravel. For cohesionless granular soils above the water table, O'Reilly & New (1982)

Table 12.1 Trough width parameter and volume loss in different soil types
(O'Reilly & New, 1982).

Soil type	Construction method	Value of K	Value of V_l
Stiff fissured clay	Shield or none	0.4–0.5	0.5–3%
Glacial deposits	Shield in free air	0.5–0.6	2–2.5%
	Shield in compressed air		1–1.25%
Recent silty clay deposits	Shield in free air	0.6–0.7	30–45%
	Shield in compressed air		5–20%

recommended a value of trough width parameter *K* between 0.2 and 0.3. However, more recent work by Marshall (2009) and Marshall et al. (2012) has shown that trough width parameter in sand is dependent on initial density and tunnel volume loss. Also, we should remember that, unlike clays, sands and gravels do not exhibit constant volume behaviour, and the volume loss at the surface or subsurface will not be the same as the volume loss at the tunnel.

In a series of geotechnical centrifuge experiments investigating the effect of tunnel size, depth and volume loss on the surface and subsurface settlement trough in sand, Marshall et al. (2012) showed that the Gaussian curve does not always provide a good fit to the settlement trough. As volume loss at the tunnel increased, the trough width decreased, indicating the development of a chimney-type failure mechanism above the tunnel.

In addition, they found that as volume loss at the tunnel increased, the volume loss at the surface increased at a faster rate, which they ascribed to the relative density of the sand. Under shear strain, the sand was found to contract initially, followed by dilation at higher shear strains. They also found that as depth increased or the size of the tunnel decreased, the trough width became narrower.

WORKED EXAMPLE 12.1   OPEN FACE TBM TUNNEL IN HOMOGENEOUS CLAY

An open-face shield mounted with a roadheader is to be used to excavate a 10 m diameter tunnel in stiff clay. The tunnel axis is 20 m below the ground surface. Based on case histories in similar geology using similar construction methods, the volume loss is predicted to be 1.5% and the trough width parameter *K* is 0.45.

Calculate the maximum surface settlement S_{max}. Then calculate the settlement at $y = 5$ m, 10 m, 15 m and 20 m.

We first need to calculate the face area *A*:

$$A = \frac{\pi D^2}{4} = \frac{\pi \times 10^2}{4} = 78.54 \text{ m}^2$$

Then we can calculate the volume loss V_s using Equation 12.4:

$$V_s = V_l A = 0.015 \times 78.54 = 1.178 \text{ m}^2$$

The trough width is given by Equation 12.3:

$$i = Kz_0 = 0.45 \times 20 = 9 \text{ m}$$

The maximum settlement S_{max} is given by Equation 12.2:

$$S_{max} = \frac{V_s}{\sqrt{2\pi}.i} = \frac{1.178}{\sqrt{2\pi} \times 9} = 0.05222 \text{ m} = 52.2 \text{ mm}$$

The settlement at $y = 5$ m is given by Equation 12.1:

$$S = S_{max}\exp\left(\frac{-y^2}{2i^2}\right) = 0.052 \times \exp\left(\frac{-(5^2)}{2 \times 9^2}\right) = 0.04475 \text{ m} = 44.8 \text{ mm}$$

Note that $-y^2$ means $-(y^2)$, so if using a spreadsheet, pay particular attention to the use of brackets. Also take care with percentages and use consistent units (m or mm, but not both).

Similarly, the settlements at the other offsets are given in Table 12.2.

Table 12.2 Worked Example 12.1 calculated settlements.

	Offset from tunnel centreline y (m)				
	0	5	10	15	20
Settlement S (mm)	52	45	28	13	4

It is not considered good practice to predict surface settlements to better than 1 mm accuracy. This is because of the considerable uncertainties involved in the estimation of trough width, volume loss and the characterisation of the geology. Therefore, final outputs of any settlement calculation should be rounded to the nearest mm.

WORKED EXAMPLE 12.2   CLOSED FACE TBM TUNNEL IN SAND

An earth pressure balance TBM is to be used to excavate a 7 m diameter tunnel in sand. The tunnel axis is 15 m below the ground surface. Based on case histories in similar geology using similar construction methods, the volume loss is predicted to be 0.5% and the trough width parameter K is 0.25.

Calculate the maximum surface settlement S_{max}. Then calculate the settlement at $y = 2.5$ m, 5 m, 7.5 m and 10 m.

We first need to calculate the face area A:

$$A = \frac{\pi D^2}{4} = \frac{\pi \times 7^2}{4} = 38.48 \text{ m}^2$$

Then we can calculate the volume loss V_s using Equation 12.4:

$$V_s = V_l A = 0.005 \times 38.48 = 0.192 \text{ m}^2$$

The trough width is given by Equation 12.3:

$$i = Kz_0 = 0.25 \times 15 = 3.75 \text{ m}$$

The maximum settlement S_{max} is given by Equation 12.2:

$$S_{max} = \frac{V_s}{\sqrt{2\pi}.i} = \frac{0.192}{\sqrt{2\pi} \times 3.75} = 0.020 \text{ m} = 20 \text{ mm}$$

The settlement at $y = 5$ m is given by Equation 12.1:

$$S = S_{max}\exp\left(\frac{-y^2}{2i^2}\right) = 0.020 \times \exp\left(\frac{-(5^2)}{2 \times 3.75^2}\right) = 0.0082 \text{ m} = 8.2 \text{ mm}$$

Similarly, the settlements at the other offsets are given in Table 12.3.

Table 12.3 Worked Example 12.2 calculated settlements.

	Offset from tunnel centreline y (m)				
	0	2.5	5	7.5	10
Settlement S (mm)	20	16	8	3	1

WORKED EXAMPLE 12.3   OPEN FACE TUNNEL IN CLAY WITH OVERLYING COARSE-GRAINED SOIL LAYER

The same tunnel in Worked Example 12.1 is to be constructed in stiff clay with an overlying sandy gravel layer.

The tunnel axis is 20 m below the ground surface. The sand layer is 8 m thick.

Based on case histories in similar geology using similar construction methods, the volume loss is predicted to be 1.5% and the trough width parameter K is 0.45 in the clay and 0.25 in the sandy gravel.

Calculate the maximum surface settlement S_{max}. Then calculate the settlement at $y = 5$ m, 10 m, 15 m and 20 m.

We first need to calculate the face area A:

$$A = \frac{\pi D^2}{4} = \frac{\pi \times 10^2}{4} = 78.54 \text{ m}^2$$

Then we can calculate the volume loss V_s using Equation 12.4:

$$V_s = V_l A = 0.015 \times 78.54 = 1.178 \text{ m}^2$$

The trough width is given by:

$$i = K_1 z_1 + K_2 z_2 = 0.45 \times 12 + 0.25 \times 8 = 5.4 + 2 = 7.4 \text{ m}$$

The maximum settlement S_{max} is given by Equation 12.2:

$$S_{max} = \frac{V_s}{\sqrt{2\pi} . i} = \frac{1.178}{\sqrt{2\pi} \times 7.4} = 0.06351 \text{ m} = 63.5 \text{ mm}$$

The settlement at $y = 5$ m is given by Equation 12.1:

$$S = S_{max} \exp\left(\frac{-y^2}{2i^2}\right) = 0.052 \times \exp\left(\frac{-(5^2)}{2 \times 7.4^2}\right) = 0.05055 \text{ m} = 50.6 \text{ mm}$$

Similarly, the settlements at the other offsets are given in Table 12.4.

Table 12.4 Worked Example 12.3 calculated settlements.

	Offset from tunnel centreline y (m)				
	0	5	10	15	20
Settlement S (mm)	64	51	26	8	2

The effect of the overlying sand layer is to make the surface settlement trough width narrower, from 9 m in Worked Example 12.1 to 7.4 m in this example. This causes a higher maximum settlement and a steeper trough, as shown in the comparison of the two troughs in Figure 12.2.

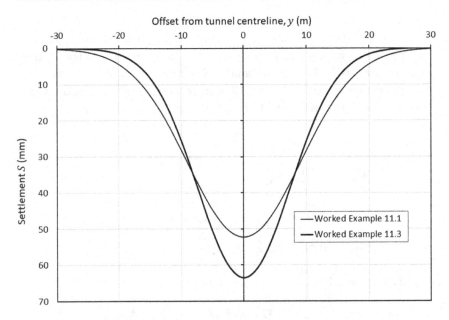

Figure 12.2 Settlement troughs from Worked Examples 12.1 and 12.3.

12.2 ESTIMATING VOLUME LOSS

Short-term volume loss arises from several sources in TBM-driven tunnels. There is ground movement towards the face of the tunnel, which will depend on support pressure and unsupported length. There is always an overcut at the front of the shield that allows the shield to steer, either created by an overcutting bead when the profile is cut by the shield itself, or by the position of the cutters on the cutterhead. This means that there will be ground closure around the shield and tailskin, adding to the volume loss. In closed face TBMs, this closure around the shield may be reduced if bentonite slurry is pumped in and pressurised around the shield. As the TBM is advanced, either the ring will be expanded against the ground using a wedgeblock segment, or grout will be introduced into the annulus between the segmental lining and the ground. The pressure of the ring expansion or the grout pressure relative to the in situ stress in the ground will also contribute to volume loss, as will any unfilled voids due to overbreak or incomplete grouting. Finally, there is deformation of the lining under the ground load, which allows more volume loss to occur.

For conventional soft ground tunnels lined with shotcrete, short-term volume loss arises from ground movement towards the face and then by deformation of the lining. Since the lining is loaded at early age, deformation of the lining will typically be greater than for a precast concrete segmental lining.

12.2.1 Estimating short-term volume loss in stiff clays

For most tunnels in clay there will be a long-term increase in volume loss. This will be fully discussed in Section 12.4. In this section we will only discuss 'short-term' volume loss, i.e. when the tunnel face has advanced sufficiently far past the section in question that ground movements have stabilised, but in a sufficiently short time period that undrained behaviour can be assumed.

The magnitude of ground movements towards a tunnel heading, and hence the volume loss, is dependent on the stability ratio N (also known as the stability number). The stability ratio was introduced in Chapter 2, and is given by:

$$N = \frac{\gamma z_0 + \sigma_s - \sigma_t}{c_u} \tag{12.5}$$

N is the stability ratio, which is dimensionless
γ is the bulk unit weight of the soil in kN/m^3
z_0 is the depth to the tunnel axis in m
σ_s is the effect at tunnel level of a surcharge at the surface (for instance consisting of a stockpile of bulk materials, traffic loads, a flexible raft foundation or a body of water) in kPa
σ_t is an internal face pressure provided by compressed air, slurry in the head of a slurry TBM or earth pressure in an EPB machine in kPa
c_u is the undrained shear strength in kPa (for guidance on what this value should be, see Chapter 2)

The relationship between stability ratio and volume loss was first proposed by Schmidt (1969). This was further developed by Mair et al. (1981) who, based on centrifuge tests by Mair (1979), related volume loss to a parameter they called 'Load Factor' LF as follows:

$$LF = \frac{N}{N_c} \tag{12.6}$$

LF is the load factor, which is dimensionless
N is the stability ratio, which is dimensionless
N_c is the critical stability ratio, which is the stability ratio at collapse, also dimensionless

The critical stability ratio N_c was defined in Chapter 2 and is usually determined from the empirical design chart produced by Mair (1979) and Kimura & Mair (1981) based on centrifuge testing of tunnel headings in overconsolidated kaolin clay. This is shown again in Figure 12.3.

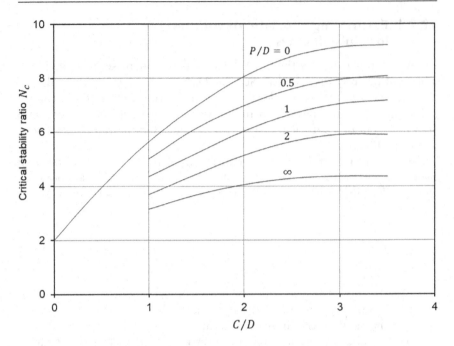

Figure 12.3 Design chart based on Mair's (1979) centrifuge tests (redrawn from Kimura & Mair, 1981).

P is the unsupported length of the idealised tunnel heading in m
C is the cover from the crown of the tunnel to the top of the clay in m
D is the excavated diameter of the tunnel in m

The load factor LF is the inverse of factor of safety and is equal to zero when the support pressure within the tunnel is exactly equal to the in situ stress in the ground and equal to 1 when the tunnel heading is on the point of collapse.

Case histories of volume losses measured above real tunnels in clay and their estimated load factors were compiled by Macklin (1999), who proposed an empirical formula to predict volume loss, as follows:

$$V_l(\%) = 0.23e^{4.4LF} \text{ for } LF \geq 0.2 \tag{12.7}$$

V_l is the volume loss as a percentage
LF is the load factor, defined in Equation 12.6

Load factors below 0.2 are not included because this would be outside the range of the centrifuge data. It is also the value of load factor below which the soil is essentially elastic.

The relationship between volume loss and load factor in Equation 12.7 is the central dashed line in Figure 12.4. Upper and lower bounds to the

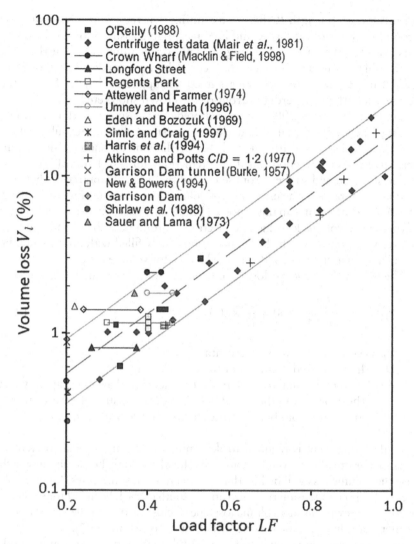

Figure 12.4 Load factor chart for volume loss prediction in overconsolidated clays (redrawn from Macklin, 1999).

case history data are also shown as solid lines. One of the outlying points from Eden & Bozozuk (1969) is not in overconsolidated clay but extremely sensitive Leda Clay, and the undrained shear strength was almost certainly overestimated by Macklin (1999).

Since we are concerned here with ground movements that are well within the ultimate limit state, then some judgement is required to determine the *P/D* ratio used for calculating the critical stability ratio and hence the load factor. Macklin (1999) recommended using the length of the shield for unsupported length P, thereby assuming that the ground will not close around it, stating that this would be a reasonable assumption for load factors less than 0.5. Where grouting does not occur immediately at the back of the shield, unsupported length P could be even longer. For shotcrete-lined tunnels, Macklin (1999) assumed unsupported length P was the advance length plus overdig in front of the crown shotcrete.

Dimmock & Mair (2007) compared predictions using Macklin's load factor approach with actual volume losses measured during the Jubilee Line Extension in London. They showed that predictions for shotcrete-lined tunnels were reasonably good, but that predictions for TBM tunnels could be underestimated and so they proposed a slightly modified approach where the load factor approach was used only for volume loss due to ground movements ahead of the face, using an unsupported length only based on excavation ahead of the shield. Volume loss due to convergence around the shield and any void between the shield and the lining not immediately filled with grout would be calculated separately and added to the load factor volume loss.

The volume loss due to closure of ground around the shield is given by:

$$V_l(\%) = \frac{\pi D \delta}{\pi D^2 / 4} \times 100 = \frac{4\delta}{D} \times 100 \tag{12.8}$$

V_l is the volume loss as a percentage
D is the excavated diameter of the tunnel in m
δ is the average annular overcut (i.e. the theoretical average gap between the ground and the shield before ground closure), plus any allowance for a void between the shield and the lining, in m

Note that judgement is required to determine the average annular overcut δ, because closure of the ground around the shield may only be partial, in which case the volume loss will be less than the overcut. In some cases, for example the Jubilee Line Extension running tunnels south of the lake in St James's Park, overbreak may result in a volume loss due to closure around the shield that is greater than the theoretical overcut (Dimmock & Mair, 2007).

An alternative approach is to use a cylindrical cavity expansion method, modelling the displacements of the soil around the shield. Mair & Taylor (1993) proposed a cylindrical cavity expansion equation for estimating displacements around a heading, which can be used for this purpose. It assumes axisymmetry and that the soil is elastic-perfectly plastic.

$$\frac{\delta}{r_0} = \frac{c_u}{2G} \left(\frac{r_0}{r} \right) \exp(N^* - 1) \tag{12.9}$$

δ is the radial displacement at radius r in m
r_0 is the radius of the cylindrical cavity, i.e. the radius of the excavation at the front of the shield, in m

c_u is the undrained shear strength in kPa

G is the shear modulus of the soil in kPa

r is the radius where we wish to calculate radial displacement in m

N^* is the stability ratio, where $N^* = \sigma_0 / c_u$

σ_0 is the initial total stress at the cavity boundary in kPa, usually assumed to be equal to the vertical total stress

c_u is the undrained shear strength in kPa

To find the closure around the shield, we take $r = r_0 = D/2$. So Equation 12.9 becomes:

$$\delta = \frac{c_u D}{4G} \exp(N^* - 1) \tag{12.10}$$

This value of δ can be used in Equation 12.8 to calculate volume loss. If Equation 12.10 gives a larger radial displacement than the average annular overcut, then take the average annular overcut as the value of δ in Equation 12.8.

Dimmock & Mair (2007) also propose that an estimate be made of the volume loss due to closure of the void between the shield and the lining. This depends on the construction method, and they say that if the void is grouted tight, or an expanded lining is expanded tightly against the ground to the full diameter of the shield, then it should be negligible.

WORKED EXAMPLE 12.4   ESTIMATING VOLUME LOSS FOR A SHOTCRETE TUNNEL IN OVERCONSOLIDATED CLAY

A shotcrete-lined tunnel with a face area A of 87 m² is to be constructed in stiff overconsolidated clay.

The centroid of the tunnel is 25 m below the ground surface. The excavation level of the crown of the tunnel is at 20 m below the ground surface.

Excavation ahead of the leading edge of the last shotcrete sprayed is 1.5 m.

The undrained shear strength of the clay is given by $c_u = 50 + 8z$ kPa, where z is the depth below the ground surface. The unit weight is $\gamma = 20$ kN/m³

Estimate the volume loss this tunnel will experience using the load factor approach.

We first need to calculate the equivalent diameter of the tunnel:

$$D = 2\sqrt{\frac{A}{\pi}} = 2\sqrt{\frac{87}{\pi}} = 10.525 \text{ m}$$

Cover $C = 20$ m, so $C/D = 20/10.525 = 1.900$ Unsupported length $P = 1.5$ m, so $P/D = 1.5/10.525 = 0.143$ Reading from the design chart in Figure 12.3, $N_c = 7.6$

The undrained shear strength varies with depth so we will conservatively take the value at $0.6z_0$ (see Section 2.2.2), i.e. at 15 m depth. Thus:

$$c_u = 50 + 8 \times 15 = 170 \text{ kPa}$$

Stability ratio is calculated using Equation 12.5:

$$N = \frac{\gamma z_0 + \sigma_s - \sigma_t}{c_u} = \frac{20 \times 25 + 0 - 0}{170} = 2.94$$

Therefore, load factor $LF = N/N_c = 2.94/7.6 = 0.39$.
Using Equation 12.7:

$$V_l(\%) = 0.23e^{4.4LF} = 0.23\exp(4.4 \times 0.39) = 1.28\%$$

WORKED EXAMPLE 12.5   ESTIMATING VOLUME LOSS FOR A TBM IN OVERCONSOLIDATED CLAY

An EPB TBM-driven tunnel with an excavation diameter of 6.5 m is to be constructed in very stiff overconsolidated clay overlain by a 5 m thick layer of Made Ground.

The tunnel axis is 18 m below the ground surface. The average face pressure to be applied (i.e. the value at axis level) is 100 kPa. Assume that no support pressure is applied around the shield.

The average annular overcut of the TBM is 40 mm. The rings are grouted immediately as they leave the tailskin, 7 m behind the cutterhead, with a grout pressure equal to the overburden pressure.

The Made Ground has a unit weight of $\gamma = 18$ kN/m³ and is assumed to have no cohesion.

The undrained shear strength of the clay is given by $c_u = 75 + 11z$ kPa, where z is the depth below the top of the clay. The unit weight is $\gamma = 20$ kN/m³. The shear modulus $G = 200c_u$.

Estimate the volume loss this tunnel will experience using the load factor approach for the tunnel face, with an allowance for closure around the shield.

Cover of clay $C = 18 - 6.5/2 - 5 = 9.75$ m, so $C/D = 9.75/6.5 = 1.5$; Unsupported length ahead of the shield $P = 0$ m, so $P/D = 0$; Reading from the design chart in Figure 12.3, $N_c = 7.0$.

The undrained shear strength varies with depth so we will conservatively take the value at 0.6z from the top of the clay (see Section 2.2.2), i.e. at 12.8 m depth, or 7.8 m depth from the top of the clay:

$$c_u = 75 + 11 \times 7.8 = 160.8 \text{ kPa}$$

The stability ratio is calculated using Equation 12.5:

$$N = \frac{\gamma z_0 + \sigma_s - \sigma_t}{c_u} = \frac{(18 \times 5 + 20 \times 13) + 0 - 100}{160.8} = \frac{250}{160.8} = 1.55$$

Therefore, load factor $LF = N/N_c = 1.55/7.0 = 0.22$.
   Using Equation 12.7:

$$V_l(\%) = 0.23e^{4.4LF} = 0.23\exp(4.4 \times 0.22) = 0.61\%$$

Next we will see if the ground closes around the shield due to elastoplastic deformation, using the cavity expansion plasticity solution from Mair & Taylor (1993), given in Equation 12.10:

$$\delta = \frac{c_u D}{4G} \exp(N^* - 1)$$

For this equation, it makes sense to use the undrained shear strength and shear modulus at tunnel axis level (13 m from top of clay):

$$c_u = 75 + 11z = 75 + 11 \times 13 = 218 \text{ kPa}$$

And:

$$G = 200c_u = 200 \times 218 = 43600 \text{ kPa}$$

We also need to calculate N^*:

$$N^* = \sigma_0/c_u$$

Now the full overburden pressure at axis level is given by:

$$\sigma_0 = 18 \times 5 + 20 \times 13 = 350 \text{ kPa}$$

Therefore:

$$N^* = \sigma_0/c_u = 350/218 = 1.61$$

So closure (radial displacement) around the shield is given by:

$$\delta = \frac{c_u D}{4G}\exp(N^* - 1) = \frac{218 \times 6.5}{4 \times 43600}\exp(1.61 - 1) = 0.015 \text{ m} = 15 \text{ mm}$$

This radial displacement is less than the average annular gap of 40 mm, so we will use it to calculate volume loss using Equation 12.8:

$$V_l(\%) = \frac{4\delta}{D} \times 100 = \frac{4 \times 0.015}{6.5} \times 100 = 0.92\%$$

Adding together the volume loss in front of the shield and around the shield gives a total volume loss of 1.53%. This is probably an overestimate because we have assumed an infinitely long cylinder when estimating ground closure around the shield, and also some of the radial deformation to equilibrium will have already occurred in front of the TBM so we are counting it twice. However, this is a better estimate than using Dimmock & Mair (2007)'s method, which assumes full closure of the annular overcut.

Since we know that the shield is not supporting the ground because the plasticity solution told us that the ground closure around the shield was less than the average annular gap, then we could also calculate the volume loss by using an unsupported length $P = 7$ m.

Using $P/D = 7/6.5 = 1.08$ and $C/D = 1.5$ in the design chart in Figure 12.3 gives $N_c = 5.2$.

The stability ratio remains the same, so load factor is now:

$$LF = N/N_c = 1.55/5.2 = 0.30$$

Using Equation 12.7:

$$V_l(\%) = 0.23e^{4.4LF} = 0.23\exp(4.4 \times 0.30) = 0.85\%$$

This is significantly less than the first method we used, which estimated a volume loss of 1.53%, but, as mentioned previously, is probably more accurate because the first method assumed an infinitely long unlined cylinder and is double-counting some of the radial deformation in the separate estimates of volume loss ahead of the TBM and around the shield.

There will be no volume loss due to a void between the shield and the lining, because grout is injected through the tailskin at a pressure equal to the overburden pressure. Also, much, if not all, of the radial convergence of the ground around the shield will be reversed by this grout pressure, so the total volume loss is a conservative estimate.

This example is intended to show that only a rough estimation of volume loss is possible, but using a variety of different tools and methods we can get a feel for what aspects of the construction method are important. In this case, the application of face pressure reduces the volume loss ahead of the cutterhead, but a significant proportion of the total volume loss is still caused by the ground closing around the shield.

12.2.2 Estimating volume loss in other soil types

Vu et al. (2016) provide a state-of-the-art review of methods of estimating volume loss at the face, around the shield, and at the tail due to grouting pressure. For volume loss at the face, they use the load factor approach described in the previous section for clays, but do not explain how to estimate volume loss at the face in drained soils even though they later present the results of this kind of calculation.

For volume loss along the shield, Vu et al. (2016) assume the annular overcut is filled with either bentonite flowing from the face, or with grout flowing from the tail. In the Netherlands, inert grouts are often used, so it does not cause problems if the grout flows forwards around the shield. For cementitious grouts this situation is strenuously avoided by employing brushes or plates on the outside of the tailskin, otherwise grout can build up around the shield and create a lot of friction.

For volume loss during grouting, Vu et al. (2016) recommend the use of a cavity expansion method proposed by Mair & Taylor (1993) for clay soils to model the cylindrical expansion or contraction caused by grout pressure relative to the overburden pressure. Again, they do not explain how this can be adapted from the undrained case to the drained case, though they do present results for sand.

Generally speaking, settlements above tunnels in sands and gravels with faces that are stable due to either cohesion or support pressure, will be small. When instability occurs, the failure is 'brittle' and ground movements increase quickly. Therefore, as long as face pressure is well controlled to maintain stability, or the ground has sufficient cohesion, volume loss in sand can be kept below 1.0% and in some cases reliably below 0.5%.

12.3 LONGITUDINAL AND TRANSIENT SURFACE SETTLEMENTS

In Section 12.1 we looked at short-term transverse settlements after construction of the tunnel. In effect, the transverse section we considered was well back from the face. In some situations, we may want to know the transient settlements as the tunnel face approaches and passes a point of interest. This was illustrated in Figure 1.6.

Attewell & Woodman (1982) showed how ground surface settlements at all points relative to the face could be estimated using a combination of a normal probability distribution function in the transverse direction and a normal cumulative distribution function in the longitudinal direction.

$$S = \frac{V_s}{\sqrt{2\pi}.i} \exp\left[\frac{-y^2}{2i^2}\right]\left\{G\left(\frac{x-x_i}{i}\right) - G\left(\frac{x-x_f}{i}\right)\right\} \qquad (12.11)$$

S is the vertical settlement at any point on the surface with coordinates (x, y), in m

V_s is the volume loss in m^2

i is the trough width or the transverse distance to the point of inflexion in m

y is the transverse distance from the centreline of the tunnel in m

x is the position along the centreline relative to the face of the tunnel in m

x_i is the tunnel starting point in m

x_f is the tunnel face position in m

G is the standard normal cumulative distribution function. In particular, $G(0) = 0.5$ and $G(\infty) = 1$. Tables of values may be found online or in statistics textbooks. In Microsoft Excel, use the NORMSDIST function.

Equation 12.11 can also be expressed in terms of the 'short-term' transverse settlement at the same offset but well back from the face, now denoted S_∞. This may be easier to calculate because we can determine the transverse settlements using Equation 12.1 and then multiply them by a function that varies depending on position relative to the face.

$$S = S_\infty\left\{G\left(\frac{x-x_i}{i}\right) - G\left(\frac{x-x_f}{i}\right)\right\} \qquad (12.12)$$

WORKED EXAMPLE 12.6   OPEN FACE SHOTCRETE-LINED TUNNEL IN HOMOGENEOUS CLAY

A shotcrete-lined tunnel is to be excavated in clay with a face area of 54 m^2. The tunnel axis is 15 m below the ground surface. Based on case histories in similar geology using similar construction methods, the volume loss is predicted to be 1% and the trough width parameter K is 0.5.

Calculate the maximum surface settlement above the face. Then calculate the settlement at $y = 5$ m, 10 m, 15 m and 20 m.

Calculate the same surface settlements at 10 m in front of the face, 5 m in front of the face, 5 m behind the face and 10 m behind the face. Assume the tunnel start position is $x_i = 0$ m and the tunnel face position is $x_f = 100$ m.

We will first calculate the volume loss V_s using Equation 12.4:

$$V_s = V_l A = 0.01 \times 54 = 0.54 \text{ m}^2$$

The trough width is given by Equation 12.3:

$$i = K z_0 = 0.5 \times 15 = 7.5 \text{ m}$$

Above the face, $x = x_f$, therefore:

$$\left(\frac{x - x_i}{i}\right) = \left(\frac{100 - 0}{7.5}\right) = 13.33$$

$$\left(\frac{x - x_f}{i}\right) = \left(\frac{100 - 100}{7.5}\right) = 0$$

Calculating, or looking up the standard normal cumulative distribution values gives:

$$G\left(\frac{x - x_i}{i}\right) = G(13.33) = 1$$

$$G\left(\frac{x - x_f}{i}\right) = G(0) = 0.5$$

The maximum settlement above the face is given by Equation 12.11:

$$S = \frac{V_s}{\sqrt{2\pi}.i} \exp\left[\frac{-y^2}{2i^2}\right]\left\{G\left(\frac{x - x_i}{i}\right) - G\left(\frac{x - x_f}{i}\right)\right\}$$

$$= 0.02872 \times \exp\left[\frac{-0}{112.5}\right]\{1 - 0.5\}$$

$$= 0.02872 \times 1\{0.5\} = 0.014 \text{ m} = 14 \text{ mm}$$

Now we can calculate the settlements along a transverse line above the face using the maximum settlement we just calculated. Firstly, for $y = 5$ m:

$$S = S_{max}\exp\left(\frac{-y^2}{2i^2}\right) = 0.014 \times \exp\left(\frac{-(5^2)}{2 \times 7.5^2}\right) = 0.012 \text{ m} = 12 \text{ mm}$$

Similarly, the settlements at the other offsets are given in Table 12.5.

Table 12.5 Worked Example 12.6 calculated settlements.

	Offset from tunnel centreline y (m)				
	0	5	10	15	20
Settlement S (mm)	14	12	6	2	0

The settlements at 5 m ahead of the face, 10 m ahead of the face, 5 m behind the face and 10 m behind the face are calculated using the following values in Table 12.6.

Table 12.6 Worked Example 12.6 longitudinal settlement factors.

	$\dfrac{x-x_i}{i}$	$G\left(\dfrac{x-x_i}{i}\right)$	$\dfrac{x-x_f}{i}$	$G\left(\dfrac{x-x_f}{i}\right)$
Above the face	13.33	1	0	0.5
5 m ahead of the face	14	1	0.67	0.75
10 m ahead of the face	14.67	1	1.33	0.91
5 m behind the face	12.67	1	−0.67	0.25
10 m behind the face	12	1	−1.33	0.09

Since the tunnel face is so far from the starting point, the function in the third column of the table is always equal to 1. The position relative to the face (the fifth column in the table) is going to be what modifies the settlement values. We can see that if we subtract these values from 1, we get smaller settlements ahead of the face, increasing as we pass the face and move behind the face, as we would expect from our understanding of ground movements as illustrated by Figure 1.6.

The full set of settlements is given in Table 12.7.

Table 12.7 Worked Example 12.6 longitudinal settlements.

	Offset from tunnel centreline y (m)				
Settlement S (mm)	0	5	10	15	20
10 m ahead of the face	3	2	1	0	0
5 m ahead of the face	7	6	3	1	0
Above the face	14	12	6	2	0
5 m behind the face	21	17	9	3	1
10 m behind the face	26	21	11	4	1

These transient surface settlements may also be presented graphically, as shown in Figure 12.5.

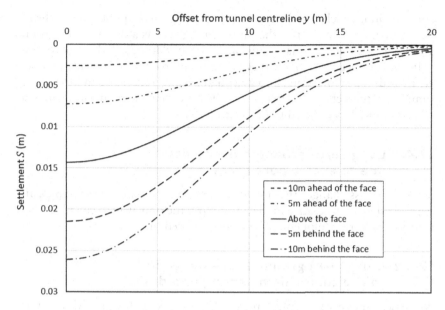

Figure 12.5 Worked Example 12.6 transverse surface settlement curves relative to face position.

It is usually assumed, as we have done so far, that the shape of the transverse settlement trough remains constant and that only the magnitude of the settlements increases as the tunnel face approaches and passes a transverse array. However, data presented by Hill & Stärk (2015, 2016) from Crossrail tunnels at Whitechapel, London, showed that the trough was wider ahead of the face, becoming narrower as the face passed the array, and then became wider again behind the face and into the long term. They also found that when a pilot tunnel was constructed first followed by an enlargement, the pilot tunnel would have a wider trough than the enlargement. The reasons for these effects are not known for certain, but may be due to the stress history of the soil and the changes in stress path direction the soil experiences as the tunnel advances.

12.4 LONG-TERM GROUND MOVEMENTS

As discussed in Chapter 1, long-term ground movements only occur in undrained soils. They can occur in drained soils if the groundwater level changes over the long term, but that is a straightforward 1D consolidation problem that we will not cover here because it is not caused by the tunnel construction.

There are two causes of long-term ground movements. The first is the excess pore pressures induced by volumetric changes and shearing during construction, which will dissipate over time to a final equilibrium or steady-state value. The second is if the tunnel lining acts as a drain on the ground around it, then this drainage will cause consolidation and thus settlement.

In low permeability clays, these long-term ground movements can take years or decades to diminish to a negligible rate. Therefore, modern urban tunnelling projects will often only allow settlement monitoring to stop when the rate has reduced to below 2 mm/year.

12.4.1 Long-term ground movements due to excess pore pressures

Since the stress changes that induce excess pore pressures are complex and depend on the constitutive model of the soil, the only way to calculate them is to use numerical modelling. This was covered in Section 7.6.2.

12.4.2 Long-term ground movements due to the tunnel lining acting as a drain

Numerical models by Wongsaroj et al. (2007 & 2013) and Mair (2008) demonstrated that in London Clay even slight changes in permeability of the lining or the ground can have a significant effect on long-term settlement magnitude and extent. Wongsaroj et al. (2007) also found that their numerical model matched the piezometer and extensometer results from the Jubilee Line Extension at St James's Park better when the London Clay was given a higher permeability in the horizontal direction compared to the vertical, which increased the lateral extent of the drainage effect.

Wongsaroj et al. (2013) defined a 'relative permeability index' RP, given as:

$$RP = \frac{k_t}{k_s} \cdot \frac{C}{t_L} \tag{12.13}$$

RP is the relative permeability index
k_t is the permeability of the lining in m/s
k_s is the permeability of the soil in m/s
C is the clay cover above the crown in m
t_L is the lining thickness in m

Laver et al. (2017) improved the model of Wongsaroj et al. (2013) to include a radial flow pattern, and derived the following equation for relative permeability index:

$$RP = \frac{Dk_t}{2k_s t_L} \ln\left(\frac{2C}{D} + 1\right) \tag{12.14}$$

k_s is now the 'average' permeability of the soil in m/s, where $k_s = \sqrt{k_v k_h}$
D is the tunnel diameter in m

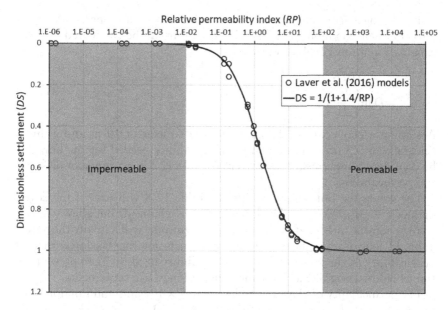

Figure 12.6 Relative permeability index and effect on long-term settlement due to consolidation (Laver et al., 2017 with permission from ASCE).

Results from Laver et al.'s 2D finite element modelling of long-term settlement at various values of *C/D* ratio and volume loss are shown in Figure 12.6. A curve with the following equation has been fitted to the data:

$$DS = \frac{1}{\left(1 + \dfrac{1.4}{RP}\right)}$$

(12.15)

DS is the 'dimensionless settlement', which is the long-term settlement above the centreline of the tunnel, normalised so that it lies on a scale of 0 to 1 where *DS* = 0 for a fully impermeable lining and *DS* = 1 for a fully permeable lining

Figure 12.6 shows that when the relative permeability index is less than 0.01, the ground-lining system behaves as though the lining were impermeable, and no long-term settlement due to drainage consolidation occurs, and when the relative permeability index is greater than 100, the ground-lining system behaves as though the lining were fully permeable and the maximum long-term settlement occurs. For relative permeability index values between 0.01 and 100, the ground is drained to a lesser extent and settlements will be intermediate. Remember that changes in mean total stress, and shear-induced dilation (in overconsolidated clays) and contraction (in normally consolidated clays) may also cause long-term heave or settlement due to consolidation processes, so even when *RP* < 0.01 there can still be long-term settlements.

The lining permeability required to achieve a relative permeability of 0.01 and hence exhibit impermeable behaviour can be calculated by rearranging Equation 12.14 as follows:

$$\frac{k_t}{k_s} = \frac{0.02t_L}{D.\ln\left(\frac{2C}{D}+1\right)} \tag{12.16}$$

Assuming a clay cover of 20 m, a tunnel diameter of 10 m and a lining thickness of 0.3 m, values for lining permeability needed for the lining to be considered impermeable are given in Table 12.8 for a range of different soil permeabilities.

The values for lining permeability in Table 12.8 will vary if the cover, diameter or thickness of the tunnel lining is changed, but they will not change significantly and the orders of magnitude will remain the same. Table 12.8 leads us to the slightly strange conclusion that the lower the soil permeability is, the lower the lining permeability needs to be to prevent the tunnel from acting like a drain and causing long-term settlements.

The permeability of clay is often less than 1×10^{-9} m/s and London Clay, for example, can have a permeability as low as 1×10^{-11} m/s (Wan & Standing, 2014), so the lining permeability needs to be at most 3.7×10^{-13} m/s and possibly as low as 3.7×10^{-15} m/s, which is approximately the permeability of a very good concrete without accounting for construction joints, shrinkage cracks or other imperfections. Therefore, if long-term settlements are to be avoided, then the tunnel probably also needs a waterproof membrane or a watertight secondary lining.

This method can be used to predict long-term settlements if the permeability of the lining is known. However, it is difficult to predict what the actual permeability of a tunnel lining will be, because it depends at least as much on joints, cracks and poor workmanship as on its inherent material properties.

If long-term settlements have been measured, then the lining permeability may be estimated 'a posteriori' (after the fact) and a database of values may be created for use in preliminary design on future projects. Laver et al. (2017) did this for the Heathrow Express Trial Tunnel (Bowers et al., 1996) and the Jubilee Line Extension at St James's Park (Nyren, 1998), and the

Table 12.8 Values of lining permeability required for lining to be considered impermeable, based on Wongsaroj et al. (2013) equation for relative permeability.

Soil permeability	Lining permeability required for RP = 0.01
1×10^{-7} m/s	3.7×10^{-11} m/s
1×10^{-8} m/s	3.7×10^{-12} m/s
1×10^{-9} m/s	3.7×10^{-13} m/s
1×10^{-10} m/s	3.7×10^{-14} m/s
1×10^{-11} m/s	3.7×10^{-15} m/s

lining permeabilities were approximately 3×10^{-11} m/s for the sprayed concrete lining of the Heathrow Express Trial Tunnel and 5×10^{-10} m/s for the expanded wedgeblock precast concrete segmental lining of the Jubilee Line Extension. Both tunnels were in London Clay, and so far there have not been any case studies analysed in this way in other types of clay.

12.5 SUBSURFACE GROUND MOVEMENTS

Subsurface settlements have also been found to follow a Gaussian settlement trough, with the trough width becoming narrower with depth. In undrained soils this can lead to some reasonably straightforward prediction methods, but in drained soils it is a little more complicated.

12.5.1 Subsurface ground movements in undrained soils

Due to the assumption of constant volume behaviour, the volume loss of a subsurface settlement trough will be equal to the volume loss at the surface. Therefore, we only need to know the trough width in order to be able to calculate the subsurface settlements.

O'Reilly & New (1982) only wrote about surface settlements, but for tunnels of different depths they found a linear correlation between tunnel depth and trough width. Assuming that the presence of the ground surface itself has little effect, then we could adapt their equation to calculate subsurface settlements:

$$i = K(z_0 - z) \tag{12.3}$$

i is the trough width or distance to the point of inflexion in m
K is the trough width parameter
z_0 is the depth to axis of the tunnel under construction in m
z is the depth to the point under consideration in m, where $z \leq z_0$. At the surface, $z = 0$ and therefore $i = Kz_0$

By using this equation, we are effectively assuming that the subsurface settlement trough width is equal to K multiplied by the height above the tunnel, where K is a constant.

For tunnels in clays, Mair et al. (1993) looked at case history and centrifuge test data of subsurface settlements and observed that the trough width could be wider close to the tunnel than predicted by Equation 12.3. Effectively, K could increase with depth. They plotted i/z_0 against z/z_0 and fitted a straight line to the data. This is shown in Figure 12.7 with more recent case studies added from Clayton et al. (2006), Cooper et al. (2002), Jones (2010), New & Bowers (1994), Nyren (1998), Standing & Selman (2001) and Wan et al. (2017b). The measurements by Cooper et al. (2002), Jones (2010) and Standing & Selman (2001) are not from instruments in

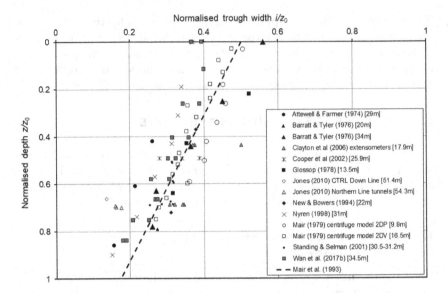

Figure 12.7 Subsurface trough width variation with depth, normalised to tunnel axis depth. Tunnel depths are given in square brackets.

boreholes, but from monitoring of third party tunnels above the tunnel being constructed. These data appear to follow the same settlement pattern as the other data, indicating that the tunnels are reasonably flexible (due to the joints in the lining) and follow the ground deformations.

The equation of the straight dashed line in Figure 12.7 was proposed by Mair et al. (1993), and is:

$$\frac{i}{z_0} = 0.175 + 0.325\left(1 - \frac{z}{z_0}\right) \tag{12.17}$$

Substituting for i using the relationship in Equation 12.3 gives:

$$K = \frac{0.175 + 0.325\left(1 - \frac{z}{z_0}\right)}{1 - \frac{z}{z_0}} \tag{12.18}$$

Note that Equation 12.17 ensures that at the surface $K = 0.5$. Mair et al. (1993) base this assumption on Rankin (1988)'s review of case histories of surface settlements in the UK and overseas, where he stated this value was appropriate for "initial practical estimation purposes". This is not appropriate for soft clays, where K at the surface can be greater than 0.5.

It is easier to see what is going on if K is plotted on a graph against normalised depth z/z_0. This is shown in Figure 12.8.

An observation we can make about Figure 12.7 and Figure 12.8 is that although Mair et al.'s relationship appears to fit the data reasonably well,

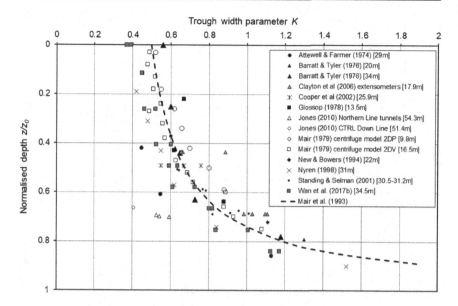

Figure 12.8 Subsurface trough width parameter K variation with normalised depth z/z_0. Tunnel depths are given in square brackets.

the deeper tunnels tend to be to the left and the shallower tunnels tend to be to the right of it. Jones (2010) speculated that by using the normalised depth z/z_0 no account is made for the real distances involved. For deep tunnels, Mair et al.'s equation would predict high K values for a much larger distance above the tunnel than for a shallow tunnel. Also, the value of K at the surface must be equal to 0.5, regardless of the depth of the tunnel. Lake et al. (1996) noted that for surface settlements the trough width parameter may be smaller than might be predicted when the tunnel is deeper than 20 m, and it does seem to be the case that deeper tunnels tend to have a K value at the surface less than 0.5.

Jones (2010) therefore plotted K against height above the tunnel $(z_0 - z)$. This is shown in Figure 12.9 with more recent data from Crossrail (Wan et al., 2017b) added, and the logarithmic curve fit adjusted to take account of the new data. The advantage of this method is that surface settlements may also be plotted on the same graph and the value of K at the surface can vary depending on the depth of the tunnel and does not always have to be constant.

The logarithmic relationship in Figure 12.9 is:

$$K = -0.261\ln(z_0 - z) + 1.286 \qquad (12.19)$$

In Figure 12.9 we can see that there is a lot of scatter in the data in the first 10 m above tunnel axis level and this is probably due to the effect of tunnel diameter on the trough width so close to the tunnel. This is also present in Figure 12.8 but is less evident due to the way the data is presented. Figure 12.10 presents the same data as Figure 12.9 but without the log scale.

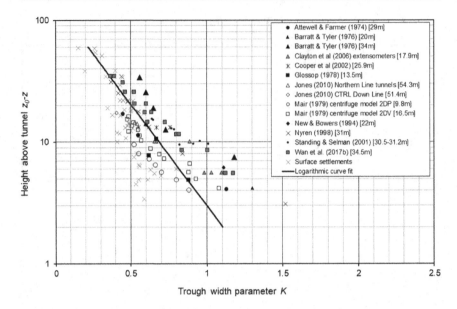

Figure 12.9 Trough width parameter *K* variation with height above tunnel axis ($z_0 - z$) on a logarithmic scale. Subsurface settlements are from case histories shown in the legend. Surface settlements are from Mair & Taylor (1997), Jones et al. (2008), Clayton et al. (2006), Jones (2010) and Wan et al. (2017a).

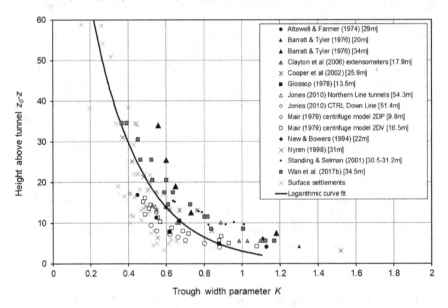

Figure 12.10 Trough width parameter *K* variation with height above tunnel axis ($z_0 - z$). Subsurface settlements are from case histories shown in the legend. Surface settlements from Mair & Taylor (1997), Jones et al. (2008), Clayton et al. (2006), Jones (2010) and Wan et al. (2017a).

There is clearly more work to be done to improve the meta-analysis, since there is still quite a large amount of scatter in the data. It is likely that the cover, the tunnel diameter and the presence of non-cohesive superficial deposits above the clay are factors, and possibly also the construction method and the volume loss.

WORKED EXAMPLE 12.7   SUBSURFACE SETTLEMENTS ABOVE A TUNNEL IN HOMOGENEOUS CLAY

A TBM with an excavated diameter of 8.5 m is to be advanced in clay. The tunnel axis is 35 m below the ground surface. Based on case histories in similar geology using similar construction methods, the volume loss is conservatively predicted to be 1%.

Calculate the trough width parameter K at 10 m, 15 m, 25 m and 35 m above tunnel axis, using three different methods:

1. Assuming K has a constant value of 0.5.
2. Using Mair et al.'s empirical relationship in Equation 12.18.
3. Using the logarithmic empirical relationship in Equation 12.19.

Then calculate the maximum settlement at each level.

We will first calculate the trough width parameter K and then trough width i using each of the methods:

Method 1:

K is constant, and i is given by $i = K(z_0 - z)$, giving the values in Table 12.9.

Knowing that the percentage volume loss is 1%, we can calculate the volume loss V_s from the excavated area as follows:

$$V_s = 1\% \times \pi \left(\frac{D^2}{4} \right) = 0.01 \times \pi \times \frac{8.5^2}{4} = 0.567 \ \text{m}^2$$

The maximum settlement can be calculated from the volume loss V_s and the trough width i by rearranging Equation 12.2:

$$S_{max} = \frac{V_s}{\sqrt{2\pi}.i}$$

The S_{max} values are given in Table 12.9. They reduce as height above the tunnel increases because the settlement trough is getting wider while the volume loss is constant.

Table 12.9 Worked Example 12.7 Method 1 calculation.

$z_0 - z$	10 m	15 m	25 m	35 m
K	0.5	0.5	0.5	0.5
i	5 m	7.5 m	12.5 m	17.5 m
S_{max}	45.3 mm	30.2 mm	18.1 mm	12.9 mm

Method 2:

For this method we are using Mair et al.'s empirical relationship in Equation 12.18:

$$K = \frac{0.175 + 0.325\left(1 - \frac{z}{z_0}\right)}{1 - \frac{z}{z_0}}$$

Table 12.10 shows the steps in the calculation row-by-row. The question gave the height above the tunnel axis of each point $z_0 - z$, so we first convert that to depth z, then we can calculate the relative depth z/z_0. Then K is calculated using Equation 12.18 (shown above). Then i and S_{max} are calculated as for Method 1.

Table 12.10 Worked Example 12.7 Method 2 calculation.

$z_0 - z$	10 m	15 m	25 m	35 m
z	25 m	20 m	10 m	0 m
z/z_0	0.714	0.571	0.286	0
K	0.938	0.733	0.570	0.5
i	9.38 m	11.00 m	14.25 m	17.50 m
S_{max}	24.1 mm	20.6 mm	15.9 mm	12.9 mm

Compared to Method 1, using Mair et al.'s equation has resulted in higher values of K, which increase closer to the tunnel, except for at the surface where in both cases $K = 0.5$. The higher values of K result in higher values of i in the subsurface. The wider trough width results in lower maximum settlement S_{max}, because the same trough volume is spread over a wider area.

Method 3:

For this method we are using the logarithmic empirical relationship in Equation 12.19, which is:

$$K = -0.261\ln(z_0 - z) + 1.286$$

Table 12.11 shows the calculation row-by-row.

Table 12.11 Worked Example 12.5 Method 3 calculation.

$z_0 - z$	10 m	15 m	25 m	35 m
K	0.685	0.579	0.446	0.358
i	6.85 m	8.69 m	11.15 m	12.53 m
S_{max}	33.0 mm	26.1 mm	20.3 mm	18.1 mm

The subsurface trough width i values are greater than for Method 1, but less than for Method 2, because this is a relatively deep tunnel. For a shallow tunnel, it could be the other way round.

At the surface, the trough width is narrower than predicted using either Method 1 or Method 2, because in this case $K < 0.5$, and for Method 1 and 2 $K = 0.5$ at the surface.

The logarithmic empirical relationship will give a surface trough width parameter $K < 0.5$ for tunnel depths greater than about 20 m, but will give values of K greater than 0.5 for tunnel depths less than 20 m.

12.5.2 Subsurface ground movements in drained soils

Much less is known about subsurface ground movements in drained coarse-grained soils. This is because the ground behaviour is much more complex. It cannot be assumed to be constant volume, and different regions of the ground will dilate or contract, depending on the soil's relative density and the current level of strain. Also, as we saw in Chapters 2 and 3, localised shear bands can form in drained soils such that the behaviour can approximate to sliding blocks at limit state, whereas failure in undrained soils is progressive and will be bounded by wide regions of shear that are better described by velocity fields.

For these reasons, trough width in drained soils will depend on the level of ground deformation that has occurred. In effect, for a given soil, trough width varies with tunnel volume loss, tending to get narrower as tunnel volume loss increases.

This behaviour is illustrated in Figure 12.11, which shows the variation of trough width parameter K with depth ratio, where z is any depth below the ground surface and z_0 is the depth of the tunnel axis, so $z/z_0 = 0$ is at the ground surface and $z/z_0 = 1$ is at tunnel axis level.

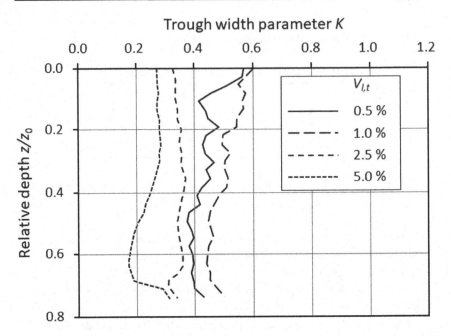

Figure 12.11 Variation of trough width parameter K with depth ratio z/z_0 in dry sand (redrawn from Marshall, 2009).

Figure 12.11 also shows that, for tunnels in sand, trough width parameter K tends to remain approximately constant or even decrease with depth, which is the opposite of what occurs in clay.

Further work by Franza et al. (2019) investigated the influence of initial density of the sand, and found that denser sand formed a ground arch more readily, and closer to the tunnel, whereas loose sand, if it could form one at all, formed a ground arch further away and was more prone to the chimney-like vertical displacements above the tunnel crown that result in a narrower settlement trough shape.

Another important thing to remember is that when we are discussing ground movements in drained soils, because the behaviour is not constant volume, the volume loss at any level in the ground can be different to the volume loss at the tunnel. For loose and medium dense sands, it is usually the case that the volume loss at the surface is significantly higher than volume loss at the tunnel, whereas for dense sands the volume loss at the surface is similar to that at depth, as shown in Figure 12.12 (Franza et al., 2020).

Predicting subsurface settlement in sand is therefore quite difficult. However, as was explored in Chapter 3, only relatively small values of effective support pressure are required to keep deformations and hence volume loss quite small. Therefore, as long as face pressure is well controlled to maintain stability, volume loss in sand can be kept below 1.0% and in some cases reliably below 0.5%. At these magnitudes of ground movements, in

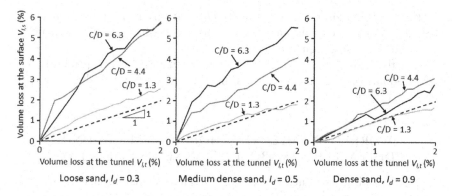

Figure 12.12 Relationship between volume loss at the tunnel and volume loss at the surface, at different C/D ratios and relative densities, in dry sand (redrawn from Franza et al., 2020).

dense sands the behaviour may be close to constant volume, whereas in looser sands the TBM driver may have to restrict ground movements at the tunnel to a much greater degree because the volume loss at the surface will be amplified, particularly for deeper tunnels (c.f. Figure 12.12).

Numerical modelling of dilatancy effects and shear band formation is very challenging. Yang (2017) implemented a version of the 'Norsand' model, which could model stress dilatancy, in PLAXIS3D, and compared it to the built-in 'Hardening Soil Model', which could not. He found that there was very little difference at 0.5% volume loss, but at 1% volume loss and higher the surface and subsurface settlement curves diverged. Therefore, fairly standard models can be used to predict ground movements as long as deformations are kept small, for example if a closed-face TBM is used with good control of face pressure.

12.6 HORIZONTAL GROUND MOVEMENTS

Up to now we have focussed on vertical ground movements, but horizontal ground movements also occur due to tunnelling and they can be just as important to surface or subsurface structures.

If we assume that ground movements are constant volume and directed towards the tunnel axis (O'Reilly & New, 1982), then it follows that horizontal displacements are given by:

$$H_{(y,z)} = \frac{-y}{(z_0 - z)} S_{(y,z)} \tag{12.20}$$

$H_{(y,z)}$ is the horizontal displacement at transverse offset y and depth z
$S_{(y,z)}$ is the vertical displacement (settlement) at transverse offset y and depth z

In this way, horizontal displacements $H_{(y,z)}$ at any position (y,z) may be calculated by first calculating a vertical displacement $S_{(y,z)}$ at the same position and then using Equation 12.20.

Note that O'Reilly & New (1982) use z rather than $(z_0 - z)$ in Equation 12.20 (their equation (7)), but define z as 'height above the axis', therefore it is equivalent to $(z_0 - z)$ here.

WORKED EXAMPLE 12.8   HORIZONTAL GROUND DISPLACEMENTS ABOVE A TUNNEL IN STIFF CLAY

A TBM with an excavated diameter of 8 m is to be driven through stiff clay. The tunnel axis is 20 m below the ground surface. Based on case histories in similar geology using similar construction methods, the volume loss is conservatively predicted to be 1.2% and the trough width parameter $K = 0.5$. Calculate the horizontal displacements at the ground surface.

We will first calculate the volume loss:

$$V_s = V_l A = V_l \pi \frac{D^2}{4} = 0.012 \times \pi \times \frac{8^2}{4} = 0.603186 \text{ m}^2$$

Next we need to calculate the trough width at the surface using Equation 12.3:

$$i = Kz_0 = 0.5 \times 20 = 10 \text{ m}$$

Then we can calculate the maximum settlement by rearranging Equation 12.2:

$$S_{max} = \frac{V_s}{\sqrt{2\pi}.i} = \frac{0.603186}{\sqrt{2\pi} \times 10} = 0.024064 \text{ m} = 24.1 \text{ mm}$$

Now we can calculate the vertical settlements using Equation 12.1:

$$S = S_{max} \exp\left(\frac{-y^2}{2i^2}\right)$$

The distance at which settlements usually become negligible is at a transverse offset $y = \pm 3i$. Therefore, we will calculate vertical settlements S in a spreadsheet with a range of y from $-3i$ to $3i$. An example calculation of S is given below for $y = -10$ m:

$$S = 0.024064 \exp\left(\frac{-(-10)^2}{2 \times 10^2}\right) = 0.024064 \exp(-0.5)$$

$$= 0.024064 \times 0.606531 = 0.0146 \text{ m} = 14.6 \text{ mm}$$

A graph of S from the spreadsheet calculation is shown in Figure 12.13.

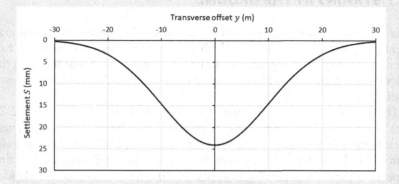

Figure 12.13 Worked Example 12.8 vertical settlements.

Next we will calculate horizontal displacements from the vertical settlements, using Equation 12.20. An example, again for $y = -10$ m, is given below:

$$H_{(y,z)} = \frac{-y}{z_0 - z} S_{(y,z)}$$

$$H_{(-10,0)} = \frac{-10}{20} \times 0.0146 = 0.0073 \text{ m} = 7.3 \text{ mm}$$

The horizontal displacements calculated in the spreadsheet are shown in Figure 12.14. Note that the displacements are positive on the left-hand side and negative on the right-hand side, as the ground is moving towards the tunnel centreline. Horizontal displacement is zero above the tunnel centreline, as the ground is moving inwards from both sides and there must be symmetry. The maximum horizontal displacement occurs at the point of inflexion of the settlement trough, i.e. at $y = -i$ and the minimum likewise occurs at $y = i$.

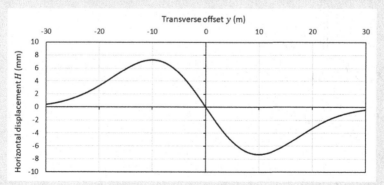

Figure 12.14 Worked Example 12.8 horizontal displacements.

12.7 STRAINS IN THE GROUND

The Gaussian settlement trough may also be used to derive settlement gradients, curvatures and strains, which are needed to predict the impact of tunnelling on utilities pipelines or buildings. This section will focus on the calculation of these derivatives. They will be used in Chapter 13 when estimating building damage.

The horizontal ground strain may be found by differentiating Equation 12.20 with respect to y. The simplest way to do this is to manually calculate changes in horizontal displacement along 1 m segments in a spreadsheet. The more elegant method is to derive an equation, which is not as straightforward as it first appears. We start by expanding Equation 12.20, because there are some hidden functions of y we need to account for:

$$H_{(y,z)} = \frac{-y}{z_0 - z} S_{(y,z)} = \frac{-y}{z_0 - z} S_{(max,z)} \exp\left(\frac{-y^2}{2i^2}\right) \tag{12.21}$$

$H_{(y,z)}$ is the horizontal displacement at transverse offset y and depth z
$S_{(y,z)}$ is the vertical displacement (settlement) at transverse offset y and depth z
$S_{(max,z)}$ is the maximum settlement at depth z
y is the transverse offset from the tunnel centreline
z is the depth below the ground surface
z_0 is the depth to the tunnel axis

Now we have to differentiate the following expression:

$$\varepsilon_h = \frac{dH_{(y,z)}}{dy} = \frac{d}{dy}\left[\frac{-y}{z_0 - z} S_{(max,z)} \exp\left(\frac{-y^2}{2i^2}\right)\right] \tag{12.22}$$

Since Equation 12.22 contains a product of two terms with y in them, we need to use the product rule, which states that:

$$\frac{duv}{dy} = u\frac{dv}{dy} + v\frac{du}{dy} \tag{12.23}$$

Let $\varepsilon_h = d(uv)/dy$, where:

$$u = \frac{-y}{z_0 - z} S_{(max,z)} \tag{12.24}$$

and:

$$v = \exp\left(\frac{-y^2}{2i^2}\right) \tag{12.25}$$

The derivative du/dy is given by:

$$\frac{du}{dy} = \frac{d}{dy}\left[\frac{-y}{z_0 - z}S_{(max,z)}\right] = \frac{-S_{(max,z)}}{z_0 - z} \tag{12.26}$$

The derivative dv/dy is given by:

$$\frac{dv}{dy} = \frac{d}{dy}\left[\exp\left(\frac{-y^2}{2i^2}\right)\right] = -\frac{1}{i^2}\cdot y\cdot\exp\left(\frac{-y^2}{2i^2}\right) \tag{12.27}$$

Therefore, given that $\varepsilon_h = duv/dy$, then:

$$\varepsilon_h = \left(\frac{-y}{z_0 - z}S_{(max,z)}\right)\cdot\left[-\frac{1}{i^2}\cdot y\cdot\exp\left(\frac{-y^2}{2i^2}\right)\right] + \exp\left(\frac{-y^2}{2i^2}\right)\cdot\left(\frac{-S_{(max,z)}}{z_0 - z}\right) \tag{12.28}$$

Which simplifies to:

$$\varepsilon_h = \left(\frac{-S_{(max,z)}}{z_0 - z}\right)\cdot\exp\left(\frac{-y^2}{2i^2}\right)\cdot\left[1 - \frac{y^2}{i^2}\right] \tag{12.29}$$

The horizontal strains calculated using Equation 12.29 for the tunnel in Worked Example 12.8 are shown in Figure 12.15. As one would expect from the distribution of horizontal displacements (c.f. Figure 12.14), there is a maximum compressive strain above the tunnel centreline and the strain becomes tensile outside the points of inflexion (the points of inflexion of the Gaussian settlement trough, not of the horizontal strain). This latter effect is because the maximum inward horizontal displacement occurs at

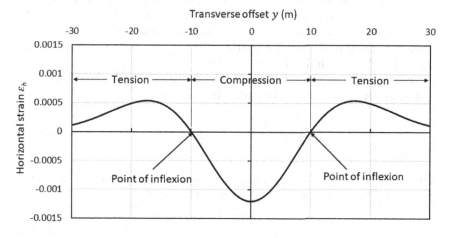

Figure 12.15 Worked Example 12.8 horizontal strains.

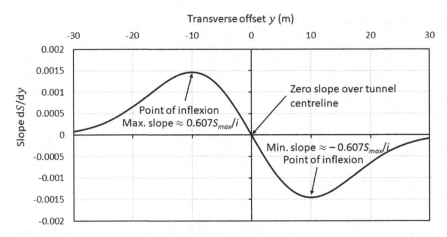

Figure 12.16 Worked Example 12.8 ground surface slope.

the points of inflexion; where horizontal displacement is increasing there is tension, and where it is decreasing there is compression.

The change in slope θ of the ground surface, or of a subsurface transverse line if we are considering the effect of tunnelling on a basement or underground utility, can be found by differentiating the vertical settlement (Equation 12.1) with respect to y.

$$\theta = \frac{dS}{dy} = \frac{d}{dy} S_{max} \exp\left(\frac{-y^2}{2i^2}\right) = \frac{-y}{i^2} S_{max} \exp\left(\frac{-y^2}{2i^2}\right) \tag{12.30}$$

The change in slope of the ground surface, calculated using Equation 12.30 for the tunnel in Worked Example 12.8, is shown in Figure 12.16. The maximum and minimum slope occur at the points of inflexion of the Gaussian settlement trough, i.e. at a transverse offset $y = \pm i$. There is zero slope over the tunnel centreline. Often slope is expressed as a fraction, for example 0.002 would be written as 1/500.

Curvature of the ground surface may also be of interest. This is the second derivative of vertical settlement with respect to y.

$$\frac{d^2 S}{dy^2} = \frac{d}{dy} \frac{-y}{i^2} S_{max} \exp\left(\frac{-y^2}{2i^2}\right) \tag{12.31}$$

We need to use the product rule again (Equation 12.23). Let $d^2S/dy^2 = C.d(uv)/dy$, where C is a constant given by:

$$C = \frac{-S_{max}}{i^2} \tag{12.32}$$

Now let:

$$u = y \tag{12.33}$$

and:

$$v = \exp\left(\frac{-y^2}{2i^2}\right) \tag{12.34}$$

Therefore:

$$\frac{du}{dy} = 1 \tag{12.35}$$

and:

$$\frac{dv}{dy} = -\frac{y}{i^2}\exp\left(\frac{-y^2}{2i^2}\right) \tag{12.36}$$

Now we use the product rule equation (Equation 12.23) to find the curvature:

$$\frac{d^2S}{dy^2} = \frac{-S_{max}}{i^2}\left\{ y \times \left[-\frac{y}{i^2}\exp\left(\frac{-y^2}{2i^2}\right)\right] + 1 \times \exp\left(\frac{-y^2}{2i^2}\right)\right\} \tag{12.37}$$

This simplifies to:

$$\frac{d^2S}{dy^2} = \frac{S_{max}}{i^2} \cdot \exp\left(\frac{-y^2}{2i^2}\right) \cdot \left[\frac{y^2}{i^2} - 1\right] \tag{12.38}$$

The curvature of the ground surface, calculated using Equation 12.38 for the tunnel in Worked Example 12.8, is shown in Figure 12.17.

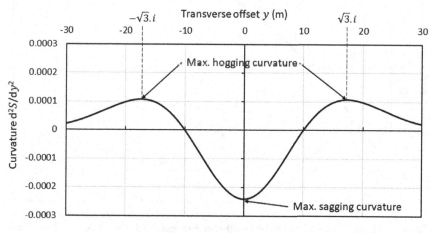

Figure 12.17 Worked Example 12.8 ground surface curvature.

The maximum sagging curvature, which is perhaps obvious if you look back at Figure 12.13 showing the settlement trough, occurs over the tunnel centreline at $y = 0$. The maximum hogging curvature occurs at $y = \pm\sqrt{3}.i$.

In Figure 12.17, sagging curvature is negative because in the coordinate system we have chosen to use the vertical axis z is pointing downwards so it represents a negative change in slope. Hogging curvature is positive because it represents a positive change in slope.

12.8 GROUND MOVEMENTS DUE TO SHAFT CONSTRUCTION

Until recently, there were very few case studies of ground movements due to shaft construction. This may be because shafts were rarely built close enough to adjacent buildings for ground movements to be a concern. However, nowadays it is increasingly common for deep, large diameter shafts to be constructed in urban areas close to existing buildings or utilities.

New & Bowers (1994) analysed data from an access shaft at Heathrow, near London, UK. Settlements and horizontal displacements due to shaft excavation were measured along two radial lines labelled 'S' and 'T', as shown in Figure 12.18. The 11 m diameter shaft was constructed by caisson-sinking and then underpinning to

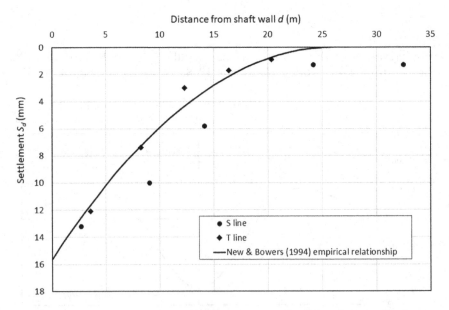

Figure 12.18 Settlements caused by Heathrow Express Trial Tunnel Access Shaft construction (redrawn from New & Bowers, 1994).

16 m depth, and then excavated in 1 m steps and lined with shotcrete to 26 m below ground level.

Also shown on Figure 12.18 is an empirical relationship proposed by New & Bowers (1994) between distance from the shaft wall d and the settlement S_d, which has the following equation:

$$S_d = \frac{\alpha.(H-d)^2}{H}$$

(12.39)

S_d is the settlement at distance d from the shaft wall in m
α is an empirical constant
H is the depth of the shaft excavation in m
d is the distance from the shaft wall in m

New & Bowers (1994) found the best fit for the empirical constant α was 0.0006. This corresponds to a maximum settlement at the shaft wall equal to 0.06% of the depth, which they said was consistent with field data for propped retaining structures in stiff fissured clays. However, this relationship probably only holds true for shafts constructed in a similar manner and in similar ground conditions to the one at Heathrow.

Faustin et al. (2018a) compared the settlements measured during excavation of the 39 m deep Crossrail Limmo Peninsular Auxiliary Shaft, constructed with sheet piles and then a shotcrete lining, with Equation 12.39 and found that the settlements close to the shaft wall were significantly underpredicted. They attributed this to the much larger diameter, which was 28 m compared to the 11 m Heathrow Access Shaft in New & Bowers (1994). Equation 12.39 does not account for shaft diameter. In contrast, the Limmo Peninsular Main Shaft in East London, of similar size to the Auxiliary Shaft but with a circular diaphragm wall installed before excavation began, had a maximum settlement close to the shaft extrados of approximately 15 mm, approximately 10 mm of which was due to dewatering and only about 5 mm was due to shaft excavation. Similarly, Schwamb et al. (2016) reported that Thames Water's Abbey Mills Shaft F in East London, 30 m diameter and 68 m deep and also of diaphragm wall construction, caused maximum surface settlements of less than 4 mm.

Therefore, it is likely that settlements due to shaft construction depend on the following factors:

- shaft depth H
- shaft diameter D
- ground conditions
- construction method, in particular the stiffness of the lining and whether it is installed before excavation (e.g. diaphragm walls) or progressively (e.g. underpinning using a segmental lining or shotcrete)

Faustin et al. (2018b) showed that pre-installed linings, such as diaphragm wall panels or secant piles, could induce as much or more settlement during installation than the subsequent excavation. These settlements were of the order of $0.02\%H$ (0.02% of the depth of the shaft), which was a similar magnitude to the settlements due to excavation in these cases. These settlements due to installation are not always presented in case studies, so it is important to remember this when using case studies to make predictions.

Faustin et al. (2018b) analysed 26 case studies of circular shaft construction in London, and very considerately placed all the data on the journal's website. They separated the shafts into two main types; shafts where the lining was pre-installed (named 'SBE', for 'support before excavation') and shafts where the lining was installed concurrently with excavation (named 'EBS', for 'excavation before support'). SBE shafts were further subdivided into two groups: those that were a jacked caisson and those that were either diaphragm wall or secant pile. Combined SBE/EBS shafts were also considered as two separate groups with the initial part of the shaft as a circular sheet pile cofferdam or a secant pile wall, or the initial part as a jacked caisson. In both cases the EBS for the lower part consisted of a shotcrete lining. This is summarised in Table 12.12.

When all this data was plotted by Faustin et al. (2018b), they found that New & Bowers' (1994) empirical relationship (Equation 12.39) did not fit well to all shaft types. For some shaft types the maximum settlement at the shaft wall could be substantially different to $0.06\%H$, and the extent of the settlements could also be more or less than a distance H from the shaft wall.

Developing this further, Equation 12.39 can be altered to the following expression without changing its quadratic nature:

$$S_d = \frac{\alpha.(\beta H - d)^2}{\beta^2 H} \tag{12.40}$$

Table 12.12 Shaft types defined by Faustin et al. (2018b).

Faustin et al. (2018b) shaft types	Description
'SBE pre-installed'	Diaphragm wall or secant pile wall shaft
'SBE jacked'	Precast concrete segments as a jacked caisson
'EBS'	Precast concrete segments or shotcrete underpinning
'Combined SBE & EBS'	Pre-installed sheet pile or secant pile walls through superficial gravels followed by shotcrete lining in the London Clay
'Combined SBE jacked & EBS'	Precast concrete segments as a jacked caisson through superficial gravels followed by shotcrete lining in the London Clay

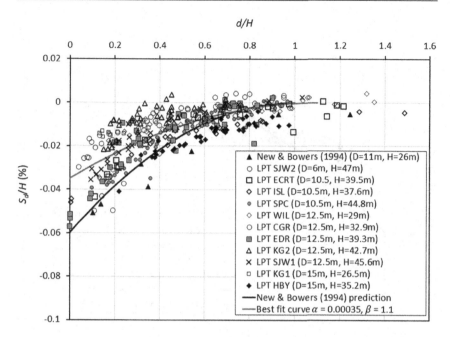

Figure 12.19 Settlements caused by shaft construction – underpinning using precast concrete segments or shotcrete (EBS only).

The distance at which settlements are zero is given by βH, and the maximum settlement at the shaft wall is still given by αH. I used regression to find the best fit values of α and β for each shaft type. When $\beta = 1$, Equation 12.40 is the same as Equation 12.39.

First let's look at similar situations to New & Bowers' case study; a jacked caisson followed by underpinning using segments or shotcrete, shown in Figure 12.19. Faustin et al. (2018b) put it into the EBS category, even though the initial part of the shaft was a jacked caisson. This was sensible because their settlements are at the higher end of the range. Also on Figure 12.19 is a solid black line labelled 'New & Bowers prediction', which is Equation 12.39, and a solid grey line labelled 'Best fit curve', which is Equation 12.40. A least squares regression method was used to find the best fit for α and β.

Note that Figure 12.19 has the axes normalised by dividing by the excavation depth H, and the settlement is expressed as a percentage. This allows shafts of different depths to be compared on the same graph and is a trick borrowed from the analysis of retained cut and basement excavations (e.g. Peck, 1969).

Figure 12.20 shows the case studies for shafts constructed using a jacked caisson of precast segments followed by shotcrete underpinning. There is a lack of data close to the shaft wall that leaves some uncertainty about the maximum settlements, but the best fit curve gives a better estimate of the extent of the settlements, to $1.35H$.

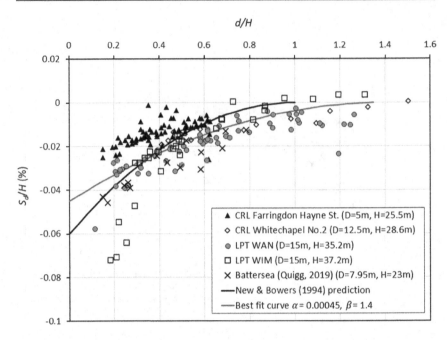

Figure 12.20 Settlements caused by shaft construction – jacked caisson followed by underpinning.

Quigg (2019) presented a case study, also in London, of a jacked caisson followed by underpinning using shotcrete, and this has been added to the dataset provided by Faustin et al. (2018b) in Figure 12.20. Quigg found that by looking at the jacked caisson settlement on its own, there was a maximum settlement of approximately 0.04%H, and for the shotcrete underpinning it was 0.06%H.

Figure 12.21 shows the settlements for secant piles or sheet piles pre-installed through the superficial gravels and made ground, followed by shotcrete underpinning.

Figure 12.22 shows the settlements for shafts constructed using a jacked caisson only. Data from Quigg (2019) for the initial jacked caisson part of the shaft only has been added to the data from Faustin et al. (2018b).

Figure 12.23 shows the settlements for shafts constructed using a pre-installed diaphragm wall or secant pile wall lining. Due to the high rigidity of this type of shaft, as the ground moves towards the excavation it can induce a small amount of heave close to the shaft wall. It was not possible to adequately fit a curve to this data. It should be borne in mind that there will be settlements due to wall installation before excavation begins, and these are of a similar magnitude to the settlements due to excavation (Faustin et al., 2018b).

The values of α and β for the different types of shaft construction are summarised in Table 12.13. As with any empirical relationship, consideration

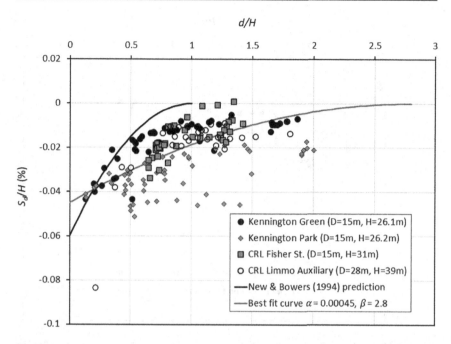

Figure 12.21 Settlements caused by shaft construction – secant piles or sheet piles followed by shotcrete underpinning.

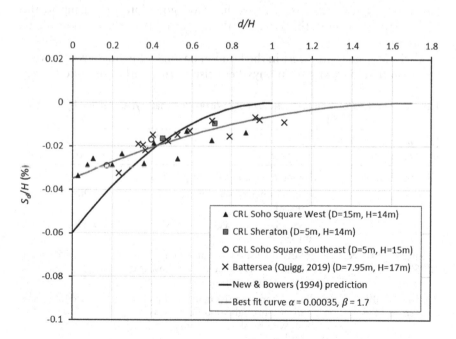

Figure 12.22 Settlements caused by shaft construction – jacked caisson only.

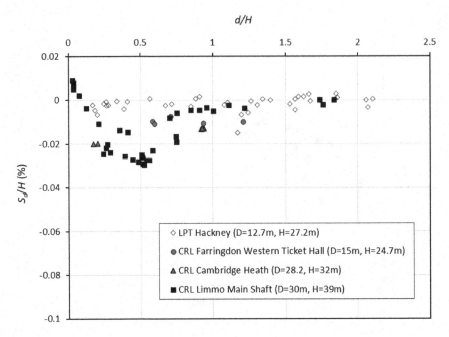

Figure 12.23 Settlements caused by shaft construction – pre-installed secant piles or diaphragm walls only.

has to be given to the case studies that have gone into producing it, the variability or scatter of the data, and the applicability of the relationship to any new situation.

All these case studies were in the London Basin. Unfortunately, case studies of shaft settlements from anywhere else in the world are quite scarce.

Table 12.13 Values of α and β found by curve-fitting for shaft types defined by Faustin et al. (2018b).

Faustin et al. (2018b) shaft types	*Description*	α	β
'SBE pre-installed'	Diaphragm wall or secant pile wall shaft	N/A	N/A
'SBE jacked'	Precast concrete segments as a jacked caisson	0.00035	1.7
'EBS'	Precast concrete segments or shotcrete underpinning	0.00035	1.1
'Combined SBE & EBS'	Pre-installed sheet pile or secant pile walls through superficial gravels followed by shotcrete lining in the London Clay	0.00045	2.8
'Combined SBE jacked & EBS'	Precast concrete segments as a jacked caisson through superficial gravels followed by shotcrete lining in the London Clay	0.00045	1.4

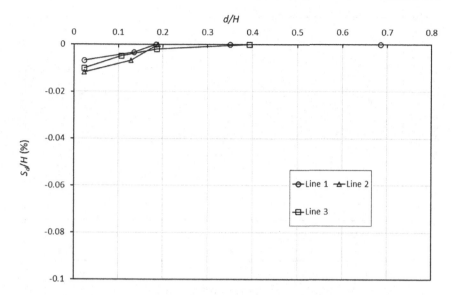

Figure 12.24 Settlements due to diaphragm wall installation and shaft excavation in predominantly coarse-grained soils (data from Muramatsu & Abe, 1996).

Muramatsu & Abe (1996) presented a case study of a 28.2 m ID circular diaphragm wall shaft in a coarse-grained soil with a maximum settlement of 0.01%H. The diaphragm walls were 98 m deep and the excavation depth was 60.3 m. Settlements were measured along three radial lines in different directions. The excavation-induced settlements were very small, even though the ground was very soft. Approximately half of the settlements shown in Figure 12.24 were due to diaphragm wall installation. These settlements are even smaller than most of those presented in Figure 12.23.

Based on what we know about shaft base heave stability in clay (c.f. Section 4.2), and the relationship between undrained stability and volume loss for tunnels demonstrated by the load factor approach in Section 12.2, we can expect that the magnitude of settlements will be larger in clays that have a lower undrained strength than London Clay, and will be smaller in stronger clays. Unfortunately, at present, we do not have the field data to confirm and quantify this.

Very few case studies of shaft construction include horizontal ground movements. Muramatsu & Abe (1996) found that maximum horizontal displacements 1.4 m from the shaft wall were between 6 and 10 mm, compared to vertical settlements of between 4 and 7 mm. Schwamb et al. (2016) installed instrumentation in and around a large diaphragm wall shaft in East London, 29 m diameter and with an excavation depth of 73 m and a diaphragm wall depth of 84 m. They also found that near-surface settlements were between 4 and 7 mm, but unlike Muramatsu & Abe (1996), the maximum horizontal movements at depth measured by inclinometers were

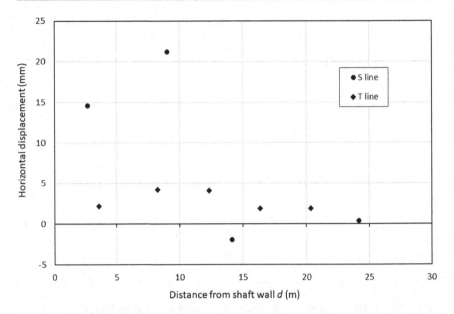

Figure 12.25 Horizontal displacements due to Heathrow Express Trial Tunnel Access Shaft construction (redrawn from New & Bowers, 1994).

less than 4 mm, and at the surface were negligible. The differences may have been due to the quality of the joints between diaphragm wall panels or the stiffness of the ground.

New & Bowers (1994) measured horizontal ground surface displacements along two radial lines 'S' and 'T', at the same locations as the vertical settlements presented earlier in Figure 12.18. These are shown in Figure 12.25. There are two anomalous displacements in the S line data, at 2.7 and 9.1 m, that were due to the proximity of heavy plant movements and shallow excavations for temporary works, so the T line data is more representative of the greenfield situation.

At present there is insufficient empirical evidence to make good predictions of horizontal ground movements due to shaft construction. More measurements are needed. It is likely that numerical modelling would need to be used if a shaft is to be constructed close to structures sensitive to horizontal displacements, such as buildings, pipelines, sewers or tunnels.

WORKED EXAMPLE 12.9   SETTLEMENTS DUE TO JACKED CAISSON SHAFT CONSTRUCTION

A 15 m diameter shaft is to be constructed using a jacked caisson segmental lining to an excavation depth of 45 m.

Estimate the maximum settlement and the settlement at 10, 20 and 30 m from the shaft wall based on case studies of similar shafts.

Look at Figure 12.22 and Table 12.13. The best fit curve to the case study data has parameters $\alpha = 0.00035$ and $\beta = 1.7$. Using Equation 12.40 we get:

$$S_d = \frac{\alpha.(\beta H - d)^2}{\beta^2 H} = \frac{0.00035(1.7 \times 45 - d)^2}{1.7^2 \times 45}$$

At $d = 0$ the maximum settlement is $S_d = \alpha H = 0.00035 \times 45 = 0.0158$ m $= 15.8$ mm

At $d = 10$ m the settlement is given by:

$$S_d = \frac{0.00035(1.7 \times 45 - 10)^2}{1.7^2 \times 45} = 0.0119 \text{ m} = 11.9 \text{ mm}$$

Similarly, at $d = 20$ m the settlement is 8.6 mm, and at $d = 30$ m the settlement is 5.8 mm. By putting Equation 12.40 into a spreadsheet, we can plot a graph of the predicted settlement, as shown in Figure 12.26.

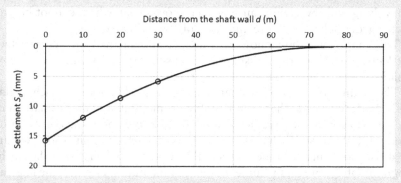

Figure 12.26 Worked Example 12.9 shaft settlements.

Further refinements to this methodology could be to take into account the diameter of the shaft D. Newhouse (2018) cites a GCG report for Crossrail that proposes the following relationship:

$$\frac{\alpha}{D} = 0.006\%/\text{m}, D \le 25 \text{ m} \tag{12.41}$$

where for values of D greater than 25 m, a constant value of $\alpha = 0.15\%$ should be used.

Newhouse (2018) showed, with reference to the same case studies that Faustin et al. (2018b) had access to, that Equation 12.41 is a conservative upper bound to the data for underpinned shafts, but is very conservative in relation to diaphragm wall or secant pile wall shafts. He then proposed alternative values for α / D, depending on construction method. Although the correlation to construction method is strong, the correlation between shaft diameter and α is weak, and this may be because the range of shaft depths and diameters is quite small for each construction method.

Newhouse (2018) then plotted all the data on a graph of α vs. D / H and proposed design lines with slope $S_{v,max} / D$. This method, however, does not take into account the effect of construction method or shaft diameter on the extent of settlements, represented by β in Equation 12.40.

An improvement could be to use Equation 12.40 with the recommended values of α for the construction method, which determines the maximum settlement at the shaft wall. Then we could relate β, which determines the extent of the settlements as a multiple of shaft excavation depth H, to shaft diameter. Setting $\beta = 0.1D$ fits the data reasonably well, and means that a shaft of 10 m diameter has an extent of settlement equal to H and a shaft of 30 m diameter has an extent equal to $3H$.

12.9 SUMMARY

Tunnelling will invariably cause ground movements, which cannot be predicted accurately in advance and can only be estimated.

This chapter has shown how to estimate transient and short-term surface settlements and horizontal displacements using the Gaussian curve approximation, both at the surface and subsurface.

The volume loss may be estimated using the load factor approach in clay soils. In drained soils there is no analogous method, and an estimate must be made based on case studies and what we know about the construction method and the soil behaviour. If a face pressure is applied and we are in dense sand then volume loss should be low and behaviour should be approximately constant volume. In loose or medium sands, volume loss can be significantly higher at the surface than at the tunnel.

The Gaussian curve approximation enables the calculation of ground strain, slope and curvature. This will be needed in the following chapter on building damage assessment.

Shafts will also cause ground movements. Ground movements due to shaft construction generally follow a quadratic relationship with a maximum at the shaft wall and settlements reducing to zero some distance away from the shaft. The depth and diameter of the shaft are also important factors.

Unfortunately there are only sufficient case studies to develop empirical relationships for shafts in the London Basin, which may not be applicable elsewhere. Shafts where the support is provided before excavation, such as secant pile or diaphragm wall shafts, cause very little ground movement, whereas shafts where the support is provided after excavation, such as underpinned shafts lined with precast concrete segments or shotcrete, cause larger ground movements. The installation of secant piles or diaphragm walls will cause ground movements before excavation inside the shaft has begun.

12.10 PROBLEMS

Q12.1.   A 10 m diameter shotcrete-lined tunnel is to be excavated in stiff clay at a depth to axis of 23 m. Assume the clay has a constant value of undrained shear strength of 125 kPa and a bulk unit weight of 20 kN/m³. Assume the unsupported length is 2 m.
   i.   Estimate the volume loss using the load factor approach.
   ii.   Estimate the trough width parameter K at the surface using Table 12.1 and then calculate trough width i at the surface.
   iii.   Calculate maximum short-term surface settlement S_{max} using the Gaussian curve equation.
   iv.   Calculate the short-term settlements at transverse offsets of 5, 10, 15, 20 and 25 m.
   v.   Assuming that 50% of the short-term settlement happens ahead of the face and 50% behind the face, calculate settlements along the tunnel centreline at 20, 15, 10, 5, –5, –10, –15 and –20 m relative to the face. Assume the tunnel is very far from its start position.

Q12.2.   A 5 m diameter EPB TBM tunnel is to be excavated in dense sand at a depth to axis of 15 m. From prior experience of driving EPB TBMs in these ground conditions, a conservative estimate of volume loss at the ground surface of 1% has been assumed.
   i.   Calculate the short-term value of maximum settlement S_{max} and trough width i, for $K = 0.25$.
   ii.   Calculate at suitable intervals and then plot a graph of the transverse short-term surface settlement trough.
   iii.   Calculate horizontal displacements at the same intervals and then plot a graph, assuming constant volume behaviour.
   iv.   Calculate horizontal strains at the same intervals and then plot a graph.
   v.   Calculate change in ground slope at the same intervals and then plot a graph.
   vi.   Calculate ground curvature at the same intervals and then plot a graph.
   vii.   Repeat (i) to (vi) for $K = 0.5$ and plot onto the same graphs. What difference does it make?

Q12.3.   A 7 m diameter EPB TBM tunnel is to be excavated in clay at a depth to axis of 17 m. The expected value of minimum face pressure at axis level is 100 kPa, with 100 kPa bentonite slurry pressure also provided around the shield, which is 6 m long. The undrained shear strength of the clay is given by $c_u = 25 + 3z$, where z is the depth below the ground surface, and the bulk unit weight is 19 kN/m^3.

i.   Estimate the volume loss using the load factor approach, taking unsupported length P as the length of the shield.

ii. Calculate the face pressure required to limit the volume loss to 1%

iii. Using a trough width parameter $K = 0.5$, calculate trough width i at the surface.

iv. Using a constant value of $K = 0.5$, calculate trough width i at 4, 8 and 12 m depth.

v.   Using Mair et al. (1993)'s empirical relationship (Equation 12.18), calculate trough width i at 0, 4, 8 and 12 m depth.

vi. Using the logarithmic empirical relationship (Equation 12.19), calculate trough width i at 0, 4, 8 and 12 m depth.

Q12.4. An 8 m diameter open face TBM with a backhoe is to driven under a railway embankment at a 45° skew. The embankment is 10 m high and consists of non-cohesive fill with a bulk unit weight of 17 kN/m^3. The natural ground below the embankment is stiff boulder clay and the tunnel is 20 m below the natural ground surface level. The undrained shear strength of the clay is given by $c_u = 50 + 8z$, where z is the depth below the natural ground surface, the shear modulus $G = 200c_u$ and the bulk unit weight is 20 kN/m^3.

i.   Assuming trough width parameter $K = 0.5$ in the stiff boulder clay and $K = 0.25$ in the fill and volume loss $V_l = 1.5\%$, calculate the surface settlements along the embankment centreline when the tunnel face is at 20, 10, 0, –10 and –20 m from the embankment centreline. Plot these five settlement troughs on a graph against distance from the tunnel centreline along the track chainage rather than transverse offset.

ii. Calculate the maximum change in track gradient for each of the five settlement troughs.

iii. Estimate the volume loss at the face using Dimmock & Mair (2007)'s load factor approach, assuming that excavation can be no more than 1.5 m ahead of the front of the shield.

iv. The average annular overcut of the shield is 20 mm and the length of the shield is 5 m. Will the ground close onto the shield? What is the volume loss due to closure of the ground around the shield?

Q12.5. A 15 m diameter shaft is to be constructed in stiff overconsolidated clay by underpinning with precast concrete segments. The final excavation depth is 30 m.

i.   Estimate the extent of the settlement, i.e. the radial distance from the shaft wall at which settlement is zero.

ii. Estimate the settlement adjacent to the shaft wall.
iii. Calculate the settlement profile at intervals of no more than 5 m from the shaft wall to the extent of settlement. Plot them on a graph.
iv. A building is 11 m away from the shaft wall. At what excavation depth would you expect settlement to begin affecting the building?
v. [Advanced] Referring back to Section 4.2 on base heave stability of shafts in clay, what factors would you expect to influence the magnitude of settlement in any type of clay soils?

REFERENCES

Atkinson, J. H. & Potts, D. M. (1977). Subsidence above shallow tunnels in soft ground. *Proc. ASCE Geot. Engrg Div.* **103**, GT4, 307–325.

Attewell, P. B. & Farmer, I. W. (1974). Ground deformations resulting from shield tunnelling in London Clay. *Canadian Geotechnical Journal* **11**, 380–395.

Attewell, P. B. & Woodman, J. P. (1982). Predicting the dynamics of ground settlement and its derivatives caused by tunnelling in soil. *Ground Engineering* **15**, No. 8, 13–22 & 36.

Barratt, D. A. & Tyler, R. G. (1976). *Measurements of ground movement and lining behaviour on the London Underground at Regent's Park*, TRRL Lab Report 684. Crowthorne, Berkshire: TRRL.

Bowers, K. H., Hiller, D. M., and New, B. M. (1996). Ground movement over three years at the Heathrow Express Trial Tunnel. *Proc. Int. Symp. on Geotechnical Aspects of Underground Construction in Soft Ground*, pp. 557–562. Rotterdam: Balkema.

Burke, H. H. (1957). Garrison dam – tunnel test section investigation. *J. Soil Mech. Found. Div. A.S.C.E.*, **83**, SM4, Paper No. 1438, 1–50.

Burland, J. B. (2001). Chapter 3: Assessment methods used in design. *Building Response to Tunnelling, Volume 1: Projects and Methods* (eds Burland, J. B., Standing, J. R. & Jardine, F. M.), CIRIA Special Publication 200, pp. 23–43. London: Thomas Telford Publishing.

Clayton, C. R. I., van der Berg, J. P. & Thomas, A. H. (2006). Monitoring and displacements at Heathrow Express Terminal 4 station tunnels. *Géotechnique* **56**, No. 5, 323–334.

Cooper, M. L., Chapman, D. N., Rogers, C. D. F. & Chan, A. H. C. (2002). Movements in the Piccadilly Line tunnels due to the Heathrow Express construction. *Géotechnique* **52**, No. 4, 243–257.

Dimmock, P. S. & Mair, R. J. (2007). Estimating volume loss for open-face tunnels in London Clay. *Proc. Instn Civ. Engrs – Geot. Engrg* **160**, GE1, 13–22.

Eden, W. J. & Bozozuk, M. (1969). Earth pressures on Ottawa outfall sewer tunnel. *Canadian Geotech. J.* **6**, No. 17, 17–32.

Faustin, N. E., Mair, R. J., Elshafie, M. Z. E. B., Menkiti, C. O. & Black, M. (2018a). Field measurements of ground movements associated with circular shaft construction. *Proc. 9th Int. Symp. on Geotechnical Aspects of Underground Construction in Soft Ground* (eds Negro, A. & Cecilio Jr., M. O.), IS-São Paulo, pp. 301–308. London: Taylor & Francis Group.

Faustin, N. E., Elshafie, M. Z. E. B. & Mair, R. J. (2018b). Case studies of circular shaft construction in London. *Proc. Instn Civ. Engrs – Geot. Engrg* **171**, No. 5, 391–404.

Franza, A., Marshall, A. M. & Zhou, B. (2019). Greenfield tunnelling in sands: the effects of soil density and relative depth. *Géotechnique* **69**, No. 4, 297–307.

Franza, A., Marshall, A. M., Zhou, B., Shirlaw, N. & Boone, S. (2020). Discussion of: Greenfield tunnelling in sands: the effects of soil density and relative depth. *Géotechnique* **70**, No. 7, 639–646.

Glossop, N. H. (1978). *Ground movements caused by tunnelling in soft soils.* Ph.D. thesis, University of Durham.

Harris, D. I., Mair, R. J., Love, J. P., Taylor, R. N. & Henderson, T. O. (1994). Observations of ground and structure movements for compensation grouting during tunnel construction at Waterloo Station. *Géotechnique* **44**, No. 4, 691–713.

Hill, N. & Stärk, A. (2015). Volume loss and long-term settlement at Kempton Court, Whitechapel. *Crossrail Project: Infrastructure design and construction – Volume 2*(eds Black, M., Dodge, C. & Yu, J.), pp. 347–385. London: ICE Publishing.

Hill, N. & Stärk, A. (2016). Long-term settlement following SCL-tunnel excavation. *Crossrail Project: Infrastructure design and construction – Volume 3* (ed. Black, M.), pp. 227–247. London: ICE Publishing.

Jones, B. D., Thomas, A. H., Hsu, Y. S. & Hilar, M. (2008). Evaluation of innovative sprayed-concrete-lined tunnelling. *Proc. Instn Civ. Engrs – Geot. Engrg* **161**, GE3, 137–149.

Jones, B. D. (2010). Low-volume-loss tunnelling for London Ring Main Extension. *Proc. Instn. Civ. Engrs – Geotech. Engrg* **163**, GE3, June, 167–185.

Kimura, T. & Mair, R. J. (1981). Centrifugal testing of model tunnels in soft clay. *Proc. 10th Int. Conf. Soil Mech. & Found. Engrg*, Stockholm, Vol. 1, pp. 319–322.

Lake, L. M., Rankin, W. J. & Hawley, J. (1996). *Prediction and effects of ground movements caused by tunnelling in soft ground beneath urban areas*, CIRIA Project Report 30. London: CIRIA.

Laver, R. G., Soga, K., Wright, P. & Jefferis, S. (2013). Permeability of aged grout around tunnels in London. *Géotechnique* **63**, No. 8, 651–660.

Laver, R., Li, Z. & Soga, K. (2017). Method to Evaluate the Long-Term Surface Movements by Tunneling in London Clay. *ASCE J. of Geotech. and Geoenv. Engineering* **143**, No. 3, 1–7.

Macklin, S. R. (1999). The prediction of volume loss due to tunnelling in over-consolidated clay based on heading geometry and stability number. *Ground Engineering*, No. 4, 30–33.

Macklin, S. R. & Field, G. R. (1998). The response of London Clay to full-face TBM tunnelling at West Ham, London. *Proc. Int. Conf. on Urban Ground Engineering*, Hong Kong, November 1998.

Mair, R. J. (1979). *Centrifuge modelling of tunnel construction in soft clay.* Ph.D. thesis, University of Cambridge.

Mair, R. J. (2008). Tunnelling and geotechnics: new horizons. *Géotechnique* **58**, No. 9, 695–736.

Mair, R. J., Gunn, M. J. & O'Reilly, M. P. (1981). Ground movements around shallow tunnels in soft clay. *Proc. 10th Int. Conf. on Soil Mechanics and Foundation Engrg*, Stockholm, 1981, Vol. 1, pp. 323–328.

Mair, R. J. & Taylor, R. N. (1993). Prediction of clay behaviour around tunnels using plasticity solutions. *Predictive Soil Mechanics - Proc. of the Wroth Memorial Symposium* (eds Houlsby, G. T. & Schofield, A. N.), St Catherine's College, Oxford, 27th-29th July 1992, pp.449–463. London: Thomas Telford.

Mair, R. J. & Taylor, R. N. (1997). Bored tunnelling in the urban environment. Theme Lecture, Plenary Session 4. *Proc. 14th Int. Conf. Soil Mechanics and Foundation Engineering*, Hamburg, Vol. 4.

Mair, R. J., Taylor, R. N. & Bracegirdle, A. (1993). Subsurface settlement profiles above tunnels in clays. *Géotechnique* 43, No. 2, 315–320.

Marshall, A. M. (2009). *Tunnelling in sand and its effect on pipelines and piles.* Ph.D. thesis, University of Cambridge.

Marshall, A. M., Farrell, R., Klar, A. & Mair, R. (2012). Tunnels in sands: the effect of size, depth and volume loss on greenfield displacements. *Géotechnique* 62, No. 5, 385–399.

Muramatsu, M. & Abe, Y. (1996). Considerations in shaft excavation and peripheral ground deformation. *Proc. Int. Symp. on Geotechnical Aspects of Underground Construction in Soft Ground* (eds Mair, R. J. & Taylor, R. N.), London, UK, pp. 173–178. Rotterdam: Balkema.

New, B. M. & Bowers, K. H. (1994). Ground movement model validation at the Heathrow Express trial tunnel. *Tunnelling '94, Proc. 7th Int. Symp. IMM and BTS*, London, UK, pp. 310–329. London: Chapman and Hall.

Newhouse, J. (2018). Ground movement due to shaft construction. *Ground Engineering*, July 2018, 26–30.

Nyren, R. J. (1998). *Field measurements above twin tunnels in clay.* Ph.D. thesis, Imperial College of Science, Technology and Medicine, London.

O'Reilly, M. P. (1988). Evaluating and predicting ground settlements caused by tunnelling in London Clay. *Proc. Tunnelling '88*, pp. 231–241. London, UK: IMM.

O'Reilly, M. P. & New, B. M. (1982). Settlements above tunnels in the United Kingdom – their magnitude and prediction. *Proc. Tunnelling '82*, pp. 173–181. London, UK: IMM

Peck, R. B. (1969). Deep excavations and tunnelling in soft ground. *Proc. 7th Int. Conf. Soil Mechanics and Foundation Engrg (7th ICSMFE)*, Mexico, State-of-the-art report, pp. 225–290.

Pound, C. (2003). Session 1 Report: Prediction of damage to buildings and other structures from tunnelling. *Response of buildings to excavation induced ground movements*, Proc. Int. Conf. held at Imperial College, London, UK, 17th-18th July 2001 (ed. Jardine, F. M.), CIRIA Special Publication 199, pp. 27–36. London: CIRIA.

Quigg, G. (2019). Battersea cable tunnel: powering regeneration in central London, UK. *Proc. Instn Civ. Engrs – Civil Engrg* 172, No. 2, 77–82, https://doi.org/10.1680/jcien.18.00020.

Rankin, W. J. (1988). Ground movements resulting from urban tunnelling: prediction and effects. *Conf. on Engineering Geology of Underground Movements*, Nottingham, Geological Society Engineering Geology Special Publication No. 5, pp. 79–92.

Sauer, G. & Lama, R. D. (1973). An application of New Austrian Tunnelling Method in difficult builtover areas in Frankfurt/Main Metro. *Proc. Symp. On Rock Mechanics and Tunnelling Problems*, Kurukshetra, India, December 1973, pp. 79–92. New Delhi: Indian Geotechnical Society.

Schmidt, B. (1969). *Settlements and ground movements associated with tunnelling in soils.* Ph.D. thesis, University of Illinois, Urbana.

Schwamb, T., Elshafie, M. Z. E. B., Soga, K. & Mair, R. J. (2016). Considerations for monitoring of deep circular excavations. *Proc. Instn Civ. Engrs – Geot. Engrg* **169**, GE6, 477–493.

Shirlaw, J. N., Doran, S. & Benjamin, B. (1988). A case study of two tunnels driven in the Singapore 'Boulder Bed' and in grouted coral sands. *Engineering Geology of Underground Movements* (eds Culshaw et al.), Geological Society Engineering Geology Special Publications Volume 5, pp. 93–103.

Simic, D. & Craig, R. N. (1997). Lisbon Metro—settlement behaviour of large diameter shield driven and NATM tunnels. *Proc. Conf. Tunnelling '97,* IMM, London.

Standing, J. R. & Selman, R. (2001). The response to tunnelling of existing tunnels at Waterloo and Westminster. *Building Response to Tunnelling,* Vol. 2 Case studies, CIRIA Special Publication 200, pp. 509–546. London: Thomas Telford.

Umney, A. R. & Heath, G. R. (1996). Recorded settlements from the DLR tunnels to Bank. *Proc. Int. Symp. on Geotechnical Aspects of Underground Construction in Soft Ground* (eds Mair, R. J. & Taylor, R. N.), London, UK, pp. 757–761. Rotterdam: Balkema.

Vu, M. N., Broere, W. & Bosch, J. (2016). Volume loss in shallow tunnelling. *Tunnelling and Underground Space Technology* **59**, 77–90.

Wan, M. S. P. & Standing, J. R. (2014). Field measurement by fully grouted vibrating wire piezometers. *Proc. Instn Civ. Engrs – Geot. Engrg.* **167**, GE6, 547–564.

Wan, M. S. P., Standing, J. R., Potts, D. M. & Burland, J. B. (2017a). Measured short-term ground surface response to EPBM tunnelling in London Clay. *Géotechnique* **67**, No. 5, 420–445.

Wan, M. S. P., Standing, J. R., Potts, D. M. & Burland, J. B. (2017b). Measured short-term subsurface ground displacements from EPBM tunnelling in London Clay. *Géotechnique* **67**, No. 9, 748–779.

Wongsaroj, J., Soga, K. & Mair, R. J. (2007). Modelling of long-term ground response to tunnelling under St James's Park, London. *Géotechnique* **57**, No. 1, 75–90.

Wongsaroj, J., Soga, K. & Mair, R. J. (2013). Tunnelling-induced consolidation settlements in London Clay. *Géotechnique* **63**, No. 13, 1103–1115.

Yang, B. (2017). *Numerical modelling of tunnelling in sand using a state parameter constitutive model.* Ph.D. thesis, University of Warwick.

Chapter 13

Estimating building damage

Buildings may be affected by ground movements induced by tunnelling in a variety of ways, and how they are affected will depend on the buildings themselves. Reinforced concrete or steel frame buildings, with concrete floor slabs and with or without infilled masonry wall panels, will behave differently to houses consisting of load-bearing brick or stone walls and timber floors. Both horizontal and vertical movements will affect buildings, as well as the change of slope of the ground and its curvature.

After working through this chapter, you will understand:

- the concept of limiting tensile strain and its correlation to building damage categories
- the stages of building damage assessment
- deflection ratio and how to calculate it
- the idealisation of a building as a simple beam and how this helps estimate maximum strains
- the combination of maximum bending and diagonal strains with horizontal strain to determine the damage category
- the use of modification factors
- what may be included in a detailed Stage 3 assessment

After working through this chapter, you will be able to:

- perform Stage 1, 2 and 2b building damage assessments for masonry walls and frame structures

It is necessary to assess buildings along the route of a tunnel for the risk of damage due to tunnelling settlements. This will usually follow a staged process. At each stage buildings are classified into one of six damage categories, from 0 to 5, as shown in Table 13.1. Categories 0, 1 and 2 are aesthetic damage, 3 and 4 are serviceability damage, and 5 is the most severe category where stability of the building will be affected and it may collapse. At the preliminary stage, very simple methods are used to conservatively assess the risk of settlement damage to all the buildings, with the aim that many of them

DOI: 10.1201/9780429470387-13

Table 13.1 Classification system for visible damage to building walls (based on Burland et al., 1977; Rankin, 1988; Burland, 2001).

Category of damage	Normal degree of severity	Description of typical damage (ease of repair is in bold) and typical crack width	Limiting tensile strain ε_{lim} (%)
0	Negligible	Hairline cracks < 0.1 mm wide.	0–0.05
1	Very slight	**Fine cracks that are easily treated during normal redecoration.** Damage generally restricted to internal wall finishes. Close inspection may reveal some cracks in external brickwork or masonry. Typical cracks < 1 mm.	0.05–0.075
2	Slight	**Cracks easily filled. Redecoration probably required. Recurrent cracks can be masked by suitable linings.** Cracks may be visible externally and **some repointing may be required to ensure weather-tightness.** Doors and windows may stick slightly. Typical cracks < 5 mm.	0.075–0.15
3	Moderate	**The cracks require some opening up and can be patched by a mason. Repointing of external brickwork and possibly a small amount of brickwork to be replaced.** Doors and windows sticking. Service pipes may fracture. Weather-tightness often impaired. Typical cracks 5–15 mm or several cracks > 3 mm.	0.15–0.3
4	Severe	**Extensive repair work involving breaking-out and replacing sections of walls, especially over doors and windows.** Windows and door frames distorted, floor sloping noticeably. Walls leaning or bulging noticeably, some loss of bearing in beams. Service pipes disrupted. Typical cracks 15–25 mm, but also depends on the number of cracks.	> 0.3
5	Very severe	**This requires a major repair job involving partial or complete rebuilding.** Beams lose bearing, walls lean badly and require shoring. Windows broken with distortion. Danger of instability. Typical cracks > 25 mm, but depends on the number of cracks.	> 0.3

can be ruled out as being at low risk. The remaining buildings will proceed to a second stage, where they will be analysed in more detail, again hoping that many of them will be moved to the low risk category. Any remaining buildings will proceed to a third, more detailed stage. Usually, sensitive structures, high rise buildings and heritage buildings will go straight to the third stage.

Structures may be considered sensitive if they contain sensitive equipment or if even minor damage would be intolerable to the user/owner or for the proper functioning of the structure (Rankin, 1988).

Building damage up to category 2 can be easily caused by a variety of environmental phenomena, such as shrinkage or thermal effects on the structure itself, or natural movements of the ground due to rising or lowering groundwater, tree root suctions, or other effects. This means that identification of the cause of any category 1 or 2 damage is difficult and could be a combination of causes, whereas category 3 damage is almost certain to be associated with ground movements due to tunnelling if they are occurring at that time. Therefore the division between category 2 and 3 is important (Burland, 2001). It is also the threshold beyond which repair work starts to become expensive and it is common practice on tunnelling projects to design underground construction methods and settlement mitigation measures so that all buildings are predicted to be in category 2 or below and are hence 'low risk'.

Building damage assessments usually follow a three stage process. Stage 1, sometimes called 'preliminary assessment', Stage 2 or 'second stage assessment' and Stage 3 or 'detailed evaluation' (Burland, 2001; Bowers & Moss, 2006). On major projects this process is often set out in the legislation used to fund and enable the project to go ahead (e.g. for CTRL see Moss & Bowers, 2006, for Crossrail see Crossrail, 2008 or DeJong et al., 2019).

13.1 STAGE 1 ASSESSMENT

Stage 1, or 'preliminary assessment', is a very simple and conservative approach. The aim is to identify buildings that are very likely to be in the low risk category and to rule them out from further studies, so that the number of buildings that require a more time-consuming Stage 2 assessment is reduced.

Two criteria are used, the predicted greenfield maximum slope and maximum settlement of the ground surface at the location of each building. If the maximum slope is less than 1/500 and the maximum settlement is less than 10 mm, then the building has negligible risk of any damage (Rankin, 1988).

It is straightforward, for a given tunnel alignment, to plot contours of surface settlement either side of the tunnel. Knowing the depth and diameter of the tunnel, and estimating the value of maximum settlement S_{max} and trough width i, the transverse offset y from the tunnel centreline to any settlement contour value S may be found by rearranging Equation 12.1 to obtain the following expression:

$$y = \sqrt{2i^2.\ln\left(\frac{S_{max}}{S}\right)} \tag{13.1}$$

  y is the transverse offset from the tunnel centreline, i.e. the horizontal distance perpendicular to the tunnel centreline, in m

i is the trough width, i.e. the horizontal distance from the tunnel centreline to the point of inflexion of the Gaussian curve, in m

S is the settlement at transverse offset y in m

S_{max} is the maximum settlement, which occurs at the tunnel centreline,
in m

Equation 12.1 can also be differentiated to obtain an equation for slope
θ at any offset distance y (c.f. Equation 12.30 in Section 12.7 for the full
derivation):

$$\theta = \frac{dS}{dy} = -\frac{y}{i^2} S_{max} \exp\left(\frac{-y^2}{2i^2}\right) \tag{13.2}$$

Equation 13.2 can only be solved for y iteratively, which can be achieved
using a solver or 'goal seek' function in a spreadsheet.

Equation 13.1 and Equation 13.2 can be used to define limits either side
of the tunnel on a set of plans of the alignment. All buildings that are outside both limits can then be excluded from further analysis.

WORKED EXAMPLE 13.1   STAGE 1
BUILDING DAMAGE ASSESSMENT

An 8 m diameter tunnel for a railway line is to be constructed using a
TBM through a dense urban area at 18 m depth to axis. The parameters
to be used for Stage 1 building damage assessment are a volume loss
$V_l = 1.5\%$ and a trough width parameter $K = 0.5$. Calculate the transverse offset y at which surface settlement $S = 10$ mm. Calculate the
transverse offset y at which surface slope $\theta = 1/500$.

First calculate volume loss V_s using:

$$V_s = V_l \frac{\pi D^2}{4} = 0.015 \times \frac{\pi \times 8^2}{4} = 0.754 \text{ m}^2$$

Now trough width i can be calculated at the surface ($z = 0$) using
Equation 12.3:

$$i = K(z_0 - z) = 0.5 \times (18 - 0) = 9 \text{ m}$$

S_{max} can be calculated using Equation 12.2 rearranged:

$$S_{max} = \frac{V_s}{\sqrt{2\pi} \cdot i} = \frac{0.754}{\sqrt{2\pi} \times 9} = 0.0334 \text{ m} = 33.4 \text{ mm}$$

We can now calculate the transverse offset from the tunnel centreline to the 10 mm settlement contour using Equation 13.1:

$$y = \sqrt{2i^2 \cdot \ln\left(\frac{S_{max}}{S}\right)} = \sqrt{2 \times 9^2 \times \ln\left(\frac{0.0334}{0.01}\right)} = 13.977 \text{ m}$$

This can be checked by inserting this value for y into Equation 12.1 and seeing that the settlement calculated is 10 mm.

A limiting slope of $1/500 = \pm 0.002$. Using Equation 13.2 in a spreadsheet, values of slope θ can be calculated every 1 m and plotted on a graph against transverse offset y, as shown in Figure 13.1. It can be seen that the limiting slope value is exceeded in two discrete areas, and this is often the case because the maximum and minimum slope values occur at the points of inflexion and the slope is always zero above the tunnel centreline.

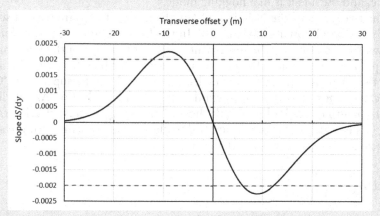

Figure 13.1 Worked Example 13.1 ground surface slope values.

In this case, since all buildings between $y = -13.977$ m and $y = 13.977$ m will be exposed to a surface settlement greater than 10 mm, we are only concerned with the outer limits of where the slope is greater than 1/500. By using a goal seek function in a spreadsheet, this occurs at $y = \pm 12.262$ m.

This can be checked by inserting $y = 12.262$ into Equation 13.2:

$$\theta = -\frac{y}{i^2}S_{max}\exp\left(\frac{-y^2}{2i^2}\right) = -\frac{12.262}{9^2} \times 0.0334 \times \exp\left(\frac{-12.262^2}{2 \times 9^2}\right) = -0.002$$

Therefore, in this case the 10 mm settlement contour will be the limiting criterion. All buildings between $y = -13.977$ m and $y = 13.977$ m will be exposed to a surface settlement greater than 10 mm, and will need to carried through to Stage 2 assessment.

13.2 STAGE 2 ASSESSMENT

In this stage, each building is assumed to deform to the greenfield settlement profile at the level of the foundations. The deflection ratio is then calculated in the hogging zone(s) and the sagging zone. The façade of the building is then assumed to deform like a simple elastic beam under a point load and this enables tensile strains to be calculated. These tensile strains are then combined with the horizontal strains in the ground using the principle of superposition to obtain resultant strains. These are then compared to the limiting tensile strain values in Table 13.1 to determine the damage category.

The deflection ratio is defined in Figure 13.2. The greenfield settlement trough is an arbitrary one for illustrative purposes. The building, shown in grey, is assumed to deform to the same shape as the greenfield settlement trough. Between the points of inflexion the building is bending in a sagging mode, and outside it is in a hogging mode.

The two parts of the building, of lengths L_{hog} and L_{sag}, are treated separately and are known as 'partitions'. In the hogging zone a straight line is drawn from a point on the settlement trough at the edge of the building to the point of inflexion. The maximum vertical distance between this line and the settlement trough is the deflection Δ_{hog}^{GF}, where the superscript 'GF' indicates that this is the greenfield deflection. This deflection Δ_{hog}^{GF} divided by the length of the building in the hogging zone L_{hog} is the deflection ratio DR_{hog}^{GF}. The same procedure is applied to the building partition in the sagging zone. The equations are therefore:

$$DR_{hog}^{GF} = \frac{\Delta_{hog}^{GF}}{L_{hog}} \tag{13.3}$$

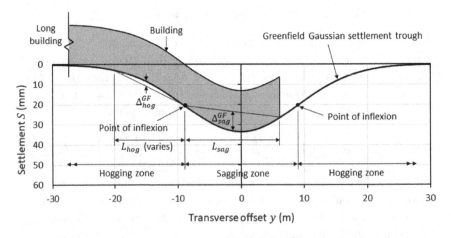

Figure 13.2 Stage 2 building damage assessment – definition of deflection ratio.

$$DR_{sag}^{GF} = \frac{\Delta_{sag}^{GF}}{L_{sag}} \qquad (13.4)$$

Note that a long building that spans beyond both points of inflexion may have two hogging zones and therefore two values of hogging deflection ratio will need to be calculated.

The maximum deflection values Δ_{hog}^{GF} and Δ_{sag}^{GF} can be difficult to calculate and are not necessarily in the centre. They can be found graphically using a scale ruler, or by using a spreadsheet to calculate the distance between the line and the settlement trough curve at, for example, every 0.5 m along the length and then to use the maximum of these values as the deflection. In practice, taking the value at the centre of the zone will be approximately correct in most cases.

The next step is to impose these deflection ratios onto the building. Burland & Wroth (1974) first proposed this should be done by modelling the building as a simple beam under the action of a point load. This will not give the exact deformation mode of the building, but this is only intended to be an approximate method, with many gross simplifications.

Figure 13.3 shows how a bearing-wall building can be idealised as a beam representing the load-bearing façade in bending and shear deformation according to the method of Burland & Wroth (1974).

Buildings will often have a height that is large relative to the 'span' in our beam idealisation, at least compared to typical beams in structural engineering. Therefore shear deflection may be significant and needs to be added to the bending deflection. The midspan deflection of a centrally-loaded simply-supported beam including both bending and shear deflection is given by the following equation (Gere & Timoshenko, 1991: p.694; or Bhatt, 1999: pp. 275–279):

$$\Delta = \frac{PL^3}{48EI}\left[1 + \frac{72EI}{5L^2 AG}\right] \qquad (13.5)$$

Δ is the midspan deflection of the beam in m

P is the applied line load in kN, uniformly applied across the width of the beam

L is the length of the beam (for sagging this would be L_{sag} and for hogging L_{hog}) in m

E is the Young's modulus of the beam in kN/m²

A is the shear area of the beam in m²

I is the second moment of area in m⁴

G is the shear modulus of the beam in kN/m²

The first term in Equation 13.5 is the deflection due to bending and the second term is the deflection due to shear. Note that Equation 13.5 is different to the one proposed by Burland & Wroth (1974), which had 18 as the coefficient of the second term in the brackets rather than 72/5. This

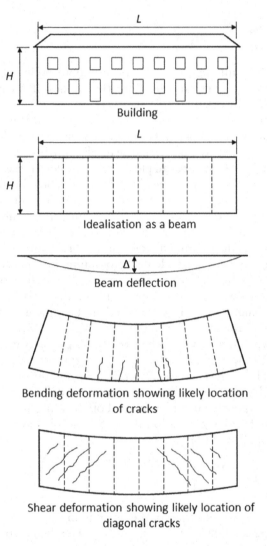

Figure 13.3 Stage 2 building damage assessment – idealisation of building as a beam (based on Burland & Wroth, 1974).

is because they assumed that the shear deflection was equal to the sum of the maximum shear strains along the centroid of the beam, but this is not the case. In reality shear stress varies parabolically with the maximum at the centroid and zero shear stress at the top and bottom of a rectangular cross-section, and shear strain varies with a cubic relationship. This is called 'warping', and a 'form factor for shear' may be calculated to take account of this, which for a rectangular section is equal to 6/5. Burland & Wroth's equation effectively assumes the form factor for shear is 3/2.

Gere & Timoshenko (1991) use the unit load method to derive Equation 13.5 and Bhatt (1999) uses a different method but arrives at the same answer. Gere & Timoshenko's method is given in Appendix C. This error in Burland & Wroth's deflection equation was first spotted by Netzel (2009).

The cross-sectional area A in Equation 13.5 is the shear area, and this is not simply the building height multiplied by its width in the tunnel longitudinal direction (unless the building consists of a solid isotropic rectangular prism!). Burland & Wroth (1974) assume that the problem is plane strain and the building properties may be expressed per unit width in the tunnel longitudinal direction, and therefore that $A = H$ in Equation 13.5, where H is the height of the building from the foundation level to the eaves, and second moment of area I is in units of m^4/m. It is important to remember this because when estimating the stiffness of more structurally complex buildings with concrete slabs, columns, masonry infill walls, and other structural features, the shear area and second moment of area will need to be calculated in a more rigorous manner.

For masonry bearing-wall structures, a masonry façade perpendicular to the tunnel alignment can be considered to act as a single plane stress rectangular section of length L_{hog} or L_{sag}, height H and of unit width. Note that we are ignoring the effect of openings in the façade, such as windows and doors. This is an unconservative assumption because stress concentrations tend to occur near the corners of openings where cracks are more likely to initiate and propagate from.

We need to rearrange Equation 13.5 so we can relate deflection ratio Δ/L to extreme fibre bending strain, while removing the point load P from the equation.

The idealised geometry of the beam is shown in Figure 13.4. The z axis is out of the page. The beam has width in the z axis direction b, height in the y axis direction H and length in the x axis direction L. A line load P is applied at midspan.

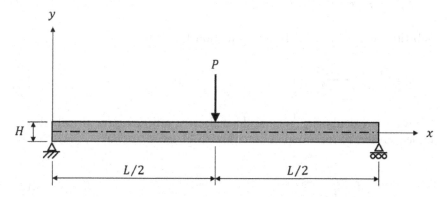

Figure 13.4 Simply-supported rectangular beam with a centrally-applied line load.

The midspan bending moment is given by:

$$M = \frac{PL}{4} \tag{13.6}$$

M is the midspan bending moment in kNm
P is a centrally-applied line load in kN
L is the span of the beam in m

The extreme fibre bending stress is given by:

$$\sigma_{bmax} = \frac{Md}{I} \tag{13.7}$$

σ_{bmax} is the extreme fibre bending stress in kPa
d is the vertical distance from the neutral axis to the extreme fibre in m
I is the second moment of area of the beam in m^4

Substituting Equation 13.6 into Equation 13.7 gives:

$$\sigma_{bmax} = \frac{PLd}{4I} \tag{13.8}$$

Therefore the extreme fibre bending strain may be given by:

$$\varepsilon_{bmax} = \frac{\sigma_{bmax}}{E} = \frac{PLd}{4EI} \tag{13.9}$$

ε_{bmax} is the extreme fibre bending strain
E is the Young's modulus of the beam in kPa

Rearranging Equation 13.9 for P gives:

$$P = \frac{4EI}{Ld}\varepsilon_{bmax} \tag{13.10}$$

Substituting Equation 13.10 into Equation 13.5 gives:

$$\Delta = \frac{L^2}{12d}\left[1 + \frac{72EI}{5L^2AG}\right]\varepsilon_{bmax} \tag{13.11}$$

We want to calculate ε_{bmax} from the deflection ratio we have already determined earlier, so rearranging Equation 13.11 gives:

$$\varepsilon_{bmax} = \frac{\Delta/L}{\dfrac{L}{12d}\left[1 + \dfrac{72EI}{5L^2AG}\right]} \tag{13.12}$$

In the sagging zone, the building's neutral axis can be assumed to be at mid-height, i.e. $d = H/2$. Also, assuming plane stress, i.e. a building of unit width, such that shear area $A = H$ and the second moment of area is also per unit width, we get the following expression for second moment of area:

$$I_{sag} = \frac{H^3}{12} \qquad (13.13)$$

For a masonry bearing-wall, rather than thinking of it as plane stress, we can just as well use the thickness of the wall t. The shear area is then given by $A = tH$ and the second moment of area is given by $I_{sag} = tH^3/12$. When inserted into Equation 13.12, thickness t cancels out and the resulting equation is the same.

In the hogging zone, the building's neutral axis is usually conservatively assumed to be at the foundation level, i.e. $d = H$, because the ground-structure interface may provide restraint (Burland & Wroth, 1974). Using the parallel axis theorem, this gives:

$$I_{hog} = \frac{H^3}{3} \qquad (13.14)$$

It is strictly speaking impossible for the neutral axis to be at the foundation level, because then the compressive stress required for equilibrium in bending is infinite (Netzel, 2009). However, many materials, such as masonry (or steel fibre reinforced concrete – see Chapter 8) that have a very low tensile strength relative to their compressive strength, will have a neutral axis very close to the compressive extreme fibre in bending. Therefore, the assumption is approximately correct in terms of estimating the strain distribution.

Now to calculate the maximum diagonal tensile strain ε_{dmax} we first need an expression relating it to the simply supported rectangular beam under point load.

The shear strain is related to the shear force by the following expression:

$$\gamma_{xy} = \frac{\alpha V}{AG} \qquad (13.15)$$

γ_{xy} is the shear strain
α is the ratio of maximum shear stress to average shear stress, which for a rectangular section is 3/2 (see Figure C.2 in Appendix C)
V is the shear force, where $V = P/2$, in kN
A is the cross-sectional area, where $A = bH$, in m^2
G is the shear modulus in kPa

Substituting for V and α in Equation 13.15 gives:

$$\gamma_{xy} = \frac{3P}{4AG} \qquad (13.16)$$

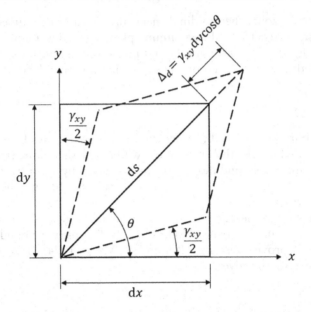

Figure 13.5 Definition of diagonal extension caused by shear strain.

Figure 13.5 defines diagonal extension, which we will call Δ_d. A rectangular element with sides of length dx and dy is deformed into a rhomboid by the action of shear stresses on the sides τ_{xy} and τ_{yx}, where because of equilibrium $\tau_{xy} = \tau_{yx}$. Angles and deformations are assumed to be small, and are greatly exaggerated in the figure. The length of the sides of the element do not change as we are considering pure shear.

To get the maximum diagonal tensile strain we use the following equation:

$$\varepsilon_{dmax} = \frac{\Delta_d}{ds} = \frac{\gamma_{xy}dy\cos\theta}{ds} \tag{13.17}$$

Now $dy/ds = \sin\theta$, therefore:

$$\varepsilon_{dmax} = \gamma_{xy}\sin\theta\cos\theta \tag{13.18}$$

We also know that the direction θ of the maximum diagonal tensile strain is 45°, because shear stresses $\tau_{xy} = \tau_{yx}$. Thus:

$$\varepsilon_{dmax} = \frac{\gamma_{xy}}{2} \tag{13.19}$$

And Equation 13.16 becomes:

$$\varepsilon_{dmax} = \frac{3P}{8AG} \tag{13.20}$$

Rearranging Equation 13.20 for P gives:

$$P = \frac{8AG}{3}\varepsilon_{dmax} \tag{13.21}$$

Substituting Equation 13.21 into Equation 13.5 gives:

$$\Delta = \frac{AGL^3}{18EI}\left[1 + \frac{72EI}{5L^2AG}\right]\varepsilon_{dmax} \tag{13.22}$$

Similarly to the bending strain we want to calculate the maximum diagonal tensile strain ε_{dmax} from the deflection ratio we have already determined earlier, so simplifying and rearranging Equation 13.22 gives:

$$\varepsilon_{dmax} = \frac{\Delta/L}{\left[\dfrac{AGL^2}{18EI} + \dfrac{4}{5}\right]} \tag{13.23}$$

Poisson's ratio v is often assumed to be equal to 0.3 for masonry (Burland & Wroth, 1974). Although it doesn't appear in Equation 13.12 or Equation 13.40, Poisson's ratio is what relates Young's modulus E and shear modulus G in the following equation (e.g. Gere & Timoshenko, 1991):

$$E = 2G(1+v) \tag{13.24}$$

E is the Young's modulus
G is the shear modulus
v is the Poisson's ratio

Assuming $v = 0.3$, then the ratio E / G, which appears in Equation 13.12 and Equation 13.23, may be replaced with the number 2.6. The E / G ratio represents the relationship between bending and shear deformation of the building as a whole. Remember that this does not take account of the openings in the wall, we are considering it to be a solid rectangular prism, so the Poisson's ratio implied by the global E/G ratio value is the same as the Poisson's ratio of the material. If we were to use a different value of E/G ratio, for example the suggested value of 12.5 for a frame structure (Burland & Wroth, 1974), then the E/G ratio would cease to be directly related to the Poisson's ratio of the material. This will become important later in the chapter.

So now, from the hogging and sagging greenfield deflection ratios, we can calculate the maximum bending tensile strain and the maximum diagonal tensile strain in the building in hogging or sagging (by applying the appropriate value of second moment of area I). We now need to use the principle of superposition to add in the effect of horizontal axial strain to obtain a resultant bending strain and a resultant diagonal strain (Boscardin & Cording, 1989; Mair et al., 1996a).

The expression for greenfield horizontal strain was derived in Section 12.7 and is given by:

$$\varepsilon_h = \left(\frac{-S_{(max,z)}}{z_0 - z} \right) \cdot \exp\left(\frac{-y^2}{2i^2} \right) \cdot \left[1 - \frac{y^2}{i^2} \right] \qquad (13.25)$$

ε_h is the greenfield horizontal strain
$S_{(max,z)}$ is the maximum settlement at depth z below the ground surface in m
z is the depth below the ground surface in m
z_0 is the depth to the tunnel axis in m
y is the transverse offset in m
i is the trough width in m

For Stage 2 assessment, it is usual practice to use the *average* horizontal strains under the building's foundation in the hogging and sagging zones, i.e. along L_{hog} and L_{sag}. This was justified by Mair et al. (1996a and 1996b) by arguing that these greenfield horizontal strains are applied to the building, where they are added to tensile strains generated by shear and bending, and the precise location of these strains is unknown. They also argued that the method is effectively empirical and predicts building damage satisfactorily, also bearing in mind that the horizontal strain induced in the building is in many cases considerably less than the greenfield horizontal strain in the ground.

Therefore, Equation 13.25 should be used at, say, 1 m intervals along the building partitions within the hogging or sagging zone to calculate horizontal strains, which are then averaged within each zone to obtain the hogging and sagging values of horizontal strain.

Since the maximum extreme fibre bending strain acts in the same direction as the horizontal axial strain, the resultant bending strain ε_{br} is simply given by:

$$\varepsilon_{br} = \varepsilon_{bmax} + \varepsilon_h \qquad (13.26)$$

ε_{br} is the resultant bending strain
ε_{bmax} is the maximum extreme fibre bending tensile strain
ε_h is the average horizontal strain along the building partition

Remember that in the sagging zone the horizontal strain will be compressive and will therefore have a beneficial effect on the resultant bending strain, and in the hogging zone it will be tensile and will have an adverse effect.

The resultant diagonal strain ε_{dr} needs to be found using Mohr's circle of strain, because the maximum diagonal tensile strain and the horizontal strain are not in the same direction and the direction and magnitude of the resultant principal strain is therefore unknown.

Two systems of strain are superposed:

1. The horizontal (axial) strain ε_h. This is in the beam's x axis direction (c.f. Figure 13.4), therefore we can say that $\varepsilon_x = \varepsilon_h$. In this system this is the major principal strain, so $\gamma_{xy} = 0$ and using Hooke's Law we also have $\varepsilon_y = -v\varepsilon_h$.
2. The maximum diagonal tensile strain ε_{dmax}. This is at 45° to the beam's x axis. This system is in pure shear, so the major principal strain is ε_{dmax} and the minor principal strain is $-\varepsilon_{dmax}$. This system can be represented by a shear strain $\gamma_{xy} = 2\varepsilon_{dmax}$.

The System 1 Mohr's circle of strain is shown in Figure 13.6.

For System 1, the variation of strain with angle θ to the horizontal $\varepsilon_{\theta,h}$ is given by the average at the centre of its Mohr's circle plus the radius multiplied by $\cos 2\theta$:

$$\varepsilon_{\theta,h} = \frac{\varepsilon_h(1-v)}{2} + \frac{\varepsilon_h(1+v)}{2}\cos 2\theta \qquad (13.27)$$

The System 2 Mohr's circle of strain is shown in Figure 13.7.

For system 2, the variation of strain with angle ϕ from the direction of principal strain ε_{dmax} is given by the average at the centre of its Mohr's circle plus the radius multiplied by $\cos 2\phi$:

$$\varepsilon_\phi = \varepsilon_{dmax}\cos 2\phi \qquad (13.28)$$

Now since we know that the diagonal strain is acting at 45° to the horizontal, we can say that:

$$\phi = \theta + 45° \qquad (13.29)$$

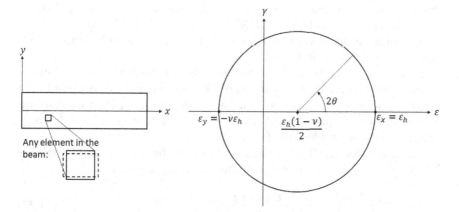

Figure 13.6 Mohr's circle of strain for System 1: horizontal strain.

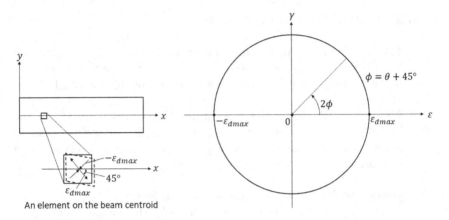

An element on the beam centroid

Figure 13.7 Mohr's circle of strain for System 2: diagonal strain.

And therefore that:

$$\cos2\phi = \cos(2\theta + 90°) = \sin2\theta \tag{13.30}$$

So we can rewrite Equation 13.28 as:

$$\varepsilon_{\theta,d} = \varepsilon_{dmax}\sin2\theta \tag{13.31}$$

Superposition, i.e. adding together, of Equations 13.27 and 13.31 gives:

$$\varepsilon_\theta = \frac{\varepsilon_h(1-v)}{2} + \frac{\varepsilon_h(1+v)}{2}\cos2\theta + \varepsilon_{dmax}\sin2\theta \tag{13.32}$$

Now we can draw a Mohr's circle of strain for the superposed strains ε_θ, but we don't yet know the direction or magnitude of the principal strain when the two are combined. The Mohr's circle is shown in Figure 13.8.

We can calculate the values of ε_θ at $\theta = 0$, 45° and 90° using Equation 13.32, and these are shown on Figure 13.8.

From Figure 13.8 we can see that the value of strain at the centre of the circle is the average of the $\theta = 0$ and $\theta = 90°$ values, and this is given in the figure. If we can also calculate the radius of the circle, then we can find the magnitude of the principal strain, which will be the resultant diagonal strain ε_{dr} that we are looking for.

Distance CB is the $\theta = 45°$ value of ε_θ minus the average strain, and distance CA is the $\theta = 0°$ value of ε_θ minus the average strain.

$$CB = \frac{\varepsilon_h(1-v)}{2} + \varepsilon_{dmax} - \frac{\varepsilon_h(1-v)}{2} = \varepsilon_{dmax} \tag{13.33}$$

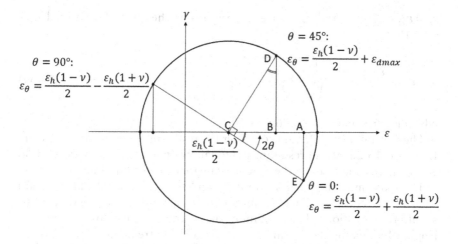

Figure 13.8 Mohr's circle of strain for superposition of System 1 and System 2.

$$CA = \frac{\varepsilon_h(1-v)}{2} + \frac{\varepsilon_h(1+v)}{2} - \frac{\varepsilon_h(1-v)}{2} = \frac{\varepsilon_h(1+v)}{2} \qquad (13.34)$$

Triangles CBD and CAE are similar triangles, therefore the radius R of the Mohr's circle may be given by:

$$R = \sqrt{CB^2 + CA^2} = \sqrt{\left(\frac{\varepsilon_h(1+v)}{2}\right)^2 + \left(\varepsilon_{dmax}\right)^2} \qquad (13.35)$$

The magnitude of the principal strain of the two superposed systems is therefore the average strain plus the radius:

$$\varepsilon_{dr} = \frac{\varepsilon_h(1-v)}{2} + \sqrt{\left(\frac{\varepsilon_h(1+v)}{2}\right)^2 + \left(\varepsilon_{dmax}\right)^2} \qquad (13.36)$$

Another way to derive the equation is to use the principal strain equation for plane strain (e.g. Gere & Timoshenko, 1991: p.438), which also applies to plane stress and is given by:

$$\varepsilon_{1,2} = \frac{\varepsilon_x + \varepsilon_y}{2} \pm \sqrt{\left(\frac{\varepsilon_x - \varepsilon_y}{2}\right)^2 + \left(\frac{\gamma_{xy}}{2}\right)^2} \qquad (13.37)$$

$\varepsilon_{1,2}$ represents the major principal strain ε_1 and the minor principal strain ε_2

Since the diagonal strain is pure shear and the horizontal (axial) strain is a principal strain (i.e. with no shear strain), we can say that

$\varepsilon_x = \varepsilon_h$, $\varepsilon_y = -v\varepsilon_h$ and $\gamma_{xy}/2 = \varepsilon_{dmax}$, and insert these into Equation 13.37 such that:

$$\varepsilon_{dr} = \frac{\varepsilon_h(1-v)}{2} + \sqrt{\left(\frac{\varepsilon_h(1+v)}{2}\right)^2 + \left(\varepsilon_{dmax}\right)^2} \tag{13.38}$$

which is the same as Equation 13.36.

The maximum tensile value of either the resultant bending strain ε_{br} from Equation 13.26 or the resultant diagonal strain ε_{dr} from Equation 13.36 will be used to determine the damage category for the building.

The maximum diagonal strain ε_{dmax} and the maximum bending strain ε_{bmax} are not considered in combination with each other because they occur at different locations. The maximum shear strain occurs close to the mid-height of a rectangular beam and is zero at the extreme fibre, whereas the maximum bending strain occurs at the extreme fibre. The distribution of shear stress is discussed in Appendix C.

**WORKED EXAMPLE 13.2   STAGE 2
BUILDING DAMAGE ASSESSMENT**

This is the same tunnel as described in Worked Example 13.1 – Stage 1 building damage assessment.

An 8 m diameter tunnel for a railway line is to be constructed using a TBM through a dense urban area at 18 m depth to axis. The parameters to be used for Stage 2 building damage assessment are a volume loss $V_l = 1.5\%$ and a trough width parameter $K = 0.5$.

A row of terraced houses is situated with its front façade transverse to the tunnel alignment, as shown in Figure 13.9. The façade is a solid double skin brick wall, i.e. with a thickness of 9 inches or 23 cm. The height from the foundation to the eaves is 5.6 m, and the total length is 31.5 m.

The settlement trough was calculated in Worked Example 13.1, where it was found that the trough width $i = 9$ m, and the maximum settlement $S_{max} = 33.4$ mm.

The building should be divided into partitions in the hogging and sagging zones, and we will treat each partition as though they were separate buildings. This is shown in Figure 13.10, where $L_{hog} = 16$ m and $L_{sag} = 15.5$ m.

Hogging zone calculation
In the hogging zone, we can calculate the settlements at $y = -25$ m and at $y = -9$ m and then calculate the values of settlement T caused by whole-body tilt between those two points, i.e. we plot a straight line between them.

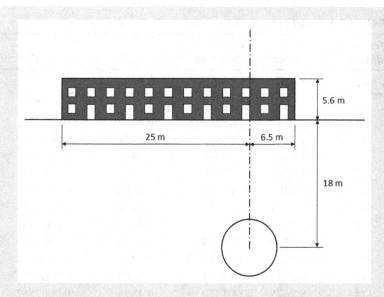

Figure 13.9 Worked Example 13.2 problem geometry for Stage 2 building damage assessment.

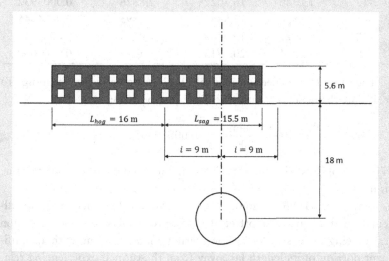

Figure 13.10 Worked Example 13.2 hogging and sagging zones.

Then we can calculate the actual Gaussian curve settlements at 1 m intervals, and for each one determine the deflection, i.e. the difference between the actual settlement and the straight line value. This is shown in Table 13.2.

Table 13.2 also has values of horizontal strain ε_h calculated using Equation 13.25, which we will need later.

Table 13.2 Worked Example 13.2 calculation of hogging deflection.

Transverse offset	Greenfield settlement	Settlement due to tilt	Deflection	Horizontal strain
y (mm)	S (mm)	T (mm)	$\Delta = S - T$ (mm)	ε_h
-25	0.7055	0.7055	0	0.0002632
-24	0.9547	1.9284	-0.9737	0.0003241
-23	1.2761	3.1512	-1.8752	0.0003921
-22	1.6846	4.3741	-2.6895	0.0004656
-21	2.1968	5.5970	-3.4002	0.0005424
-20	2.8294	6.8198	-3.9904	0.0006191
-19	3.5996	8.0427	-4.4431	0.0006913
-18	4.5231	9.2655	-4.7424	0.0007539
-17	5.6139	10.4884	-4.8745	0.0008009
-16	6.8823	11.7113	-4.8289	0.0008261
-15	8.3338	12.9341	-4.6004	0.0008231
-14	9.9675	14.1570	-4.1895	0.0007862
-13	11.7752	15.3798	-3.6046	0.0007107
-12	13.7401	16.6027	-2.8626	0.0005937
-11	15.8361	17.8256	-1.9895	0.0004345
-10	18.0279	19.0484	-1.0205	0.0002349
-9	20.2713	20.2713	0	0.0000000

The maximum deflection in Table 13.2 is shown in grey shading. To calculate the deflection ratio, we use Equation 13.3:

$$DR_{hog}^{GF} = \frac{\Delta_{hog}^{GF}}{L_{hog}} = \frac{-4.8744536}{16000} = -0.000304653$$

The sign of the deflection ratio is usually ignored and the absolute value is used.

Mair et al. (1996a) recommended cutting off the deflection ratio calculation at a transverse offset of $\pm 2.5i$, presumably to avoid deflection ratio being made smaller by considering a long building with a significant portion of its length far away from the tunnel, experiencing relatively little settlement. In this case, however, cutting off the calculation at $2.5i$ ($y = 22.5$ m) would give a significantly lower value of deflection ratio than the one calculated here.

Netzel (2009) also noted that considering the length of the building outside the cut-off point often results in a larger deflection ratio. Also if L/H is larger the maximum bending tensile strain will be larger (c.f. Equation 13.12). In one example he shows that considering the length of the building beyond the cut-off can increase maximum tensile strain by 75%. However, Netzel (2009) did not consider that a longer building will result in a lower average horizontal strain and potentially a lower resultant strain. Usually, for a building that extends a long distance into the hogging zone, the maximum resultant strain is at a maximum when the building partition is cut-off somewhere between $\pm2.5i$ and $\pm3i$.

We now need to know the properties of the building and idealise it as a rectangular isotropic beam of unit width.

For brick or masonry, Burland & Wroth (1974) recommended using a Poisson's ratio of 0.3, such that $E/G = 2.6$.

The neutral axis in hogging is assumed to be at the foundation level of the structure, and therefore the distance of the extreme fibre in tension from the neutral axis $d = H = 5.6$ m.

The second moment of area in hogging is given by Equation 13.14:

$$I_{hog} = \frac{H^3}{3} = \frac{5.6^3}{3} = 58.5387 \ \text{m}^4/\text{m}$$

The cross-sectional area of the building modelled as a beam of unit width is given by:

$$A = 1 \times H = 1 \times 5.6 = 5.6 \ \text{m}^2/\text{m}$$

The maximum bending tensile strain may be calculated using Equation 13.12:

$$\varepsilon_{bmax} = \frac{\Delta/L}{\dfrac{L}{12d}\left[1+\dfrac{72EI}{5L^2AG}\right]} = \frac{0.000304653}{\dfrac{16}{12\times5.6}\left[1+\dfrac{72\times58.5387}{5\times16^2\times5.6}\times2.6\right]} = 0.0005060 = 0.050\%$$

The maximum diagonal tensile strain may be calculated using Equation 13.23:

$$\varepsilon_{dmax} = \frac{\Delta/L}{\left[\dfrac{AGL^2}{18EI}+\dfrac{4}{5}\right]} = \frac{0.000304653}{\left[\dfrac{5.6\times16^2}{18\times58.5387}\cdot\dfrac{1}{2.6}+\dfrac{4}{5}\right]} = 0.0002302 = 0.023\%$$

The average horizontal strain is the average of the values in the right-hand column of Table 13.2, which is 0.0005448 or 0.054%.

The resultant bending strain is given by Equation 13.26:

$$\varepsilon_{br} = \varepsilon_{bmax} + \varepsilon_h = 0.0005060 + 0.0005448 = 0.0010508 = 0.105\%$$

The resultant diagonal strain is given by Equation 13.38:

$$\varepsilon_{dr} = \frac{\varepsilon_b(1-v)}{2} + \sqrt{\left(\frac{\varepsilon_b(1+v)}{2}\right)^2 + (\varepsilon_{dmax})^2} = \frac{0.0005448(1-0.3)}{2}$$

$$+ \sqrt{\left(\frac{0.0005448(1+0.3)}{2}\right)^2 + 0.0002302^2} = 0.0006130 = 0.061\%$$

The maximum of the resultant bending strain and the resultant diagonal strain is used to determine the damage category. This is the resultant bending strain, which is 0.105%. Referring to Table 13.1, the hogging zone of this building is in damage category 2 – 'slight'.

Sagging zone calculation

In the sagging zone, we can calculate the settlements at $y = -9$ m and at $y = 6.5$ m and then calculate the values of settlement T caused by whole-body tilt between those two points, i.e. we plot a straight line between them. Then we can calculate the actual Gaussian curve settlements at 1 m intervals, and for each one determine the deflection, i.e. the difference between the actual settlement and the straight line value. This is shown in Table 13.3.

Table 13.3 Worked Example 13.2 Calculation of sagging deflection.

Transverse offset	Greenfield settlement	Settlement due to tilt	Deflection	Horizontal strain
y	S (mm)	T (mm)	$\Delta = S - T$ (mm)	ε_b
-9	20.2713	20.2713	0.0000	0.0000000
-8	22.5141	20.6247	1.8894	-0.0002625
-7	24.6984	20.9781	3.7202	-0.0005421
-6	26.7620	21.3315	5.4305	-0.0008260
-5	28.6423	21.6849	6.9574	-0.0011001
-4	30.2786	22.0384	8.2402	-0.0013499
-3	31.6156	22.3918	9.2238	-0.0015613
-2	32.6066	22.7452	9.8614	-0.0017220
-1	33.2160	23.0986	10.1174	-0.0018226
0	33.4217	23.4520	9.9697	-0.0018568
1	33.2160	23.8054	9.4106	-0.0018226
2	32.6066	24.1588	8.4478	-0.0017220
3	31.6156	24.5122	7.1033	-0.0015613
4	30.2786	24.8657	5.4129	-0.0013499
5	28.6423	25.2191	3.4232	-0.0011001
6	26.7620	25.5725	1.1895	-0.0008260
6.5	25.7492	25.7492	0.0000	-0.0006843

Table 13.3 also has values of horizontal strain calculated using Equation 13.25, which we will need later.

The maximum deflection in Table 13.3 is highlighted in grey shading. To calculate the deflection ratio, we use Equation 13.4:

$$DR_{sag}^{GF} = \frac{\Delta_{sag}^{GF}}{L_{sag}} = \frac{10.1174}{15500} = 0.000652738$$

We now need to know the properties of the building and idealise it as a rectangular isotropic beam of unit width.

We will again use a Poisson's ratio of 0.3, such that $E/G = 2.6$.

The neutral axis in sagging is assumed to be at the mid-height of the structure, and therefore the distance of the extreme fibre in tension from the neutral axis $d = H/2 = 5.6/2 = 2.8$ m.

The second moment of area in sagging is given by Equation 13.13:

$$I_{sag} = \frac{H^3}{12} = \frac{5.6^3}{12} = 14.6347 \text{ m}^4/\text{m}$$

The cross-sectional area of the building modelled as a beam of unit width is again given by:

$$A = 1 \times H = 1 \times 5.6 = 5.6 \text{ m}$$

The maximum bending tensile strain may be calculated using Equation 13.12:

$$\varepsilon_{bmax} = \frac{\Delta/L}{\dfrac{L}{12d}\left[1 + \dfrac{72EI}{5L^2AG}\right]} = \frac{0.000652738}{\dfrac{15.5}{12 \times 2.8}\left[1 + \dfrac{72 \times 14.6347}{5 \times 15.5^2 \times 5.6} \times 2.6\right]} = 0.0010055 = 0.101\%$$

The maximum diagonal tensile strain may be calculated using Equation 13.23:

$$\varepsilon_{dmax} = \frac{\Delta/L}{\left[\dfrac{AGL^2}{18EI} + \dfrac{4}{5}\right]} = \frac{0.000652738}{\left[\dfrac{5.6 \times 15.5^2}{18 \times 14.6347} \cdot \dfrac{1}{2.6} + \dfrac{4}{5}\right]} = 0.0002361 = 0.024\%$$

The average horizontal strain is the average of the values in the right-hand column of Table 13.3, which is −0.0011829 or −0.118%. Note that it is negative because the horizontal ground strain is compressive in the sagging zone.

The resultant bending strain is given by Equation 13.26:

$$\varepsilon_{br} = \varepsilon_{bmax} + \varepsilon_h = 0.0010055 - 0.0011829 = -0.0001774 = -0.018\%$$

The resultant diagonal strain is given by Equation 13.38:

$$\varepsilon_{dr} = \frac{\varepsilon_h(1-v)}{2} + \sqrt{\left(\frac{\varepsilon_h(1+v)}{2}\right)^2 + \left(\varepsilon_{dmax}\right)^2} = \frac{-0.0011829(1-0.3)}{2}$$

$$+ \sqrt{\left(\frac{-0.0011829(1+0.3)}{2}\right)^2 + 0.0002361^2} = 0.0003903 = 0.039\%$$

The maximum of the resultant bending strain and the resultant diagonal strain is used to determine the damage category. The resultant bending strain is compressive, so it is the resultant diagonal strain that is critical, at 0.039%. Referring to Table 13.1, the sagging zone of this building is in damage category 1 – 'very slight'.

Since both the hogging and the sagging zone are in damage category 2 or below, the building can be considered at low risk of damage due to settlement and no further analysis will be required.

13.2.1 Generic Stage 2 assessment

Harris & Franzius (2006) proposed a generic form of Stage 2 assessment as a more efficient means of assessing large numbers of buildings. Major urban tunnelling projects may have hundreds or even thousands of buildings that progress from Stage 1 to Stage 2 and it is costly and time-consuming to produce a Stage 2 assessment and report for each one. In this method, no real buildings are considered but instead a wide range of generic building lengths and positions are analysed.

Representative transverse sections are produced, each with a representative surface settlement trough. Along each section, a large number of building geometries are analysed using the standard Stage 2 assessment method and the highest maximum tensile strain is determined.

This method is limited to buildings with foundations at or close to the surface, since it is the surface settlement trough that is used in the analysis. Harris & Franzius (2006) suggest a limit of $z_f/z_0 < 0.2$, where z_f is the foundation depth and z_0 is the tunnel axis depth, which in most cases means a single basement, or no basement, is acceptable. Buildings with two or more basement levels or piled foundations have to be assessed separately.

Using this method, many ordinary buildings over smaller diameter bored tunnels may be dismissed, and it is usually buildings close to large caverns, for instance crossovers, turnouts or underground stations, that will be taken forward for more detailed analysis.

13.3 STAGE 2B ASSESSMENT

A further refinement to the Stage 2 assessment may be added to filter out even more buildings prior to the relatively expensive Stage 3, and I have called this 'Stage 2b'. It has been known, at least since the experiences of the Jubilee Line Extension in London, that applying the greenfield deflection ratios and horizontal strains to a building is very conservative, particularly with respect to the horizontal strain (Viggiani & Standing, 2001), and particularly when dealing with buildings on continuous concrete strip or raft foundations. For example, a recent set of detailed field measurements in Bologna, Italy by Farrell et al. (2014) showed that horizontal strains measured in two buildings of different stiffnesses were negligible, even though the settlements were quite large with a volume loss of 5.1% causing a maximum settlement of approximately 200 mm. Numerical modelling by Goh & Mair (2012) found that if a building's axial stiffness EA is greater than 10^5 kN/m, which is less than the stiffness of most buildings, then the horizontal strain induced in the building is effectively zero.

Based on finite element modelling of tunnelling in a greenfield situation and of tunnelling under a building modelled as a simple beam, Potts & Addenbrooke (1997) proposed modification factors to reduce the deflection ratios and horizontal strains. These were defined as follows:

$$M^{DRsag} = \frac{DR_{sag}}{DR_{sag}^{GF}} \tag{13.39}$$

$$M^{DRhog} = \frac{DR_{hog}}{DR_{hog}^{GF}} \tag{13.40}$$

$$M^{ehc} = \frac{\varepsilon_{hc}}{\varepsilon_{hc}^{GF}} \tag{13.41}$$

$$M^{eht} = \frac{\varepsilon_{ht}}{\varepsilon_{ht}^{GF}} \tag{13.42}$$

M^{DRsag} and M^{DRhog} are the modification factors for deflection ratio in sagging and hogging, respectively

DR_{sag} and DR_{hog} are the deflection ratios at the foundation level of the building in the sagging and hogging zone, respectively, including the effect of the building on the ground movements

DR_{sag}^{GF} and DR_{hog}^{GF} are the greenfield deflection ratios of the ground at the foundation level of the building in the sagging and hogging zone, respectively

M^{ehc} and M^{eht} are the modification factors for horizontal compressive strain and horizontal tensile strain, respectively

ε_{hc} and ε_{ht} are the maximum horizontal compressive and tensile strains, respectively, at the foundation level of the building, including the effect of the building on ground movements

ε_{hc}^{GF} and ε_{ht}^{GF} are the maximum greenfield horizontal compressive and tensile strains, respectively, of the ground at the foundation level of the building

The relative stiffness of the structure-soil system was characterised by two relative stiffness factors, one for bending stiffness and one for axial stiffness. By calculating the relative stiffness values for the structure-soil system, the four modification factors could be read off from a design chart. They could then be used to convert greenfield deflection ratios and greenfield horizontal strains into values that take account of the structure-soil interaction.

In theory, the deflection ratio of a building will depend on both its bending stiffness and its axial stiffness. However, Goh & Mair (2008) showed that when the axial stiffness is high and in the range corresponding to actual buildings, the modification factor for deflection ratio depends mainly on bending stiffness.

After performing a large number of further analyses taking into account building length in the longitudinal direction, the effect of tunnel depth, the influence of interface roughness between the building and the ground, the number of floor slabs in the building, 3D tunnel construction and the effect of building weight, Franzius et al. (2006a) modified the relative stiffness expressions, and these are the ones we will use here. If you would like to see a comparison of the relative stiffness methods of Potts & Addenbrooke (1997), Franzius et al. (2006a) and Goh & Mair (2008), then refer to DeJong et al. (2019).

The relative bending stiffness is given by:

$$\rho_{mod}^* = \frac{EI}{E_s z_0 B^2 L} \tag{13.43}$$

ρ_{mod}^* is the modified relative bending stiffness (modified compared to the original proposed by Potts & Addenbrooke, 1997). It is dimensionless.

EI is the bending stiffness of the building, where E is the Young's modulus of the building material and I is the second moment of area of the building idealised as a beam.

E_s is the secant Young's modulus of the soil at depth $z_0/2$, at 0.01% axial strain in a triaxial compression test.

z_0 is the depth to the tunnel axis.

B is the total building width perpendicular to tunnelling.

L is the building length in the direction of tunnelling.

Note that the definitions of B and L are different to what we have used so far in this chapter, but this is the notation used in Potts & Addenbrooke (1997) and Franzius et al. (2006a).

The relative axial stiffness is given by:

$$\alpha^*_{mod} = \frac{EA}{E_s BL} \tag{13.44}$$

α^*_{mod} is the modified relative axial stiffness. It is dimensionless.

EA is the axial stiffness of the building, where E is the Young's modulus of the building material and A is the cross-sectional area of the building structure.

L is the length of the building in the tunnel longitudinal direction.

Note that Goh & Mair (2012) suggested new expressions for the relative bending and axial stiffness that included the total length of the building in the transverse direction B, such that the charts had only one design curve that was not dependent on e/B. However, this was for deep multipropped excavations and it is unclear whether this method would be appropriate for tunnels.

The design procedure for a Stage 2b assessment is as follows:

1. Predict greenfield settlement at the building foundation level using a Gaussian curve with assumed values of volume loss and trough width.
2. From the Gaussian curve, calculate greenfield deflection ratios in sagging and hogging (DR^{GF}_{sag} and DR^{GF}_{hog}), and maximum greenfield horizontal tensile and compressive strains (ε^{GF}_{hc} and ε^{GF}_{ht}).
3. Evaluate the EI and EA values for the building and obtain modification factors from the design charts (Figures 13.11 and 13.12). The e/B ratio is needed. Eccentricity e is the perpendicular horizontal distance from the centreline of the tunnel to the midpoint of the building width B.
4. Multiply the greenfield values of deflection ratios and maximum horizontal strains to obtain the modified values.
5. Use the modified values of deflection ratios and maximum horizontal strains to calculate resultant diagonal and bending strains and thereby obtain the damage category for the building using the limiting tensile strains in Table 13.1. If it falls within damage category 0, 1 or 2 then it will not have to be carried forward to Stage 3 assessment.

A limitation of this approach is that the modelling by Potts & Addenbrooke (1997) and Franzius et al. (2006a) used the same small strain stiffness model for London Clay, with a target volume loss of 1.5%. It is not known if these modification factors are appropriate for soils with different properties, e.g. drained cohesionless coarse-grained soils, or normally-consolidated clays or silts, or when the tunnelling method achieves significantly smaller or larger volume losses. Farrell et al. (2014) found that the design lines of Franzius et al. (2006a) provided an upper bound to actual observed modification factors for deflection ratio in hogging and sagging, both for

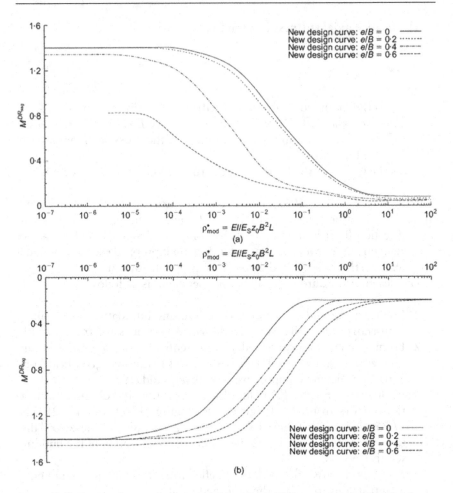

Figure 13.11 Design charts relating relative bending stiffness ρ^*_{mod} to modification factor in (a) sagging M^{DRsag} and in (b) hogging M^{DRhog} (Franzius et al., 2006a).

centrifuge model experiments in dry silica sand and for field measurements in Bologna. The ground conditions in Bologna consisted of overconsolidated highly stratified fluvial deposits with layers of silty clays and clayey silts, interbedded with lenses of sandy silt and silty sand. As mentioned previously, the volume loss for this 12 m diameter tunnel was very high at 5.1%.

Although compared to 3D finite element analyses of an advancing tunnel in Franzius et al. (2006a) as well as 2D plane strain finite element analyses, the modification factors only produce modified deflection ratios and horizontal strains for after the tunnel has passed, i.e. the short-term plane strain situation. As the tunnel face approaches, one would expect the nearer parts of the building to move downwards and towards the face more than the further away parts of the building. This will cause the building to twist,

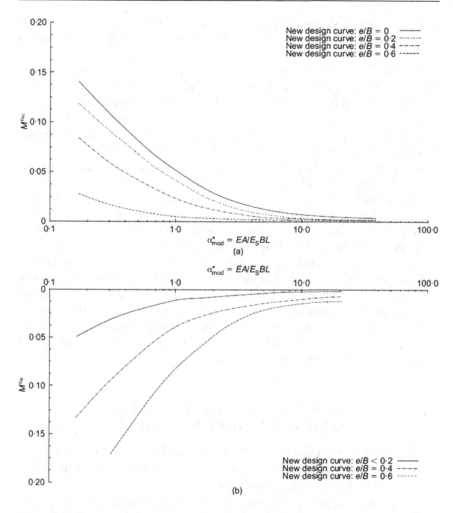

Figure 13.12 Design charts relating relative axial stiffness α^*_{mod} to modification factor in (a) compression $M^{\varepsilon hc}$ and (b) tension $M^{\varepsilon ht}$ (Franzius et al., 2006a).

first towards the tunnel as it approaches, and then in the opposite direction as it passes beyond (i.e. the incremental ground and structure movements will be predominantly directed towards the tunnel face at every advance step). This was investigated by Franzius et al. (2006b), where modification factors for twist were proposed.

A further limitation is that the modelling by Potts & Addenbrooke (1997) and Franzius et al. (2006a and 2006b) only examined buildings that are square to the tunnel centreline, i.e. not skewed at an arbitrary angle. In reality, most buildings will not be perpendicular or parallel to the tunnel centreline. Skew should be expected to cause permanent twist of the building. This is discussed to some degree by Franzius et al. (2006b) and field data of skewed building twist can be found in Standing (2001) for Elizabeth House in London, the data being later analysed by Franzius (2004).

Giardina et al. (2018) performed a large parametric study comparing the relative stiffness methods of Potts & Addenbrooke (1997), Franzius et al. (2006a) and Goh & Mair (2012) for a large number of building configurations, and found they gave very different values of modification factors in both hogging and sagging, and hence very different predictions of strain. Therefore these methods should be used with some caution.

At the end of the day, after all these sophisticated numerical and laboratory models and field measurements have been used to calculate these modification factors, we are going to apply them to greenfield deflection ratio and horizontal strain. We will then use these modified values of deflection ratio and horizontal strain in the simple beam model, with all its limitations. There is no way that the simple beam model can calculate the true strains in a building or accurately predict where or when damage will occur. It is, however, a simple well-established method of assessing risk that is very widely used and accepted. Perhaps one day, if enough field measurements are available, we can move to a purely empirical method of Stage 2 building damage assessment. Alternatively, with increasing computer power it may be possible to undertake a generic Stage 2 assessment using large numbers of numerical models.

WORKED EXAMPLE 13.3   STAGE 2B BUILDING DAMAGE ASSESSMENT

This is the same tunnel as described in Worked Example 13.1 – Stage 1 building damage assessment, that we also used for Worked Example 13.2 – Stage 2 building damage assessment.

An 8 m diameter tunnel for a railway line is to be constructed using a TBM through a dense urban area at 18 m depth to axis. The parameters to be used for Stage 2 building damage assessment are a volume loss $V_l =$ 1.5% and a trough width parameter $K = 0.5$.

A row of terraced houses is situated with its front façade transverse to the tunnel alignment, as shown in Figure 13.9. The façade is a solid double skin brick wall, i.e. with a thickness of 9 inches or 23 cm. The height from the foundation to the eaves is 5.6 m, and the total length is 31.5 m.

The secant ground stiffness determined at 0.01% axial strain in a triaxial compression test from a sample taken from a depth midway between the surface and the tunnel axis is 100 MPa.

The Young's modulus of the brick wall may be assumed to be 10 GPa and Poisson's ratio 0.3.

The greenfield hogging and sagging deflection ratios were calculated in Worked Example 13.2 and had the following values:

$$DR_{hog}^{GF} = \frac{\Delta_{hog}^{GF}}{L_{hog}} = 0.000304653$$

$$DR_{sag}^{GF} = \frac{\Delta_{sag}^{GF}}{L_{sag}} = 0.000652738$$

Hogging zone calculation
Using the design charts in Figure 13.11 requires the calculation of the relative bending stiffness using Equation 13.43:

$$\rho_{mod}^{*} = \frac{EI}{E_s z_0 B^2 L}$$

For unit wall thickness, the length of the building in the tunnel longitudinal direction $L = 1$.

The building's length in the transverse direction $B = 31.5$ m, and we will use $I = I_{hog} = 58.5387$ m^4/m that we calculated in Worked Example 13.2.

$$\rho_{mod}^{*} = \frac{E}{E_s} \cdot \frac{I_{hog}}{z_0 B^2} = \frac{10 \times 10^3}{100} \cdot \frac{58.5387}{18 \times 31.5^2} = 0.3278$$

We will use the lower chart in Figure 13.11, for which we need to know the value of e/B. Now e is the transverse distance of the midpoint of the building from the tunnel centreline, which is 9.25 m. Therefore:

$$\frac{e}{B} = \frac{9.25}{31.5} = 0.29$$

Reading off from the design chart gives $M^{DRhog} = 0.27$.

Therefore the hogging deflection ratio may be calculated using Equation 13.40:

$$DR_{hog} = M^{DRhog} \times DR_{hog}^{GF} = 0.27 \times 0.000304653 = 0.00008226$$

The maximum bending tensile strain and the maximum diagonal tensile strain can now be calculated using the modified value of deflection ratio. By inspection of the equations, they will be multiplied by 0.27 giving:

$$\varepsilon_{bmax} = 0.27 \times 0.0005060 = 0.0001366 = 0.014\%$$

$$\varepsilon_{dmax} = 0.27 \times 0.0002302 = 0.00006215 = 0.006\%$$

The horizontal strain also needs to modified, and first we need to calculate the relative axial stiffness using Equation 13.44:

$$\alpha^*_{mod} = \frac{EA}{E_s BL}$$

Again we assume unit wall thickness so $A = H$ and $L = 1$, such that:

$$\alpha^*_{mod} = \frac{EA}{E_s BL} = \frac{10 \times 10^3}{100} \cdot \frac{5.6}{31.5 \times 1} = 17.78$$

Reading off from the lower design chart in Figure 13.12 gives $M^{eht} = 0.005$. This is applied to the maximum greenfield horizontal tensile strain (remember that for the standard Stage 2 analysis we used the average horizontal strain). The maximum greenfield horizontal tensile strain we previously calculated in Table 13.2 is $\varepsilon^{GF}_{ht} = 0.0008261$.

$$\varepsilon_{ht} = M^{eht} \times \varepsilon^{GF}_{ht} = 0.005 \times 0.0008261 = 4.13 \times 10^{-6} = 4.13 \times 10^{-4} \%$$

The resultant bending strain may now be calculated using Equation 13.26:

$$\varepsilon_{br} = \varepsilon_{bmax} + \varepsilon_h = 0.0001366 + 4.13 \times 10^{-6} = 0.000141 = 0.0141\%$$

The resultant diagonal strain is given by Equation 13.38:

$$\varepsilon_{dr} = \frac{\varepsilon_h(1-v)}{2} + \sqrt{\left(\frac{\varepsilon_h(1+v)}{2}\right)^2 + \left(\varepsilon_{dmax}\right)^2} = \frac{4.13 \times 10^{-6}(1-0.3)}{2}$$

$$+ \sqrt{\left(\frac{4.13 \times 10^{-6}(1+0.3)}{2}\right)^2 + 0.00006215^2} = 0.00006366 = 0.0064\%$$

The maximum tensile strain in the building is therefore the resultant bending strain of 0.0141%. Referring to Table 13.1, the hogging zone of this building is in damage category 0 – 'negligible'. In Worked Example 13.2, without the modification factors, the damage category in the hogging zone was 2 – 'slight', so this is a considerable improvement.

Sagging zone calculation
Everything is as for hogging, but we will use $I = I_{sag} = 14.6347$ m⁴/m that we calculated in Worked Example 13.2.

$$\rho^*_{mod} = \frac{E}{E_s} \cdot \frac{I_{sag}}{z_0 B^2} = \frac{10 \times 10^3}{100} \cdot \frac{14.6347}{18 \times 31.5^2} = 0.0819$$

We will use the upper chart in Figure 13.11, and as for the hogging calculation, $e/B = 0.29$.

Reading off from the design chart gives $M^{DRsag} = 0.29$.

Therefore the sagging deflection ratio may be calculated using Equation 13.39:

$$DR_{sag} = M^{DRsag} \times DR_{sag}^{GF} = 0.29 \times 0.000652738 = 0.000189294$$

The maximum bending tensile strain and the maximum diagonal tensile strain can now be calculated using the modified value of deflection ratio. By inspection of the equations, they will be multiplied by 0.29 giving:

$$\varepsilon_{bmax} = 0.29 \times 0.0010055 = 0.00029160 = 0.029\%$$

$$\varepsilon_{dmax} = 0.29 \times 0.0002361 = 0.00006847 = 0.007\%$$

The horizontal strain also needs to modified. The relative axial stiffness is the same as for hogging, i.e. $\alpha^*_{mod} = 17.78$.

Reading off from the upper design chart in Figure 13.12 gives $M^{ehc} = 0.002$. This is applied to the maximum greenfield horizontal compressive strain (remember that for the standard Stage 2 analysis we used the average horizontal strain). The maximum greenfield horizontal compressive strain we previously calculated in Table 13.3 is $\varepsilon_{hc}^{GF} = -0.0018568$.

$$\varepsilon_{hc} = M^{ehc} \times \varepsilon_{hc}^{GF} = 0.002 \times (-0.0018568) = -3.714 \times 10^{-6} = -3.714 \times 10^{-4}\%$$

The resultant bending strain may now be calculated using Equation 13.26:

$$\varepsilon_{br} = \varepsilon_{bmax} + \varepsilon_h = 0.00029160 - 3.714 \times 10^{-6} = 0.0002879 = 0.029\%$$

The resultant diagonal strain is given by Equation 13.38:

$$\varepsilon_{dr} = \frac{\varepsilon_h(1-v)}{2} + \sqrt{\left(\frac{\varepsilon_h(1+v)}{2}\right)^2 + (\varepsilon_{dmax})^2} = \frac{(-3.714 \times 10^{-6})(1-0.3)}{2}$$

$$+ \sqrt{\left(\frac{(-3.714 \times 10^{-6})(1+0.3)}{2}\right)^2 + 0.00006847^2} = 0.0000672 = 0.007\%$$

The maximum tensile strain in the sagging zone of the building is therefore the resultant bending strain of 0.029%. Referring to Table 13.1, the sagging zone of this building is therefore in damage category 0 – 'negligible'.

In Worked Example 13.2, without the modification factors, the damage category in the sagging zone was 1 – 'very slight', so this is an improvement.

Note that in the sagging zone, the horizontal strain is compressive and is therefore beneficial, so using the modification factor for horizontal strain will be reducing this beneficial effect.

13.3.1 Buildings with discrete pad foundations

So far we have assumed that the buildings have a continuous strip or raft foundation in contact with the ground. It is not clear how buildings with discrete footings will respond to ground movements induced by tunnelling.

Goh & Mair (2014) presented a case study of two 2-storey reinforced concrete frame structures in Singapore, with individual footings founded on short timber piles estimated to be 6 to 9 m long. They found that the horizontal strains in the building were much higher than for structures with continuous footings. In addition, the highest horizontal strains occurred where the columns were not connected together at ground floor level. Where ground beams were present between columns, horizontal strains were much smaller. This was corroborated by finite element modelling.

13.3.2 Steel or reinforced concrete frame buildings

So far we have only considered buildings whose response is dominated by load-bearing wall behaviour, which are typically masonry houses. Another common form is a frame structure. These typically consist of columns and beams made from steel or reinforced concrete, with reinforced concrete floor slabs. The façades may have brick infill or cladding. Therefore, the building stiffness predominantly comes from the floor slabs, rather than the walls.

The same procedure may be applied as for masonry walls, but when the axial and bending stiffness are calculated, the floor slabs are considered. We need to find a way to add together the stiffness of various structural elements to obtain the stiffness of the building as a whole.

Potts & Addenbrooke (1997) suggested that the bending stiffness of a framed structure may be estimated by using the parallel axis theorem, namely:

$$(EI)_{structure} = E \sum_{i=1}^{n+1} \left(I_{slab} + A_{slab} d_{slab}^2 \right) \tag{13.45}$$

$(EI)_{structure}$ is the bending stiffness of the overall structure modelled as a single beam

I_{slab} is the second moment of area of an individual slab i

A_{slab} is the cross-sectional area of an individual slab i

d_{slab} is the vertical distance from the overall structure's neutral axis to the individual slab's neutral axis

n is the number of storeys, such that the building has $n + 1$ slabs

The axial stiffness is likewise given by:

$$(EA)_{structure} = E \sum_{i=1}^{n+1} (A_{slab}) \qquad (13.46)$$

$(EA)_{structure}$ is the axial stiffness of the overall structure modelled as a single beam

Potts & Addenbrooke (1997) acknowledged that the parallel axis theorem would tend to overestimate stiffness, as it is assumed that the slabs act compositely with transfer of shear between each slab. However, it is unlikely that there would be sufficient shear walls and moment connections between columns and beams to make this happen, except perhaps in a partial manner.

In making predictions of building response for Elizabeth House, Waterloo, during tunnelling for the Jubilee Line Extension, Mair & Taylor (2001) assumed that the floor slabs and 1.4 m thick raft foundation all contributed to overall bending stiffness, but did not act compositely. They assumed that all the slabs, including the raft foundation, were the same thickness, i.e. 0.3 m thick, using the following equation:

$$(EI)_{structure} = E \sum_{i=1}^{n+1} (I_{slab}) \qquad (13.47)$$

Since this was a 10-storey structure with two basement levels, it probably seemed unreasonable to use all the slabs in the building for the axial stiffness, so only the lower basement, basement and ground floor slabs were included with 1.4 m, 0.3 m and 0.3 m thickness, respectively (though the reason for doing this was never explained by Mair & Taylor, 2001). The predictions agreed well with actual measured building movements (Standing, 2001).

Goh & Mair (2014) tried both methods of estimating the bending stiffness of frame structures modelled using finite element analysis and compared the computed modification factors to the simple beam method, as shown in Figure 13.13. Using the parallel axis theorem (Equation 13.46) significantly underestimated the modification factor, while summing the bending stiffness (Equation 13.47) gave a realistic upper bound estimate of modification factor.

Based on finite element modelling, Goh & Mair (2014) also proposed correction factors for the stiffening effect of columns on beams, which can

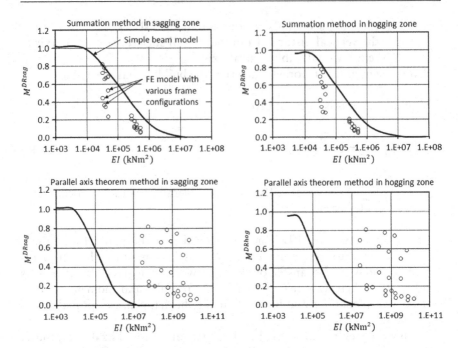

Figure 13.13 Modification factor from a simple beam model compared with finite element modelling of various configurations of frame structure with bending stiffness calculated either using the summation method or the parallel axis theorem (redrawn from Goh & Mair, 2014).

be used to make a better estimate of the modification factor when using the simple beam model. Meyerhof (1953) put forward an approach to estimate the bending stiffness of a building taking account of columns, infill panels and load-bearing walls without openings. Melis & Rodriguez Ortiz (2001) developed a method to take account of the contribution of walls and their proportion of openings, floors, and foundation slabs with either a rigid or hinged connection to the superstructure. The second moment of area is calculated relative to the neutral axis of the basement. Giardina et al. (2018) provided a useful review of all these methods.

In summary, these methods are ways of estimating the stiffness of a framed building, i.e. the second moment of area I and the cross-sectional area A, for use in either a Stage 2 or 2b analysis:

- In the case of a Stage 2 analysis, the calculated values of second moment of area I and cross-sectional area A are assigned to the beam, and the greenfield deflection ratio and horizontal strain are imposed to obtain a value of maximum tensile strain, which can then be used to determine the damage category.
- In a Stage 2b analysis, the stiffness of the building is one of the inputs needed to calculate the relative stiffness and hence obtain the

modification factors. Then, the greenfield deflection ratio and horizontal strain, multiplied by the modification factors, are applied to the beam to obtain a value of maximum tensile strain. The second moment of area I and cross-sectional area A of the beam will be the same as that determined in the relative stiffness calculation.

In both types of analysis, we also need to know the position of the neutral axis in order to use the bending strain equation (c.f. Equation 13.12), for both hogging and sagging. Since reinforced concrete or steel frame structures can support tension forces as well as compression forces (unlike masonry), then it seems sensible that the neutral axis should be the neutral axis of the structure, which is likely to be close to the mid-height, or a bit below, since basement and ground floor slabs are likely to be thicker. Often it is assumed to be at mid-height because of the difficulty in accounting for all structural details. Unlike for masonry walls, for frame structures the same neutral axis position should be used in both hogging and sagging.

Yet another aspect to bear in mind is that a frame structure cannot be assumed to be isotropic in the way we assumed a load-bearing masonry wall was isotropic, as it is an assembly of structural elements acting together in a partially composite manner. Therefore, the E/G ratio will not be directly dependent on the Poisson's ratio. Burland & Wroth (1974) argued that because a frame structure has a relatively low stiffness in shear and a reasonable degree of tensile restraint, then an E/G ratio of 12.5 should be used. This has a large influence on the maximum tensile strains calculated using Equation 13.12 and Equation 13.23. In most cases the maximum diagonal tensile strain will be larger than the maximum bending tensile strain and they will both be smaller compared to values obtained when $E/G = 2.6$.

It shouldn't be assumed that the behaviour of a framed building will be the same as a simple beam or that the damage to a framed building can be described using this method (Giardina et al., 2018). It is a very simple method used to make a conservative estimate of the level of risk of potential damage.

WORKED EXAMPLE 13.4 STAGE 2B BUILDING DAMAGE ASSESSMENT FOR A FRAME STRUCTURE

This is the same tunnel as described in Worked Example 13.1 – Stage 1 building damage assessment, that we also used for Worked Example 13.2 – Stage 2 building damage assessment and Worked Example 13.3 – Stage 2b building damage assessment.

An 8 m diameter tunnel for a railway line is to be constructed using a TBM through a dense urban area at 18 m depth to axis. The parameters to be used for Stage 2b building damage assessment are a volume loss $V_l = 1.5\%$ and a trough width parameter $K = 0.5$.

A reinforced concrete frame building is over the tunnel alignment, as shown in Figure 13.14. It has 4 storeys with no basement. All the storeys are 3.5 m high. The ground floor has a reinforced concrete raft foundation 1.0 m thick. The other floor slabs are all 0.3 m thick.

The secant ground stiffness determined at 0.01% axial strain in a triaxial compression test from a sample taken from a depth midway between the surface and the tunnel axis is 100 MPa.

The Young's modulus of the concrete is 30 GPa and Poisson's ratio 0.2.

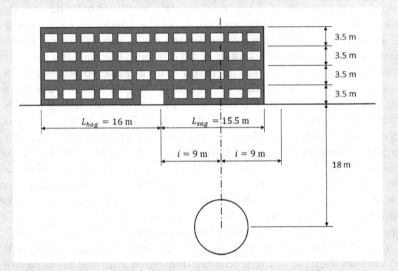

Figure 13.14 Worked Example 13.4 – Cross-section through building and tunnel.

Since the building has the same length and position relative to the tunnel as Worked Example 13.2, the greenfield hogging and sagging deflection ratios were calculated previously and had the following values:

$$DR_{hog}^{GF} = \frac{\Delta_{hog}^{GF}}{L_{hog}} = 0.000304653$$

$$DR_{sag}^{GF} = \frac{\Delta_{sag}^{GF}}{L_{sag}} = 0.000652738$$

We will assume that the building stiffness depends only on the reinforced concrete slabs, and ignore the contribution of columns, shear walls and other structural features.

Assuming that the overall second moment of area of the structure may be approximated by summing the individual slabs, using Equation 13.47 for a unit width gives:

$$I_{structure} = \sum_{i=1}^{n+1}(I_{slab}) = \frac{1 \times 1.0^3}{12} + 4\left(\frac{1 \times 0.3^3}{12}\right) = 0.0833 + 0.0090 = 0.0923 \text{ m}^4/\text{m}$$

Note that the contribution of the raft is much more than the sum of the four floor slabs above. Comparing this value of second moment of area with those calculated in Worked Example 13.2, this value is much lower than for a solid brick wall and here we are not calculating separate values of second moment of area for hogging and sagging.

We also need to know the cross-sectional area, again per unit width. We will make the same assumption as Mair & Taylor (2001) and count only the slabs at ground level or below, which in our case means just the raft:

$$A = 1.0 \times 1.0 = 1.0 \text{ m}^2/\text{m}$$

Using the design charts in Figure 13.11 requires the calculation of the relative bending stiffness using Equation 13.43:

$$\rho^*_{mod} = \frac{EI}{E_s z_0 B^2 L}$$

For unit thickness, the length of the building in the tunnel longitudinal direction $L = 1$.

The building's length in the transverse direction $B = 31.5$ m, so we get:

$$\rho^*_{mod} = \frac{E}{E_s} \cdot \frac{I_{structure}}{z_0 B^2} = \frac{30 \times 10^3}{100} \cdot \frac{0.0923}{18 \times 31.5^2} = 1.550 \times 10^{-3}$$

We will use the charts in Figure 13.11, for which we need to know the value of e/B. Now e is the transverse distance of the midpoint of the building from the tunnel centreline, which is 9.25 m. Therefore:

$$\frac{e}{B} = \frac{9.25}{31.5} = 0.29$$

Reading off from the design charts gives $M^{DRhog} = 1.26$ and $M^{DRsag} = 1.02$.

The hogging deflection ratio may be calculated using Equation 13.40:

$$DR_{hog} = M^{DRhog} \times DR^{GF}_{hog} = 1.26 \times 0.000304653 = 3.839 \times 10^{-4}$$

And the sagging deflection ratio may be calculated using Equation 13.39:

$$DR_{sag} = M^{DRsag} \times DR_{sag}^{GF} = 1.02 \times 0.000652738 = 6.658 \times 10^{-4}$$

The distance from the structure's neutral axis to the extreme fibre in bending is assumed to be the distance from the overall structure's neutral axis to the top floor slab in hogging and to the ground floor slab in sagging.

To find the height of the neutral axis above ground level d_{sag}, take moments of area about ground level:

$$0.3 \times 3.5 + 0.3 \times 7 + 0.3 \times 10.5 + 0.3 \times 14 = (1 + 4 \times 0.3) d_{sag}$$

$$\therefore d_{sag} = 4.77 \text{ m}$$

Now the vertical distance from the neutral axis to the extreme fibre in hogging will be given by:

$$d_{hog} = H - d_{sag} = 14 - 4.77 = 9.23 \text{ m}$$

Similarly, the relative axial stiffness is calculated using Equation 13.44:

$$\alpha_{mod}^* = \frac{EA}{E_s BL}$$

Again we assume unit thickness $L = 1$, such that:

$$\alpha_{mod}^* = \frac{EA}{E_s BL} = \frac{30 \times 10^3}{100} \cdot \frac{1}{31.5 \times 1} = 9.524$$

Reading off from the design charts in Figure 13.12 gives $M^{\varepsilon ht} = 0.006$ and $M^{\varepsilon hc} = 0.002$. These modification factors will be applied to the maximum greenfield horizontal strain in the hogging and sagging zone.

The modified horizontal strain in the hogging zone is given by:

$$\varepsilon_{ht} = M^{\varepsilon ht} \times \varepsilon_{ht}^{GF} = 0.006 \times 0.0008261 = 4.957 \times 10^{-6} = 4.957 \times 10^{-4}\%$$

And the modified horizontal strain in the sagging zone is given by:

$$\varepsilon_{hc} = M^{\varepsilon hc} \times \varepsilon_{hc}^{GF} = 0.002 \times (-0.0018568) = -3.714 \times 10^{-6} = -3.714 \times 10^{-4}\%$$

Hogging zone calculation
The maximum bending tensile strain and the maximum diagonal tensile strain can now be calculated using the modified value of

deflection ratio. Taking $E/G = 12.5$, the maximum bending tensile strain in hogging is given by:

$$\varepsilon_{bmax} = \frac{DR_{hog}}{\dfrac{L_{hog}}{12d_{hog}}\left[1+\dfrac{72EI}{5L^2AG}\right]} = \frac{3.839\times10^{-4}}{\dfrac{16}{12\times9.23}\left[1+\dfrac{72\times0.0923}{5\times16^2\times1.0}\times12.5\right]}$$

$$= 0.002496 = 0.250\%$$

The maximum diagonal tensile strain in hogging is given by:

$$\varepsilon_{dmax} = \frac{DR_{hog}}{\left[\dfrac{AGL_{hog}^{2}}{18EI}+\dfrac{4}{5}\right]} = \frac{3.839\times10^{-4}}{\left[\dfrac{1.0\times16^2}{18\times0.0923}\cdot\dfrac{1}{12.5}+\dfrac{4}{5}\right]} = 0.000029 = 0.003\%$$

The resultant bending strain may now be calculated using Equation 13.26:

$$\varepsilon_{br} = \varepsilon_{bmax} + \varepsilon_{ht} = 0.002496 + 4.957\times10^{-6} = 0.002501 = 0.250\%$$

The resultant diagonal strain is given by Equation 13.38:

$$\varepsilon_{dr} = \frac{\varepsilon_{ht}(1-v)}{2} + \sqrt{\left(\frac{\varepsilon_{ht}(1+v)}{2}\right)^2 + \left(\varepsilon_{dmax}\right)^2}$$

It may be unclear what the value of Poisson's ratio v should be – the value for concrete or the value for the structure as a whole. If we calculate the value for the structure as a whole, then based on $E/G = 12.5$ we get $v = 5.25$. A typical value for concrete is 0.2.

Although maximum diagonal tensile strain ε_{dmax} and horizontal strain ε_{ht} have been calculated based on the strain distribution of the structure as a whole, when we are calculating the resultant we are effectively superposing the strains at a single point in the structure so we should use the Poisson's ratio of the material, which is 0.2 in this case.

$$\varepsilon_{dr} = \frac{\varepsilon_{ht}(1-v)}{2} + \sqrt{\left(\frac{\varepsilon_{ht}(1+v)}{2}\right)^2 + \left(\varepsilon_{dmax}\right)^2} = \frac{4.957\times10^{-6}(1-0.2)}{2}$$

$$+ \sqrt{\left(\frac{4.957\times10^{-6}(1+0.2)}{2}\right)^2 + 0.000029^2} = 0.00003113 = 0.003\%$$

The maximum tensile strain in the building is therefore the resultant bending strain of 0.250%. Referring to Table 13.1, the hogging zone of this building is in damage category 3 – 'moderate'.

Sagging zone calculation

Taking $E/G = 12.5$, the maximum bending tensile strain in sagging is given by:

$$\varepsilon_{bmax} = \frac{DR_{sag}}{\dfrac{L_{sag}}{12d_{sag}}\left[1 + \dfrac{72EI}{5L^2AG}\right]} = \frac{6.658 \times 10^{-4}}{\dfrac{16}{12 \times 4.77}\left[1 + \dfrac{72 \times 0.0923}{5 \times 15.5^2 \times 1.0} \times 12.5\right]} = 0.002228 = 0.223\%$$

The maximum diagonal tensile strain in sagging is given by:

$$\varepsilon_{dmax} = \frac{DR_{sag}}{\left[\dfrac{AGL_{sag}^2}{18EI} + \dfrac{4}{5}\right]} = \frac{6.658 \times 10^{-4}}{\left[\dfrac{1.0 \times 15.5^2}{18 \times 0.0923} \cdot \dfrac{1}{12.5} + \dfrac{4}{5}\right]} = 0.000054 = 0.005\%$$

The resultant bending strain may now be calculated using Equation 13.26:

$$\varepsilon_{br} = \varepsilon_{bmax} + \varepsilon_{hc} = 0.002228 + -3.714 \times 10^{-6} = 0.002224 = 0.222\%$$

The resultant diagonal strain is given by Equation 13.38:

$$\varepsilon_{dr} = \frac{\varepsilon_{hc}(1-v)}{2} + \sqrt{\left(\frac{\varepsilon_{hc}(1+v)}{2}\right)^2 + \left(\varepsilon_{dmax}\right)^2} = \frac{-3.714 \times 10^{-6}(1-0.2)}{2}$$

$$+ \sqrt{\left(\frac{-3.714 \times 10^{-6}(1+0.2)}{2}\right)^2 + 0.000054^2} = 0.00005256 = 0.005\%$$

The maximum tensile strain in the building is therefore the resultant bending strain of 0.222%. Referring to Table 13.1, the sagging zone partition of this building is in damage category 3 – 'moderate'.

We categorise the building according to the worst damage category of the hogging and sagging zone partitions. Therefore, the building is in damage category 3 – 'moderate' and needs to be taken forward to Stage 3 assessment for more detailed survey and analysis.

13.4 STAGE 3 ASSESSMENT

A Stage 3 assessment is a refinement of the preceding analysis, and must take account of the individual building's structural details and the tunnelling method (Burland, 2001), as opposed to Stage 2, which may be approached generically (Harris & Franzius, 2006). Sometimes Stage 3 is approached as a series of sub-steps, each sub-step increasing the sophistication of the building and tunnel model (Crossrail, 2008). It is assumed that at each sub-step the maximum tensile strain will decrease and the damage category may also decrease. If the damage category reduces to 2 or below, then the assessment is stopped.

If deemed necessary, a building condition survey is undertaken by a structural engineer to determine the structural form and condition of the building (Crossrail, 2008). They will look for evidence of previous movements and damage, the type of foundations, and the form of the structure. Buildings that have already suffered previous damage may be less able to cope with additional movements. Buildings on continuous foundations such as strip footings or concrete raft foundations will be less vulnerable to differential movements. Buildings with structural continuity such as steel or reinforced concrete frame structures are less likely to suffer damage than load-bearing brick or masonry wall structures (Burland, 2001).

Giardina et al. (2020) defined four main areas where the tunnel-soil-building model may be refined: excavation, propagation of displacements from the tunnel boundary to the ground surface, the effect of soil-structure interaction, and the vulnerability of the building to the imposed displacements.

13.4.1 Tunnel excavation

Up to now, we have imposed empirical greenfield settlements on the building. By looking into the tunnel excavation method in detail, a lower value of volume loss may be found to be justified. Or perhaps by looking at the particular ground conditions at the location of the building, a lower value of volume loss or a larger trough width may be justified.

If a numerical model is used, then the geometry of the problem may be made more realistic by including soil properties that vary with stratum and with depth, by modelling the depth and skew of the building's foundation relative to the tunnel, and by including details of the tunnel excavation, such as face pressure, shield gap or grout pressure. Refer to Chapter 7, on numerical modelling, for more details.

13.4.2 Soil-structure interaction

As we have seen in Stage 2b, where finite element modelling by Potts & Addenbrooke (1997) and Franzius et al. (2006a) was used to determine modification factors to take account of soil-structure interaction, a numerical model that includes the building could be used to calculate building strains and thereby determine the damage category. Usually the building is modelled as a beam in 2D or using shell elements in 3D with equivalent elastic properties (Goh & Mair, 2012). This enables the interaction of the bending, shear and axial stiffness of the building with the ground settlements to be taken into account, as well as the building's self weight.

In a slight variation, Pickhaver et al. (2010) used a mesh of equivalent elastic Timoshenko beams to model the walls of a building in a 3D finite element analysis, including internal walls. They found that, compared to nonlinear numerical models of masonry panels with and without openings, the beams replicated the behaviour well in sagging, but required the use of a reduced flexural stiffness to replicate behaviour in hogging.

Using a centrifuge model with a building modelled by two layers of tiny mortared bricks, Farrell & Mair (2012) found that the horizontal strain transferred to buildings through their foundations is negligible, as has been found by others. When the results were compared to the modification factors produced by Franzius et al. (2006a), agreement was poor but the modification factors did provide an upper bound. This means that buildings that proceed from Stage 2b analysis are likely to have reduced strains predicted in Stage 3 analyses when more sophisticated methods are used.

Farrell & Mair (2012) also found that model buildings that respond rigidly do so by redistributing their weight. Therefore, corners or edges of the building may embed into the ground, as has been found during monitoring of real buildings affected by tunnelling (Farrell et al., 2014), and by centrifuge models of buildings on sand by Ritter et al. (2017). Although this is likely to reduce building distortion, it could cause serious problems for utility connections entering the building, which may experience a rapid change of displacement between the adjacent ground and the building.

In their numerical analyses with the building modelled by an elastic beam, Goh & Mair (2012) noted that when a building's axial stiffness is low, the transfer of horizontal strains from the ground to the building is determined by the interface properties between the ground and the building in the model. When the interface coefficient was 0.5, the modification factor also tended towards 0.5 as building axial stiffness decreased. If the interface coefficient were 1.0, meaning that the nodes of the building elements in the model were tied rigidly to the ground elements, then the modification factor would tend towards 1.0.

13.4.3 Modelling the building

There are a large number of numerical modelling studies of building response to settlement, but since very few of them have been compared to real or scale model buildings, they will not all be discussed here.

Giardina et al. (2012) built a physical 1/10th scale model of a masonry façade and applied a hogging settlement profile to its base. They found that the presence of door and window openings had an influence on the crack distribution. Corners of openings create stress concentrations and tend to be where cracks initiate. This cannot be taken into account in the Stage 2 analysis, and a building façade with large openings may end up with a higher damage category when modelled in more detail in Stage 3. Giardina et al. (2013) built a 2D continuum numerical model of the physical scale model in Giardina et al. (2012), and found it was able to accurately reproduce the façade deformation and cracking pattern. For masonry buildings, therefore, it is of crucial importance to model the initiation and propagation of cracking in the structure.

Giardina et al. (2019) presented the results of physical and numerical modelling of another masonry façade, this time of a 1/20ᵗʰ scale palace in Brescia, Italy. Two types of numerical models were compared with the experimental results: a continuum finite element model and a discrete element model. The discrete element model allowed the mortared joints between masonry blocks to be modelled explicitly, with crack patterns following the mortar joints. In the continuum model the joints could not be modelled, so a total strain rotating crack model was used, which distributes the localised damage over a certain crack bandwidth of the finite elements. The stress-strain relationship for the façade was elastic in compression and linear softening in tension, determined by the fracture energy. The crack patterns and deformations of both the models were similar to the experiment.

Giardina et al. (2015) used finite element analyses to perform a sensitivity study to investigate the importance of the interface between a building and the soil, the building's stiffness and the building's weight, and compared the results with centrifuge models. They found that it was important to enable a gap to form between the building and the soil. The results were also sensitive to building stiffness, and also it was found to be important to model the building's weight.

Ritter et al. (2017) investigated the effect of openings in a façade in centrifuge models of 3D printed buildings on sand. They found that the area of openings affected the structure response to settlement, especially in the hogging zone, where the building behaved more flexibly. In addition, the area of the openings reduced the axial stiffness such that more horizontal ground strain was transferred to the building. It therefore seems important that openings in façades should be included in any physical or numerical model.

In summary, numerical models of masonry structures affected by settlement need to include cracking, whether this is approximated in a continuum model, or done explicitly in a discrete element model. The layout and size of openings also needs to be included, as well as the building's weight. Interface elements between the building and the ground are needed to allow a gap to form if it wants to.

13.5 SUMMARY

The concept of limiting tensile strain and its correlation to building damage categories has been described.

There are three stages to building damage assessment. Each stage is intended to be progressively more sophisticated and less conservative, with the aim that structures at low risk of damage can be ruled out, reducing the number of buildings that require detailed analysis.

Stage 1 analysis involves calculation of contours of 10 mm settlement and 1:500 slope and classifying all buildings outside of both contours as at low risk of damage. All buildings within either of the contours will go on to Stage 2 assessment.

Stage 2 assessment takes the greenfield deflection ratio and horizontal strain and imposes them on each building using the simple beam method. The building is separated into partitions in the hogging and sagging zones, and cut-off at a distance of 2.5i or 3i from the tunnel centreline. The simple beam method models the building as a rectangular isotropic beam and allows the calculation of resultant bending strain and resultant diagonal strain. The greater of these in the hogging and sagging partitions of the building are taken as the limiting tensile strain and give us the damage category for the building.

Stage 2b assessment involves the use of a relative stiffness method to allow for soil-structure interaction between the building and the ground. Modification factors from Franzius et al. (2006a) can be read from design charts based on numerical modelling, and are applied to the greenfield deflection ratios and horizontal strains. This generally provides a less conservative estimate of damage category compared to Stage 2.

Stage 3 assessment must take account of the individual building's structural details and the tunnelling method. It is usually approached in a series of sub-steps that introduce layers of complexity and refine the prediction. As well as desk studies and surveys of the building itself, Stage 3 analysis may include numerical modelling of the structure, with structural details, façade openings and material behaviour characterised. For masonry structures the model should include cracking. Generally speaking, Stage 3 should reduce the predicted damage, but can in some cases result in greater predicted damage.

13.6 PROBLEMS

Q13.1. This question is about Stage 1 building damage assessment. A 6.5 m diameter tunnel will be driven under an urban area at a depth to axis of 17 m. Assume the volume loss is 2% and the trough width parameter K is 0.3.

   i. Calculate the trough width i and the maximum settlement S_{max}.
   ii. Calculate the transverse offset from the tunnel centreline at which the settlement is 10 mm.
   iii. Calculate the transverse offset from the tunnel centreline at which the slope is 1:500.
   iv. Give the transverse offset within which buildings would proceed to Stage 2 assessment.

Q13.2. This question is about Stage 1 building damage assessment. A 72 m² face area shotcrete tunnel will be driven in a straight line under an urban area at a constant level for 120 m length. Due to the surface level changing, the depth to axis varies between 17 and 25 m from one end of the tunnel to the other. Assume the volume loss is 1.5% and the trough width parameter K is 0.45.

i. Calculate the trough width i and the maximum settlement S_{max} at 17 and 25 m.

ii. Calculate the transverse offset from the tunnel centreline at which the settlement is 10 mm for each case.

iii. Calculate the transverse offset from the tunnel centreline at which the slope is 1:500 for each case.

iv. Draw the transverse offset within which buildings would proceed to Stage 2 assessment as a contour on a plan of the tunnel.

Q13.3. This question is about Stage 2 building damage assessment. A 4.5 m diameter tunnel is to be driven under a brick masonry building at a depth to axis of 12 m. The building façade is perpendicular to the tunnel centreline and it has strip foundations at 1 m below ground level. The façade is 8 m high from the foundation level to the eaves, and extends 25 m either side of the tunnel centreline. Assume the volume loss is 1.5% and the trough width parameter K is 0.5.

i. Calculate the trough width i and the maximum settlement S_{max}.

ii. Separate the building into hogging and sagging partitions, with the hogging partitions cut-off at $2.5i$.

iii. Calculate the hogging and sagging deflection ratios.

iv. Calculate the average horizontal strain in the hogging partition and the sagging partition.

v. Calculate the maximum bending tensile strain and the maximum diagonal tensile strain in the hogging partition.

vi. Calculate the resultant bending strain and resultant diagonal strain in the hogging partition and assign the building to a damage category.

vii. Calculate the maximum bending tensile strain and the maximum diagonal tensile strain in the sagging partition.

viii. Calculate the resultant bending strain and resultant diagonal strain in the sagging partition and assign the building to a damage category. What is the overall damage category for the building?

Q13.4. This question is about Stage 2 building damage assessment. A 7 m diameter tunnel is to be driven under a reinforced concrete frame building at a depth to axis of 20 m. The building is perpendicular to the tunnel centreline, is 30 m long and is centred over the tunnel alignment. It has a single level basement with a 1.2 m thick reinforced concrete raft foundation slab at 5 m below ground level. The building has a 0.4 m thick ground floor slab at ground level and three more 0.3 m thick floor slabs at 3.5 m centres above. Assume the volume loss is 1.5% and the trough width parameter K is 0.5.

   i. Calculate the trough width i and the maximum settlement S_{max} at the foundation level.
  ii. Separate the building into hogging and sagging partitions, with the hogging partitions cut-off at $2.5i$ if necessary, and calculate the hogging and sagging deflection ratios.
 iii. Calculate the average horizontal strain in the hogging partition and the sagging partition.
  iv. Calculate the maximum bending tensile strain and the maximum diagonal tensile strain in the hogging partition.
   v. Calculate the resultant bending strain and resultant diagonal strain in the hogging partition.
  vi. Calculate the maximum bending tensile strain and the maximum diagonal tensile strain in the sagging partition.
 vii. Calculate the resultant bending strain and resultant diagonal strain in the sagging partition. What is the overall damage category for the building?

Q13.5. This question is about Stage 2b building damage assessment and uses the same geometry of tunnel and building described in Q13.4. Assume the reinforced concrete slabs have a Young's modulus $E = 30$ GPa and assume the secant ground stiffness determined at 0.01% axial strain in a triaxial compression test from a sample taken from a depth midway between the surface and the tunnel axis $E_s = 150$ MPa.

   i. For the tunnel and building described in Q13.4, calculate the relative bending stiffness and relative axial stiffness of the building.
  ii. Using the relative stiffness values, use the design charts from Franzius et al. (2006a) to find values for modification factors.
 iii. Apply the modification factors to the deflection ratios and maximum horizontal strains in the hogging and sagging partitions.
  iv. Calculate the maximum bending tensile strain and the maximum diagonal tensile strain in the hogging and sagging partitions.
   v. Calculate the resultant bending strain and resultant diagonal strain in the hogging and sagging partitions. What is the overall damage category for the building?

REFERENCES

Bhatt, P. (1999). *Structures*. New Jersey: Prentice Hall.
Boscardin, M. D. & Cording, E. J. (1989). Building response to excavation-induced settlement. *ASCE J. Geot. Eng.* 115, No. 1, 1–21.

Bowers, K. H. & Moss, N. A. (2006). Settlement due to tunnelling on the CTRL London Tunnels. *Proc. 5th Int. Symp. on Geotechnical Aspects of Underground Construction in Soft Ground* (eds Bakker, K. J., Bezuijen, A., Broere, W. & Kwast, E. A.), Amsterdam, The Netherlands, 15th–17th June 2005, pp.203–209. London: Taylor & Francis.

Burland, J. B. (2001). Assessment methods used in design. *Building response to tunnelling, volume 1: projects and methods* (eds Burland, J. B., Standing, J. R. & Jardine, F. M.), CIRIA Special Publication 200, pp.23–43. London: Thomas Telford Publishing.

Burland, J. B., Broms, B. B. & de Mello, V. F. B. (1977). Behaviour of foundations and structures. *Proc. 9th Int. Conf. Soil Mechanics and Foundation Engineering*, Tokyo, Japan, Session 2, pp. 495–546.

Burland, J. B. & Wroth, C. P. (1974). Settlement of buildings and associated damage. State of the art review. *Conf. on Settlement of Structures*, Cambridge, pp. 611–654. London: Pentech Press.

Crossrail (2008). *Crossrail Information Paper D12 – Ground Settlement*. London, UK: CLRL.

DeJong, M. J., Giardina, G., Chalmers, B., Lazarus, D., Ashworth, D. & Mair, R. J. (2019). Impact of the Crossrail tunnelling project on masonry buildings with shallow foundations. *Proc. Instn Civ. Engrs – Geot. Engrg* 172, No. 5, 402–416.

Farrell, R. P. & Mair, R. J. (2012). Centrifuge modelling of the response of buildings to tunnelling. *Proc. 7th Int. Symp. on Geotechnical Aspects of Underground Construction in Soft Ground* (ed. Viggiani, G.), Rome, Italy, 17th–19th May 2011, pp. 343–351. London: Taylor & Francis Group.

Farrell, R., Mair, R., Sciotti, A. & Pigorini, A. (2014). Building response to tunnelling. *Soils Found.* 54, No. 3, 269–279.

Franzius, J. N. (2004). *The behaviour of buildings due to tunnel induced subsidence*. PhD thesis, Imperial College, University of London, 2004.

Franzius, J. N., Potts, D. M. & Burland, J. B. (2006a). The response of surface structures to tunnel construction. *Proc. Instn Civ. Engrs – Geot. Engrg* 159, GE1, 3–17.

Franzius, J. N., Potts, D. M. & Burland, J. B. (2006b). Twist behaviour of buildings due to tunnel induced ground movement. *Proc. 5th Int. Symp. on Geotechnical Aspects of Underground Construction in Soft Ground* (eds Bakker, K. J., Bezuijen, A., Broere, W. & Kwast, E. A.), Amsterdam, The Netherlands, 15th–17th June 2005, pp. 107–113. London: Taylor & Francis Group.

Gere, J. M. & Timoshenko, S. P. (1991). *Mechanics of materials*, 3rd S. I. edition. London: Chapman & Hall.

Giardina, G., DeJong, M. J., Chalmers, B., Ormond, B. & Mair, R. J. (2018). A comparison of current analytical methods for predicting soil-structure interaction due to tunnelling. *Tunn. Undergr. Space Technol.* 79, 319–335.

Giardina, G., DeJong, M. J. & Mair, R. J. (2015). Interaction between surface structures and tunnelling in sand: Centrifuge and computational modelling. *Tunn. Undergr. Space Technol.* 50, 465–478.

Giardina, G., Losacco, N., DeJong, M. J., Viggiani, G. M. B. & Mair, R. J. (2020). Effect of soil models on the prediction of tunnelling-induced deformations of structures. *Proc. Instn Civ. Engrs – Geot. Engrg* 173, No.5, 379–397.

Giardina, G., Marini, A., Hendriks, M. A. N., Rots, J. G., Rizzardini, F. & Giuriani, E. (2012). Experimental analysis of a masonry façade subject to tunnelling-induced settlement. *Eng. Struct.* 45, 421–434.

Giardina, G., Marini, A., Riva, P. & Giuriani, E. (2019). Analysis of a scaled stone masonry facade subjected to differential settlements. *Int. J. Architect. Heritage*, published online, DOI: 10.1080/15583058.2019.1617911.

Giardina, G., van de Graaf, A. V., Hendriks, M. A. N., Rots, J. G. & Marini, A. (2013). Numerical analysis of a masonry façade subject to tunnelling-induced settlements. *Eng. Struct.* 54, 234–247.

Goh, K.H. & Mair, R.J. (2008). Response of a building under excavation-induced ground movements, *Proc. Int. Conf. on Deep Excavations*, 10th–12th Nov 2008, Singapore.

Goh, K.H. & Mair, R.J. (2012). The response of buildings to movements induced by deep excavations. *Proc. 7th Int. Symp. on Geotechnical aspects of underground construction in soft ground* (ed. Viggiani, G.), Rome, Italy, 17th–19th May 2011, pp. 903–910. London: Taylor & Francis Group.

Goh, K. H. & Mair, R. J. (2014). The response of framed buildings to excavation-induced movements. *Soils Found.* 54, No. 3, 250–268.

Harris, D. I. & Franzius, J. N. (2006). Settlement assessment of running tunnels – a generic approach. *Proc. 5th Int. Symp. on geotechnical aspects of underground construction in soft ground* (eds Bakker, K. J., Bezuijen, A., Broere, W. & Kwast, E. A.), Amsterdam, The Netherlands, 15th–17th June 2005, pp. 225–230. London: Taylor & Francis Group.

Mair, R. J. & Taylor, R. N. (2001). Elizabeth House: settlement predictions. *Building response to tunnelling*, vol. 1 projects and methods, CIRIA Special Publication 200, pp. 195–215. London: Thomas Telford.

Mair, R. J., Taylor, R. N. & Burland, J. N. (1996a). Prediction of ground movements and assessment of risk of building damage due to bored tunnelling. *Proc. Int. Symp. on geotechnical aspects of underground construction in soft ground* (eds Mair, R. J. & Taylor, R. N.), London, UK, pp. 713–718. Rotterdam: Balkema.

Mair, R. J., Taylor, R. N. & Burland, J. N. (1996b). Discussion: Reply to discussion by M. P. O'Reilly. *Proc. Int. Symp. on geotechnical aspects of underground construction in soft ground* (eds Mair, R. J. & Taylor, R. N.), London, UK, pp. 765–766. Rotterdam: Balkema.

Melis, M. & Rodriguez Ortiz, J. (2001). Consideration of the stiffness of buildings in the estimation of subsidence damage by EPB tunnelling in the Madrid subway. *Response of buildings to excavation induced ground movements*, Proc. Int. Conf. held at Imperial College, London, UK, 17th–18th July 2001 (ed. Jardine, F. M.), CIRIA Special Publication 199. London: CIRIA.

Meyerhof, G. G. (1953). Some recent foundation research and its application to design. *Struct. Eng.* 31, 151–167.

Moss, N. A. & Bowers, K. H. (2006). The effect of new tunnel construction under existing metro tunnels. *Proc. 5th Int. Symp. on Geotechnical Aspects of Underground Construction in Soft Ground* (eds Bakker, K. J., Bezuijen, A., Broere, W. & Kwast, E. A.), Amsterdam, The Netherlands, 15th–17th June 2005, pp. 151–157. London: Taylor & Francis.

Netzel, H. D. (2009). *Building response due to ground movements.* PhD thesis, Technische Universiteit Delft. Delft, The Netherlands: Delft University Press.

Pickhaver, J. A., Burd, H. J. & Houlsby, G. T. (2010). An equivalent beam method to model masonry buildings in 3D finite element analysis. *Comput. Struct.* **88**, 1049–1063.

Potts, D. M. & Addenbrooke, T. I. (1997). A structure's influence on tunnelling-induced ground movements. *Proc. Instn Civ. Engrs – Geot. Engrg* **125**, April, 109–125.

Rankin, W. J. (1988). Ground movements resulting from urban tunnelling: prediction and effects. *Conf. on Engineering Geology of Underground Movements*, Nottingham, Geological Society Engineering Geology Special Publication No. 5, pp. 79–92.

Ritter, S., Giardina, G., DeJong, M. J. & Mair, R. J. (2017). Influence of building characteristics on tunnelling-induced ground movements. *Géotechnique* **67**, No. 10, 926–937.

Standing, J. R. (2001). Elizabeth House, Waterloo. *Building response to tunnelling*, vol. 2 case studies, CIRIA Special Publication 200, pp. 547–612. London: Thomas Telford.

Viggiani, G. & Standing, J. R. (2001). The Treasury. *Building response to tunnelling*, vol. 2 case studies, CIRIA Special Publication 200, pp. 401–432. London: Thomas Telford.

Appendix A: Derivation of wedge-prism method

Anagnostou & Kovári (1994, 1996a, 1996b) used a limit equilibrium solution they attributed to Horn (1961) and developed it to allow direct calculation of the required support pressure in any situation above or below the water table, with or without seepage, for tunnels where the unsupported length $P = 0$. The geometry is shown in Figure A.1. This solution is potentially much more useful for practical situations involving closed-face tunnelling machines, although whether it represents the true collapse load is not guaranteed because the geometry is a simplification. Jancsecz & Steiner (1994) stated that use on several slurry tunnel boring machine (TBM) projects has validated the wedge-prism method, as does Broere (1998). Anagnostou & Kovári (1996a) compared results from the wedge-prism method with centrifuge tests in dry sand by Chambon & Corté (1994), and Messerli et al. (2010) ran their own 1 g model tests of a tunnel in dry sand, and in both cases a good agreement was found.

The wedge ABCDEF is acted on by:

1. its own weight
2. resultant normal and shear forces along failure surfaces BCF, ADE and ABFE
3. the resulting support force acting on surface ABCD
4. a vertical silo pressure produced by the rectangular prism above the surface CDEF

The angle of the wedge, labelled ω in Figure A.1, is unknown. It may be found by iteration, either by finding the value that results in the maximum support force, or by fixing the support force and finding the value that results in the lowest factor of safety.

Groundwater pressure is assumed to be hydrostatic, and effective stress is used throughout the calculations.

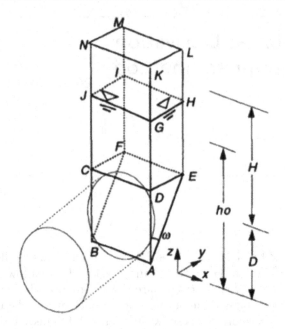

Figure A.1 Wedge and prism model (Anagnostou & Kovári, 1996b).

In reality, the shear stresses on the failure surfaces depend on the horizontal effective stress acting normal to each surface, and this cannot be calculated easily. Anagnostou & Kovári (1994) recommended using a constant ratio of horizontal to vertical effective stress of 0.8.

FINDING THE VERTICAL PRESSURE ON SURFACE CDEF

The vertical pressure on CDEF is found by applying Janssen's silo formula to the prism above the water table (HIJGKLMN), and then to the prism below the water table (CDEFHIJG).

Using Jancsecz & Steiner (1994)'s definition initially as it is simpler:

Perimeter of CDEF = perimeter of JGHI = U

Area of CDEF = area of JGHI = F

The forces on a slice of a prism silo, excluding weight, are shown in Figure A.2.

The weight of each slice is given by:

$$dW = F\gamma dz \qquad (A.1)$$

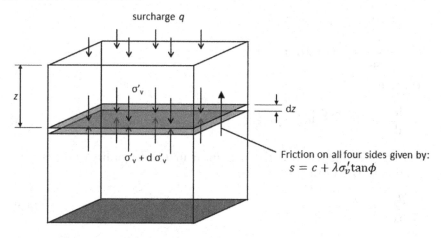

Figure A.2 Prism silo representation.

where γ is the unit weight of the soil, which would be the dry unit weight above the water table and the submerged unit weight below the water table, and λ is the horizontal to vertical effective stress ratio. Note that the symbol λ is used rather than K_0 because K_0 refers specifically to the ratio of horizontal to vertical *in situ* effective stress at rest. In this situation we are finding the point of mass failure of the soil, so it is very disturbed and K_0 no longer exists.

The vertical force on the top of the slice is given by $F\sigma_v$. The upwards vertical force on the bottom of the slice is given by $F(\sigma_v + \mathrm{d}\sigma_v)$.

The resultant friction resistance force around the sides of each slice is given by:

$$\mathrm{d}S = U\mathrm{d}z\left(c + \lambda\sigma_v' \tan\phi\right) \tag{A.2}$$

Vertical equilibrium of the slice gives:

$$\mathrm{d}W = F\left(\sigma_v' + \mathrm{d}\sigma_v'\right) - F\sigma_v' + \mathrm{d}S \tag{A.3}$$

This can be written in full as:

$$F\gamma\mathrm{d}z = F\left(\sigma_v' + \mathrm{d}\sigma_v'\right) - F\sigma_v' + U\mathrm{d}z\left(c + \lambda\sigma_v' \tan\phi\right) \tag{A.4}$$

Which simplifies to:

$$F\gamma\mathrm{d}z = F\mathrm{d}\sigma_v' + U\mathrm{d}z\left(c + \lambda\sigma_v' \tan\phi\right) \tag{A.5}$$

Dividing across by $\mathrm{d}z$ gives:

$$F\gamma = F\frac{\mathrm{d}\sigma_v'}{\mathrm{d}z} + U\left(c + \lambda\sigma_v' \tan\phi\right) \tag{A.6}$$

Dividing across by F gives:

$$\gamma = \frac{d\sigma'_v}{dz} + \frac{U}{F}(c + \lambda\sigma'_v \tan\phi) \tag{A.7}$$

Rearranging:

$$\frac{d\sigma'_v}{dz} + \left(\frac{U\lambda\tan\phi}{F}\right)\sigma'_v = \gamma - \frac{U}{F}c \tag{A.8}$$

Since σ'_v is a function of z, this is a linear first-order ordinary differential equation (ODE) of the form:

$$\frac{dy}{dx} + Py = Q \tag{A.9}$$

where, $y = \sigma'_v$, $x = z$, $P = \left(\frac{U\lambda\tan\phi}{F}\right)$, and $Q = \gamma - \frac{U}{F}c$.

We need to integrate to solve this ODE, but this is not straightforward. The solution to this form of ODE may be found if we can find a function $F(x)$ such that $dF/dx = PF(x)$. Or, using the names of variables and constants in our specific situation, a function $F(z)$ such that $dF/dz = PF(z)$, with P defined as above. Try using:

$$F(z) = \exp\left(\frac{U\lambda\tan\phi z}{F}\right) \tag{A.10}$$

Then:

$$\frac{1}{F(z)}\left(\frac{d}{dz}(F(z)\sigma'_v(z))\right) = Q \tag{A.11}$$

Substituting $F(z)$ and Q according to the definitions in Equation A.9 gives:

$$\frac{1}{\exp\left(\frac{U\lambda\tan\phi z}{F}\right)}\left\{\frac{d}{dz}\left(\exp\left(\frac{U\lambda\tan\phi z}{F}\right)\sigma'_v\right)\right\} = \gamma - \frac{U}{F}c \tag{A.12}$$

Solving this gives:

$$\frac{1}{\exp\left(\frac{U\lambda\tan\phi z}{F}\right)}\left(\exp\left(\frac{U\lambda\tan\phi z}{F}\right)\frac{d\sigma'_v}{dz} + \frac{U\lambda\tan\phi}{F}\exp\left(\frac{U\lambda\tan\phi z}{F}\right)\sigma'_v\right) = \gamma - \frac{U}{F}c \tag{A.13}$$

Which can be simplified to:

$$\frac{d\sigma'_v}{dz} + \exp\left(\frac{U\lambda\tan\phi}{F}\right)\sigma'_v = \gamma - \frac{U}{F}c \tag{A.14}$$

This is our original ODE, proving that Equation A.12 is equivalent to Equation A.8.

Equation A.12 is useful to us because we can integrate it, as follows. First rearrange Equation A.12:

$$\frac{d}{dz}\left\{\exp\left(\frac{U\lambda\tan\phi z}{F}\right)\sigma'_v\right\} = \left(\gamma - \frac{U}{F}c\right)\exp\left(\frac{U\lambda\tan\phi z}{F}\right) \tag{A.15}$$

Now integrate with respect to dz:

$$\exp\left(\frac{U\lambda\tan\phi z}{F}\right)\sigma'_v = \left(\gamma - \frac{U}{F}c\right)\int\exp\left(\frac{U\lambda\tan\phi z}{F}\right)dz \tag{A.16}$$

This gives:

$$\exp\left(\frac{U\lambda\tan\phi z}{F}\right)\sigma'_v = \frac{\left(\gamma - \dfrac{U}{F}c\right)\exp\left(\dfrac{U\lambda\tan\phi z}{F}\right)}{\dfrac{U\lambda\tan\phi}{F}} + C \tag{A.17}$$

where C is an integration constant. To find C we use the boundary condition $\sigma'_v = q$ at $z = 0$:

$$q = \frac{\left(\gamma - \dfrac{U}{F}c\right)}{\dfrac{U\lambda\tan\phi}{F}} + C \tag{A.18}$$

Therefore:

$$C = q - \frac{\left(\gamma - \dfrac{U}{F}c\right)}{\dfrac{U\lambda\tan\phi}{F}} \tag{A.19}$$

Putting this expression for C into Equation A.17 gives:

$$\exp\left(\frac{U\lambda\tan\phi z}{F}\right)\sigma'_v = \frac{\left(\gamma - \dfrac{U}{F}c\right)\exp\left(\dfrac{U\lambda\tan\phi z}{F}\right)}{\dfrac{U\lambda\tan\phi}{F}} + q - \frac{\left(\gamma - \dfrac{U}{F}c\right)}{\dfrac{U\lambda\tan\phi}{F}} \tag{A.20}$$

Rearranging:

$$\sigma'_v = q\exp\left(\frac{-U\lambda\tan\phi z}{F}\right) + \frac{\left(\gamma - \dfrac{U}{F}c\right)}{\dfrac{U\lambda\tan\phi}{F}}\left(1 - \exp\left(\frac{-U\lambda\tan\phi z}{F}\right)\right) \tag{A.21}$$

Simplifying:

$$\sigma'_v = \frac{\frac{F}{U}\gamma - c}{\lambda\tan\phi}\left(1 - \exp\left(\frac{-U\lambda\tan\phi z}{F}\right)\right) + q\exp\left(\frac{-U\lambda\tan\phi z}{F}\right) \tag{A.22}$$

Equation A.22 is the Janssen silo equation used by Jancsecz & Steiner (1994) in their paper to calculate the vertical effective stress on the top of the wedge. Anagnostou & Kovári (1994) use a slightly different notation, where the ratio of area to perimeter $F/U = r$, which may also be expressed in terms of the wedge slope angle ω (cf. Figure A.1) as follows:

$$r = \frac{0.5D\tan\omega}{1 + \tan\omega} \tag{A.23}$$

Substituting F/U for r in Equation A.22 and applying $z = (H - H_w)$ for the prism above the water table (HIJGKLMN) gives the following expression:

$$\sigma'_v = \frac{\gamma_d r - c}{\lambda\tan\phi}\left(1 - \exp\left(-\lambda\tan\phi(H - H_w)/r\right)\right) + q\exp\left(-\lambda\tan\phi(H - H_w)/r\right) \tag{A.24}$$

Anagnostou & Kovári (1994) do not include a surface surcharge in their equations, so for now we will assume that $q = 0$. Therefore the vertical effective stress on surface GHIJ may be given by:

$$\sigma'_{v,GHIJ} = \frac{\gamma_d r - c}{\lambda\tan\phi}\left(1 - \exp\left(-\lambda\tan\phi(H - H_w)/r\right)\right) \tag{A.25}$$

The prism below the water table has the following equation for the vertical effective stress on surface CDEF:

$$\sigma'_{v,CDEF} = \frac{\gamma' r - c}{\lambda\tan\phi}\left(1 - \exp\left(-\lambda\tan\phi H_w/r\right)\right) + q\exp\left(-\lambda\tan\phi H_w/r\right) \tag{A.26}$$

For the lower prism, $q = \sigma'_{v,GHIJ}$, which gives:

$$\sigma'_{v,CDEF} = \frac{\gamma' r - c}{\lambda\tan\phi}\left(1 - \exp\left(-\lambda\tan\phi H_w/r\right)\right)$$
$$+ \frac{\gamma_d r - c}{\lambda\tan\phi}\left(1 - \exp\left(-\lambda\tan\phi(H - H_w)/r\right)\right)\left(\exp\left(-\lambda\tan\phi H_w/r\right)\right) \tag{A.27}$$

This simplifies to:

$$\sigma'_{v,CDEF} = \frac{\gamma'r - c}{\lambda\tan\phi}\left(1 - \exp\left(-\lambda\tan\phi H_w/r\right)\right)$$

$$+ \frac{\gamma_d r - c}{\lambda\tan\phi}\left(\exp\left(-\lambda\tan\phi H_w/r\right) - \exp\left(-\lambda\tan\phi H/r\right)\right) \tag{A.28}$$

Equation A.28 is exactly the same as Equation (2) in Anagnostou & Kovári (1994).

If we wanted to include a surface surcharge, we could repeat this process using Equation A.24 instead of Equation A.25 and we would get:

$$\sigma'_{v,CDEF} = \frac{\gamma'r - c}{\lambda\tan\phi}\left(1 - \exp\left(-\lambda\tan\phi H_w/r\right)\right)$$

$$+ \frac{\gamma_d r - c}{\lambda\tan\phi}\left(\exp\left(-\lambda\tan\phi H_w/r\right) - \exp\left(-\lambda\tan\phi H/r\right)\right) \tag{A.29}$$

$$+ q\exp\left(-\lambda\tan\phi H/r\right)$$

FINDING THE SLIDING FRICTION ON SURFACES ADE AND BCF

Anagnostou & Kovári (1994) assume that the vertical effective stress on these surfaces is made up of two components, which are illustrated in Figure A.3, and is according to the German standard for slurry walls (DIN 4126, 1986). The first component is the interface vertical effective stress $\sigma'_{v,CDEF}$, which is assumed to decrease linearly from $\sigma'_{v,CDEF}$ at the top, to zero at the bottom. The second component is the self-weight of the soil, which is assumed to increase linearly from zero at the top to γD at the bottom.

Note that later papers by the same authors (Anagnostou & Kovári, 1996a, 1996b) assume the square ABCD is of equal area to the circular tunnel face, and so is not of side D but is slightly smaller. Broere (1998) states that this makes the required support force about 3% smaller, so is not significant for most practical purposes. When using these calculations, just make D^2 equal to the face area for a more accurate solution.

The shear stress τ_ω is given by:

$$\tau_\omega = c + \lambda_\omega \sigma'_z \tan\phi \tag{A.30}$$

To compensate for uncertainties in the distribution of vertical effective stress, which some studies have found to be overestimated by the DIN 4126 method, Anagnostou & Kovári (1994) recommended using a reduced

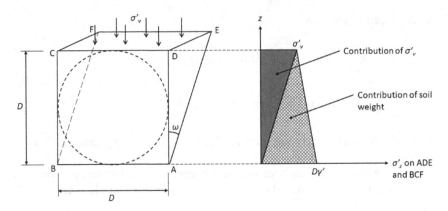

Figure A.3 Anagnostou & Kovári (1994, 1996a & 1996b)'s assumptions for vertical effective stress on wedge surfaces ADE and BCF.

value of the horizontal to vertical effective stress ratio $\lambda_\omega = 0.4$. A later paper (Anagnostou, 2012) used a method of slices to investigate what the true value of λ_ω should be, and found that $\lambda_\omega = 0.4$ was reasonably conservative.

The mean shear stress is found by using the assumption of the triangular distribution of σ'_v and $D\gamma'$ in Figure A.3, as well as the triangular shape of ADE or BCF. This triangular distribution is an assumption and alternative assumptions investigated by Broere (2001), and Anagnostou (2012) shows how there is no need to make an assumption of the distribution of vertical stress if a method of slices is used, dividing the wedge into infinitesimal horizontal slices.

Because of the triangular shape of ADE/BCF the average occurs at distance $D/3$ from the top. For σ'_v the distribution decreases linearly from $\sigma'_{v,CDEF}$ at the top of the wedge, to zero at the bottom, such that the average will be $2/3 \times \sigma'_{v,CDEF}$. Similarly, for $D\gamma'$ the average will be $1/3 \times D\gamma'$. Therefore:

$$\overline{\tau_\omega} = c + \lambda_\omega \left(\frac{1}{3}\gamma'D + \frac{2}{3}\sigma'_{v,CDEF} \right) \frac{\tan\phi}{v} \tag{A.31}$$

where v is the factor of safety. When calculating the minimum support pressure required to prevent failure, use $v = 1$.

FINDING THE SLIDING FRICTION ON ABEF

Anagnostou & Kovári (1994) do not describe how the sliding friction on the wedge slope surface ABEF is calculated, but a very sensible way of doing this is set out by Dias & Bezuijen (2015) and this will be explained

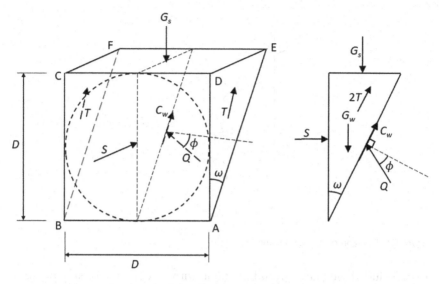

Figure A.4 Forces on the wedge viewed in 3D (left) and in 2D cross-section (right).

in the following diagrams and equations. The forces are shown in Figure A.4,[1] where:

S is the effective support force required to maintain equilibrium (this is what we want to calculate)

T are the friction and cohesion forces on the sides of the wedge, assumed parallel to the wedge slope ABEF

G_s is the effective vertical force due to effective silo pressure on CDEF

Q is the effective reaction force due to friction on the wedge slope ABEF

C_w is the resistance due to cohesion on the wedge slope ABEF

G_w is the effective weight of the wedge

Kolymbas (2008) points out that the horizontal component of friction on the surface CDEF is ignored by "most authors". Certainly Jancsecz & Steiner (1994), Anagnostou & Kovári (1994, 1996a, 1996b), Broere (1998) and Dias & Bezuijen (2015) do not include it, though none of them explain why it is neglected. The reason for neglecting it is that the wedge and prism are not really separate moving rigid blocks, even though we may consider them separately to calculate the vertical stress, but they should be considered as a single rigid plastic block moving downwards. If friction on CDEF

---

[1] Note that both Jancsecz & Steiner (1994) and Broere (1998) have errors in their wedge forces diagrams. The forces on the wedge slope ABEF can either be represented by a shear force parallel to the slope and a normal force perpendicular to the slope, or by a resultant force at an angle ϕ to the normal. They both show a resultant at an angle ϕ to the normal *and* a shear force parallel to the slope.

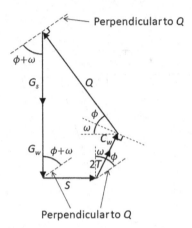

Figure A.5 Polygon of forces on the wedge.

were included then faces would be stable with no cohesion or support pressure, which is clearly unrealistic.

The effective reaction force Q and the effective support force S are unknown. Using the method described by Dias & Bezuijen (2015), if we solve for force equilibrium in the direction perpendicular to Q, then it will not appear in the equation and we can more easily calculate S. Q is orientated at an angle ϕ to the slope normal. The best way to visualise the force equilibrium is to use a 'polygon of forces', where the forces have to create a closed polygon if they are in equilibrium, as shown in Figure A.5.

The equilibrium equation perpendicular to Q may be expressed as:

$$\left(G_w + G_s\right)\cos\left(\phi + \omega\right) = \left(2T + C_w\right)\cos\phi + S\sin\left(\phi + \omega\right) \tag{A.32}$$

Rearranging for S gives:

$$S = \frac{G_w + G_s}{\tan\left(\phi + \omega\right)} - \frac{\left(2T + C_w\right)\cos\phi}{\sin\left(\phi + \omega\right)} \tag{A.33}$$

T may be calculated by multiplying the mean shear stress calculated in Equation A.31 by the area of ADE or BCF, which is given by $0.5D^2\tan\omega$:

$$T = \overline{\tau_\omega} \times \left(0.5D^2\tan\omega\right) = \lambda_\omega\left(\frac{1}{3}\gamma'D + \frac{2}{3}\sigma'_{v,CDEF}\right)\frac{\tan\phi}{v}\left(0.5D^2\tan\omega\right) \tag{A.34}$$

G_s may be calculated by multiplying $\sigma'_{v,CDEF}$ calculated in Equation A.28 or Equation A.29 by the area of CDEF as follows:

$$G_s = \sigma'_{v,CDEF}D^2\tan\omega \tag{A.35}$$

G_w is simply the effective weight of the wedge, given by:

$$G_w = 0.5\gamma'D^3 \tan\omega \tag{A.36}$$

C_w is the resistance force due to cohesion on the wedge slope, which is the cohesion c multiplied by the surface area of ABEF, given by:

$$C_w = \frac{cD^2}{\cos\omega} \tag{A.37}$$

S is the effective support force required to maintain equilibrium of the wedge and prism. The required minimum effective support pressure, usually denoted by s', may be found by dividing S by the area of ABCD, which is D^2.

CALCULATING VALUES FOR THE F_0 NOMOGRAM

If the same assumptions are used that Anagnostou & Kovári (1996a, 1996b) used to create their nomograms, i.e. $\gamma_d/\gamma' = 1.6$, $\lambda = 0.8$ and $\lambda_\omega = 0.4$, then we can calculate values of s' and compare these to the nomograms.
    For $H = 8$ m, $D = 8$ m, $H_w = 4$ m, $\gamma_d = 16$ kN/m³, $\gamma' = 10$ kN/m³, $\phi = 15°$ and $c = 0$ kPa:

$$H/D = 1$$

$$h_0 = 1.5D$$

The maximum value of S is found when $\omega = 40°$. At this value:

$r = 1.82502$	[Equation A.23]
σ'_v above the water table = 51.06769 kPa	[Equation A.25]
$\sigma'_{v,CDEF} = 63.84038$ kPa	[Equation A.28]
$\overline{\tau_\omega} = 7.41972$ kPa	[Equation A.31]
$T = 199.23$ kN	[Equation A.34]
$G_s = 3428.38$ kN	[Equation A.35]
$G_w = 2148.10$ kN	[Equation A.36]
$S = 3434.84$ kN	[Equation A.33]
$s' = S/D^2 = 53.6694$ kPa	
$F_0 = s'/\gamma'D = 0.671$	

Further values can be found in the same way, for $\phi = 20, 25, 30, 35°$, for $H/D = 1, 2, 5$, and for $h_0 = 1.5D$ and $h_0 = H + D$, and exact agreement with the nomograms can be found, except that the values found for $h_0 = 1.5D$

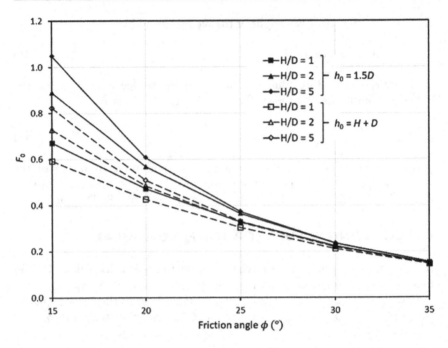

Figure A.6 F_0 nomogram based on wedge-prism calculations presented in this appendix.

exactly match those in the nomograms for $h_0 = H + D$, and vice-versa. The lines were mislabelled in the papers by Anagnostou & Kovári (1996a & 1996b), as confirmed by Anagnostou (2019). The results are shown in Figure A.6.

CALCULATING VALUES FOR THE F_1 NOMOGRAM

If the same assumptions are used that Anagnostou & Kovári (1996a, 1996b) used to create their nomograms, i.e. $\gamma_d/\gamma' = 1.6$, $\lambda = 0.8$ and $\lambda_\omega = 0.4$, then we can calculate values of s' and from Equation 3.2 we can calculate F_1. We are assuming no groundwater flow, so only the F_0 and F_1 terms are active. Therefore:

$$s' = F_0 \gamma' D - F_1 c \tag{A.38}$$

For the same situation we used to calculate F_0 above, i.e. $H = 8$ m, $D = 8$ m, $H_w = 4$ m, $\gamma_d = 16$ kN/m³, $\gamma' = 10$ kN/m³, $\phi = 15°$ and $c = 5$ kPa:

$$H/D = 1$$

$$h_0 = 1.5D$$

The maximum value of S is found when $\omega = 40°$. At this value:

$r = 1.82502$	[Equation A.23]
σ'_v above the water table $= 42.32333$ kPa	[Equation A.25]
$\sigma'_{v,CDEF} = 49.62981$ kPa	[Equation A.28]
$\overline{\tau_\omega} = 11.40433$ kPa	[Equation A.31]
$T = 306.22$ kN	[Equation A.34]
$G_s = 3428.38$ kN	[Equation A.35]
$G_w = 2148.10$ kN	[Equation A.36]
$C_w = 417.73$ kN	[Equation A.37]
$S = 2155.58$ kN	[Equation A.33]
$s' = S/D^2 = 33.6809$ kPa	

Now from the definition of F_1 (i.e. Equation A.38):

$$F_1 = \frac{\left(F_0 \gamma' D - s'\right)}{c} = \frac{\left(53.6694 - 33.6809\right)}{5} = 3.9977 \tag{A.39}$$

Further values can be found in the same way, for $\phi = 20, 25, 30, 35°$, for $H/D = 1, 2, 5$, and for $h_0 = 1.5D$ and $h_0 = H + D$, and exact agreement with

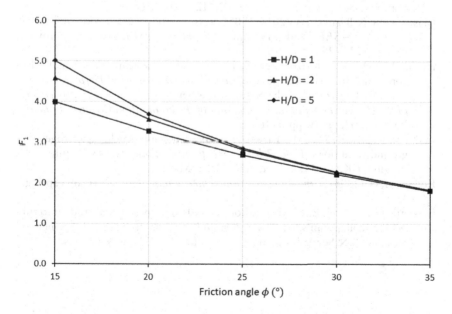

Figure A.7 F_1 nomogram based on wedge-prism calculations presented in this appendix.

the nomograms can be found. The values for $h_0 = 1.5D$ are the same as for $h_0 = H + D$, indicating that the position of the water table does not affect F_1. The results are shown in Figure A.7 and are the same as the values in the papers by Anagnostou & Kovári (1996a, 1996b).

REFERENCES

Anagnostou, G. (2012). The contribution of horizontal arching to tunnel face stability. *geotechnik* **35**, Heft 1, 34–44.

Anagnostou, G. (2019). *Personal communication.*

Anagnostou, G. & Kovári, K. (1994). The face stability of slurry shield-driven tunnels. *Tunn. Deep Space* **9**, No. 2, 165–174.

Anagnostou, G. & Kovári, K. (1996a). Face stability conditions with earth-pressure-balanced shields. *Tunn. Undergr. Space Technol.* **11**, No. 2, 165–173.

Anagnostou, G. & Kovári, K. (1996b). Face stability in slurry and EPB shield tunnelling. *Geotechnical aspects of underground construction in soft ground* (eds Mair, R. J. & Taylor, R. N.), pp. 453–458. Rotterdam, The Netherlands: Balkema.

Broere, W. (1998). Face stability calculation for a slurry shield in heterogeneous soft soils. *Proc. Tunnels and Metropolises* (eds Negro Jr, A. & Ferreira, A. A.), Sao Paulo, Brazil, pp. 215–218. Rotterdam: Balkema.

Broere, W. (2001). *Tunnel face stability & new CPT applications.* Ph.D. thesis, Technical University of Delft.

Chambon, J. F. & Corté, J. F. (1994). Shallow tunnels in cohesionless soil: stability of tunnel face. *J. Geotech. Eng. ASCE* **120**, No. 7, 1150–1163.

Dias, T. G. S. & Bezuijen, A. (2015). TBM pressure models: calculation tools. *Proc. WTC2015 – SEE Tunnel: Promoting Tunneling in SEE Region*, Dubrovnik, Croatia, 22nd–28th May 2015.

Horn, M. (1961). Horizontaler Erddruck auf senkrechte Abschlussflächen von Tunnelröhren. *Landeskonferenz der Ungarischen Tiefbauindustrie*, pp. 7–16. Horizontal earth pressure on perpendicular tunnel face. *Proceedings of the Hungarian National Conference of the Foundation Engineer Industry*, Budapest, Hungary, pp. 7–16.

Jancsecz, S., Steiner, W., 1994. Face support for a large mix-shield in heterogeneous ground conditions. *Proc. 7th Int. Symp. Tunnelling '94*, 5th–7th July 1994, London, UK, pp. 531–549. London, UK: IMM.

Kolymbas, D. (2008). *Tunnelling and tunnel mechanics.* Berlin, Germany: Springer Verlag.

Messerli, J., Pimentel, E. & Anagnostou, G. (2010). Experimental study into tunnel face collapse in sand. *Physical modelling in geotechnics* (eds Springman, S., Laue, J. & Seward, L.), pp. 575–580. London: Taylor & Francis.

Appendix B: Details from derivation of Curtis–Muir Wood equations

INTEGRATION BY PARTS

For calculation of N, the following trigonometric expressions need to integrated:

$$A = \int_0^{\frac{\pi}{2}} \cos 2\theta \sin\theta \; d\theta \tag{B.1}$$

$$B = \int_0^{\frac{\pi}{2}} \sin 2\theta \cos\theta \; d\theta \tag{B.2}$$

Use the following integration by parts rule:

$$\int uv \; d\theta = u \int v \; d\theta - \int \frac{du}{d\theta} \left(\int v \; d\theta \right) d\theta \tag{B.3}$$

Starting with A, let $u = \cos 2\theta$ and $v = \sin\theta$. First calculate:

$$\frac{du}{d\theta} = \frac{d}{d\theta} \cos 2\theta = -2\sin 2\theta \tag{B.4}$$

And:

$$\int v \; d\theta = \int \sin\theta \; d\theta = -\cos\theta \tag{B.5}$$

Now using Equation (B.3):

$$A = \int_0^{\frac{\pi}{2}} \cos 2\theta \sin\theta \; d\theta = \left[\cos 2\theta (-\cos\theta) - \int (-2\sin 2\theta)(-\cos\theta) d\theta \right]_0^{\frac{\pi}{2}} \quad (B.6)$$

This simplifies to:

$$A = \left[-\cos 2\theta \cos\theta - 2\int \sin 2\theta \cos\theta \, d\theta \right]_0^{\frac{\pi}{2}} \quad (B.7)$$

Now let's calculate B. Let $u = \sin 2\theta$ and $v = \cos\theta$. First calculate:

$$\frac{du}{d\theta} = \frac{d}{d\theta} \sin 2\theta = 2\cos 2\theta \quad (B.8)$$

And:

$$\int v \; d\theta = \int \cos\theta \; d\theta = \sin\theta \quad (B.9)$$

Now using Equation (B.3):

$$B = \int_0^{\frac{\pi}{2}} \sin 2\theta \cos\theta \; d\theta = \left[\sin 2\theta \sin\theta - \int 2\cos 2\theta \sin\theta \, d\theta \right]_0^{\frac{\pi}{2}} \quad (B.10)$$

This simplifies to:

$$B = \left[\sin 2\theta \sin\theta - 2\int \cos 2\theta \sin\theta \, d\theta \right]_0^{\frac{\pi}{2}} \quad (B.11)$$

And can be expressed as:

$$B = \left[\sin 2\theta \sin\theta - 2A \right]_0^{\frac{\pi}{2}} \quad (B.12)$$

Similarly Equation (B.7) can be expressed as:

$$A = \left[-\cos 2\theta \cos\theta - 2B \right]_0^{\frac{\pi}{2}} \quad (B.13)$$

If we substitute for B in Equation (B.13), we get:

$$A = \left[-\cos 2\theta \cos \theta - 2\left(\sin 2\theta \sin \theta - 2A \right) \right]_0^{\frac{\pi}{2}}$$

(B.14)

This simplifies to:

$$A = \left[-\cos 2\theta \cos \theta - 2 \sin 2\theta \sin \theta + 4A \right]_0^{\frac{\pi}{2}}$$

(B.15)

Then:

$$A = \frac{1}{3} \left[\cos 2\theta \cos \theta + 2 \sin 2\theta \sin \theta \right]_0^{\frac{\pi}{2}}$$

(B.16)

Applying the integration limits gives:

$$A = \frac{1}{3} \left\{ (0+0) - (1+0) \right\} = -\frac{1}{3}$$

(B.17)

Substituting this value for A into Equation (B.12):

$$B = \left[\sin 2\theta \sin \theta \right]_0^{\frac{\pi}{2}} + \frac{2}{3} = \left\{ (0) - (0) \right\} + \frac{2}{3} = \frac{2}{3}$$

(B.18)

For the calculation of M, the following trigonometric expressions need to be integrated:

$$A = \int_0^{\frac{\pi}{4}} \cos 2\theta \sin \theta \ d\theta$$

(B.19)

$$B = \int_0^{\frac{\pi}{4}} \sin 2\theta \cos \theta \ d\theta$$

(B.20)

$$C = \int_0^{\frac{\pi}{4}} \sin 2\theta \ d\theta$$

(B.21)

A and B turn out to have the same values as before when the integration was between 0 and $\pi/2$, i.e. $-1/3$ and $2/3$ respectively. The value for C is given by:

$$C = \int_0^{\frac{\pi}{4}} \sin 2\theta \; d\theta = \left[-\frac{1}{2}\cos 2\theta \right]_0^{\frac{\pi}{4}} = \left\{ (0) - \left(-\frac{1}{2} \right) \right\} = \frac{1}{2} \qquad \text{(B.22)}$$

Quod erat demonstrandum.
    Return to Section 6.3.4.

Appendix C: Derivation of the deflection of a rectangular simply-supported beam under a point load

This is the derivation of Equation 13.5 in Chapter 13, which gives the mid-span deflection of a simply supported rectangular beam with a centrally-applied point load. The equation is reproduced here:

$$\Delta = \frac{PL^3}{48EI}\left[1 + \frac{72EI}{5L^2AG}\right] \qquad (13.5)$$

The idealised geometry is shown in Figure C.1. The z axis is out of the page. The beam has width in the z axis direction b, height in the y axis direction H and length in the x axis direction L. A point load P is applied at midspan.

The cross-sectional area A of the beam is given by:

$$A = bH \qquad (C.1)$$

The shear force V from $x = 0$ to $x = L/2$ is given by:

$$V = \frac{P}{2} \qquad (C.2)$$

The bending moment M from $x = 0$ to $x = L/2$ is given by:

$$M = \frac{Px}{2} \qquad (C.3)$$

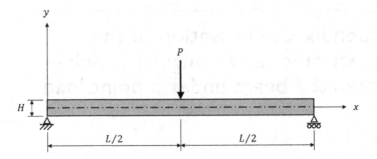

Figure C.1 Simply-supported rectangular beam with a centrally-applied point load.

BENDING DEFLECTION

The deflection at midspan due to bending moment may be obtained from simple beam theory (also known as 'Euler-Bernoulli beam theory'). If u, v, w denote displacements in the x, y, z directions then the slope θ is given by:

$$\theta = \frac{dv}{dx} = -\frac{1}{EI}\int M \, dx \qquad (C.4)$$

    E is the Young's modulus of the beam
    I is the second moment of area

Substituting for M in Equation C.4 using the expression in Equation C.3 gives:

$$\theta = -\frac{1}{EI}\int \frac{Px}{2} \, dx = -\frac{1}{EI}\left[\frac{Px^2}{4} + C_1\right] \qquad (C.5)$$

    C_1 is an integration constant

At midspan $x = L/2$ and due to symmetry the slope $\theta = 0$. Hence $C_1 = -PL^2/16$ and we get:

$$\theta = \frac{dv}{dx} = -\frac{1}{EI}\left[\frac{Px^2}{4} - \frac{PL^2}{16}\right] \qquad (C.6)$$

The slope θ, for small angles, is equal to dv/dx. If we integrate dv/dx from $x = 0$ to $x = L/2$ with respect to x we will get the midspan deflection due to bending Δ_{bmax}.

$$\Delta_{bmax} = \int \frac{dv}{dx}dx = -\frac{1}{EI}\int_0^{L/2}\left[\frac{Px^2}{4} - \frac{PL^2}{16}\right]dx = -\frac{1}{EI}\left[\frac{Px^3}{12} - \frac{PL^2 x}{16} + C_2\right]_0^{L/2} \qquad (C.7)$$

    C_2 is an integration constant

At the support, when $x = 0$, $\Delta_{bmax} = 0$, $\therefore C_2 = 0$. Therefore:

$$\Delta_{bmax} = -\frac{1}{EI}\left[\left\{\frac{PL^3}{96} - \frac{PL^3}{32} + 0\right\} - \{0\}\right] = \frac{PL^3}{48EI} \tag{C.8}$$

SHEAR STRESS

Burland & Wroth (1974), based on Timoshenko (1957), assumed that simple beam theory could also be applied to shear deflection. However, this will not be accurate because the 'shear area', the effective area of a section participating in the shear deformation, is not equal to the cross-section area (e.g. Bhatt, 1999: pp. 275–279). This effect is largely to do with warping, caused by the parabolic shear stress distribution, which induces a cubic displacement profile as shown in Figure C.2.

When simple beam theory is applied to shear deformations, the shear stress is determined on the basis of equilibrium only and compatibility is not considered (Iyer, 2005). Simple beam theory assumes that plane sections remain plane, but they do not.

The shear stress distribution, as shown in Figure C.2, is parabolic. This is because $\tau_{xy} = \tau_{yx}$ and the shear stress on the top and bottom surfaces of the beam must be zero. The maximum shear stress occurs at the centroid of the section and is equal to 1.5 times the average shear stress. This causes the shear strain distribution to also be parabolic, and this leads to a cubic displacement profile, i.e. warping, as seen in the last diagram of Figure C.2.

The average shear stress is given by:

$$\tau_{ave} = \frac{V}{A} = \frac{V}{bH} \tag{C.9}$$

τ_{ave} is the average shear stress in a cross-section
V is the shear force acting at the cross-section
b is the width of the section (in the z axis direction)
H is the height of the section (in the y axis direction)
A is the area of the cross-section, where $A = bH$

By considering equilibrium of an element in the cross-section of the beam, the following formula for shear stress, known as the 'shear formula' may be obtained:

$$\tau = \frac{VQ}{Ib} \tag{C.10}$$

τ is the shear stress
Q is the first moment of area
I is the second moment of area

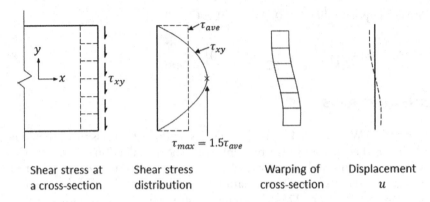

Shear stress at   Shear stress   Warping of   Displacement
a cross-section   distribution   cross-section u

Figure C.2 Shear warping in a rectangular cross-section beam.

The shear stress at distance y in the cross-section may be calculated using Figure C.3.

The shaded area A_1 in Figure C.3 is given by:

$$A_1 = b\left(\frac{H}{2} - y\right)$$ (C.11)

And the distance of the centroid of the shaded area from the neutral axis is:

$$y_1 = \frac{\left(\dfrac{H}{2} + y\right)}{2} = \frac{H}{4} + \frac{y}{2}$$ (C.12)

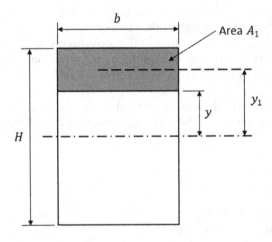

Figure C.3 First moment of area calculation.

The first moment of area Q for the shaded area is its area multiplied by the distance of its centroid from the neutral axis:

$$Q = A_1 y_1 = b\left(\frac{H}{2} - y\right)\left(\frac{H}{4} + \frac{y}{2}\right) = \frac{b}{2}\left(\frac{H^2}{4} - y^2\right) \tag{C.13}$$

Substituting for Q in the shear formula (Equation C.10) gives:

$$\tau = \frac{V}{Ib}\left[\frac{b}{2}\left(\frac{H^2}{4} - y^2\right)\right] = \frac{V}{2I}\left(\frac{H^2}{4} - y^2\right) \tag{C.14}$$

Equation C.14 is quadratic and defines the parabolic distribution of shear stress shown in Figure C.2.

UNIT LOAD METHOD TO CALCULATE SHEAR DEFLECTION – PART I

The simple beam method assumes that the maximum shear stress is related to shear strain via the shear modulus, and hence deflection may be calculated based on integrating shear strains along the centroid. Therefore, the form factor for shear is effectively 1.5. However, using closed form solutions and finite element analysis, Renton (1991), Schramm et al. (1994), Pilkey (2003) and Iyer (2005) have all found that form factor should be approximately 1.2 for rectangular beams where $b < H$. Therefore, the simple beam method is not accurate.

Gere & Timoshenko (1991, pp. 691–694) provided a derivation for shear deflection for a uniformly loaded beam with simple supports, using the unit load method. This is reproduced here but with the load changed to a point load to obtain the correct equation for shear deflection for our case. Bhatt (1999) used a different method of derivation but arrived at the same result.

The unit load method is based on the principle of virtual work, where the external work is equal to the internal work. A unit load is applied to the structure, in our case as a point load at the midspan, such that:

$$W_{ext} = 1.\Delta \tag{C.15}$$

W_{ext} is the external work
1 is the unit load
Δ is the midspan deflection due to the actual loads, which is what we want to calculate

The unit load produces reactions at the supports and stress resultants within the beam, in our case a moment M_U and a shear force V_U. These stress resultants perform the internal virtual work due to this virtual deformation.

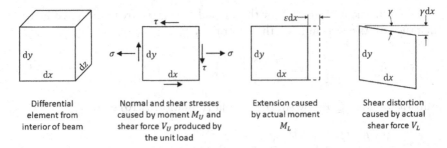

Differential element from interior of beam

Normal and shear stresses caused by moment M_U and shear force V_U produced by the unit load

Extension caused by actual moment M_L

Shear distortion caused by actual shear force V_L

Figure C.4 Virtual flexural and shearing deformations of an element of the beam.

However, and this is the clever bit, we choose the virtual deformations to be the same as the actual deformations that occur in the real beam with the real point load. The virtual deformations of an internal element of the beam are shown in Figure C.4.

The normal stress σ may be obtained from the flexure formula:

$$\sigma = \frac{M_U y}{I} \tag{C.16}$$

And the shear stress τ may be obtained from the shear formula:

$$\tau = \frac{V_U Q}{Ib} \tag{C.17}$$

The extension strain ε due to the actual moment in the structure M_L is given by:

$$\varepsilon = \frac{M_L y}{EI} \tag{C.18}$$

And the shear strain γ due to the actual shear force in the structure V_L is given by:

$$\gamma = \frac{V_L Q}{GIb} \tag{C.19}$$

The internal virtual work of the stresses σ and τ acting on the sides of the differential element is:

$$dW_{int} = (\sigma\, dy\, dz)(\varepsilon\, dx) + (\tau\, dy\, dz)(\gamma\, dx) \tag{C.20}$$

Substituting for σ, τ, ε and γ from Equations C.16 to C.19 gives:

$$dW_{int} = \frac{M_U M_L y^2}{EI^2}\, dx\, dy\, dz + \frac{V_U V_L Q^2}{GI^2 b^2}\, dx\, dy\, dz \tag{C.21}$$

Integrating Equation C.21 through the whole volume of the beam gives:

$$W_{int} = \int \frac{M_U M_L y^2}{EI^2} dx\,dy\,dz + \int \frac{V_U V_L Q^2}{GI^2 b^2} dx\,dy\,dz \qquad (C.22)$$

At any given cross-section of the beam, M_U, M_L, V_U, V_L, E, G and I are constants, so we can simplify the integrals to an integration over the cross-sectional area (denoted A) and an integration over the length of the beam (denoted L):

$$W_{int} = \int_L \frac{M_U M_L}{EI^2} \left[\int_A y^2 dy dz \right] dx + \int_L \frac{V_U V_L}{GI^2} \left[\int_A \frac{Q^2}{b^2} dy dz \right] dx \qquad (C.23)$$

By the definition of the second moment of area I, we know that:

$$\left[\int_A y^2 dy dz \right] = I \qquad (C.24)$$

The second bracketed integral in Equation C.23 is also solely a property of the cross-section dimensions of the beam. We can therefore define a property called the 'form factor for shear' f_s as follows:

$$f_s = \frac{A}{I^2} \int_A \frac{Q^2}{b^2} dy dz \qquad (C.25)$$

The form factor for shear is dimensionless. Replacing the two bracketed integrals in Equation C.23 with I and $f_s I^2 / A$ respectively we get:

$$W_{int} = \int_L \frac{M_U M_L}{EI} dx + \int_L \frac{f_s V_U V_L}{GA} dx \qquad (C.26)$$

By equating internal work in Equation C.26 with external work in Equation C.15 we obtain the equation of the unit load method:

$$\Delta = \int \frac{M_U M_L}{EI} dx + \int \frac{f_s V_U V_L}{GA} dx \qquad (C.27)$$

FORM FACTOR FOR SHEAR FOR A RECTANGULAR SECTION

The form factor for shear f_s must be evaluated for every shape of cross-section, using Equation C.25. For a rectangular section of width b and height H, the first moment of area Q is given by Equation C.13 as:

$$Q = \frac{b}{2}\left(\frac{H^2}{4} - y^2\right) \qquad (C.28)$$

We can also calculate A/I^2 as:

$$\frac{A}{I^2} = \frac{bH}{\left(\frac{bH^3}{12}\right)^2} = \frac{144}{bH^5} \qquad (C.29)$$

And Q^2 may be obtained by squaring Equation C.28:

$$Q^2 = \left[\frac{b}{2}\left(\frac{H^2}{4} - y^2\right)\right]^2 = \left[\frac{bH^2}{8} - \frac{by^2}{2}\right]^2 = \frac{b^2H^4}{64} - \frac{b^2H^2y^2}{8} + \frac{b^2y^4}{4} \quad (C.30)$$

And Q^2/b^2 is given by:

$$\frac{Q^2}{b^2} = \frac{H^4}{64} - \frac{H^2y^2}{8} + \frac{y^4}{4} \qquad (C.31)$$

Now inserting these values into Equation C.25 gives:

$$f_s = \frac{A}{I^2}\int_A \frac{Q^2}{b^2}\,dydz = \frac{144}{bH^5}\int_{-H/2}^{H/2}\left(\frac{H^4}{64} - \frac{H^2y^2}{8} + \frac{y^4}{4}\right)b.dy \qquad (C.32)$$

Integrating:

$$f_s = \frac{144b}{bH^5}\left[\frac{H^4y}{64} - \frac{H^2y^3}{24} + \frac{y^5}{20}\right]_{-H/2}^{H/2} \qquad (C.33)$$

Now applying the limits:

$$f_s = \frac{144b}{bH^5}\left[\left(\frac{H^5}{128} - \frac{H^5}{192} + \frac{H^5}{640}\right) - \left(-\frac{H^5}{128} + \frac{H^5}{192} - \frac{H^5}{640}\right)\right] \qquad (C.34)$$

Simplifying:

$$f_s = \frac{144}{H^5}\left(\frac{H^5}{64} - \frac{H^5}{96} + \frac{H^5}{320}\right) = \frac{6}{5} \qquad (C.35)$$

As mentioned previously, form factor for shear is dimensionless, and for a rectangular section we have just proved it is equal to 1.2. This is the same value as found by other sources using a variety of different methods (Renton, 1991; Schramm et al., 1994; Bhatt, 1999; Pilkey, 2003; and Iyer, 2005). Some of those methods also considered the variation of shear stress across the width of the beam due to Poisson's ratio effects, but for $b < H$ the effect is negligible. Since for buildings the thickness of the wall will always be less than the height, this is a satisfactory assumption.

UNIT LOAD METHOD TO CALCULATE SHEAR DEFLECTION – PART 2

Continuing with the unit load method, the actual stress resultants from $x = 0$ to $x = L/2$ are given by:

$$M_L = \frac{Px}{2} \tag{C.36}$$

$$V_L = \frac{P}{2} \tag{C.37}$$

For a unit load, i.e. $P = 1$, the stress resultants are:

$$M_U = \frac{1x}{2} \tag{C.38}$$

$$V_L = \frac{1}{2} \tag{C.39}$$

Inserting these stress resultants into the unit load equation gives:

$$\Delta = \int \frac{M_U M_L}{EI} dx + \int \frac{f_s V_U V_L}{GA} dx = \frac{2}{EI} \int_0^{L/2} \frac{Px^2}{4} dx + \frac{2f_s}{GA} \int_0^{L/2} \frac{P}{4} dx \tag{C.40}$$

Note that the terms are multiplied by 2 because the integration is only from 0 to $L/2$, i.e. only half the beam. We did this because the moment and shear force equations we have defined are only valid from 0 to $L/2$.
   Integrating:

$$\Delta = \frac{2}{EI} \left[\frac{Px^3}{12} \right]_0^{L/2} + \frac{2f_s}{GA} \left[\frac{P}{4} \right]_0^{L/2} = \frac{2}{EI} \left(\frac{PL^3}{96} \right) + \frac{2f_s}{GA} \left(\frac{PL}{8} \right) \tag{C.41}$$

Simplifying:

$$\Delta = \frac{PL^3}{48EI}\left(1 + \frac{12f_sEI}{GAL^2}\right) \tag{C.42}$$

Equation C.42 incorporates both bending deflection in the first term, and shear deflection in the second term and is appropriate for any cross-section simply-supported beam with a centrally-applied point load. If we substitute for $f_s = 6/5$, which is the value calculated for rectangular section beams, we get the following equation, which is Equation 13.5 in Chapter 13:

$$\Delta = \frac{PL^3}{48EI}\left(1 + \frac{72EI}{5GAL^2}\right) \tag{13.5}$$

Quod erat demonstrandum.

COMPARISON WITH BURLAND & WROTH (1974)

If we compare Equation C.42 with the following equation from Burland & Wroth (1974), we can see that the correct shear deflection we have just derived has a lower coefficient of 72/5 (= 14.4) compared to 18 in Burland & Wroth (1974).

$$\Delta = \frac{PL^3}{48EI}\left(1 + \frac{18EI}{L^2HG}\right) \tag{C.43}$$

Burland & Wroth (1974) also assumed $A = H$, i.e. that the calculation is per unit width. This is fine as long as the second moment of area I is also per unit width (i.e. $b = 1$ m), and the point load P is also expressed per unit width. This is useful because it means we do not have to know the thickness of the wall. Alternatively, we can include the wall thickness and Equation 16.43 can be expressed as:

$$\Delta = \frac{PL^3}{48EI}\left(1 + \frac{18EI}{L^2AG}\right) \tag{C.44}$$

Effectively, the form factor for shear f_s is equal to 1.5 in Burland & Wroth's equation (Equation C.43). Shear deflections calculated using Equation 13.5 will be 1.2/1.5 = 0.8 times smaller than those calculated using Burland & Wroth's equation. This means that higher strains will be generated as a higher point load is needed to achieve the same greenfield deflection ratio.

Our new, correct version of this equation (Equation 13.5) was used in Chapter 13 to derive the following equations for the maximum bending strain and the maximum diagonal strain:

$$\varepsilon_{bmax} = \frac{\Delta / L}{\frac{L}{12d}\left[1 + \frac{72EI}{5L^2 AG}\right]} \tag{13.12}$$

$$\varepsilon_{dmax} = \frac{\Delta / L}{\left[\frac{AGL^2}{18EI} + \frac{4}{5}\right]} \tag{13.23}$$

As L/H varies, the relative importance of maximum diagonal and bending strains will vary. Since the maximum tensile strain will be the greater of either the maximum diagonal strain or the maximum bending strain, it can be of interest to know which is greater. If we assume the following so that the graphs we will plot will be similar to those of Burland & Wroth (1974):

Unit width, $b = 1$ and $A = H$.
Neutral axis at foundation level for hogging, so that $I = H^3/3$ and $d = H$.
Take Poisson's ratio as 0.3, such that $E/G = 2.6$.

Then we can simplify the equations to:

$$\frac{\Delta/L}{\varepsilon_{bmax}} = \frac{L}{12H} + 1.04\frac{H}{L} \tag{C.45}$$

$$\frac{\Delta/L}{\varepsilon_{dmax}} = \frac{L^2}{15.6H^2} + \frac{4}{5} \tag{C.46}$$

Burland & Wroth (1974) also did this and derived the following equations:

$$\frac{\Delta/L}{\varepsilon_{bmax}} = \frac{L}{12H} + 1.3\frac{H}{L} \tag{C.47}$$

$$\frac{\Delta/L}{\varepsilon_{dmax}} = \frac{L^2}{15.6H^2} + 1 \tag{C.48}$$

By setting either $\varepsilon_{bmax} = \varepsilon_{lim}$ or $\varepsilon_{dmax} = \varepsilon_{lim}$ we can now plot these equations against L/H, as Burland & Wroth (1974) did. This is shown in Figure C.5.

Figure C.5 shows that the corrected equations give lower values of $(\Delta/L)/\varepsilon_{lim}$ at all values of L/H and for both bending strains and diagonal strains. This is because at the same deflection ratio Δ/L the corrected equations will give higher values of ε_{bmax} and ε_{dmax}.

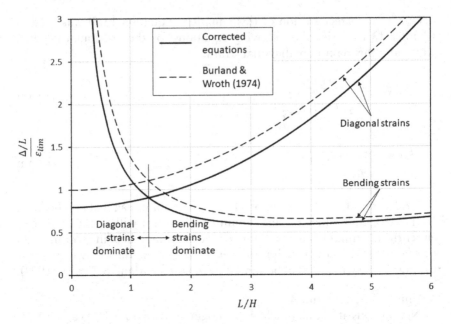

Figure C.5 Relationship between $(\Delta/L)/\varepsilon_{lim}$ and L/H for buildings modelled as isotropic rectangular cross-section beams in hogging with the neutral axis at foundation level.

Figure C.5 also shows that the corrected equations do not change the point, at exactly $L/H = 1.3$, where diagonal strains cease to dominate and bending strains begin to dominate.

We can also look at sagging if we assume the following:

Unit width, $b = 1$ and $A = H$.
Neutral axis at centroid, so that $I = H^3/12$ and $d = H/2$.
Take Poisson's ratio as 0.3, such that $E/G = 2.6$.

Then we can simplify the equations to:

$$\frac{\Delta/L}{\varepsilon_{bmax}} = \frac{L}{6H} + 052\frac{H}{L} \tag{C.49}$$

$$\frac{\Delta/L}{\varepsilon_{dmax}} = \frac{L^2}{3.9H^2} + \frac{4}{5} \tag{C.50}$$

Burland & Wroth (1974) also did this and derived the following equations:

$$\frac{\Delta/L}{\varepsilon_{bmax}} = 0.167\frac{L}{H} + 0.65\frac{H}{L} \tag{C.51}$$

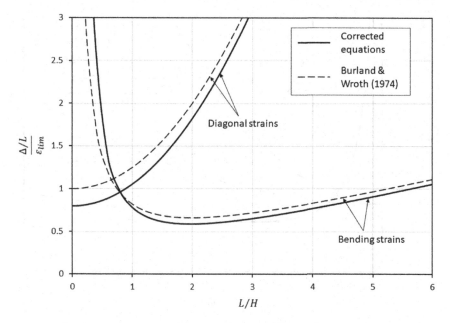

Figure C.6 Relationship between $(\Delta/L)/\varepsilon_{lim}$ and **L/H** for buildings modelled as iso-tropic rectangular cross-section beams in sagging with the neutral axis at mid-height.

$$\frac{\Delta/L}{\varepsilon_{dmax}} = 0.25\frac{L^2}{H^2} + 1 \qquad\qquad\qquad (C.52)$$

These are plotted on Figure C.6, which shows that the corrected equa-tions give lower values of $(\Delta/L)/\varepsilon_{lim}$ at all values of L/H and for both bending strains and diagonal strains. This was also the case for hogging, but here in sagging the difference is much more marked. The points at which diagonal strains cease to dominate and bending strains begin to dominate are further left than for the hogging case and are not at the same value of L/H for the corrected equations and Burland & Wroth (1974).

THE EFFECT OF VARYING E/G RATIO

In Figures C.5 and C.6, we assumed that Poisson's ratio $v = 0.3$ and therefore $E/G = 2.6$. However, the effective E/G ratio can vary a lot in real buildings. Burland & Wroth (1974) describe situations where the longitudinal stiffness of the 'building as a beam' can be much higher than the shear stiffness, for instance a building with lots of openings in

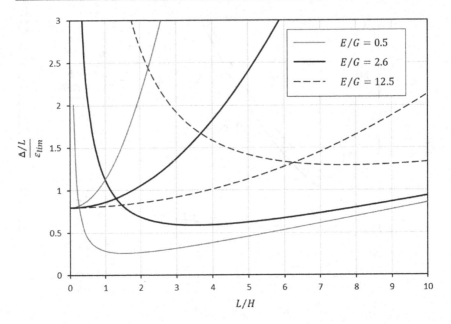

Figure C.7 The effect of varying *E/G* ratio for buildings in hogging.

it, and situations where the shear stiffness can be higher than the longitudinal stiffness, for instance a wall made of precast concrete elements connected with dowels.

Burland & Wroth (1974) used limiting values of $E/G = 12.5$ for a building with high longitudinal stiffness but low shear stiffness, and $E/G = 0.5$ for a building with low longitudinal stiffness but high shear stiffness. These values have been used with our corrected equations, and the limiting tensile strains are shown in Figure C.7 for hogging only.

Figure C.7 shows that the best situation for minimising tensile strains is to have a building that is flexible in shear, i.e. to have a high E/G ratio. Also of note in Figure C.7 is that as E/G increases, the diagonal strains become more important than bending strains over a larger range of L/H, up to around $L/H = 6.3$.

Figure C.8 shows the effect of varying E/G ratio for buildings in sagging. Similar to Figure C6.7, it shows that the best situation for minimising tensile strains is to have a building that is flexible in shear, i.e. to have a high E/G ratio. When $E/G = 0.5$, bending dominates at all except very low values of L/H. When $E/G = 12.5$, diagonal strains dominate up to around $L/H = 3.2$.

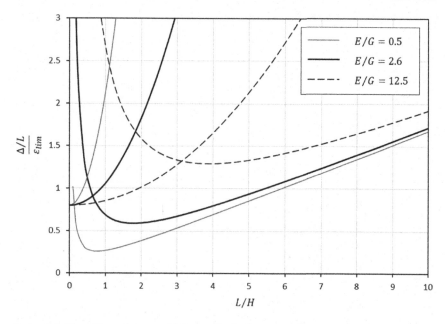

Figure C.8 The effect of varying *E/G* ratio for buildings in sagging.

REFERENCES

Bhatt, P. (1999). *Structures.* London, UK: Addison-Wesley Longman Ltd.

Burland, J. B. & Wroth, C. P. (1974). Settlement of buildings and associated damage – state of the art review. *Conf. on Settlement of Structures, Cambridge,* pp. 611–654. London, UK: Pentech Press.

Gere, J. M. & Timoshenko, S. P. (1991). *Mechanics of materials,* 3rd S. I. Edition. London, UK: Chapman & Hall.

Iyer, H. (2005). *The effects of shear deformation in rectangular and wide flange sections.* MS thesis, Virginia Polytechnic Institute and State University.

Pilkey, W. D. (2003). *Analysis and design of elastic beams.* New York, USA: John Wiley and Sons.

Renton, J. D. (1991). Generalized beam theory applied to shear stiffness. *Int. J. Solids Struct.* **27,** 1955–1967.

Schramm, U., Kitis, L., Kang, W., & Pilkey, W. D. (1994). On the shear deformation coefficient in beam theory. *Finite Elem. Anal. Des.* **16,** 141–162.

Timoshenko, S. (1957). *Strength of materials – Part I,* 3rd edition. London, UK: D van Nostrand.

Index

Printed in the United States
by Baker & Taylor Publisher Services